AF411457

Feature Papers in Horticulturae

Feature Papers in Horticulturae

Editor

Douglas D. Archbold

MDPI • Basel • Beijing • Wuhan • Barcelona • Belgrade • Manchester • Tokyo • Cluj • Tianjin

Editor
Douglas D. Archbold
Department of Horticulture,
University of Kentucky
USA

Editorial Office
MDPI
St. Alban-Anlage 66
4052 Basel, Switzerland

This is a reprint of articles from the Special Issue published online in the open access journal *Horticulturae* (ISSN 2311-7524) (available at: https://www.mdpi.com/journal/horticulturae/special_issues/Feature_Papers_Horticulturae).

For citation purposes, cite each article independently as indicated on the article page online and as indicated below:

LastName, A.A.; LastName, B.B.; LastName, C.C. Article Title. *Journal Name* **Year**, *Volume Number*, Page Range.

ISBN 978-3-0365-1992-0 (Hbk)
ISBN 978-3-0365-1993-7 (PDF)

Contents

About the Editor

Douglas D. Archbold Professor Emeritus. Retired in 2020. Performed research in fruit crop physiology and taught undergraduate classes in fruit and vegetable production and graduate classes on phytohormones and stress physiology over a 38 year career

Preface to "Feature Papers in Horticulturae"

The goal of this Special Issue is to highlight frontier research in basic to applied horticulture among selected works published in Horticulturae in 2020. Exhibiting diversity in topic areas, the research can generally be classified as follows: (1) improving the sustainability of horticultural crop production systems is key for the future; (2) integrating new technologies and new crops into existing production systems is growing; (3) managing fruit set and ripening remain challenges, especially concerning climate change; (4) postharvest storage and handling techniques continue to evolve; and (5) expanding our knowledge of unique and underutilized species is key to their conservation and horticultural use.

Douglas D. Archbold
Editor

Editorial

Special Issue: Feature Papers 2020

Douglas D. Archbold

Department of Horticulture, University of Kentucky, N318 Agricultural Sciences North, Lexington, KY 40546, USA; darchbol@uky.edu

The goal of this Special Issue is to highlight, through selected works, frontier research in basic to applied horticulture among those published in Horticulturae in 2020. Exhibiting a diversity in topic areas, the following papers are noteworthy.

1. Improving the Sustainability of Horticultural Crop Production Systems Is Key for the Future

The maintenance of soil quality and the reclamation of marginal soils are urgent priorities in sustainable systems. Interest in the use of biochar, a carbon-rich, porous material thought to improve various soil properties, has increased. Tenic et al. [1] reviewed the literature and reported that the source of organic material, or 'feedstock', used in biochar production and different parameters of pyrolysis determine its chemical and physical properties. The incorporation of biochar impacts soil–water relations and soil health, and has been shown to have an overall positive impact on crop yield. However, the pre-existing physical, chemical, and biological soil properties influence the outcome and the effects of long-term field application of biochar on the soil microcosm need to be understood.

Miceli et al. [2] showed that tomato and sweet pepper seedlings suffered negative effects of salinity on plant growth, relative water content, and stomatal conductance. However, the foliar application of GA_3 was successful in increasing salinity tolerance of tomato seedlings up to 25 mM NaCl and up to 50 mM NaCl by sweet pepper seedlings. GA_3 treatment could represent a sustainable strategy, enabling the use of saline water in vegetable nurseries when it is the only water quality available.

Strawberry growers have used different materials to mulch, but the most widely used is black polyethylene mulch films which are not biodegradable. The increasing focus on environmentally sustainable agricultural practices requires a transition to biodegradable mulch films. Ten biodegradable mulch films were used to compare their effectiveness to black polyethylene in covering the soil during a cultivation cycle of strawberry by Giordano et al. [3]. Several of the biodegradable films were comparable to black polyethylene in their effects on yield and fruit number, indicating that sustainable alternatives to black polyethylene may serve as replacements.

The effect of extracts of *Nerium oleander*, *Eucalyptus chamadulonsis* and *Citrullus colocynthis* against bacterial spot disease in tomato and the induction of resistance were assessed by Abo-Elyousr et al. [4]. Application of the plant extracts at 15% (v/v) to tomato plants significantly reduced disease severity, significantly increased total phenol and salicylic acid content, and in some instances significantly increased peroxidase activity and polyphenol oxidase after infection with the causal agent. These plant extracts showed promising antibacterial activity and could become an effective tool in integrated management programs for sustainable tomato bacterial spot control.

In sustainable cropping systems, the management of herbivorous arthropods is a challenge to tomato production. The development of resistant cultivars is critical for a transition to a sustainable system. The host selection of *Tetranychus urticae*, *Bemisia tabaci*, and *Tuta absoluta* was evaluated by de Oliveira et al. [5], characterizing their preference for high zingiberene content (HZC) or low zingiberene content (LZC) tomato leaves. Tomato genotypes with HZC showed repellency to pests and induced a nonpreference for

Citation: Archbold, D.D. Special Issue: Feature Papers 2020. *Horticulturae* **2021**, 7, 121. https://doi.org/10.3390/horticulturae7060121

Received: 18 May 2021
Accepted: 19 May 2021
Published: 21 May 2021

oviposition. Genotypes selected for HZC may be considered sources of resistance genes to arthropod pests for tomato breeding programs, and therefore have excellent potential for the development of resistant cultivars for sustainable cropping systems.

2. Integrating New Technologies and New Crops into Existing Production Systems Is Growing

The level of agricultural productivity in sub-Saharan Africa remains far below the global average, due in part to the scarce use of production- and process-enhancing technologies. Steffens et al. [6] studied the driving forces and effects of adopting innovative agricultural technologies in food value chains in the region and determined that access to credit, experience of environmental shocks and social capital were the main drivers for the adoption of improved practices. Thus, the promotion of social capital and access to financial capital can be pivotal in enhancing the adoption of innovative agricultural technologies.

The European market for fresh Asian vegetables is expected to expand across the EU member states, and the introduction of new vegetable crops has enormous potential according to Hong and Gruda [7]. Demand for diversified, nutritious, and exotic vegetables has been increasing, and four Asian vegetables, Korean ginseng sprout, Korean cabbage, Coastal hog fennel and Japanese (Chinese or Korean) angelica tree, were discussed. All possess several health benefits, are increasingly in demand, are easy to cultivate, and align with current trends of the European vegetable market. Thus, studies of production systems of these and other Asian vegetables in different European environments are required to expand their production and market availability.

3. Managing Fruit Set and Ripening Remain Challenges Especially in Relation to Climate Change

Domingues et al. [8] observed that 'Valencia' orange ripening and quality characteristics were affected by the type of rootstock on which the scion was grafted, and by a rootstock/scion interaction with a tropical versus temperate climate, giving growers something they may need to consider when planting new orchards.

Climate change associated with a warm autumn often hampers the development of the coloration of many fruits including late ripening apple varieties in New Zealand. Funke and Blanke [9] found that an exposure of at least two weeks of reflective mulch was sufficient for enhancing coloration for outside-, inside- and down-facing sides of the fruit of 'Fuji' and 'Pacific Rose' apple cultivars, increasing the portion of fruit harvested in the first pick and improving fruit storability and export quality, potentially increasing financial returns to growers. The short exposure time was considered surprising, but could be cost effective in commercial settings.

Youssef and Roberto [10] determined the incidence and possible causal pathogen(s) of premature apple fruit drop (PAFD) in Egypt, and also assessed some fungicides for controlling the disease organisms, in order to promote a sustainable system in orchards. Phytopathogenic fungi were isolated from the dropped apple fruit, and four fungicides were tested against the diseases in vitro and under naturally occurring infections in the field. All of the fungicides, applied at fruit set, significantly reduced PAFD in the field and could be useful in commercial settings.

Plums can suffer from small fruit size, premature fruit drop and alternate bearing. Lammerich et al. [11] demonstrated that European plum fruit size could be improved by either mechanical (Bonn/Baum at 380 rpm at 5 km/h) or chemical thinning with either ammonium thiosulfate or ethephon, or a combination of both, increasing likely financial returns to growers.

4. Postharvest Storage and Handling Techniques Continue to Evolve

Youssef et al. [12] compared the efficacy of different types of SO_2-generating pads on the incidence of gray mold band on the physicochemical properties of the quality of seeded table grapes grown under protected cultivation. The SO_2-generating pads with a dual release of 5 or 8 g completely inhibited the development of gray mold at all evaluation times. A high reduction of the disease incidence was also achieved by using a slow release of 4 g. In addition, the SO_2-generating pads did not alter the physicochemical properties of the grapes. Thus, these types of pads should be considered for the effective control of the gray mold of table grapes grown under protected cultivation, while maintaining grape quality.

Raspberries are a rich source of bioactive phytochemicals, but these can be altered by postharvest storage and processing techniques before human consumption. In an in-depth review, Piccolo et al. [13] reported that the content of bioactive phytochemicals is relatively stable during cold (5 °C) or frozen storage. Processing techniques such as juicing or drying negatively affect bioactive phytochemical content. Among drying techniques, hot air (oven) drying alters the content of bioactive compounds the most. For this reason, new drying technologies such as microwave and heat pumps have been developed. These novel techniques are more successful in retaining bioactive phytochemicals with respect to conventional hot air drying.

5. Expanding Our Knowledge of Unique and Underutilized Species Is Key to Their Conservation and Horticultural Use

Conservation of unique plant species has significant ecological and horticultural implications. Marler and Calonje [14] observed that male Cycas and Zamia plants produced more branches than female plants, and cycad species with determinate female strobili produced more branches on female plants than species with indeterminate female strobili. Horticultural and conservation decisions may be improved with this sexual dimorphism knowledge.

The literature containing which chemical elements are found in cycad leaves was reviewed by Deloso et al. [15] to determine the range in values of concentrations reported for essential and beneficial elements. The leaf element concentrations were influenced by biotic factors such as plant size, leaf age, and leaflet position on the rachis; by environmental factors such as incident light and soil nutrient concentrations within the root zone. These influential factors were missing from many reports, rendering the results ambiguous and comparisons among studies difficult. Future research should include the addition of more taxa, more in situ locations, the influence of season, and the influence of herbivory to understand more fully leaf nutrition for cycads.

Fruit and seed shape are important taxonomic characteristics providing information on ecological, nutritional, and developmental aspects, but their use requires quantification. Del Pozo et al. [16] proposed a method for seed shape quantification based on the comparison of the bidimensional images of the seeds with geometric figures. The diversity of the seed shape in the Arecaceae makes this family a good model system to study the application of geometric models in morphology.

Fragaria vesca L. has become a model species for the genomic studies relevant to important crop plant species in the Rosaceae family, but generating large numbers of plants from non-runner-producing genotypes is slow. Sarker et al. [17] developed an in vitro protocol that illustrated that in vitro culture of shoot axillary bud explants could generate high numbers of clonal shoots from a single seedling plant in vitro.

Conflicts of Interest: The authors declare no conflict of interest.

References

1. Tenic, E.; Ghogare, R.; Dhingra, A. Biochar—A Panacea for Agriculture or Just Carbon? *Horticulture* **2020**, *6*, 37. [CrossRef]
2. Miceli, A.; Vetrano, F.; Moncada, A. Effects of Foliar Application of Gibberellic Acid on the Salt Tolerance of Tomato and Sweet Pepper Transplants. *Horticulture* **2020**, *6*, 93. [CrossRef]
3. Giordano, M.; Amoroso, C.G.; El-Nakhel, C.; Rouphael, Y.; De Pascale, S.; Cirillo, C. An Appraisal of Biodegradable Mulch Films with Respect to Strawberry Crop Performance and Fruit Quality. *Horticulture* **2020**, *6*, 48. [CrossRef]
4. Abo-Elyousr, K.A.M.; Almasoudi, N.M.; Abdelmagid, A.W.M.; Roberto, S.R.; Youssef, K. Plant Extract Treatments Induce Resistance to Bacterial Spot by Tomato Plants for a Sustainable System. *Horticulture* **2020**, *6*, 36. [CrossRef]

5.   De Oliveira, J.R.F.; De Resende, J.T.V.; Filho, R.B.D.L.; Roberto, S.R.; Da Silva, P.R.; Rech, C.; Nardi, C. Tomato Breeding for Sustainable Crop Systems: High Levels of Zingiberene Providing Resistance to Multiple Arthropods. *Horticulture* **2020**, *6*, 34. [CrossRef]

6.   Steffens, J.; Brüssow, K.; Grote, U. A Strategic Approach to Value Chain Upgrading—Adopting Innovations and Their Impacts on Farm Households in Tanzania. *Horticulture* **2020**, *6*, 32. [CrossRef]

7.   Hong, J.; Gruda, N.S. The Potential of Introduction of Asian Vegetables in Europe. *Horticulture* **2020**, *6*, 38. [CrossRef]

8.   Domingues, A.R.; Marcolini, C.D.M.; Gonçalves, C.H.D.S.; Gonçalves, L.S.A.; Roberto, S.R.; Carlos, E.F. Fruit Ripening Development of 'Valencia' Orange Trees Grafted on Different 'Trifoliata' Hybrid Rootstocks. *Horticulture* **2020**, *7*, 3. [CrossRef]

9.   Funke, K.; Blanke, M. Spatial and Temporal Enhancement of Colour Development in Apples Subjected to Reflective Material in the Southern Hemisphere. *Horticulture* **2020**, *7*, 2. [CrossRef]

10. Youssef, K.; Roberto, S.R. Premature Apple Fruit Drop: Associated Fungal Species and Attempted Management Solutions. *Horticulture* **2020**, *6*, 31. [CrossRef]

11. Lammerich, S.; Kunz, A.; Damerow, L.; Blanke, M. Mechanical Crop Load Management (CLM) Improves Fruit Quality and Reduces Fruit Drop and Alternate Bearing in European Plum (*Prunus domestica* L.). *Horticulture* **2020**, *6*, 52. [CrossRef]

12. Youssef, K.; Junior, O.J.C.; Mühlbeier, D.T.; Roberto, S.R. Sulphur Dioxide Pads Can Reduce Gray Mold While Maintaining the Quality of Clamshell-Packaged 'BRS Nubia' Seeded Table Grapes Grown under Protected Cultivation. *Horticulture* **2020**, *6*, 20. [CrossRef]

13. Piccolo, E.; Garcìa, L.M.; Landi, M.; Guidi, L.; Massai, R.; Remorini, D. Influences of Postharvest Storage and Processing Techniques on Antioxidant and Nutraceutical Properties of *Rubus idaeus* L.: A Mini-Review. *Horticulture* **2020**, *6*, 105. [CrossRef]

14. Marler, T.; Calonje, M. Stem Branching of Cycad Plants Informs Horticulture and Conservation Decisions. *Horticulture* **2020**, *6*, 65. [CrossRef]

15. Deloso, B.E.; Krishnapillai, M.V.; Ferreras, U.F.; Lindström, A.J.; Calonje, M.; Marler, T.E. Chemical Element Concentrations of Cycad Leaves: Do We Know Enough? *Horticulture* **2020**, *6*, 85. [CrossRef]

16. del Pozo, D.G.; Martín-Gómez, J.; Tocino, Á.; Cervantes, E. Seed Geometry in the Arecaceae. *Horticulture* **2020**, *6*, 64. [CrossRef]

17. Sarker, B.C.; Archbold, D.D.; Geneve, R.L.; Kester, S.T. Rapid In Vitro Multiplication of Non-Running *Fragaria vesca* Genotypes from Seedling Shoot Axillary Bud Explants. *Horticulture* **2020**, *6*, 51. [CrossRef]

 *horticulturae*

Review

Biochar—A Panacea for Agriculture or Just Carbon?

Elvir Tenic [†], Rishikesh Ghogare [†] and Amit Dhingra *

Department of Horticulture, Washington State University, Pullman, WA 99164, USA; elvir.tenic@wsu.edu (E.T.);
rishikesh.ghogare@wsu.edu (R.G.)
* Correspondence: adhingra@wsu.edu
† These authors contributed equally to this work.

Received: 18 May 2020; Accepted: 29 June 2020; Published: 3 July 2020

Abstract: The sustainable production of food faces formidable challenges. Foremost is the availability of arable soils, which have been ravaged by the overuse of fertilizers and detrimental soil management techniques. The maintenance of soil quality and reclamation of marginal soils are urgent priorities. The use of biochar, a carbon-rich, porous material thought to improve various soil properties, is gaining interest. Biochar (BC) is produced through the thermochemical decomposition of organic matter in a process known as pyrolysis. Importantly, the source of organic material, or 'feedstock', used in this process and different parameters of pyrolysis determine the chemical and physical properties of biochar. The incorporation of BC impacts soil–water relations and soil health, and it has been shown to have an overall positive impact on crop yield; however, pre-existing physical, chemical, and biological soil properties influence the outcome. The effects of long-term field application of BC and how it influences the soil microcosm also need to be understood. This literature review, including a focused meta-analysis, summarizes the key outcomes of BC studies and identifies critical research areas for future investigations. This knowledge will facilitate the predictable enhancement of crop productivity and meaningful carbon sequestration.

Keywords: agronomy; sustainability; organic fertilizer; crop productivity; soil acidification; soil organic matter; pyrolysis; microbial activity

---

1. Introduction

"If you desire peace, cultivate justice, but at the same time cultivate the fields to produce more bread; otherwise there will be no peace" Norman Borlaug, Oslo, Norway, December 11, 1970. Nobel lecture.

What was prevalent in the 1960s holds true yet again—the world stands at a threshold where the availability of food is threatened, albeit for reasons different than six decades ago. The changing climate, deteriorating land and water conditions, and loss of biodiversity present unprecedented challenges for humankind [1]. At present, greenhouse gas (GHG) emissions are increasing rapidly, with carbon dioxide (CO_2) levels rising more than 3% annually since the 2000s. These GHG discharges have a drastic impact on the climate, despite global efforts to reduce the emissions over the last few decades [2]. As a step toward reducing GHG emissions, more than 100 countries signed and ratified the Paris Climate Agreement, aiming to limit the increase in global temperature to 1.5–2 °C over the next 30 years [3]. The achievement of this target requires the swift adoption of carbon-neutral and carbon-negative technologies to limit the global GHG emissions to approximately 9.8 gigatons of carbon [4,5]. Several approaches are being considered for CO_2 removal from the atmosphere, such as the adoption of bioenergy, direct carbon capture, afforestation and reforestation, the modification of agricultural practices, the use of bioenergy, and the direct infusion of recalcitrant carbon into the soil using biochar (BC) [6–9]. The longer-term sequestration of carbon into the soil using biochar is one of the potential carbon-negative approaches. As soils store twice as much carbon compared to atmospheric reserves and for longer periods, it has been hypothesized that increasing global soil organic matter stocks by 4 per 1000 (or 0.4%) per year in agricultural land can offset 30% of global greenhouse emission [4].

The agricultural and industrial revolutions, combined with unsustainable farming practices, have significantly affected global soil health. This is mainly a consequence of the type of fertilizer used in crop production. Earlier practices of using manure and compost replenished the soil organic matter (SOM) on a regular basis. However, the use of petroleum-derived chemical fertilizers is detrimental to SOM as they enhance the accumulation of salt and reduce microbial diversity. Fertilizers derived from the Haber–Bosch process contribute to more than 1% of total global CO_2 emissions [10,11]. Soil health has further declined with the gradual acidification of arable lands and continual soil erosion negatively affecting crop yields throughout the world. While the use of compost and manure to enhance and maintain SOM is an option, it presents limitations due to the accumulation of organic pollutants, increased pathogen pressure, and leaching of excess nutrients into waterways, leading to eutrophication [12,13].

There is a long history of enriching soils with recalcitrant carbon practiced by indigenous farmers in different parts of the world. Black-earth-like anthropogenic soils known as 'Terra Preta' have been discovered in several regions of South America and Japan. These dark soils were amended with charcoal-like substances, generally referred to as biochar (BC), and possibly other amendments such as manure, which conferred enhanced fertility to the soil [14,15]. Chemical analysis revealed that the BC-treated areas contained 70 times more carbon than the surrounding soils, demonstrating its long half-life [16]. The enhanced fertility of these soils most likely resulted from increased SOM, higher pH, higher water-holding capacity, and high nutrient-holding capacity [16–18]. Due to the potential advantages of 'Terra Preta', several global efforts are afoot to recreate such soils. Biochar represents an organic soil amendment that improves soil quality for agricultural production [19].

There are various studies reporting the impact of BC on plant growth and development; however, the results have remained inconsistent [20]. There are several types of biochar, produced using various pyrolysis parameters and feedstocks. A standardized biochar with predictable physical, biological, and chemical properties, which is beneficial to plant growth and development, remains to be developed [21]. Furthermore, a positive impact on growth and development does not directly translate to improved yields. In fact, there are reports indicating that enhanced growth and development compromises the plant defense systems. Therefore, this review discusses and summarizes the various aspects of how biochar impacts soil quality and crop productivity. The discussion concludes with a summary of some of the areas that need to be addressed to enable the widespread use of biochar in environmental, economic, agricultural, and ecological contexts.

2. Biochar and Soil

2.1. Biochar

Biochar (BC) is a carbon (C)-rich, porous material produced during the process of pyrolysis, which involves the thermochemical decomposition of organic matter in an oxygen-limited environment. Any feedstock, such as forest residue, agricultural by-products, and waste biomass can be converted into liquid fuels, gasses, and BC. The properties and yields of BC are highly variable depending on the rate of pyrolysis (fast/slow), feedstock, pyrolysis temperature, and retention time. Generally, slow pyrolysis with a heating rate of 5–20 °C per minute with higher residence time results in higher BC yield [22,23]. Fast pyrolysis with a higher heating rate (>100 °C/min) and lower residence time results in a higher yield of liquid fuel and reduced BC output [24]. Due to the complex nature of pyrolysis and diversity of feedstock, the final chemical and physical properties of BC vary. For example, a recent meta-analysis concluded that BC produced at higher temperature (600–699 °C) had a higher pH of approximately 9 compared to BC produced at lower temperatures (300–399 °C) with an approximate pH of 5 [25]. This observation was supported by another recent meta-analysis [26]. The higher reaction temperatures reduce the amount of aliphatic carbons, oxygenated functional groups, cation exchange capacity (CEC), and total content of N, H, and O. However, a higher temperature of production resulted in increased pH, amount of C fixed, total ash content, total C, and surface area of BC [25,27,28]. Ultimately, the bulk property and surface characteristics of any BC is determined by the feedstock source along with the pyrolysis parameters [23,29]. There remains a critical need to understand the characteristics of BCs produced from different feedstock, pyrolysis parameters, and the resulting relative impact on soil. In the following sections, recent research on BCs has been collated, and the effects of various BC regimens on soil physical properties, soil–water relations, soil organic matter, microbial activity, soil tilth and nutrient status, pH, crop productivity, biotic stresses, and abiotic stresses have been discussed.

2.2. Impact of Biochar on Soil

Physical Properties

Physical properties of soil, such as bulk density, porosity, and water retention are important variables that impact plant growth and development. Human intervention in agricultural practices causes soil compaction, which is one of the key factors affecting plant growth [30]. Soil texture also plays a key role. Soil compaction above $1.7\ \text{g cm}^{-3}$ results in restricted root growth and limits access to water and nutrients [31]. As a consequence, the yields of many crops such as soybean and corn have been shown to be negatively impacted [32,33]. The threshold bulk density for impact on root growth varies, with clayey soils having a lower bulk density threshold.

Amending soils with biochar increases soil porosity while decreasing soil bulk density, which aids in the transport of water, nutrients, and gases. These alterations encourage root formation and increased microbial respiration [26].

A meta-analysis reported that the addition of BC to soil reduced the bulk density of the soil by an average of 7.6% and increased its water-holding capacity and porosity by 15.1% and 8.4%, respectively [34]. Similar results were reported in another meta-analysis where the average bulk density was decreased by 12% [35]. Fifteen of the 17 studies conducted in 2019 reported that biochar effectively reduced soil bulk density and increased the porosity and available water content (Table 1). However, there were two studies that reported either no effect on bulk density after the addition of BC [36] or an insignificant decrease [37]. It was also observed that a larger average BC particle size was more effective in reducing the bulk density of sandy loam soil than sandy soil [38]. In the case of sandy soil, bulk density significantly decreased and water-holding capacity was significantly increased with the addition of BC with small particle size [38]. A majority of the recent studies used biochar produced from agricultural residue and woody residue. Generally, a positive effect on the physical properties of soil was reported (Table 1).

Table 1. A summary of recent studies related to the impact of biochar on physical properties of soil. The soil types listed in this table correspond to the types reported in the original publication. The usage of term 'slightly' in any of the categories indicates a change in mean values. BC: biochar.

Exp. Type	BC Feedstock	Pyrolysis Temp. (°C)	Soil Type	Bulk Density	Available Water Content	Total Porosity	BC Application Rates	Ref.
Lab	Agricultural residues	450	Loamy	Slightly decreased	Slightly increased	Increased	0.46% (W/W)	[37]
			Sandy	Slightly decreased	Slightly increased	Increased		
Lab	Woody residues	620	Sandy loam	Decreased	Increased	N/A	1%, 5%, 10% and 20% (V/V)	[38]
			Sandy	Decreased	Increased			
Field	N/A	N/A	N/A	Decreased	Increased	Increased	5 and 10 tons ha^{-1}	[39]
Field	Sewage sludge	700–850	Loamy sand	Decreased	N/A	N/A	20, 40, and 60 tons ha^{-1}	[40]
Lab	Agricultural residues	450	Sandy loam	Decreased	No effect	Increased	27.5 tons ha^{-1}	[41]
			Clay loam	Decreased	Increased	Increased		
Field	Agricultural residues	200–600	Loam sand	Decreased	N/A	Increased	10, 25, and 50 tons ha^{-1}	[42]
Field	Agricultural residues	360	Sandy loam	Decreased	N/A	Increased	4.5 and 9 tons ha^{-1}	[43]
Lab	Agricultural residues	300 and 700	Desert	Decreased	Increased	Increased	5% (W/W)	[44]
Field	Woody residues	N/A	N/A	Decreased	Increased	Increased	55 tons ha^{-1}	[45]
Lab	Agricultural residues	350–650	Sandy	N/A	Increased	N/A	1%, 2%, 3%, and 4% (W/W)	[46]
Lab	Woody residues	350	Sandy loam	Decreased	Increased	Increased	2%, 4%, and 6% (W/W)	[47]
Field	Woody residues	500	Silt loam	Decreased	Increased	Increased	24 and 46 tons ha^{-1}	[48]
Field	Agricultural residues	550–600	Clay loam	Decreased	N/A	Increased	10, 20, and 30 tons ha^{-1}	[49]
Yard	Agricultural residues	400–450	Planosol	Decreased	Increased	Increased	1%, 2%, and 3% (W/W)	[50]
Field	Agricultural residues	550	Haplic Luvisol	Decreased	Increased	Increased	10 and 20 tons ha^{-1}	[51]
Lab	Forest residue	450	Desert sandy	Decreased	Increased	Increased	39.5, 58.7, and 65 tons ha^{-1}	[52]
Field	Agricultural residues	550	Sandy clay loam	No difference	Decreased	No difference	5.5, 16.5, and 33 tons ha^{-1}	[36]
Field	Woody residues	580	Luvisol	Decreased	Increased	Increased	25- and 50-tons ha^{-1}	[53]

A most recent meta-analysis showed an average increase in soil porosity by 6.27%, decrease in bulk density by 7.47%, and increase in water-holding capacity by 9.82% [54]. However, biochar derived from softwood and walnut shell did not affect soil porosity or water retention over a period of six years in silty clay soil. It was suggested that the effect on soil porosity and water retention was temporary until the pores of biochar were occluded with clay or soil organic matter (SOM) [55]. Woody biomass derived-biochar was shown to have no effect on soil porosity or water retention after four years of amendment [56].

Comparably, in the case of soil bulk density, a majority of the studies reported a decrease after BC amendment [34,35,54]. There were also a few reports where no significant decrease in bulk density was observed [57–60]. A majority of the data from previous meta-analyses and reports indicate that the addition of BC to a coarse, textured soil had a larger positive impact on soil physical properties compared to clay textured soil [34,35,54,61,62].

Biochar produced from wheat straw (550–600 °C) incorporated with clay loamy soil improved its physical properties and enhanced the yield of wheat when irrigated with saline water [49]. The biochar amendment decreased soil bulk density by 5.5–11.6% and increased porosity by 35.4–49.5%. The biochar amendment also seemed to mitigate soil sodicity and also increased total NPK (nitrogen, phosphorous, potassium) availability in mixed soil layer. This resulted in the improvement of wheat yield by 8.6% and 8.4% at the BC application rates of 10 and 20 t/ha, respectively [49]. However, at the application rate of 30 t/ha^{-1}, the improvement in yield was the lowest (2.2%), which was probably due to high salinity and the immobilization of N. This study suggests that for saline irrigation in clay loamy soil, the optimal application rate of BC produced from agricultural residue should be between 10 and 20 tha^{-1} [49]. These studies imply that initial soil characteristics, along with BC application rate and type, determine the final changes in the physical properties of the soil.

2.3. Soil–Water Relations

Accessible fresh water supplies are becoming increasingly limited, and 70% of available fresh water supports crop irrigation [63]. Although biochar holds promise for improved hydrological functions, there are differing schools of thought regarding the role of BC in improving the long- term water-holding capacity of soil [64]. BC amendment has been reported to increase rainfall absorption and soil water-holding capacity, particularly in non-irrigated production regions [65–67]. However, the pre-existing physical and biochemical characteristics of the soil and the wide array of BC production parameters (feedstock inputs, pyrolysis temperatures, application methods, and geographical variables) ultimately determine the BC's impact on water-holding capacity. In order to probe the influence of BC on water dynamics, initial experiments were performed with soil columns in greenhouses with the addition of farm or potting soils. Field studies are now becoming prevalent in peer-reviewed literature, particularly within the last 10 years.

The identification of key features that contribute to improved water retention could lead to an expanded role for BC in crop production. Overall, it was determined that feedstock selection and pyrolysis temperature, the most predictive variables impacting water status, impact BC surface chemistry and porosity, the latter of which is a major contributor to the water-holding capacity of BC [68,69]. Pore saturation is highly dependent on BC surface chemistry, which is affected by pyrolysis temperature. An increase in pyrolysis temperature volatilizes organic elements and thermally cracks the biomass, thereby rendering hydrophobic compounds more hydrophilic and increasing the overall BC porosity [70]. Conversely, BCs produced via low-temperature pyrolysis exhibit negative capillary pressure, inhibiting the hydration of the pore space [71].

Comparative analysis of Fourier transform infrared (FTIR) spectroscopy data collected from nine different feedstocks pyrolyzed at 250 °C, 500 °C, and 700 °C revealed the relationship between BC surface chemistry and hydrophobicity [72]. The spectrometer data indicated that the functional group C=O in carboxylic acid was present only in the BC obtained from pyrolysis at 250 °C, making it hydrophobic. BC produced at 500 °C and 700 °C were deemed more appropriate for improving soil water status. A significant correlation was identified between low pyrolysis temperature (<300 °C) and surface functional groups (specifically acidic moieties), and increased hydrophobicity contributing to low water retention was reported [68,73]. Other factors, including cation exchange capacity, play a role along with the variables of surface groups and porosity in determining the hydrophobic properties for each specific BC [74].

Considerable variation in total pore volumes was reported in BCs produced at 400 °C, 600 °C, and 800 °C from various feedstocks. Wood-based BC possessed a comparatively higher range of micropores (5–30 μm), and although the number of micropores decreased with increasing pyrolysis temperatures, this BC still retained relatively large pore volumes overall due to pyrogenic micropores. In contrast, the pore volumes of BC derived from poultry manure and agricultural wastewater sludge were smaller, indicating that these feedstocks may not be suitable for improving water retention in amended soils [75]. While BC amendment imparts large increases in porosity, permeability, and moisture retention in clay soils, these affects are diminished in silt loam soils [76,77]. The particle size of BC had a clear impact on soil bulk density, with a linear decrease in bulk density of sandy soil observed when large-particle-containing hardwood BC (620 °C) was used. Smaller BC particles increased water-holding capacity compared to larger BC particles [38], which was possibly due to the increased microporosity resulting from higher pyrolysis temperatures. Despite this, the addition of BC at 25 Mg ha^{-1} to sandy soils did not result in increased water retention. In a study with *Miscanthus giganteus* residue-derived BC (450 °C), the increased porosity of larger BC particles proved beneficial for soil water retention, while smaller BC particles under 0.15 mm retained water too well, thereby strongly reducing its bioavailability [37].

BC was reported to increase the water-holding capacity in coarse and medium textured soils by an average of 51% and 13%, respectively [78]. This was attributed to a higher abundance of soil micropores resulting from the intrinsic microporosity of BC. However, a reduction in water-holding capacity was reported in fine-textured soils, which was possibly due to the overall decrease in micropores or occlusion of existing pores. Field studies of high-porosity BCs derived from softwood (600–700 °C) and walnut shell (900 °C) reported a temporary improvement of water-holding capacity; however, no long-term improvement was seen in BC-amended silty clay loam soils subjected to a corn–tomato rotation with conventional or organic production regimes [55]. Plant-available water in fine-textured soils could be enhanced through the management or manipulation of hydrophobic properties of BC, thereby improving BC–soil interactions [78]. For example, it has been reported that grapevine feedstocks subjected to low pyrolysis temperatures (approximately 400 °C) yield BC with a 23% higher available water content in clay soils [79].

2.4. Soil Tilth and Nutrient Status

Defining management approaches to increase the productivity of agricultural soils remains a priority as food demand increases and arable farmland decreases [80]. As a mineral-rich organic material, BC can be incorporated into agricultural soils, potentially serving as a slow-releasing fertilizer, positively affecting soil tilth and enhancing the nutrient status of agricultural soils [81–83]. The basis for this potential use lies in the unique porosity of BC, its facilitation of chemical and physical interactions between nutrients and the carbon material, and its strong intrinsic sorption properties. Due to the large surface area, porous microstructures, and negative surface charge, BC enhances nutrient retention in the soil. Furthermore, the nutrient retention properties of BC may significantly reduce irrigation or the rainfall-induced leaching of water-soluble minerals [66,84]. The slow desorption of the BC-sequestered nutrient elements may supply a steady rate of nutrient delivery, thereby alleviating the need for excessive fertilizer use. Together, these agronomic benefits to soil health may also mitigate freshwater eutrophication that results from fertilizer runoff, prevent pesticide contamination, and reduce the risk of environmental damage [85–87].

While composition varies based on feedstock and pyrolysis parameters, a universal characteristic of BC is that it is carbon-dense, which facilitates the retention of necessary plant nutrients such as N, P,

K, Mg, Fe, and Ca [88–91]. Depending on the soil status and existing nutrient deficiencies, BCs can be custom-manufactured to replenish depleted nutrients. It has been demonstrated that BCs derived from different feedstocks possess variable amounts of beneficial plant nutrients [47,66,91–94]. The general characteristics of three major BC feedstock sources are as follows:

- Organic waste feedstocks, such as animal manure and sewage sludge-derived BC, are rich in potassium and phosphorus, low in C levels, and low in surface area; additionally, eggshell-derived BC is elevated in calcium levels
- Wood-based BC is high in organic matter and surface area, while low in CEC and N, P, and K levels
- Crop residue-derived BC properties reside somewhere in between those of the two previous categories, with specific crops producing BC with different properties (e.g., wheat and rice BC is high in silicon content; soybean BC is high in N).

These feedstocks can be blended in appropriate ratios to produce BCs with desired nutrient and/or mineral profiles. Further modifications, including the alteration of pyrolysis parameters, physical alterations, chemical modifications, and BC-mediated composting have been discussed to aid in the customization of BC to ameliorate detrimental soil aspects [91].

Soil pH and the abundance/availability of important plant nutrients such as phosphorus (P) and nitrogen (N) are positively affected in BC-amended soils. Limiting the pyrolysis temperatures to less than 700 °C enhances the levels of P and N in BC, both of which can be lost at higher temperatures due to the volatilization and transformation of NH_4^+ to heterocyclic-N [92]. Wood-derived BC (450–550 °C) applied at 20 t/ha significantly improved the bioavailability of P in sandy soils, which is an effect that was primarily attributed to the perturbation of abiotic processes (adsorption/desorption of P, altered redox potentials, development of organomineral aggregations) [95]. BC nitrogen levels are correlated closely with the original source of the char; feedstocks high protein biomass, such as grasses, generate BCs with higher N levels (approximately 10% by weight), while wood-derived BCs tend to be N-poor (approximately 1% by weight) [92]. While several individual studies show that wheat BC (450 °C) applications increased total soil N [96], other studies found significant decreases of soil NH_4^+ and NO_3^- following BC addition. The latter outcome was likely due to the inherent recalcitrance of the small amount of extractable inorganic N and organic N present. In the studies reporting increased N, this increase could be attributed to a heightened abundance of recruited microorganisms, which assist in the degradation of soil organic nitrogen [97].

In addition to improving mineral nutrient retention, BC has a role in the amelioration of soil erosion and the improvement of overall soil structure [98,99]. A study utilizing hardwood (600 °C) BC at 15 and 30 t/ ha concentrations to amend clay-rich soils in incubation containers demonstrated improved soil aggregate structure and soil stabilization [100]. This is likely due to the interaction of carboxylic and phenolic functional groups on the BC surface, resulting in the formation of cation bridges and consequent BC–mineral complexes [101]. For example, microaggregates observed to form upon the incorporation of hardwood-derived BC (700 °C) into soil with application rates of 2.5% or 5% correlated with a 50–64% decrease in soil loss, respectively [102]. An additional study with oak wood-derived BC applied at a rate of 10 Mg/ha provided further evidence for the stabilizing effects of BC, with significant decreases in soil loss of almost 20% observed in a simulated rainfall experiment. In addition to improving soil retention, BC appeared to reduce the impact force from rainfall, thereby facilitating the reduction of particle detachment [103].

2.5. Soil Acidification

The expanding global incidence of soil acidification is concerning, with acidic soils (pH < 5.5) currently accounting for approximately 50% of arable land [104,105]. The excessively low pH of acidic soil results in reduced productivity and decreased crop fertility. The main causes of soil acidification include the use of ammonia-based fertilizers and low nitrogen-use efficiency. In soil, ammonia fertilizers are converted to nitrates and hydrogen ions. The hydrogen ions that are left over following the uptake of nitrates by crops or after nitrate leaching increase the soil acidity [106]. The removal of crop residue also accelerates soil acidification. An excessive reduction of pH leads to the increased solubility of soil-bound aluminum; thus, soil acidification generally leads to aluminum (Al) toxicity [105]. Aluminum toxicity, in turn, leads to deficiencies in phosphorus, calcium, magnesium, and potassium cations and contributes to impaired root growth.

Current strategies to alleviate soil acidification include liming, the application of crop residue, and the use of industrial products; however, these methods have several disadvantages. For example, liming material elicits a disproportionately strong effect on top surface soils in comparison with lower layers. This method is also costly, due to the high transportation costs of liming material [106,107]. The application of industrial products can lead to heavy metal toxicity [108]. Similarly, the excessive application of organic material may lead to both heavy metal accumulation and eutrophication, the latter resulting from augmented concentrations of nitrogen and phosphorus [26,109]. Hence, biochar, which is naturally alkaline, is a potential solution to the problem of soil acidity.

Various studies have validated the effectiveness of BC in reducing soil acidity [110–112], and a linear correlation of biochar alkalinity with the resulting soil pH has been established [105]. Furthermore, it has been demonstrated that the increased pH-buffering capacity of BC-amended soils is due to a BC-derived increase in cation exchange capacity [113]. The carbonates and oxides of cations such as Ca, K, Mg, Na, and Si formed during pyrolysis are known to react with dissolved Al and hydrogen ions in soil, leading to increased soil pH and decreased Al uptake by the plants [114]. Previous meta-analyses and individual studies have concluded that in imparting increased buffering capacity, BC amendment can increase the soil pH by >2.0 units [115]. Not surprisingly, the original feedstock material plays a key role in determining the final pH of BC. For example, BC generated from manure has higher alkalinity, pH-buffering capacity, and propensity for the alleviation of Al toxicity, compared to crop residue-derived BC. Thus, the former would be more suitable for extremely acidic soils [116]. Soils exhibiting Al toxicity could be reclaimed via BC amendment, the ash content of which would precipitate Al_3^+ to less toxic $Al(OH)_3$ and $Al(OH)_4$ [117]. Furthermore, carboxyl and other organic functional groups on the BC surface would provide additional sites for Al_3^+ binding [117]. Functional groups such as COO^- and O^- also contribute to the alkalinity of biochar through reaction with free H^+ ions [27,118].

In soil, H^+ is produced through the aerobic conversion of ammonia to nitrate. Experimental results have demonstrated that BC amendment leads to decreased soil nitrification through the adsorption of NH_3 and NH_4 onto the BC surface. Soil amendment with wheat straw-derived BC (500 °C) led to reduced nitrification in cadmium-contaminated Ferralsol soil by decreasing soil acidity (Table 2) [119]. Similarly, amendment with pig manure-derived BC (300 °C) resulted in decreased soil acidification and increased cation exchange capacity [120,121], and crop residue-derived BC (500 °C) led to improved rice growth, yield, and soil nutrient availability in acidified soil [122]. Collectively, information from the literature has established that carbon content, nutrient availability, and alkalinity are highest when BC is generated from manure feedstock, intermediate when generated from crop residue feedstock, and lowest when generated from woody plants-based feedstock (Table 2). Finally, biochar produced at higher temperature has higher pH and might be more suitable for countering soil acidity.

Table 2. Selected recent studies documenting the impact of biochar on acidified soils.

Exp. Type	Soil Type	BC Feedstock	Pyrolysis Temperature (°C)	Effect of Biochar Amendment	BC Application Rates	Reference
Lab	Ultisols	Crop residue	400	Inhibited soil re-acidification and increased pH buffering capacity	3% (W/W)	[105]
Lab	Ferrosol	Crop residue	450	Promoted nitrification and inhibited re-acidification of Cd-contaminated soils	3% (W/W)	[119]
Lab	Sandy	Pig manure and poultry litter	300	Decreased soil acidification and increased cation exchange capacity	0.5 %, 1%, and 2% (W/W)	[120]
Lab	Ultisols / Oxisol	Crop residue	400	Increased soil pH buffering capacity and increased the resistance of soils to re-acidification	3% and 5% (W/W)	[123]
Lab	Ultisols	Crop residue	400	Increased soil pH, neutralized soil acidity, increased soil pH-buffering capacity, and increased resistance of soils to re-acidification	1% and 3% (W/W)	[124]
Lab	N/A	Crop residue	500	Biochar significantly promoted rice growth and the yield increased in acidified soil	2% (W/W)	[122]
Lab	Oxisols	Crop residue	N/A	Alleviated soil acidification	1%, 2% and 5% (W/W)	[125]
Lab	Loamy sand	Sewage sludge	300	Reduced soil acidification	0.5%, 1% and 2% (W/W)	[126]

3. Biochemical Properties

3.1. Soil Organic Matter (SOM)

Soil organic matter comprises the total organic carbon in a soil and is the main determinant of overall soil fertility. SOM components consist of plant residue, animal waste, microbial populations, and active and stable organic matter in soil. SOM contributes to soil fertility by serving as a nutrient source for crops and microbes, causes soil aggregation, and improves water retention and nutrient exchange. It also helps to reduce soil compaction and surface crusting. It has been reported that the impact of biochar on SOM depends on the following variables [127–129]:

1. Type of biomass used for production of BC
2. Pyrolysis temperature
3. Pre-existing SOM levels in the soil

Amending soils with biochar often results in alterations in C cycling and mineralization, and this effect is known as 'priming'. Previous studies have reported both positive and negative effects of priming. Grass-derived BC produced at lower temperatures (250 °C and 400 °C) resulted in positive priming resulting in increased C mineralization. However, BC produced at higher temperatures (525 °C and 600 °C) from hardwood resulted in negative priming [128]. It was hypothesized that negative priming resulted from the organic matter binding to the biochar and thereby becoming unavailable to microbial and enzymatic action.

Analogous results were observed in a study that showed that crop residue-derived biochar produced at lower temperature (300–550 °C) generally resulted in positive priming when applied to arable and fallow soils; however, in the case of grassland soils, the effect was negative [127]. In another study, which aimed to analyze the short-term effect of biochar on SOM, BC produced from woody feedstocks at lower temperatures (350 °C) had greater positive priming during 0–13 days of biochar application both in low and high pH clay loam soil [130]. The extent of positive priming was reduced for low and high pH clay loam soil when BC produced at higher temperature (700 °C) was used. The addition of fresh labile substrate, such as rye grass, to BC produced at both high and low temperatures further increased priming and mineralization [130]. If the goal is to sequester carbon, rapid mineralization caused in conjunction with low temperature-derived BC results in carbon loss, necessitating BC reapplication. The application of high temperature-derived BC can also be used to reduce the priming effect and aide in carbon sequestration.

The addition of 3% (w/w) BC prepared from forest residues at 550 °C has been reported to delay the decomposition of SOM and reduce N mineralization when added to acidic red loam soil [131]. However, some studies did not find BC to contribute to SOM decomposition [132,133]. A reduction in priming and a 16% reduction of SOM decomposition was reported when crop residue-derived BC pyrolyzed via gasification at 1200 °C was added to sandy loam soil [134]. The reduction may be due to a shift in the preference of the microbial community for biochar as a C source [135]. There have been various studies in which BC produced at lower temperature ranging from 450 to 550 °C stimulated positive priming when added to sandy loam (Table 3). Overall, BC promoted increases in C sequestration, organic carbon retention, SOM, mineralization, phosphorous and potassium content, and plant biomass [136–139]. Conversely, one study reported a decrease in soil microbial biomass and SOM mineralization when crop residue-derived BC (450 °C) was applied to sandy loam soil [140].

Table 3. Selected recent studies summarizing the effect of biochar on soil organic matter (SOM).

Exp. Type	Soil Type	BC Feedstock	Pyrolysis Temperature (°C)	Effect of Biochar Amendment	BC Application Rates	Reference
Lab	Acidic red loam	Forest residue	550	Decomposition of SOC(soil organic carbon) declined and reduced mineralization of SOM	1% and 3% (W/W)	[131]
Field	Sandy loam	Crop residue	Gasification at 1200	Reduced SOM degradation by 16%.	30 tons ha^{-1}	[134]
Lab	Podzol Antric	Woody biomass	550	Increased the SOM mineralization	1% (W/W)	[141]
Field	Sandy	Crop residue	350	Increased soil organic matter and N	5% (W/W)	[142]
Field	Sandy	Crop residue	450	Decreased organic matter and N content	5% (W/W)	[142]
Field	N/A	Sewage sludge biochar	N/A	SOM are increased	16.5 tons ha^{-1}	[136]
Field	Plaggic Anthrosols	Crop residue	350	Positive priming	N/A	[137]
Field	Silt loams	Woody biomass	900	Increased soil organic matter, soil pH, phosphorus, potassium, sulfur, and the shoot and root biomass of wheat	12, 24.6, and 49.3 tons ha^{-1}	[138]
Field	Sandy loam	Crop residue	450 to 500	Decrease of SOM mineralization, reduce soil microbial biomass	5.5 tons ha^{-1}	[140]
Lab	Sandy loam	Woody biomass	450	Increased organic carbon retention and promoted carbon sequestration	2%, 5%, and 10% (W/W)	[139]
Field	Sandy loam	Crop residue	360	SOC increased after biochar application and did not contribute to soil aggregation	4.5 and 9 tons ha^{-1} year^{-1}	[43]
Lab	Sandy loam	Crop residue	600	Significantly increased SOM, microbial respiration, and microbial biomass	0.5% and 1% (W/W)	[143]

3.2. Microbial Activity

Considerable emphasis has been placed on the topic of microbial dynamics in agricultural systems and their role in crop productivity. The health and diversity of soil microbial populations as a function of agro-ecosystem well-being has diverse implications for water-use efficiency, soil structure and stability, nutrient cycling, disease resistance, and eventual crop productivity [144,145]. While other organic amendments are only stable for relatively short periods in the soil environment, BC is more stable and remains in the soil for hundreds to thousands of years, as it is not easily degraded, and it could support soil microbial communities for an extended period of time with reduced inputs [66].

The diverse and specific physiochemical characteristics of BC that influence soil microbial composition are increased labile carbon, pH, surface area for colonization, and water content in amended soils. BC addition induces remodeling of the microbial diversity and community structure of the soil; however, the changes are highly variable and dependent on the individual soil properties [146,147]. It was reported that low pyrolysis temperature BCs (>350 °C) harbor a greater number of organic residues and are commonly characterized by lower pH. In contrast, at high temperatures (<600 °C), the abundance of organic moieties contributes to the production of a higher pH BC. It was concluded that pyrolysis temperature (and the BC-related characteristics associated with temperature) is the single most important factor that determines how the microbial communities are influenced [148]. Overall, there is a consensus that BCs foster the growth and maintenance of soil microbial communities [95,149–151].

3.2.1. Fungi

In terms of their abundance and diversity, both beneficial [152–154] and detrimental [95,155–157] effects of BC on fungal communities have been reported. In comparison to bacteria, fungi respond differently to organic and inorganic treatments. Soil bacteria act as better indicators of soil fertility than soil fungi [158]. The mechanisms for improved fungal diversity and abundance appear to be correlated more with the physical microstructure of BC and the recalcitrant organic carbon than other factors. This was demonstrated in a study where corn straw BC (500 °C) derived aqueous extractable substances and organic extractable substances, and the remaining solid BC were tested [159]. It has been hypothesized that BC addition preferably fosters bacterial communities over fungal communities. The bacteria may starve the fungi of C and therefore outcompete them [155].

It was demonstrated that fungal diversity was lowered in soybean and rice straw BC (500 °C) soils compared to controls, although individual order, family, genus, and species level fungal communities were affected differently [160]. These outcomes could be a result of the "unbalanced competition" theory. This theory describes the phenomena of saprotrophs exponentially increasing their abundance due to the easily mineralizable carbon found in BC, therefore leading to an overall decrease of other fungal groups and potentially suppressing their abundance and diversity [161]. Other speculations underlying decreased fungal diversity and population include the high levels of organic compounds, mineral elements, and higher soil pH due to BC amendment [162].

3.2.2. Bacteria

The microbial community consisting of bacteria tend to respond positively to BC, as several studies have reported a significant increase in abundance and diversity, after BC application, especially in the rhizosphere soil [162–165]. For example, an increase in specific bacterial families and species such as phosphorous solubilizing [166], nitrifiers [167], and N-fixing and denitrifiers [168] was reported with *Malus pumila* woodchip BC (500 °C), *Eucalyptus saligna* hardwood BC (550 °C), and sugar maple wood BC (400 °C) soil amendment, respectively. Additional studies found only modest or no differences [140,159,168,169]. The change in the composition of bacterial communities after the incorporation of BC in soil is highly dependent on the pre-existing bacterial community, soil type, and overall BC characteristics.

Generally, the Gram-negative bacterial community is favored in the nutrient-enriched BC-amended soils and initially predominates the soil environment since it performs specific and narrow functions. They outcompete Gram-positive bacteria that rely on recalcitrant C as their main energy source. Gram-positive bacteria become the dominant bacteria type over time due to BC's ability to form stable aggregates with soil organic matter (SOM) [148]. Utilizing sugarcane-derived BC (450 °C), it was found that bacterial populations increased significantly while fungal populations were significantly reduced in heavy metal-contaminated soils. This was possibly due to the enhanced heavy metal immobilization by the BC addition, although other factors may have contributed to the observations [162]. Similar results with wood (fir, cedar) BC (450–550 °C) indicated a significant shift toward a bacteria-dominated microbial community in a short-term study (3 months) and was attributed to the increased release of labile C from the BC or stable SOM–BC aggregates [95]. A study utilizing bamboo BC (500 °C) provided further support for the concept that bacteria are more sensitive to BC compared to the fungal community, which was mostly due to increased pH with increasing BC addition [170]. These results indicate that alkaline conditions due to BC amendment (liming effect) favor and promote bacterial growth and may inhibit fungal growth.

The high complexity of BC–soil interactions and microbial community dynamics leaves many 'gray areas' in this field that require further investigation. However, assessing the long-term effects of BC-amended soils and microbial population diversity and activity are highly recommended for future studies. Research has indicated that the BC surface and pores can be inundated with plant exudates and dead cells, inorganic and organic complexes, and larger soil microorganisms. These factors may reduce the total available space for microbial colonization of aged BCs over time [171].

3.3. Abiotic and Biotic Stressors

3.3.1. Heavy Metals

Soil contamination with organic and inorganic toxins increases environmental and agricultural risks and poses a threat to both plants and humans. Efforts to develop remediation processes that bind the contaminants, limit their mobility and bioavailability, and foster improved soil health are ongoing. Currently, organic materials such as charcoal, soot, kerogen and activated carbon are used as amendments for limiting and reducing the bioavailability of multiple soil contaminants [172,173]. The organic contaminants have been shown to sorb preferentially to the carbonaceous fractions present in soil, limiting their bioavailability [174]. BC has also been shown to reduce the bioavailability of heavy metal contaminants. Several studies analyzed the effect of BC amendment on soils contaminated with heavy metals such as arsenic, cadmium (Cd), copper (Cu), nickel (Ni), lead (Pb), and zinc (Zn) [175–177]. The high surface area of BCs results in more effective contaminant binding; however, one of the recent meta-analyses pointed out that the pyrolysis temperature at which a given BC is produced influences its remediation efficiency and type of contaminant that can be removed [175]. A majority of studies tested the effect of BC on Cd pollution and concluded that the higher BC surface area had a smaller effect on Cd bioavailability [175]. The BCs produced at lower temperatures (300–500 °C) have a higher density of functional groups, while BC produced at higher temperatures results in a larger surface area and lower density of functional groups. Another study revealed that BC produced from wheat straw at 450 °C, with a higher density of functional groups, was more effective in treating Cd and Pd-contaminated soil [178]. However, BC produced at higher temperature is also more alkaline and results in the immobilization of heavy metals in acidic soil via the liming effect. The addition of rice and wheat straw-derived BC in soils contaminated with Pb, Cu, Cd, and Zn led to a reduced mobility and bioavailability of the heavy metals, resulting in increased yields and a decreased enrichment of heavy metals in the tested plants [176]. Biochar is also used for the remediation of soil from contaminated sites due to rapid industrialization. It has been recently demonstrated that BC derived from pine wood was able to reduce the bioavailability of Cd, Pb, and Zn in metalloid-contaminated soils at a smelting site and promoted plant growth [179]. Biochar derived from hardwood (600 °C) was shown to be effective in reducing Ni and Zn by 83–93% in a historically polluted site in the United Kingdom [179].

The effect of sewage sludge BC pyrolyzed at 330–500 °C on alleviating heavy metal toxicity was evaluated. It was observed that with the increase in pyrolysis temperature, the availability of heavy metal in tropical soils was decreased [180]. This might be due to the increased pH, pore size, and surface area in the BC produced at higher temperatures leading to the formation of carbonates, sulfates, phosphates, and metal hydroxides [180]. Due to the reduction in the bioavailability of heavy metal cations, maize yields increased in BC-amended soil in comparison to NPK-fertilized soil [180]. Irrigation with untreated wastewater leads to the accumulation of lead, cadmium, zinc, and iron, which can be taken up by plants or leach into ground water, adversely affecting plant growth and human health. In a recent study, the effect of plantain peel-derived biochar (450–500 °C) on potato yield was studied in sandy soil irrigated with wastewater [181]. This BC regime resulted in the adsorption of soil Cd and Zn and the reduction of the Cd level by 69% and 33% in tuber flesh [181].

A summary of selected studies reporting the effect of BC on the alleviation of heavy metal toxicity is presented in Table 4. Most of the listed studies were carried out in pots. Therefore, large-scale field studies are required to understand the interactions among a particular biochar, soil type, and contaminant. To use biochar for soil remediation, the specific soil and biochar properties must be taken into consideration. Some studies show that certain BC amendment results in high heavy metal immobilization. However, if the mechanism of immobilization is only physical adsorption or cation exchange, these BCs may not be suitable for long-term remediation due to weakly bound metals. BCs that immobilize heavy metals through precipitation or complex formation should be used for long-term remediation.

Table 4. Selected recent studies reporting the effect of biochars (BCs) derived from various feedstocks on heavy metal remediation in different types of soils.

Exp. Type	Soil Type	BC Feedstock	Pyrolysis Temperature (°C)	Effect of Biochar Amendment	BC Application Rates	Reference
Lab	N/A	Wood, bamboo, rice straw, and walnut shell	500	Reduced Zn, Cd, Cu, and Pb solubility	5% (W/W)	[182]
Lab	Aridisols	Woodchip-derived biochar	300	Reduced extractable Cd, Pd, Ni, and Cu. Improved antioxidant enzyme activity. Increased rapeseed fresh shoot biomass, fresh root biomass, total chlorophyll, total pigments, carotenoids, and lycopene concentration	1% and 2% (W/W)	[183]
Lab	Sandy loam soil	Wood derived biochar	350–500	Reduction in the accumulation of Cu and Zn in spinach	5% and 10% (W/W)	[184]
Lab	N/A	Switchgrass and poultry litter	700	Decreased the Zn, Cd, and Pb bio-accessibility	0.5%, 1.0%, 2.0%, and 4.0% (W/W)	[185]
Lab	Paddy soil	Wheat straw	450	Reduced soil Cd bioavailability	5% and 15% (W/W)	[186]
Lab	Clay soil	Corncob biochar	600	Reduced lead leaching	5% (W/W)	[187]
Lab	N/A	Wheat straw	350–650	Lower temperature BC led to increased Zn (II) and Cd (II) immobilization acidic condition, and higher temperature BC led to increased Zn (II) and Cd (II) immobilization alkaline condition	N/A	[188]
Lab	N/A	Manure	500	Promoted Zn and Cd precipitation and reduced total Cd and Zn concentrations in switchgrass shoots and roots	0%, 2.5%, and 5%, (W/W)	[189]
		Poultry litter	500			
		Lodgepole pine	500–700	Reduced Zn concentration in roots		
Lab	N/A	Rice husk biochar	550	Decreased leaching of Cd, Cu, Pb, and Zn	0.5%, 1%, and 2% (W/W)	[190]
		Maple leaf biochar				
Lab	Stagnic Phaeozem	Pine wood	N/A	Decreased heavy metal accumulation in above-ground parts of *Hordeum vulgare*	2.5% (W/W)	[191]

3.3.2. Salt

Salt stress is known to negatively affect soil properties, plant development, and crop productivity due to disturbed soil structure, soil organic matter, microbial activity, and C:N ratio. Salt stress causes oxidative stress in plants, down-regulating antioxidant enzyme activity [192]. Due to excessive salinization, the sodium ions bind to cation exchange sites in soil, causing poor crop growth and yields. Although saline soil can be reclaimed by washing or excessively irrigating with water to remove excessive salts, it is neither economically nor physically feasible for large fields [193]. On the other hand, sodic soils require treatment with other cations such as calcium to remove excess sodium from cation exchange sites followed by the leaching of sodium [193]. The application of organic amendments such as manure or compost has been shown to improve soil fertility by reducing salt stress. In saline soil, the organic amendment improves soil porosity, leading to the leaching of excess salt. In sodic soil, organic amendments might help by improving the physical characteristics of soil, such as triggering cation exchange with the calcium present in organic amendment and Na present in soil.

Several studies have reported the positive impact of BC on the nutrient status, conductivity, and improved physical and chemical properties of soil. The variable amount of plant nutrients present in the BC can compensate for the nutrient deficiency and improve the fertility of saline soils. For example, sodification raises soil pH, thereby limiting the bioavailability of P. In such soils, BC can act as a P source and improve its availability, aiding plant growth [194].

It has been demonstrated that the mixture of hardwood and softwood biochar produced at 500 °C, when mixed with sandy loam soil and irrigated with a saline solution, improved the yield of potato, maize, and wheat [195–197]. It was also shown that BC was able to reduce the Na^+/K^+ ratio in the xylem sap of wheat and potato and reduced Na concentration in maize xylem [197]. When BC produced from wheat straw (350–550 °C) was combined with poultry manure and incorporated into Aqui-Entisol soils, a decrease in Na uptake was observed, leading to increased biomass in maize and an increase of yield in wheat [194,198]. Similarly, rice straw-derived BC (600 °C) alleviated salt stress in paddy soil. There was a significant reduction in bulk density, electrical conductivity, exchangeable Na, and exchangeable chlorine ions in the soil, creating favorable conditions for rice seedling growth [199]. A selection of recent reports in which BC addition was reported to alleviate salt stress is summarized in Table 5.

A majority of the reports support the role of BC in improving soil health, plant growth, and the biological properties of soil. It has been reported that BC adsorbs Na salt and improves plant growth; however, salt-affected land is only considered reclaimed if the Na salts are removed. Therefore, the repeated application of BC might have negative consequences in a case where an increased accumulation of Na-salt bound BC aggravates the salinization problem [193]. There is a need for a better understanding of how different BCs interact with different types of salt-affected soils prior to the prescription of any recommendations.

Table 5. Selected recent studies reporting the effect of biochar on salt stressed soils.

Exp. Type	Soil Type	BC Feedstock	Pyrolysis Temp. (°C)	Effect of Biochar Amendment	BC Application Rates	References
Lab	Loam clay	Rice straw	300–600	Reduced bulk density, electrical conductivity, exchangeable Na^+ and Cl^-. Reduced salt accumulation in rice seedlings.	0.3% (W/W)	[199]
Field	N/A	Citrus wood	N/A	Improved plant growth and productivity. Improved nutrient concentration in soil, dehydration tolerance, and water retention.	5 and 10 tons ha^{-1}	[200]
Lab	Coastal soil	Wood chips	600	Improved photosynthetic performance and alleviated oxidative damage and salt stress.	5% (W/W)	[201]
Lab	Sandy clay loam	Rice straw	450	Mitigated oxidative and salt stress. Reduced Cd and Na concentration in plant.	3% and 5% (W/W)	[202]
Lab	N/A	Maple residues	560	Improved plant growth and xylem structure. Reduced salinity and plant stress hormones.	5% and 10% (W/W)	[203]
Lab	N/A	Rice straw	300	Increased seed germination rates of cowpea. Increased photosynthetic efficiency and photosynthetic pigments.	N/A	[204]

3.3.3. Biotic Stress

Several recent reports have emerged showing BC to aid plants in countering biotic stresses. It has been suggested that BC-mediated nutrient retention, adsorption, pH adjustment, and increased water holding provides plants with the capacity to respond to pathogens and to counter the effect of toxic metabolites generated by plants [205].

The severity of gray mold, powdery mildew, and anthracnose on strawberry plants was evaluated in the presence of 3% (w/w) citrus-derived BC (450 °C). It was observed that greenhouse waste-derived biochar when mixed with coconut fiber/peat reduced the severity of gray mold after disease challenge by 74% in mature strawberry plants and by 53% in young strawberry plantlets. Post-disease challenge, anthracnose severity was reduced by 39–49% and powdery mildew severity was reduced by 68% [206]. Both citrus wood and greenhouse waste-derived BC reduced gray mold severity as well. It was observed that BC application induced the expression of genes related to the systematic acquired resistance and induced systemic pathways, which might have contributed to the reduction in disease severity [206]. The ability of biochar to absorb pathogenic cell wall-degrading enzymes and toxic metabolites produced by soil pathogen *Fusarium oxysporum* was also tested with tomato seedlings [205]. The tomato seedlings were treated with 3% BC produced from eucalyptus wood chips and greenhouse pepper plant waste pyrolyzed at 350 °C and 600 °C. It was observed that seedlings exposed to enzymes from *Fusarium oxysporum* and toxic metabolites without BC developed severe disease-like symptoms, whereas those symptoms were significantly reduced in the seedlings grown with BC amendment [205]. The exact mechanism of interaction with BC is still unclear; however, it was observed that a majority of the fungal enzymes that were immobilized by BC through adsorption were deactivated [205]. A commercial-scale study conducted over a period of 3 years tested the effect of BC on growth and disease resistance in *Capsicum annuum* L. (sweet pepper) [207]. Pepper seedlings were planted in four combinations of sandy soil amended with biochars produced from greenhouse pepper plant waste and eucalyptus chips. During the first year of growth, it was observed that greenhouse waste-derived BC (450 °C) reduced the severity of powdery mildew by almost 50% 168 days post-planting in comparison with controls. In the second year of the study, the greenhouse waste BC (350 °C and 450 °C) showed the highest pepper yield compared to the other treatments and the control [207], in addition to a significant reduction in powdery mildew severity. The incidence of plants affected by broad leaf mite was also reduced when amended with greenhouse waste BC. A comparable trend was observed in the third-year trial. Powdery mildew severity was reduced by 25% in both greenhouse waste and eucalyptus wood chip-derived BC after 160 days of growth [207]. Biochars produced from greenhouse waste (350 °C) and eucalyptus chips (600 °C) were shown to be effective in suppressing crown and root rot in tomato caused by *Fusarium oxysporum* f. sp. *radicis lycopersici* [208]. The application of greenhouse waste BC at 0.5%, 1%, and 3% reduced disease severity by 72%. The eucalyptus chip BC also reduced disease severity by 44% compared to the control plants [208]. There are also some reports where no significant effect of BC on soil-borne pathogen suppression was observed [209–211].

The number of studies exploring the role of BC in pathogen suppression is significantly less than other organic amendments such as compost, peat, and crop residue. Hence, additional studies are needed in order to understand the mechanism behind the ability of biochar to suppress pathogens and to be able to prescribe biochar regimens as safe and effective amendment strategies for the improvement of plant resistance to soil-borne pathogens.

4. Impact of Biochar on Crop Production

Increasing crop yields to feed a burgeoning population is a daunting task in the face of a myriad abiotic and biotic challenges, including the reduction of arable farmlands and increased plant stressors due to the changing climate [212–214]. These issues are especially important in organic production systems where the average crop yields are 5–34% lower compared to conventional farming [215–218]. The use of biochar in soil remediation can be a useful strategy, especially in degraded soils [219,220]. Furthermore, the potential to significantly reduce the organic yield gap through better fertilization regimes has been proposed, suggesting an expanded role for nutrient-rich biochars [221].

A meta-analysis of BC effects on plant productivity concluded that BC use holds promise as a method to increase crop yields and could further promote ecosystem services and carbon storage [91,222–224]. It was noted that increased soil N, P, K, the reduction of soil acidity due to the liming effect of BC, and improved water relations contributed to various soil and crop responses. In this review, a comprehensive literature search was performed in the Google Scholar search engine with the search terms "biochar crop productivity yield" for the years 2017–2019, which yielded 330 entries. These entries were further parsed using minimal criteria terms—BC feedstock source, pyrolysis temp, retention time, and soil type. The second round reduced the number of entries to 18. The results of the literature search are summarized in Table 6.

Table 6. Impact of BC on crop productivity summarized from a comprehensive literature search. Soil types listed in the table correspond to the types reported in the original studies. CEC: cation exchange capacity.

Crop Productivity				Soil type, Experiment Type, Length	Biochar Feedstock	Pyrolysis Temp °C, Residence Time, Application Rate	References
Crop Tested	Productivity	Beneficial	Detrimental				
Cherry tomato (*Solanum lycopersicum*)	Bamboo BC increased tomato yields	Both BCs improved tomato quality with increased total sugars	Rice husk BC did not improve total N %	Clay loamy / *Greenhouse* / Short-term ≤ 1 year	Rice husk and bamboo	500 / 1 h / 2% and 5% (w/w)	[225]
Lettuce (*Lactuca sativa*)	For both soils, BC rates of 20 and 30 t/ha^{-1} significantly increased above-ground biomass	Effective fertilizer for lettuce production at least for two growing cycles	Biosolid BC could increase harmful soil elements such as heavy metals	Silty loam and sandy loam / *Greenhouse* / Short-term ≤ 1 year	Fecal matter	450 / 1 h / 10, 20, and 30 t/ha	[226]
Chrysanthemum (*Glebionis coronaria*, cv. 'Crown Daisy') Leaf lettuce	3% BC significantly decreased yields No effect	BC increased WHC(water holding capacity) and SOM	Higher BC application reduced plant productivity	Pedocals, silt-clay / *Greenhouse* / Short-term ≤ 1 year	Peanut shells	350 / 3 h / 0%, 1.5%, 3%, and 5% (w/w) = to 0, 37.5, 75, and 125 t/ha in the field	[227]
Beans	Bean yields were significantly reduced with BC application	Increased germination rate in BC-amended soils	Significant decreases in some macro and micronutrients	Krome loamy / *Greenhouse* / Short-term ≤ 1 year	*Melaleuca quinquenervia* (Broad-leaved paperbark) hardwood	350 / 7 h / 2% and 5% (w/w)	[228]
Wheat (cv. 'Yecora Rojo')	300 °C BC with NPK increased yields	Increased soil water retention and decreased bulk density	BC alone decreased yields with BC produced at higher temp° (400, 500, 600 °C)	Loamy sand / *Greenhouse* / Short-term ≤ 1 year	Date palm tree residues	300, 400, 500, and 600 / 4 h / 8 t/ha	[229]
Potatoes (*Solanum tuberosum* L., cv. 'Russet Burbank')	No significant differences in yield	BC increased soil CEC	BC had no effect on leaf greenness rate or photosystem activity	Sandy / *Field Study* / Long-term, 2 years	Green plantain peels	450–500 / 18–25 min / 13.5 t/ha (1% w/w)	[230]
Tomato & Maize (*Zea mays*)	BC does not have a significant long-term effect on yield	Increased K$^+$, Ca^{2+}, and PO4-P in the soil in year 2	Delayed nutrient availability from BC and short-lived effects	Rincon silty clay loam / *Field Study* / Long-term, 4 years	Walnut shells	900 / 1–2 h / 10 t/ha	[231]
Winter wheat (cv. 'Xiaoyan no. 22')	Low levels (1%, 2%) of BC had a positive effect on wheat yields	Total nitrogen and SOC increased with BC applications	Under drought conditions, BC addition decreased the availability of nutrients	Silty-clay / *Outdoor pot study* / Short-term ≤ 1 year	Apple wood	450 / 8 h / 1%, 2%, 4%, and 6% (w/w)	[232]
Maize	BC and fertilizer led to a significant increase in maize yield	BC improved soil water-holding capacity	BC alone had no effect on maize yields	Sandy clay loam / *Field Study* / Short-term ≤ 1 year	Maize cobs	500 / 1 h / 20 t/ha	[233]
Chinese cabbage (*Brassica rapa*)	BC significantly improved crop yields	BC increased soil pH and CEC	BC did not affect the soil bulk density and porosity	Loamy / *Field Study* / Short-term ≤ 1 year	Barley straw	400 / 1 h / 10 t/ha	[234]

Table 6. *Cont.*

Crop Productivity				Soil type, Experiment Type, Length	Biochar Feedstock	Pyrolysis Temp °C, Residence Time, Application Rate	References
Crop Tested	Productivity	Beneficial	Detrimental				
Radish (*Raphanus sativus* L. cv. French Breakfast)	Increased yields in second year	Reduced bulk density and increased porosity, moisture content, soil pH	No effect on first-year growth	Alfisol or Luvisol	Local hardwoods (*Parkis biglosa, Khaya senegalensis, Prosopis africana* and *Terminalia glaucescens*)	580	[53]
				Field Study		24 h	
				Long-term, 2 years		25 and 50 t/ha	
Rice (cv. 'Naveen')	Increased grain yield up to 24%	Increased total organic C in soils	Microbial carbon use efficiency decreased due to BC addition	Sandy clay loam	Rice husk	350	[235]
				Field Study		6 h	
				Long-term, 3 years		0.5, 1, 2, 4, 8, 10 t/ha	
Maize (cv. 'hybrid LG 6030')	Increased corn yields	Increased P levels during the two years of cultivation	BC was unable to supply the necessary K for further crop production	Red-Yellow Latosol with clayey texture	Sewage sludge	300 and 500	[236]
				Field Study		30 min	
				Long-term, 2 years		15 Mg/ha	
Okra (*Abelmoschus esculentus* L., cv. 'OH-397')	Increased yields vs. controls	Significant increase in SOC and microbial activity	Lower benefit cost ratios for BC compared to controls	Inceptisol with sandy loam texture	Mixed local hardwoods	450	[237]
				Field Study		4 h	
				Long-term, 2 years		5 t/ha	
Rice (*Oryza sativa* L.) & Wheat (*Triticum* ssp.)	Not affected	BC amendment increased the soil water-holding capacity, soil nutrients, and SOC	Short-term effects and BC alone did not increase yields	Hydragric Anthrosol, sandy	Wheat straw	350–550	[166]
				Field Study		2–3 h	
				Long-term, 6 years		20 and 40 t/ha	
Sunflower (*Helianthus annuus* L., cv. 'Embrapa 122/V2000')	Sunflower seed and oil yield declined	Increased levels of most soil minerals and total carbon levels	Nitrogen levels in leaves and the nitrogen uptake of the entire plant decreased with biochar application	Dark red soil, Typic Hapludalfs	Sugarcane bagasse and sunflower residues	500–600	[238]
				Field Study		1 h	
				Short-term ≤ 1 year		1% (w/w)	
Spring barley (*Hordeum vulgare* L.)	Increased yields with BC + NPK	Increased soil water status in BC amended soils in the first year; increased soil carbon status	BC only decreased yields for both crops compared to control NPK plants	Sandy loamy silt; calcareous Chernozem on loess	Hardwood	550	[239]
				Field Study		2 h	
Sunflower	No difference vs. controls			Long-term, 2 years		72 t/ha	
Rice (*Oryza sativa* L.) & Wheat (*Triticum* ssp.)	Not affected	BC amendment increased the soil water-holding capacity, soil nutrients, and SOC	Short-term effects and BC alone did not increase yields	Hydragric Anthrosol, sandy	Wheat straw	350–550	[166]
				Field Study		2–3 h	
				Long-term, 6 years		20 and 40 t/ha	
Cauliflower (*Brassica oleracea*, cv. 'Desire')	No significant improvement in crop yield	No negative effects to crop productivity or soil quality	Soil moisture and bulk density not affected by BC additions	Ferralsol	Woody Eucalyptus 'Blue Mallee'	550	[240]
Pea (*Pisum sativum*, cv. 'Ashton')				*Field Study*		30 min	
Broccoli 'Ironman'				Short-term, 1 year		10 t/ha	

In terms of productivity alone, a majority of the studies reported a beneficial impact of BC on crop yields [53,225,226,229,232–237]. Experimental plants included lettuce, cabbage, radish, tomato, wheat, rice, maize, and okra. Soils were amended with BC derived from major feedstock sources such as hardwood, manure, and crop residues. Positive results from this mixture of plants and biochars indicate a theoretical system to 'mix and match' crop with BC for optimal productivity. Interestingly, none of the studies included perennial plant species in the experimental design. That is another area where the impact of BC remains to be assessed.

Due to the range of tested soil conditions, many factors altered by BC amendment were implicated in reported yield gains. For instance, lettuce yields were positively influenced with 20 and 30 t/ha fecal-derived BC (450 °C) [226]. Mineral-enriched BC proved to be an effective fertilizer for two growing cycles in the greenhouse pot study. Additional experiments in greenhouses with leafy crops proved that significant yield increases are possible with BC soil amendment [20,241]. Two studies with wheat showed increased yields as a result of soils amended with 1–2% apple wood-derived BC (450 °C) due to increased nitrogen levels [232] and increased soil water retention with 8 t/ha date palm tree residue-derived BC (300–600 °C) [229]. The increase in soil organic carbon and the stimulatory effect on microbial communities raised rice yields in soil amended with rice husk BC (350 °C) [235] and okra yields amended with hardwood-derived BC (450 °C) [237]. In addition to reporting increases in yields, these studies also discussed the limitations of field applications of BC.

BC contains key plant nutrients, although at a low level as demonstrated by several studies, and it may have led to the lack of a complete plant nutrient profile in the soils to obtain a desirable increase in yields [225]. Multiple studies reported mixed results in terms of crop production [227,239] or described no effect [166,230,231,240]. Soil nutrient content and CEC were improved with BC amendment but were short-lived and resulted in comparable crop productivity compared to controls in studies with rice and wheat growing in wheat straw BC (350–550 °C) [166], potatoes with green plantain peel-derived BC (450–500 °C) [230], and tomatoes and maize with walnut shell-derived BC (900 °C) [231]. The growth of Spring barley and sunflowers was tested with hardwood-derived BC (550 °C) at 72 t/ha. The treatments increased barley yields but had no effect on sunflower productivity [239]. The BC-only amendment did increase soil water status and carbon levels; however, increased barley productivity was noted only when BC was mixed with NPK compared to NPK-only controls. While increased water-holding capacity and soil carbon levels with peanut shell-derived BC (350 °C) were also reported [227], these alterations did not lead to any effect on lettuce yields.

Undesirable effects on crop productivity following BC soil amendment were also reported in two of the studies [228,238]. Although beans demonstrated an increased rate of germination in BC amended soils, their yields were significantly reduced with hardwood BC (350 °C) application at 2% and 5% [228]. Other studies with legumes reported a gain in yield when grown in BC-amended soils. The yields of mash bean improved with sugarcane bagasse BC (350 °C), with and without chemical fertilizer, due to the increased SOC, total N, and decreased bulk density. Importantly, nitrogen fixation increased by 83% in the biochar-only treatment due to higher nodule numbers [242]. Additionally, fava bean growth with wheat straw-derived BC (500 °C) amendment applied at a 2.5% w/w rate in addition to saltwater irrigation led to significantly increased dry seed yield compared to controls, which was mainly attributed to the high salt sorption capacity of BC [243]. The higher nutrient content in the crop residue-derived BCs reported above may have helped elevate yields compared to controls, while the already nutrient poor hardwood-derived BC may have reduced bean yields. While BC can be a source of nutrients, the complex interactions in the soil environment may have reduced the capacity of available nutrients in the soils inflicting significant yield losses [244]. Additional studies are required to develop a more comprehensive model of BC effect on legume production.

Other factors potentially responsible for lower productivity include soil nutrient deficiencies found with sugarcane bagasse and sunflower-derived BC-amended (500–600 °C) soils [238]. As a result of decreased nitrogen uptake with increasing BC application, sunflower seed and oil yield saw a significant decrease. The 1% field application of the BCs may have increased specific communities of bacteria and enhanced certain enzyme activity such as urease, which is an important enzyme in soil nitrogen status, as reported by a field study with the addition of sugarcane bagasse biochar (SCBC) [162]. However, fungal communities suffered due to SCBC addition, and final yields of *Brassica chinesis* L. (pak choi) were reduced compared to controls. It was found that a 4% application rate of SCBC supported normal plant growth and increased sugar and cane yields [245]. The SOC, soil–water related properties, and nutrient levels were enhanced by SCBC, leading to increased plant productivity. Further research is needed to identify BCs appropriate for specific plant species and initial soil characteristics for improved plant growth and development.

Although crop responses were generally positive, the high variability within the listed studies makes it difficult to draw any broad conclusions except that the type and application rate of BC will require customization. The benefits of BC application mainly consist of increased water-holding potential, better nutrient cycling, and increased soil carbon reserves. This may lead to no effect or only minimally increasing yields in the short term, but further testing in the field should illuminate the effects of long-term BC amendment on crop yields [246]. Other regions of industrial agriculture and tropical environments may show a more pronounced BC effect and may be better at exploiting the advantages of BC. BC application in marginal soils will likely lead to increased crop productivity by increasing the overall soil fertility through pH and CEC adjustments, better water retention, and increased microbial activity [247,248]. Nevertheless, considerable caution should be observed when using extremely heavy rates of BC. The elevated risk of heavy metal contamination due to feedstocks rich in accumulated metals or other phytotoxic compounds could decrease crop productivity with increasing BC applications [249,250].

The overall conclusion is that BC application is favorable for improving crop productivity sustainably. Certain agricultural systems require different inputs to achieve higher crop yields, and designing BC to meet those specific needs could lead to optimized production methods and products.

5. Conclusions and Future Opportunities

Analysis of the published literature supports the role of BC as one of the many viable solutions to soil-related challenges of food production in the face of persistent global issues. While it is not a panacea, the humble porous carbon-composed BC has the ability to physically, biologically, and chemically alter soil properties, which has multifarious consequences. There is an opportunity for carbon sequestration and establishing carbon negative cycles with the expanded use of BC. Countering deteriorating soil health due to industrial agriculture, BC amendment can help support higher crop productivity and contribute to improving global food security. The overall impact of BC can be increased by its application on highly weathered and marginal soils that are characterized by depleted nutrients levels, reduced water retention, and the lack of a competent soil structure.

Future climate models indicate that water stress will be a key driver of reduced crop productivity. BC has been shown to improve soil water-holding capacity, making it a potential candidate for alleviating water stress [251–253]. Finding the right permutation of BC feedstock, application rate, and crop variety are vital for improving agricultural production and reducing the related carbon footprint. Initial progress in understanding the effects of BC use has led to the efforts of producing 'designer biochars' which promise to exploit the positive properties and dampen negative effects mainly through feedstock selection but also through chemical, physical, or natural alterations of the BC [91,254–256].

Various studies and published meta-analyses on BC have pointed out the benefits of BC soil amendment; however, there are still several areas that require attention and resolution. Important aspects that affect crop productivity include feedstock sources, various BC production methods, initial soil characteristics, crop variety, and experimental conditions. Others have also reported irregular effects of BC on crop productivity due to these variables [257]. Since it is an irreversible decision to amend soils with BC, the various impacts of augmenting soils are important considerations before conducting field applications [258].

Several articles were not included in the comprehensive literature review due to the omission of key variables that are critical for assessing the overall impact of BC studies. The conclusions of this review are that standardizing BC experiments is vitally important to hypothesis testing and replicating studies to move the research field forward. All future BC articles should include meticulous descriptions of biochar feedstock sources, pyrolysis temperature, and retention time. Not covered in this review, but still significant, are the various economic factors to be considered before undertaking any large-scale BC applications, including the production or acquisition of BC, shipping and transporting, and the time and labor required for field application [237,259]. Additionally, the lack of consistent responses of microbial communities to BC amendment highlights a knowledge gap regarding the mechanism by which microbes interact with BC. BC has the potential to decrease the bioavailability and efficiency of some herbicides, which is yet another variable that needs consideration [260]. Further, analysis of the published literature leads to the conclusion that the following areas need further investigation:

- Biomes underrepresented in the current biochar-associated literature, such as forests and perennial crops (the vast majority of BC studies are directed toward temperate and tropical areas);
- Effects of biochar on non-model crop species (present studies primarily focus on model organisms such as tomato, maize, rice, and wheat);
- Evaluation of BC in field studies to build upon the extensive greenhouse studies;
- Develop an understanding of the highly complex interactions between different soil types, different biochar types, and their impact on plant productivity
- Assessment of biochar-amended soil microbial activity through meta-genomics approaches;
- Longer-term experiments to understand characteristics of 'aged' BC to assess its temporally evolving properties in soils;
- Development of cost-effective ways to minimize environmental impacts by incorporating organic fertilizer amendments such as BC.

The multidimensional and complex interactions between inherent soil properties, variable biochar properties depending on the type of feedstock used, the genetic background of the plant, and the limited amount of available empirical data make it almost impossible to predict the outcome of BC amendment. This is an obvious conclusion of this and several previous studies. Knowledge generated from the above-mentioned areas of investigation is expected to enable the large-scale utilization of biochars in agriculture, representing a step toward establishing carbon negative ecosystems.

Author Contributions: Conceptualization—A.D., E.T. and R.G.; Resources—A.D.; Writing—review and editing—E.T., R.G., A.D.; Supervision—A.D.; Funding acquisition—A.D. All authors have read and agreed to the published version of the manuscript.

Funding: This work was supported in part by the USDA-NIFA Hatch Grant WNP00011 to A.D.

Acknowledgments: The authors thank Seanna Hewitt, Karen Adams, and Richard Sharpe for their critical reading of the manuscript and their feedback.

Conflicts of Interest: A.D. serves as a consultant for AgEnergy Solutions—A biochar production startup company based in Spokane, WA, USA. AgEnergy Solutions had no role in the design of the study; in the collection, analyses, or interpretation of data; in the writing of the manuscript, or in the decision to publish the results.

Abbreviations

BC	biochar
SOM	soil organic matter
GHG	greenhouse gas
CEC	cation exchange capacity
FTIR	Fourier transform infrared spectroscopy
SBBC	sugarcane bagasse biochar
C	carbon
N	nitrogen
P	phosphorous
K	potassium
Ca	calcium
Mg	magnesium
Fe	iron
Na	sodium
Si	silicon
Cd	cadmium
Cu	copper
Ni	nickel
Pb	lead
Zn	zinc

References

1. Foley, J.A.; Ramankutty, N.; Brauman, K.A.; Cassidy, E.S.; Gerber, J.S.; Johnston, M.; Mueller, N.D.; O'Connell, C.; Ray, D.K.; West, P.C.; et al. Solutions for a cultivated planet. *Nature* **2011**, *478*, 337–342. [CrossRef]
2. Raupach, M.R.; Marland, G.; Ciais, P.; Le Quéré, C.; Canadell, J.G.; Klepper, G.; Field, C.B. Global and regional drivers of accelerating CO_2 emissions. *Proc. Natl. Acad. Sci. USA* **2007**, *104*, 10288–10293. [CrossRef] [PubMed]
3. The Paris Agreement|UNFCCC. Available online: https://unfccc.int/process-and-meetings/the-paris-agreement/the-paris-agreement (accessed on 21 March 2020).
4. Minasny, B.; Malone, B.P.; McBratney, A.B.; Angers, D.A.; Arrouays, D.; Chambers, A.; Chaplot, V.; Chen, Z.-S.; Cheng, K.; Das, B.S. Soil carbon 4 per mille. *Geoderma* **2017**, *292*, 59–86. [CrossRef]
5. Meinshausen, M.; Meinshausen, N.; Hare, W.; Raper, S.C.B.; Frieler, K.; Knutti, R.; Frame, D.J.; Allen, M.R. Greenhouse-gas emission targets for limiting global warming to 2 C. *Nature* **2009**, *458*, 1158–1162. [CrossRef]
6. Smith, P. Soil carbon sequestration and biochar as negative emission technologies. *Glob. Chang. Biol.* **2016**, *22*, 1315–1324. [CrossRef]
7. Canadell, J.G.; Raupach, M.R. Managing forests for climate change mitigation. *Science* **2008**, *320*, 1456–1457. [CrossRef]
8. Creutzig, F.; Ravindranath, N.H.; Berndes, G.; Bolwig, S.; Bright, R.; Cherubini, F.; Chum, H.; Corbera, E.; Delucchi, M.; Faaij, A. Bioenergy and climate change mitigation: An assessment. *GCB Bioenergy* **2015**, *7*, 916–944. [CrossRef]
9. Socolow, R.; Desmond, M.; Aines, R.; Blackstock, J.; Bolland, O.; Kaarsberg, T.; Lewis, N.; Mazzotti, M.; Pfeffer, A.; Sawyer, K. *Direct Air Capture of CO_2 with Chemicals: A Technology Assessment for the APS Panel on Public Affairs*; American Physical Society, 2011. Available online: https://www.aps.org/policy/reports/assessments/upload/dac2011.pdf (accessed on 21 March 2020).
10. Atafar, Z.; Mesdaghinia, A.; Nouri, J.; Homaee, M.; Yunesian, M.; Ahmadimoghaddam, M.; Mahvi, A.H. Effect of fertilizer application on soil heavy metal concentration. *Environ. Monit. Assess.* **2010**, *160*, 83. [CrossRef] [PubMed]
11. Wang, J.; Yu, L.; Hu, L.; Chen, G.; Xin, H.; Feng, X. Ambient ammonia synthesis via palladium-catalyzed electrohydrogenation of dinitrogen at low overpotential. *Nat. Commun.* **2018**, *9*, 1–7. [CrossRef] [PubMed]

12. Clarke, B.O.; Smith, S.R. Review of "emerging" organic contaminants in biosolids and assessment of international research priorities for the agricultural use of biosolids. *Environ. Int.* **2011**, *37*, 226–247. [CrossRef]

13. Hamilton, K.A.; Ahmed, W.; Rauh, E.; Rock, C.; McLain, J.; Muenich, R.L. Comparing microbial risks from multiple sustainable waste streams applied for agricultural use: Biosolids, manure, and diverted urine. *Curr. Opin. Environ. Sci. Health* **2020**, *14*, 37–50. [CrossRef]

14. Smith, N.J.H. The Amazon River Forest: A Natural History of Plants, Animals, and People. In *OUP Catalogue*; Oxford University Press: Oxford, UK, 1999.

15. Skjemstad, J.O.; Reicosky, D.C.; Wilts, A.R.; McGowan, J.A. Charcoal carbon in US agricultural soils. *Soil Sci. Soc. Am. J.* **2002**, *66*, 1249–1255. [CrossRef]

16. Glaser, B.; Haumaier, L.; Guggenberger, G.; Zech, W. The 'Terra Preta' phenomenon: A model for sustainable agriculture in the humid tropics. *Naturwissenschaften* **2001**, *88*, 37–41. [CrossRef]

17. Zech, W.; Haumaier, L.; Reinhold, H. Ecological aspects of soil organic matter in tropical land use. *Humic Subst. Soil Crop Sci. Sel. Read.* **1990**, 187–202. [CrossRef]

18. Smith, N.J.H. Anthrosols and human carrying capacity in Amazonia. *Ann. Assoc. Am. Geogr.* **1980**, *70*, 553–566. [CrossRef]

19. Woolf, D.; Amonette, J.E.; Street-Perrott, F.A.; Lehmann, J.; Joseph, S. Sustainable biochar to mitigate global climate change. *Nat. Commun.* **2010**, *1*, 56. [CrossRef]

20. Trupiano, D.; Cocozza, C.; Baronti, S.; Amendola, C.; Vaccari, F.P.; Lustrato, G.; Di Lonardo, S.; Fantasma, F.; Tognetti, R.; Scippa, G.S. The Effects of Biochar and Its Combination with Compost on Lettuce (*Lactuca sativa* L.) Growth, Soil Properties, and Soil Microbial Activity and Abundance. *Int. J. Agron.* **2017**, *2017*, 3158207. [CrossRef]

21. Ahmed, M.B.; Zhou, J.L.; Ngo, H.H.; Guo, W. Insight into biochar properties and its cost analysis. *Biomass Bioenergy* **2016**, *84*, 76–86. [CrossRef]

22. Al Arni, S. Comparison of slow and fast pyrolysis for converting biomass into fuel. *Renew. Energy* **2018**, *124*, 197–201. [CrossRef]

23. Suliman, W.; Harsh, J.B.; Abu-Lail, N.I.; Fortuna, A.-M.; Dallmeyer, I.; Garcia-Perez, M. Influence of feedstock source and pyrolysis temperature on biochar bulk and surface properties. *Biomass Bioenergy* **2016**, *84*, 37–48. [CrossRef]

24. Fu, P.; Hu, S.; Xinag, J.; Sun, L.; Yang, T.; Zhang, A.; Wang, Y.; Chen, G. Effects of Pyrolysis Temperature on Characteristics of Porosity in Biomass Chars. In Proceedings of the 2009 International Conference on Energy and Environment Technology(IEEE), Guilin, China, 16–18 October 2009; Volume 1, pp. 109–112.

25. Lehmann, J.; Joseph, S. *Biochar for Environmental Management: Science, Technology and Implementation*; Routledge: Abingdon, UK, 2015; ISBN 1134489536.

26. Dai, Z.; Zhang, X.; Tang, C.; Muhammad, N.; Wu, J.; Brookes, P.C.; Xu, J. Potential role of biochars in decreasing soil acidification—A critical review. *Sci. Total Environ.* **2017**, *581*, 601–611. [CrossRef] [PubMed]

27. Yuan, J.-H.; Xu, R.-K.; Zhang, H. The forms of alkalis in the biochar produced from crop residues at different temperatures. *Bioresour. Technol.* **2011**, *102*, 3488–3497. [CrossRef] [PubMed]

28. Dai, Z.; Brookes, P.C.; He, Y.; Xu, J. Increased agronomic and environmental value provided by biochars with varied physiochemical properties derived from swine manure blended with rice straw. *J. Agric. Food Chem.* **2014**, *62*, 10623–10631. [CrossRef] [PubMed]

29. Zhao, L.; Cao, X.; Mašek, O.; Zimmerman, A. Heterogeneity of biochar properties as a function of feedstock sources and production temperatures. *J. Hazard. Mater.* **2013**, *256*, 1–9. [CrossRef]

30. Flowers, M.D.; Lal, R. Axle load and tillage effects on soil physical properties and soybean grain yield on a mollic ochraqualf in northwest Ohio. *Soil Tillage Res.* **1998**, *48*, 21–35. [CrossRef]

31. Bruand, A.; Gilkes, R.J. Subsoil bulk density and organic carbon stock in relation to land use for a Western Australian Sodosol. *Soil Res.* **2002**, *40*, 999–1010. [CrossRef]

32. Beutler, A.N.; Centurion, J.F.; da Silva, A.P.; da Centurion, M.A.P.C.; Leonel, C.L.; da Freddi, O.S. Soil compaction by machine traffic and least limiting water range related to soybean yield. *Pesqui. Agropecu. Bras.* **2008**, *43*, 1591–1600. [CrossRef]

33. Ramazan, M.; Khan, G.D.; Hanif, M.; Ali, S. Impact of soil compaction on root length and yield of corn (Zea mays) under irrigated condition. *Middle-East J. Sci. Res.* **2012**, *11*, 382–385.

34. Omondi, M.O.; Xia, X.; Nahayo, A.; Liu, X.; Korai, P.K.; Pan, G. Quantification of biochar effects on soil hydrological properties using meta-analysis of literature data. *Geoderma* **2016**, *274*, 28–34. [CrossRef]

35. Blanco-Canqui, H. Biochar and soil physical properties. *Soil Sci. Soc. Am. J.* **2017**, *81*, 687–711. [CrossRef]

36. Obour, P.B.; Danso, E.O.; Yakubu, A.; Abenney-Mickson, S.; Sabi, E.B.; Darrah, Y.K.; Arthur, E. Water Retention, Air Exchange and Pore Structure Characteristics after Three Years of Rice Straw Biochar Application to an Acrisol. *Soil Sci. Soc. Am. J.* **2019**, *83*, 1664–1671. [CrossRef]

37. de Jesus Duarte, S.; Glaser, B.; Pellegrino Cerri, C. Effect of Biochar Particle Size on Physical, Hydrological and Chemical Properties of Loamy and Sandy Tropical Soils. *Agronomy* **2019**, *9*, 165. [CrossRef]

38. Verheijen, F.G.A.; Zhuravel, A.; Silva, F.C.; Amaro, A.; Ben-Hur, M.; Keizer, J.J. The influence of biochar particle size and concentration on bulk density and maximum water holding capacity of sandy vs sandy loam soil in a column experiment. *Geoderma* **2019**, *347*, 194–202. [CrossRef]

39. Rahim, H.U.; Mian, I.A.; Arif, M.; Rahim, Z.U.; Ahmad, S.; Khan, Z.; Zada, L.; Khan, M.A.; Haris, M. 3. Residual effect of biochar and summer legumes on soil physical properties and wheat growth. *Pure Appl. Biol.* **2019**, *8*, 16–26. [CrossRef]

40. You, J.; Sun, L.; Liu, X.; Hu, X.; Xu, Q. Effects of sewage sludge biochar on soil characteristics and crop yield in Loamy sand soil. *Pol. J. Environ. Stud.* **2019**, *28*. [CrossRef]

41. Duarte, S.D.J.; Glaser, B.; De Lima, R.P.; Cerri, C.P. Chemical, Physical, and Hydraulic Properties as Affected by One Year of Miscanthus Biochar Interaction with Sandy and Loamy Tropical Soils. *Soil Syst.* **2019**, *3*, 24. [CrossRef]

42. Nikravesh, I.; Boroomandnasab, S.; Abd Ali Naseri, A.S.M. Wheat Straw Biochar Application to Loam-Sand Soil: Impact on Yield Components of Summer Maize and Some Soil Properties. *World* **2019**, *8*, 54–60.

43. Zhou, H.; Fang, H.; Zhang, Q.; Wang, Q.; Chen, C.; Mooney, S.J.; Peng, X.; Du, Z. Biochar enhances soil hydraulic function but not soil aggregation in a sandy loam. *Eur. J. Soil Sci.* **2019**, *70*, 291–300. [CrossRef]

44. Zhang, L.; Jing, Y.; Chen, G.; Wang, X.; Zhang, R. Improvement of physical and hydraulic properties of desert soil with amendment of different biochars. *J. Soils Sediments* **2019**, *19*, 2984–2996. [CrossRef]

45. Kätterer, T.; Roobroeck, D.; Andrén, O.; Kimutai, G.; Karltun, E.; Kirchmann, H.; Nyberg, G.; Vanlauwe, B.; de Nowina, K.R. Biochar addition persistently increased soil fertility and yields in maize-soybean rotations over 10 years in sub-humid regions of Kenya. *Field Crop. Res.* **2019**, *235*, 18–26. [CrossRef]

46. Badawi, M.A. Production of Biochar from Date Palm Fronds and its Effects on Soil Properties. *By-Prod. Palm Trees Appl.* **2019**, *11*, 159.

47. Tanure, M.M.C.; da Costa, L.M.; Huiz, H.A.; Fernandes, R.B.A.; Cecon, P.R.; Pereira Junior, J.D.; da Luz, J.M.R. Soil water retention, physiological characteristics, and growth of maize plants in response to biochar application to soil. *Soil Tillage Res.* **2019**, *192*, 164–173. [CrossRef]

48. Lentz, R.D.; Ippolito, J.A.; Lehrsch, G.A. Biochar, manure, and sawdust alter long-term water retention dynamics in degraded soil. *Soil Sci. Soc. Am. J.* **2019**, *83*, 1491–1501. [CrossRef]

49. Huang, M.; Zhang, Z.; Zhai, Y.; Lu, P.; Zhu, C. Effect of straw biochar on soil properties and wheat production under saline water irrigation. *Agronomy* **2019**, *9*, 457. [CrossRef]

50. Xiu, L.; Zhang, W.; Sun, Y.; Wu, D.; Meng, J.; Chen, W. Effects of biochar and straw returning on the key cultivation limitations of Albic soil and soybean growth over 2 years. *Catena* **2019**, *173*, 481–493. [CrossRef]

51. Horák, J.; Šimanský, V.; Igaz, D. Biochar and biochar with N fertilizer impact on soil physical properties in a silty loam haplic luvisol. *J. Ecol. Eng.* **2019**, *20*, 31–38. [CrossRef]

52. Baiamonte, G.; Crescimanno, G.; Parrino, F.; De Pasquale, C. Effect of biochar on the physical and structural properties of a sandy soil. *Catena* **2019**, *175*, 294–303. [CrossRef]

53. Adekiya, A.O.; Agbede, T.M.; Aboyeji, C.M.; Dunsin, O.; Simeon, V.T. Effects of biochar and poultry manure on soil characteristics and the yield of radish. *Sci. Hortic.* **2019**, *243*, 457–463. [CrossRef]

54. Dai, Y.; Zheng, H.; Jiang, Z.; Xing, B. Combined effects of biochar properties and soil conditions on plant growth: A meta-analysis. *Sci. Total Environ.* **2020**, *713*, 136635. [CrossRef]

55. Wang, D.; Li, C.; Parikh, S.J.; Scow, K.M. Impact of biochar on water retention of two agricultural soils—A multi-scale analysis. *Geoderma* **2019**, *340*, 185–191. [CrossRef]

56. Hardie, M.; Clothier, B.; Bound, S.; Oliver, G.; Close, D. Does biochar influence soil physical properties and soil water availability? *Plant Soil* **2014**, *376*, 347–361. [CrossRef]

57. Rogovska, N.; Laird, D.A.; Rathke, S.J.; Karlen, D.L. Biochar impact on Midwestern Mollisols and maize nutrient availability. *Geoderma* **2014**, *230*, 340–347. [CrossRef]

58. Usowicz, B.; Lipiec, J.; Łukowski, M.; Marczewski, W.; Usowicz, J. The effect of biochar application on thermal properties and albedo of loess soil under grassland and fallow. *Soil Tillage Res.* **2016**, *164*, 45–51. [CrossRef]

59. Xiao, Q.; Zhu, L.-X.; Zhang, H.-P.; Li, X.-Y.; Shen, Y.-F.; Li, S.-Q. Soil amendment with biochar increases maize yields in a semi-arid region by improving soil quality and root growth. *Crop Pasture Sci.* **2016**, *67*, 495–507. [CrossRef]

60. Pratiwi, E.P.A.; Shinogi, Y. Rice husk biochar application to paddy soil and its effects on soil physical properties, plant growth, and methane emission. *Paddy Water Environ.* **2016**, *14*, 521–532. [CrossRef]

61. Obia, A.; Mulder, J.; Martinsen, V.; Cornelissen, G.; Børresen, T. In situ effects of biochar on aggregation, water retention and porosity in light-textured tropical soils. *Soil Tillage Res.* **2016**, *155*, 35–44. [CrossRef]

62. Liu, Z.; Dugan, B.; Masiello, C.A.; Barnes, R.T.; Gallagher, M.E.; Gonnermann, H. Impacts of biochar concentration and particle size on hydraulic conductivity and DOC leaching of biochar–sand mixtures. *J. Hydrol.* **2016**, *533*, 461–472. [CrossRef]

63. Chen, B.; Han, M.Y.; Peng, K.; Zhou, S.L.; Shao, L.; Wu, X.F.; Wei, W.D.; Liu, S.Y.; Li, Z.; Li, J.S.; et al. Global land-water nexus: Agricultural land and freshwater use embodied in worldwide supply chains. *Sci. Total Environ.* **2018**, *613*, 931–943. [CrossRef]

64. Fischer, B.M.C.; Manzoni, S.; Morillas, L.; Garcia, M.; Johnson, M.S.; Lyon, S.W. Improving agricultural water use efficiency with biochar—A synthesis of biochar effects on water storage and fluxes across scales. *Sci. Total Environ.* **2019**, *657*, 853–862. [CrossRef]

65. Basso, A.S.; Miguez, F.E.; Laird, D.A.; Horton, R.; Westgate, M. Assessing potential of biochar for increasing water-holding capacity of sandy soils. *GCB Bioenergy* **2013**, *5*, 132–143. [CrossRef]

66. Laghari, M.; Naidu, R.; Xiao, B.; Hu, Z.; Mirjat, M.S.; Hu, M.; Kandhro, M.N.; Chen, Z.; Guo, D.; Jogi, Q.; et al. Recent developments in biochar as an effective tool for agricultural soil management: A review. *J. Sci. Food Agric.* **2016**, *96*, 4840–4849. [CrossRef] [PubMed]

67. Obia, A.; Cornelissen, G.; Martinsen, V.; Smebye, A.B.; Mulder, J. Conservation tillage and biochar improve soil water content and moderate soil temperature in a tropical Acrisol. *Soil Tillage Res.* **2020**, *197*. [CrossRef]

68. Kinney, T.J.; Masiello, C.A.; Dugan, B.; Hockaday, W.C.; Dean, M.R.; Zygourakis, K.; Barnes, R.T. Hydrologic properties of biochars produced at different temperatures. *Biomass Bioenergy* **2012**, *41*, 34–43. [CrossRef]

69. Lei, O.; Zhang, R. Effects of biochars derived from different feedstocks and pyrolysis temperatures on soil physical and hydraulic properties. *J. Soils Sediments* **2013**, *13*, 1561–1572. [CrossRef]

70. Rafiq, M.K.; Bachmann, R.T.; Rafiq, M.T.; Shang, Z.; Joseph, S.; Long, R. Influence of Pyrolysis Temperature on Physico-Chemical Properties of Corn Stover (Zea mays L.) Biochar and Feasibility for Carbon Capture and Energy Balance. *PLoS ONE* **2016**, *11*, e0156894. [CrossRef]

71. Gray, M.; Johnson, M.G.; Dragila, M.I.; Kleber, M. Water uptake in biochars: The roles of porosity and hydrophobicity. *Biomass Bioenergy* **2014**, *61*, 196–205. [CrossRef]

72. Mao, J.; Zhang, K.; Chen, B. Linking hydrophobicity of biochar to the water repellency and water holding capacity of biochar-amended soil. *Environ. Pollut.* **2019**, *253*, 779–789. [CrossRef]

73. Suliman, W.; Harsh, J.B.; Abu-Lail, N.I.; Fortuna, A.M.; Dallmeyer, I.; Garcia-Pérez, M. The role of biochar porosity and surface functionality in augmenting hydrologic properties of a sandy soil. *Sci. Total Environ.* **2017**, *574*, 139–147. [CrossRef]

74. Batista, E.M.C.C.; Shultz, J.; Matos, T.T.S.; Fornari, M.R.; Ferreira, T.M.; Szpoganicz, B.; De Freitas, R.A.; Mangrich, A.S. Effect of surface and porosity of biochar on water holding capacity aiming indirectly at preservation of the Amazon biome. *Sci. Rep.* **2018**, *8*, 1–9. [CrossRef]

75. Kameyama, K.; Miyamoto, T.; Iwata, Y. The Preliminary Study of Water-Retention Related Properties of Biochar Produced from Various Feedstock at Different Pyrolysis Temperatures. *Materials* **2019**, *12*, 1732. [CrossRef]

76. Barnes, R.T.; Gallagher, M.E.; Masiello, C.A.; Liu, Z.; Dugan, B. Biochar-induced changes in soil hydraulic conductivity and dissolved nutrient fluxes constrained by laboratory experiments. *PLoS ONE* **2014**, *9*. [CrossRef] [PubMed]

77. Aller, D.; Rathke, S.; Laird, D.; Cruse, R.; Hatfield, J. Impacts of fresh and aged biochars on plant available water and water use efficiency. *Geoderma* **2017**, *307*, 114–121. [CrossRef]

78. Razzaghi, F.; Obour, P.B.; Arthur, E. Does biochar improve soil water retention? A systematic review and meta-analysis. *Geoderma* **2019**. [CrossRef]

79. Marshall, J.; Muhlack, R.; Morton, B.J.; Dunnigan, L.; Chittleborough, D.; Kwong, C.W. Pyrolysis Temperature Effects on Biochar–Water Interactions and Application for Improved Water Holding Capacity in Vineyard Soils. *Soil Syst.* **2019**, *3*, 27. [CrossRef]

80. Zhang, X.; Cai, X. Climate change impacts on global agricultural land availability. *Environ. Res. Lett.* **2011**, *6*, 014014. [CrossRef]

81. Nguyen, T.T.N.; Xu, C.-Y.; Tahmasbian, I.; Che, R.; Xu, Z.; Zhou, X.; Wallace, H.M.; Bai, S.H. Effects of biochar on soil available inorganic nitrogen: A review and meta-analysis. *Geoderma* **2017**, *288*, 79–96. [CrossRef]

82. Gupta, R.K.; Hussain, A.; Sooch, S.S.; Kang, J.S.; Sharma, S.; Dheri, G.S. Rice straw biochar improves soil fertility, growth, and yield of rice-wheat system on a sandy loam soil. *Exp. Agric.* **2019**, *56*, 118–131. [CrossRef]

83. Khan, M.; Fatima, K.; Ahmad, R.; Younas, R.; Rizwan, M.; Azam, M.; Abadin, Z.; ul Ali, S. Comparative effect of mesquite biochar, farmyard manure, and chemical fertilizers on soil fertility and growth of onion (*Allium cepa* L.). *Arab. J. Geosci.* **2019**, *12*, 1–7. [CrossRef]

84. Borchard, N.; Schirrmann, M.; Cayuela, M.L.; Kammann, C.; Wrage-Mönnig, N.; Estavillo, J.M.; Fuertes-Mendizábal, T.; Sigua, G.; Spokas, K.; Ippolito, J.A.; et al. Biochar, soil and land-use interactions that reduce nitrate leaching and N_2O emissions: A meta-analysis. *Sci. Total Environ.* **2019**, *651*, 2354–2364. [CrossRef]

85. Yao, Y.; Gao, B.; Zhang, M.; Inyang, M.; Zimmerman, A.R. Effect of biochar amendment on sorption and leaching of nitrate, ammonium, and phosphate in a sandy soil. *Chemosphere* **2012**, *89*, 1467–1471. [CrossRef]

86. Chen, Q.; Qin, J.; Sun, P.; Cheng, Z.; Shen, G. Cow dung-derived engineered biochar for reclaiming phosphate from aqueous solution and its validation as slow-release fertilizer in soil-crop system. *J. Clean. Prod.* **2018**, *172*, 2009–2018. [CrossRef]

87. Qiu, G.; Zhao, Y.; Wang, H.; Tan, X.; Chen, F.; Hu, X. Biochar synthesized via pyrolysis of Broussonetia papyrifera leaves: Mechanisms and potential applications for phosphate removal. *Environ. Sci. Pollut. Res.* **2019**, *26*, 6565–6575. [CrossRef] [PubMed]

88. Waqas, M.; Nizami, A.S.; Aburiazaiza, A.S.; Barakat, M.A.; Ismail, I.M.I.; Rashid, M.I. Optimization of food waste compost with the use of biochar. *J. Environ. Manag.* **2018**, *216*, 70–81. [CrossRef] [PubMed]

89. Solaiman, Z.M.; Abbott, L.K.; Murphy, D.V. Biochar phosphorus concentration dictates mycorrhizal colonisation, plant growth and soil phosphorus cycling. *Sci. Rep.* **2019**, *9*, 5062. [CrossRef] [PubMed]

90. Panwar, N.L.; Pawar, A.; Salvi, B.L. Comprehensive review on production and utilization of biochar. *SN Appl. Sci.* **2019**, *1*, 168. [CrossRef]

91. Jiang, Z.; Lian, F.; Wang, Z.; Xing, B. The role of biochars in sustainable crop production and soil resiliency. *J. Exp. Bot.* **2020**, *71*, 520–542. [CrossRef]

92. Xiao, X.; Chen, B.; Chen, Z.; Zhu, L.; Schnoor, J.L. Insight into Multiple and Multilevel Structures of Biochars and Their Potential Environmental Applications: A Critical Review. *Environ. Sci. Technol.* **2018**, *52*, 5027–5047. [CrossRef]

93. Liu, W.-J.; Jiang, H.; Yu, H.-Q. Development of Biochar-Based Functional Materials: Toward a Sustainable Platform Carbon Material. *Chem. Rev.* **2015**, *115*, 12251–12285. [CrossRef]

94. Billa, S.F.; Angwafo, T.E.; Ngome, A.F. Agro-environmental characterization of biochar issued from crop wastes in the humid forest zone of Cameroon. *Int. J. Recycl. Org. Waste Agric.* **2019**, *8*. [CrossRef]

95. Gao, S.; DeLuca, T.H. Wood biochar impacts soil phosphorus dynamics and microbial communities in organically-managed croplands. *Soil Biol. Biochem.* **2018**, *126*, 144–150. [CrossRef]

96. Zhang, R.; Zhang, Y.; Song, L.; Song, X.; Hänninen, H.; Wu, J. Biochar enhances nut quality of Torreya grandis and soil fertility under simulated nitrogen deposition. *For. Ecol. Manag.* **2017**, *391*, 321–329. [CrossRef]

97. Liu, Q.; Zhang, Y.; Liu, B.; Amonette, J.E.; Lin, Z.; Liu, G.; Ambus, P.; Xie, Z. How does biochar influence soil N cycle? A meta-analysis. *Plant Soil* **2018**, *426*, 211–225. [CrossRef]

98. Joseph, U.E.; Toluwase, A.O.; Kehinde, E.O.; Omasan, E.E.; Tolulope, A.Y.; George, O.O.; Zhao, C.; Hongyan, W. Effect of biochar on soil structure and storage of soil organic carbon and nitrogen in the aggregate fractions of an Albic soil. *Arch. Agron. Soil Sci.* **2019**, 1–12. [CrossRef]

99. Khademalrasoul, A.; Kuhn, N.J.; Elsgaard, L.; Hu, Y.; Iversen, B.V.; Heckrath, G. Short-term Effects of Biochar Application on Soil Loss During a Rainfall-Runoff Simulation. *Soil Sci.* **2019**, *184*, 17–24. [CrossRef]

100. Soinne, H.; Hovi, J.; Tammeorg, P.; Turtola, E. Effect of biochar on phosphorus sorption and clay soil aggregate stability. *Geoderma* **2014**, *219*, 162–167. [CrossRef]
101. Lin, Y.; Munroe, P.; Joseph, S.; Kimber, S.; Van Zwieten, L. Nanoscale organo-mineral reactions of biochars in ferrosol: An investigation using microscopy. *Plant Soil* **2012**, *357*, 369–380. [CrossRef]
102. Jien, S.-H.; Wang, C.-S. Effects of biochar on soil properties and erosion potential in a highly weathered soil. *Catena* **2013**, *110*, 225–233. [CrossRef]
103. Lee, S.S.; Shah, H.S.; Awad, Y.M.; Kumar, S.; Ok, Y.S. Synergy effects of biochar and polyacrylamide on plants growth and soil erosion control. *Environ. Earth Sci.* **2015**, *74*, 2463–2473. [CrossRef]
104. Inyang, M.; Dickenson, E. The potential role of biochar in the removal of organic and microbial contaminants from potable and reuse water: A review. *Chemosphere* **2015**, *134*, 232–240. [CrossRef]
105. Shi, R.; Ni, N.; Nkoh, J.N.; Li, J.; Xu, R.; Qian, W. Beneficial dual role of biochars in inhibiting soil acidification resulting from nitrification. *Chemosphere* **2019**, *234*, 43–51. [CrossRef]
106. Wang, J.; Sun, N.; Xu, M.; Wang, S.; Zhang, J.; Cai, Z.; Cheng, Y. The influence of long-term animal manure and crop residue application on abiotic and biotic N immobilization in an acidified agricultural soil. *Geoderma* **2019**, *337*, 710–717. [CrossRef]
107. Conyers, M.K.; Mullen, C.L.; Scott, B.J.; Poile, G.J.; Braysher, B.D. Long-term benefits of limestone applications to soil properties and to cereal crop yields in southern and central New South Wales. *Aust. J. Exp. Agric.* **2003**, *43*, 71–78. [CrossRef]
108. Guodao, Y.J.L.L.L. Research on Soil Acidification and Acidic Soil's Melioration. *J. South China Univ. Trop. Agric.* **2006**, *1*, 21–28.
109. Zhang, H.; Dao, T.H.; Basta, N.T.; Dayton, E.A.; Daniel, T.C. Remediation techniques for manure nutrient loaded soils. *Am. Soc. Agric. Biol. Eng.* **2006**. [CrossRef]
110. Palviainen, M.; Berninger, F.; Bruckman, V.J.; Köster, K.; de Assumpção, C.R.M.; Aaltonen, H.; Makita, N.; Mishra, A.; Kulmala, L.; Adamczyk, B.; et al. Effects of biochar on carbon and nitrogen fluxes in boreal forest soil. *Plant Soil* **2018**, *425*, 71–85. [CrossRef]
111. Wu, D.; Senbayram, M.; Zang, H.; Ugurlar, F.; Aydemir, S.; Brüggemann, N.; Kuzyakov, Y.; Bol, R.; Blagodatskaya, E. Effect of biochar origin and soil pH on greenhouse gas emissions from sandy and clay soils. *Appl. Soil Ecol.* **2018**, *129*, 121–127. [CrossRef]
112. Si, L.; Xie, Y.; Ma, Q.; Wu, L. The Short-Term Effects of Rice Straw Biochar, Nitrogen and Phosphorus Fertilizer on Rice Yield and Soil Properties in a Cold Waterlogged Paddy Field. *Sustainability* **2018**, *10*, 537. [CrossRef]
113. Xu, R.; Zhao, A.; Yuan, J.; Jiang, J. pH buffering capacity of acid soils from tropical and subtropical regions of China as influenced by incorporation of crop straw biochars. *J. Soils Sediments* **2012**, *12*, 494–502. [CrossRef]
114. Novak, J.M.; Busscher, W.J.; Laird, D.L.; Ahmedna, M.; Watts, D.W.; Niandou, M.A.S. Impact of biochar amendment on fertility of a southeastern coastal plain soil. *Soil Sci.* **2009**, *174*, 105–112. [CrossRef]
115. Jeffery, S.; Verheijen, F.G.A.; van der Velde, M.; Bastos, A.C. A quantitative review of the effects of biochar application to soils on crop productivity using meta-analysis. *Agric. Ecosyst. Environ.* **2011**, *144*, 175–187. [CrossRef]
116. Verheijen, F.; Jeffery, S.; Bastos, A.C.; Van der Velde, M.; Diafas, I. *Biochar Application to Soils—A Critical Scientific Review of Effects on Soil Properties, Processes and Functions*; European Commission Publication Office: Luxembourg, 2010; Volume 24099, p. 162.
117. Qian, L.; Chen, B. Dual role of biochars as adsorbents for aluminum: The effects of oxygen-containing organic components and the scattering of silicate particles. *Environ. Sci. Technol.* **2013**, *47*, 8759–8768. [CrossRef]
118. Fidel, R.B.; Laird, D.A.; Thompson, M.L.; Lawrinenko, M. Characterization and quantification of biochar alkalinity. *Chemosphere* **2017**, *167*, 367–373. [CrossRef] [PubMed]
119. Zhao, H.; Yu, L.; Yu, M.; Afzal, M.; Dai, Z.; Brookes, P.; Xu, J. Nitrogen combined with biochar changed the feedback mechanism between soil nitrification and Cd availability in an acidic soil. *J. Hazard. Mater.* **2019**, *390*, 121631. [CrossRef]
120. Gondek, K.; Mierzwa-Hersztek, M.; Kopeć, M.; Sikora, J.; Głąb, T.; Szczurowska, K. Influence of Biochar Application on Reduced Acidification of Sandy Soil, Increased Cation Exchange Capacity, and the Content of Available Forms of K, Mg, and P. *Pol. J. Environ. Stud.* **2019**, *28*, 1–9. [CrossRef]
121. Taghizadeh-Toosi, A.; Clough, T.J.; Sherlock, R.R.; Condron, L.M. Biochar adsorbed ammonia is bioavailable. *Plant Soil* **2012**, *350*, 57–69. [CrossRef]

122. Zhang, Y.; Chen, H.; Ji, G.; Zhang, Y.; Xiang, J.; Anwar, S.; Zhu, D. Effect of Rice-straw Biochar Application on Rice (*Oryza sativa* L.) Root Growth and Nitrogen Utilization in Acidified Paddy Soil. *Int. J. Agric. Biol.* **2018**, *20*, 2529–2536.

123. Shi, R.; Li, J.; Jiang, J.; Kamran, M.A.; Xu, R.; Qian, W. Incorporation of corn straw biochar inhibited the re-acidification of four acidic soils derived from different parent materials. *Environ. Sci. Pollut. Res.* **2018**, *25*, 9662–9672. [CrossRef]

124. Shi, R.; Hong, Z.; Li, J.; Jiang, J.; Kamran, M.A.; Xu, R.; Qian, W. Peanut straw biochar increases the resistance of two Ultisols derived from different parent materials to acidification: A mechanism study. *J. Environ. Manag.* **2018**, *210*, 171–179. [CrossRef]

125. Wang, J.; Zhang, B.; Tian, Y.; Zhang, H.; Cheng, Y.; Zhang, J. A soil management strategy for ameliorating soil acidification and reducing nitrification in tea plantations. *Eur. J. Soil Biol.* **2018**, *88*, 36–40. [CrossRef]

126. Mierzwa-Hersztek, M.; Gondek, K.; Klimkowicz-Pawlas, A.; Baran, A.; Bajda, T. Sewage sludge biochars management—Ecotoxicity, mobility of heavy metals, and soil microbial biomass. *Environ. Toxicol. Chem.* **2018**, *37*, 1197–1207. [CrossRef]

127. Cross, A.; Sohi, S.P. The priming potential of biochar products in relation to labile carbon contents and soil organic matter status. *Soil Biol. Biochem.* **2011**, *43*, 2127–2134. [CrossRef]

128. Zimmerman, A.R.; Gao, B.; Ahn, M.-Y. Positive and negative carbon mineralization priming effects among a variety of biochar-amended soils. *Soil Biol. Biochem.* **2011**, *43*, 1169–1179. [CrossRef]

129. Zhao, S.; Ta, N.; Li, Z.; Yang, Y.; Zhang, X.; Liu, D.; Zhang, A.; Wang, X. Varying pyrolysis temperature impacts application effects of biochar on soil labile organic carbon and humic fractions. *Appl. Soil Ecol.* **2018**, *123*, 484–493. [CrossRef]

130. Luo, Y.; Durenkamp, M.; De Nobili, M.; Lin, Q.; Brookes, P.C. Short term soil priming effects and the mineralisation of biochar following its incorporation to soils of different pH. *Soil Biol. Biochem.* **2011**, *43*, 2304–2314. [CrossRef]

131. Li, Y.; Zhou, C.; Qiu, Y.; Tigabu, M.; Ma, X. Effects of biochar and litter on carbon and nitrogen mineralization and soil microbial community structure in a China fir plantation. *J. For. Res.* **2019**, *30*, 1913–1923. [CrossRef]

132. Cui, J.; Ge, T.; Kuzyakov, Y.; Nie, M.; Fang, C.; Tang, B.; Zhou, C. Interactions between biochar and litter priming: A three-source 14C and δ13C partitioning study. *Soil Biol. Biochem.* **2017**, *104*, 49–58. [CrossRef]

133. Maestrini, B.; Herrmann, A.M.; Nannipieri, P.; Schmidt, M.W.I.; Abiven, S. Ryegrass-derived pyrogenic organic matter changes organic carbon and nitrogen mineralization in a temperate forest soil. *Soil Biol. Biochem.* **2014**, *69*, 291–301. [CrossRef]

134. Ventura, M.; Alberti, G.; Panzacchi, P.; Delle Vedove, G.; Miglietta, F.; Tonon, G. Biochar mineralization and priming effect in a poplar short rotation coppice from a 3-year field experiment. *Biol. Fertil. Soils* **2019**, *55*, 67–78. [CrossRef]

135. Kuzyakov, Y.; Friedel, J.K.; Stahr, K. Review of mechanisms and quantification of priming effects. *Soil Biol. Biochem.* **2000**, *32*, 1485–1498. [CrossRef]

136. De Figueiredo, C.C.; Farias, W.M.; de Melo, B.A.; Chagas, J.K.M.; Vale, A.T.; Coser, T.R. Labile and stable pools of organic matter in soil amended with sewage sludge biochar. *Arch. Agron. Soil Sci.* **2019**, *65*, 770–781. [CrossRef]

137. Ji, X.; Abakumov, E.; Xie, X.; Wei, D.; Tang, R.; Ding, J.; Cheng, Y.; Chen, J. Preferential Alternatives to Returning All Crop Residues as Biochar to the Crop Field? A Three-Source 13C and 14C Partitioning Study. *J. Agric. Food Chem.* **2019**, *67*, 11322–11330. [CrossRef] [PubMed]

138. Bista, P.; Ghimire, R.; Machado, S.; Pritchett, L. Biochar Effects on Soil Properties and Wheat Biomass vary with Fertility Management. *Agronomy* **2019**, *9*, 623. [CrossRef]

139. Schofield, H.K.; Pettitt, T.R.; Tappin, A.D.; Rollinson, G.K.; Fitzsimons, M.F. Biochar incorporation increased nitrogen and carbon retention in a waste-derived soil. *Sci. Total Environ.* **2019**, *690*, 1228–1236. [CrossRef] [PubMed]

140. Andrés, P.; Rosell-Melé, A.; Colomer-Ventura, F.; Denef, K.; Cotrufo, M.F.; Riba, M.; Alcañiz, J.M. Belowground biota responses to maize biochar addition to the soil of a Mediterranean vineyard. *Sci. Total Environ.* **2019**, *660*, 1522–1532. [CrossRef] [PubMed]

141. Orlova, N.; Abakumov, E.; Orlova, E.; Yakkonen, K.; Shahnazarova, V. Soil organic matter alteration under biochar amendment: Study in the incubation experiment on the Podzol soils of the Leningrad region (Russia). *J. Soils Sediments* **2019**, *19*, 2708–2716. [CrossRef]

142. Kammoun-Rigane, M.; Hlel, H.; Medhioub, K. Biochars Induced Changes in the Physicochemical Characteristics of Technosols: Effects of Feedstock and Pyrolysis Temperature. In *Recent Advances in Geo-Environmental Engineering, Geomechanics and Geotechnics, and Geohazards*; Springer: Berlin/Heidelberg, Germany, 2019; pp. 109–111.

143. Raiesi, F.; Khadem, A. Short-term effects of maize residue biochar on kinetic and thermodynamic parameters of soil β-glucosidase. *Biochar* **2019**, *1*, 213–227. [CrossRef]

144. Lehman, R.; Cambardella, C.; Stott, D.; Acosta-Martinez, V.; Manter, D.; Buyer, J.; Maul, J.; Smith, J.; Collins, H.; Halvorson, J.; et al. Understanding and Enhancing Soil Biological Health: The Solution for Reversing Soil Degradation. *Sustainability* **2015**, *7*, 988–1027. [CrossRef]

145. Morrow, J.G.; Huggins, D.R.; Carpenter-Boggs, L.A.; Reganold, J.P. Evaluating Measures to Assess Soil Health in Long-Term Agroecosystem Trials. *Soil Sci. Soc. Am. J.* **2016**, *80*, 450. [CrossRef]

146. Muhammad, N.; Dai, Z.; Xiao, K.; Meng, J.; Brookes, P.C.; Liu, X.; Wang, H.; Wu, J.; Xu, J. Changes in microbial community structure due to biochars generated from different feedstocks and their relationships with soil chemical properties. *Geoderma* **2014**, *226*, 270–278. [CrossRef]

147. Wang, D.; Zhang, N.; Tang, H.; Adams, J.M.; Sun, B.; Liang, Y. Straw biochar strengthens the life strategies and network of rhizosphere fungi in manure fertilized soils. *Soil Ecol. Lett.* **2019**, *1*, 72–84. [CrossRef]

148. Zhang, L.; Jing, Y.; Xiang, Y.; Zhang, R.; Lu, H. Responses of soil microbial community structure changes and activities to biochar addition: A meta-analysis. *Sci. Total Environ.* **2018**, *643*, 926–935. [CrossRef] [PubMed]

149. Palansooriya, K.N.; Wong, J.T.F.; Hashimoto, Y.; Huang, L.; Rinklebe, J.; Chang, S.X.; Bolan, N.; Wang, H.; Ok, Y.S. Response of microbial communities to biochar-amended soils: A critical review. *Biochar* **2019**, *1*, 3–22. [CrossRef]

150. Song, D.; Xi, X.; Zheng, Q.; Liang, G.; Zhou, W.; Wang, X. Soil nutrient and microbial activity responses to two years after maize straw biochar application in a calcareous soil. *Ecotoxicol. Environ. Saf.* **2019**, *180*, 348–356. [CrossRef] [PubMed]

151. Chen, J.; Sun, X.; Zheng, J.; Zhang, X.; Liu, X.; Bian, R.; Li, L.; Cheng, K.; Zheng, J.; Pan, G. Biochar amendment changes temperature sensitivity of soil respiration and composition of microbial communities 3 years after incorporation in an organic carbon-poor dry cropland soil. *Biol. Fertil. Soils* **2018**, *54*, 175–188. [CrossRef]

152. Liu, Z.; Zhu, M.; Wang, J.; Liu, X.; Guo, W.; Zheng, J.; Bian, R.; Wang, G.; Zhang, X.; Cheng, K.; et al. The responses of soil organic carbon mineralization and microbial communities to fresh and aged biochar soil amendments. *GCB Bioenergy* **2019**, *11*, 1408–1420. [CrossRef]

153. Luo, Y.; Dungait, J.A.J.; Zhao, X.; Brookes, P.C.; Durenkamp, M.; Li, G.; Lin, Q. Pyrolysis temperature during biochar production alters its subsequent utilization by microorganisms in an acid arable soil. *Land Degrad. Dev.* **2018**, *29*, 2183–2188. [CrossRef]

154. Du, Y.; Wu, J.; Anane, P.-S.; Wu, Y.; Wang, C.; Liu, S. Effects of Different Biochars on Physicochemical Properties and Fungal Communities of Black Soil. *Pol. J. Environ. Stud.* **2019**, *28*, 3125–3132. [CrossRef]

155. Bonanomi, G.; Ippolito, F.; Cesarano, G.; Vinale, F.; Lombardi, N.; Crasto, A.; Woo, S.L.; Scala, F. Biochar chemistry defined by 13C-CPMAS NMR explains opposite effects on soilborne microbes and crop plants. *Appl. Soil Ecol.* **2018**, *124*, 351–361. [CrossRef]

156. Zhang, Y.; Liu, Y.; Zhang, G.; Guo, X.; Sun, Z.; Li, T. The Effects of Rice Straw and Biochar Applications on the Microbial Community in a Soil with a History of Continuous Tomato Planting History. *Agronomy* **2018**, *8*, 65. [CrossRef]

157. Paymaneh, Z.; Gryndler, M.; Konvalinková, T.; Benada, O.; Borovička, J.; Bukovská, P.; Püschel, D.; Řezáčová, V.; Sarcheshmehpour, M.; Jansa, J. Soil Matrix Determines the Outcome of Interaction Between Mycorrhizal Symbiosis and Biochar for Andropogon gerardii Growth and Nutrition. *Front. Microbiol.* **2018**, *9*, 2862. [CrossRef]

158. Dangi, S.; Gao, S.; Duan, Y.; Wang, D. Soil microbial community structure affected by biochar and fertilizer sources. *Appl. Soil Ecol.* **2019**. [CrossRef]

159. Li, Y.; Yang, Y.; Shen, F.; Tian, D.; Zeng, Y.; Yang, G.; Zhang, Y.; Deng, S. Partitioning biochar properties to elucidate their contributions to bacterial and fungal community composition of purple soil. *Sci. Total Environ.* **2019**, *648*, 1333–1341. [CrossRef] [PubMed]

160. Zheng, N.; Yu, Y.; Shi, W.; Yao, H. Biochar suppresses N_2O emissions and alters microbial communities in an acidic tea soil. *Environ. Sci. Pollut. Res.* **2019**, 1–10. [CrossRef] [PubMed]

161. Dai, Z.; Enders, A.; Rodrigues, J.L.M.; Hanley, K.L.; Brookes, P.C.; Xu, J.; Lehmann, J. Soil fungal taxonomic and functional community composition as affected by biochar properties. *Soil Biol. Biochem.* **2018**, *126*, 159–167. [CrossRef]

162. Nie, C.; Yang, X.; Niazi, N.K.; Xu, X.; Wen, Y.; Rinklebe, J.; Ok, Y.S.; Xu, S.; Wang, H. Impact of sugarcane bagasse-derived biochar on heavy metal availability and microbial activity: A field study. *Chemosphere* **2018**, *200*, 274–282. [CrossRef] [PubMed]

163. Liu, Y.; Zhu, J.; Ye, C.; Zhu, P.; Ba, Q.; Pang, J.; Shu, L. Effects of biochar application on the abundance and community composition of denitrifying bacteria in a reclaimed soil from coal mining subsidence area. *Sci. Total Environ.* **2018**, *625*, 1218–1224. [CrossRef]

164. Yu, J.; Deem, L.M.; Crow, S.E.; Deenik, J.L.; Penton, C.R. Biochar application influences microbial assemblage complexity and composition due to soil and bioenergy crop type interactions. *Soil Biol. Biochem.* **2018**, *117*, 97–107. [CrossRef]

165. Liu, X.; Li, J.; Yu, L.; Pan, H.; Liu, H.; Liu, Y.; Di, H.; Li, Y.; Xu, J. Simultaneous measurement of bacterial abundance and composition in response to biochar in soybean field soil using 16S rRNA gene sequencing. *Land Degrad. Dev.* **2018**, *29*, 2172–2182. [CrossRef]

166. Liu, X.; Zhou, J.; Chi, Z.; Zheng, J.; Li, L.; Zhang, X.; Zheng, J.; Cheng, K.; Bian, R.; Pan, G. Biochar provided limited benefits for rice yield and greenhouse gas mitigation six years following an amendment in a fertile rice paddy. *Catena* **2019**, *179*, 20–28. [CrossRef]

167. Nguyen, T.T.N.; Wallace, H.M.; Xu, C.-Y.; (Van) Zwieten, L.; Weng, Z.H.; Xu, Z.; Che, R.; Tahmasbian, I.; Hu, H.-W.; Bai, S.H. The effects of short term, long term and reapplication of biochar on soil bacteria. *Sci. Total Environ.* **2018**, *636*, 142–151. [CrossRef]

168. Cole, E.J.; Zandvakili, O.R.; Blanchard, J.; Xing, B.; Hashemi, M.; Etemadi, F. Investigating responses of soil bacterial community composition to hardwood biochar amendment using high-throughput PCR sequencing. *Appl. Soil Ecol.* **2019**, *136*, 80–85. [CrossRef]

169. Song, Y.; Li, X.; Xu, M.; Jiao, W.; Bian, Y.; Yang, X.; Gu, C.; Wang, F.; Jiang, X. Does Biochar Induce Similar Successions of Microbial Community Structures Among Different Soils? *Bull. Environ. Contam. Toxicol.* **2019**, *103*, 642–650. [CrossRef] [PubMed]

170. Ge, X.; Cao, Y.; Zhou, B.; Wang, X.; Yang, Z.; Li, M.-H. Biochar addition increases subsurface soil microbial biomass but has limited effects on soil CO_2 emissions in subtropical moso bamboo plantations. *Appl. Soil Ecol.* **2019**, *142*, 155–165. [CrossRef]

171. Han Weng, Z.; Van Zwieten, L.; Singh, B.P.; Tavakkoli, E.; Joseph, S.; Macdonald, L.M.; Rose, T.J.; Rose, M.T.; Kimber, S.W.L.; Morris, S.; et al. Biochar built soil carbon over a decade by stabilizing rhizodeposits. *Nat. Clim. Chang.* **2017**, *7*, 371–376. [CrossRef]

172. Brändli, R.C.; Hartnik, T.; Henriksen, T.; Cornelissen, G. Sorption of native polyaromatic hydrocarbons (PAH) to black carbon and amended activated carbon in soil. *Chemosphere* **2008**, *73*, 1805–1810. [CrossRef]

173. Jonker, M.T.O.; van der Heijden, S.A.; Kreitinger, J.P.; Hawthorne, S.B. Predicting PAH bioaccumulation and toxicity in earthworms exposed to manufactured gas plant soils with solid-phase microextraction. *Environ. Sci. Technol.* **2007**, *41*, 7472–7478. [CrossRef]

174. Kreitinger, J.P.; Quiñones-Rivera, A.; Neuhauser, E.F.; Alexander, M.; Hawthorne, S.B. Supercritical carbon dioxide extraction as a predictor of polycyclic aromatic hydrocarbon bioaccumulation and toxicity by earthworms in manufactured-gas plant site soils. *Environ. Toxicol. Chem. Int. J.* **2007**, *26*, 1809–1817. [CrossRef]

175. O'Connor, D.; Peng, T.; Zhang, J.; Tsang, D.C.W.; Alessi, D.S.; Shen, Z.; Bolan, N.S.; Hou, D. Biochar application for the remediation of heavy metal polluted land: A review of in situ field trials. *Sci. Total Environ.* **2018**, *619*, 815–826. [CrossRef]

176. He, L.; Zhong, H.; Liu, G.; Dai, Z.; Brookes, P.C.; Xu, J. Remediation of heavy metal contaminated soils by biochar: Mechanisms, potential risks and applications in China. *Environ. Pollut.* **2019**, *252*, 846–855. [CrossRef]

177. Beesley, L.; Moreno-Jiménez, E.; Gomez-Eyles, J.L. Effects of biochar and greenwaste compost amendments on mobility, bioavailability and toxicity of inorganic and organic contaminants in a multi-element polluted soil. *Environ. Pollut.* **2010**, *158*, 2282–2287. [CrossRef]

178. Cui, L.; Pan, G.; Li, L.; Bian, R.; Liu, X.; Yan, J.; Quan, G.; Ding, C.; Chen, T.; Liu, Y. Continuous immobilization of cadmium and lead in biochar amended contaminated paddy soil: A five-year field experiment. *Ecol. Eng.* **2016**, *93*, 1–8. [CrossRef]

179. Lomaglio, T.; Hattab-Hambli, N.; Miard, F.; Lebrun, M.; Nandillon, R.; Trupiano, D.; Scippa, G.S.; Gauthier, A.; Motelica-Heino, M.; Bourgerie, S. Cd, Pb, and Zn mobility and (bio) availability in contaminated soils from a former smelting site amended with biochar. *Environ. Sci. Pollut. Res.* **2018**, *25*, 25744–25756. [CrossRef] [PubMed]

180. De Figueiredo, C.C.; Chagas, J.K.M.; da Silva, J.; Paz-Ferreiro, J. Short-term effects of a sewage sludge biochar amendment on total and available heavy metal content of a tropical soil. *Geoderma* **2019**, *344*, 31–39. [CrossRef]

181. Nzediegwu, C.; Prasher, S.; Elsayed, E.; Dhiman, J.; Mawof, A.; Patel, R. Effect of biochar on heavy metal accumulation in potatoes from wastewater irrigation. *J. Environ. Manag.* **2019**, *232*, 153–164. [CrossRef] [PubMed]

182. Wang, Y.; Zhong, B.; Shafi, M.; Ma, J.; Guo, J.; Wu, J.; Ye, Z.; Liu, D.; Jin, H. Effects of biochar on growth, and heavy metals accumulation of moso bamboo (Phyllostachy pubescens), soil physical properties, and heavy metals solubility in soil. *Chemosphere* **2019**, *219*, 510–516. [CrossRef] [PubMed]

183. Kamran, M.; Malik, Z.; Parveen, A.; Huang, L.; Riaz, M.; Bashir, S.; Mustafa, A.; Abbasi, G.H.; Xue, B.; Ali, U. Ameliorative effects of biochar on rapeseed (*Brassica napus* L.) growth and heavy metal immobilization in soil irrigated with untreated wastewater. *J. Plant Growth Regul.* **2020**, *39*, 266–281. [CrossRef]

184. Ghori, S.A.; Gul, S.; Tahir, S.; Sohail, M.; Batool, S.; Shahwani, M.N.; Bano, G. Wood-derived biochar influences nutrient use efficiency of heavy metals in spinach (Spinacia oleracea) under groundwater and wastewater irrigation. *J. Environ. Eng. Landsc. Manag.* **2019**, *27*, 144–152. [CrossRef]

185. Antonangelo, J.A.; Zhang, H. Heavy metal phytoavailability in a contaminated soil of northeastern Oklahoma as affected by biochar amendment. *Environ. Sci. Pollut. Res.* **2019**, *26*, 33582–33593. [CrossRef]

186. Cui, L.; Noerpel, M.R.; Scheckel, K.G.; Ippolito, J.A. Wheat straw biochar reduces environmental cadmium bioavailability. *Environ. Int.* **2019**, *126*, 69–75. [CrossRef]

187. Shen, Z.; Zhang, J.; Hou, D.; Tsang, D.C.W.; Ok, Y.S.; Alessi, D.S. Synthesis of MgO-coated corncob biochar and its application in lead stabilization in a soil washing residue. *Environ. Int.* **2019**, *122*, 357–362. [CrossRef]

188. Qian, T.-T.; Wu, P.; Qin, Q.-Y.; Huang, Y.-N.; Wang, Y.-J.; Zhou, D.-M. Screening of wheat straw biochars for the remediation of soils polluted with Zn (II) and Cd (II). *J. Hazard. Mater.* **2019**, *362*, 311–317. [CrossRef] [PubMed]

189. Novak, J.M.; Ippolito, J.A.; Watts, D.W.; Sigua, G.C.; Ducey, T.F.; Johnson, M.G. Biochar compost blends facilitate switchgrass growth in mine soils by reducing Cd and Zn bioavailability. *Biochar* **2019**, *1*, 97–114. [CrossRef]

190. Derakhshan-Nejad, Z.; Jung, M.C. Remediation of multi-metal contaminated soil using biochars from rice husk and maple leaves. *J. Mater. Cycles Waste Manag.* **2019**, *21*, 457–468. [CrossRef]

191. Andrey, G.; Rajput, V.; Tatiana, M.; Saglara, M.; Svetlana, S.; Igor, K.; Grigoryeva, T.V.; Vasily, C.; Iraida, A.; Vladislav, Z. The role of biochar-microbe interaction in alleviating heavy metal toxicity in *Hordeum vulgare* L. grown in highly polluted soils. *Appl. Geochem.* **2019**, *104*, 93–101. [CrossRef]

192. Schaeffer, S.; Koepke, T.; Dhingra, A. Tobacco: A Model Plant for Understanding the Mechanism of Abiotic Stress Tolerance. In *Improving Crop Resistance to Abiotic Stress*; Wiley-VCH Verlag GmbH & Co. KGaA: Weinheim, Germany, 2012; Volume 2, pp. 1169–1201.

193. Dahlawi, S.; Naeem, A.; Rengel, Z.; Naidu, R. Biochar application for the remediation of salt-affected soils: Challenges and opportunities. *Sci. Total Environ.* **2018**, *625*, 320–335.

194. Lashari, M.S.; Liu, Y.; Li, L.; Pan, W.; Fu, J.; Pan, G.; Zheng, J.; Zheng, J.; Zhang, X.; Yu, X. Effects of amendment of biochar-manure compost in conjunction with pyroligneous solution on soil quality and wheat yield of a salt-stressed cropland from Central China Great Plain. *Field Crop. Res.* **2013**, *144*, 113–118. [CrossRef]

195. Akhtar, S.S.; Andersen, M.N.; Liu, F. Biochar mitigates salinity stress in potato. *J. Agron. Crop Sci.* **2015**, *201*, 368–378. [CrossRef]

196. Akhtar, S.S.; Andersen, M.N.; Liu, F. Residual effects of biochar on improving growth, physiology and yield of wheat under salt stress. *Agric. Water Manag.* **2015**, *158*, 61–68. [CrossRef]

197. Akhtar, S.S.; Andersen, M.N.; Naveed, M.; Zahir, Z.A.; Liu, F. Interactive effect of biochar and plant growth-promoting bacterial endophytes on ameliorating salinity stress in maize. *Funct. Plant Biol.* **2015**, *42*, 770–781. [CrossRef]

198. Lashari, M.S.; Ye, Y.; Ji, H.; Li, L.; Kibue, G.W.; Lu, H.; Zheng, J.; Pan, G. Biochar–manure compost in conjunction with pyroligneous solution alleviated salt stress and improved leaf bioactivity of maize in a saline soil from central China: A 2-year field experiment. *J. Sci. Food Agric.* **2015**, *95*, 1321–1327. [CrossRef]

199. Zhang, J.; Bai, Z.; Huang, J.; Hussain, S.; Zhao, F.; Zhu, C.; Zhu, L.; Cao, X.; Jin, Q. Biochar alleviated the salt stress of induced saline paddy soil and improved the biochemical characteristics of rice seedlings differing in salt tolerance. *Soil Tillage Res.* **2019**, *195*, 104372. [CrossRef]

200. Abd El-Mageed, T.A.; Rady, M.M.; Taha, R.S.; Abd El Azeam, S.; Simpson, C.R.; Semida, W.M. Effects of integrated use of residual sulfur-enhanced biochar with effective microorganisms on soil properties, plant growth and short-term productivity of Capsicum annuum under salt stress. *Sci. Hortic.* **2020**, *261*, 108930. [CrossRef]

201. Mingyi, H.; Zhang, Z.; Zhu, C.; Zhai, Y.; Peirong, L. Effect of biochar on sweet corn and soil salinity under conjunctive irrigation with brackish water in coastal saline soil. *Sci. Hortic.* **2019**, *250*, 405–413. [CrossRef]

202. Hafeez, F.; Rizwan, M.; Saqib, M.; Yasmeen, T.; Ali, S.; Abbas, T.; Zia-ur-Rehman, M.; Qayyum, M. Residual effect of biochar on growth, antioxidant defence and cadmium (Cd) accumulation in rice in a Cd contaminated saline soil. *Pak. J. Agric. Sci.* **2018**, *56*. [CrossRef]

203. Nikpour-Rashidabad, N.; Tavasolee, A.; Torabian, S.; Farhangi-Abriz, S. The effect of biochar on the physiological, morphological and anatomical characteristics of mung bean roots after exposure to salt stress. *Arch. Biol. Sci.* **2019**, *71*, 14. [CrossRef]

204. Aboshanab, W.; Osman, M.; Nessim, A.; Mohsen, A.; Elsaka, M. Evaluation of Biochar as a Soil Amendment for Alleviating the Harmful Effect of Salinity on *Vigna unguiculata* (L.) Walp. *Egipt. J. Bot.* **2019**, *59*, 617–631.

205. Jaiswal, A.K.; Frenkel, O.; Tsechansky, L.; Elad, Y.; Graber, E.R. Immobilization and deactivation of pathogenic enzymes and toxic metabolites by biochar: A possible mechanism involved in soilborne disease suppression. *Soil Biol. Biochem.* **2018**, *121*, 59–66. [CrossRef]

206. Harel, Y.M.; Elad, Y.; Rav-David, D.; Borenstein, M.; Shulchani, R.; Lew, B.; Graber, E.R. Biochar mediates systemic response of strawberry to foliar fungal pathogens. *Plant Soil* **2012**, *357*, 245–257. [CrossRef]

207. Kumar, A.; Elad, Y.; Tsechansky, L.; Abrol, V.; Lew, B.; Offenbach, R.; Graber, E.R. Biochar potential in intensive cultivation of *Capsicum annuum* L.(sweet pepper): Crop yield and plant protection. *J. Sci. Food Agric.* **2018**, *98*, 495–503. [CrossRef]

208. Jaiswal, A.K.; Elad, Y.; Paudel, I.; Graber, E.R.; Cytryn, E.; Frenkel, O. Linking the belowground microbial composition, diversity and activity to soilborne disease suppression and growth promotion of tomato amended with biochar. *Sci. Rep.* **2017**, *7*, 44382. [CrossRef]

209. Knox, O.G.G.; Oghoro, C.O.; Burnett, F.J.; Fountaine, J.M. Biochar increases soil pH, but is as ineffective as liming at controlling clubroot. *J. Plant Pathol.* **2015**, *97*, 149–152.

210. Bonanomi, G.; Ippolito, F.; Scala, F. A "black" future for plant pathology? Biochar as a new soil amendment for controlling plant diseases. *J. Plant Pathol.* **2015**, *97*, 223–234.

211. Jaiswal, A.K.; Graber, E.R.; Elad, Y.; Frenkel, O. Biochar as a management tool for soilborne diseases affecting early stage nursery seedling production. *Crop Prot.* **2019**, *120*, 34–42. [CrossRef]

212. Lechenet, M.; Dessaint, F.; Py, G.; Makowski, D.; Munier-Jolain, N. Reducing pesticide use while preserving crop productivity and profitability on arable farms. *Nat. Plants* **2017**, *3*, 17008. [CrossRef] [PubMed]

213. Elijah, O.; Rahman, T.A.; Orikumhi, I.; Leow, C.Y.; Hindia, M.H.D.N. An Overview of Internet of Things (IoT) and Data Analytics in Agriculture: Benefits and Challenges. *IEEE Internet Things J.* **2018**, *5*, 3758–3773. [CrossRef]

214. Prost, L.; Berthet, E.T.A.; Cerf, M.; Jeuffroy, M.-H.; Labatut, J.; Meynard, J.-M. Innovative design for agriculture in the move towards sustainability: Scientific challenges. *Res. Eng. Des.* **2017**, *28*, 119–129. [CrossRef]

215. Reganold, J.P.; Wachter, J.M. Organic agriculture in the twenty-first century. *Nat. Plants* **2016**, *2*, 15221. [CrossRef]

216. Kravchenko, A.N.; Snapp, S.S.; Robertson, G.P. Field-scale experiments reveal persistent yield gaps in low-input and organic cropping systems. *Proc. Natl. Acad. Sci. USA* **2017**, *114*, 926–931. [CrossRef]

217. De Ponti, T.; Rijk, B.; Van Ittersum, M.K. The crop yield gap between organic and conventional agriculture. *Agric. Syst.* **2012**, *108*, 1–9. [CrossRef]

218. Sharpe, R.M.; Gustafson, L.; Hewitt, S.; Kilian, B.; Crabb, J.; Hendrickson, C.; Jiwan, D.; Andrews, P.; Dhingra, A. Concomitant phytonutrient and transcriptome analysis of mature fruit and leaf tissues of tomato (Solanum lycopersicum L. Cv. Oregon Spring) grown using organic and conventional fertilizer. *PLoS ONE* **2020**, *15*, e0227429. [CrossRef]

219. Zheng, H.; Wang, X.; Chen, L.; Wang, Z.; Xia, Y.; Zhang, Y.; Wang, H.; Luo, X.; Xing, B. Enhanced growth of halophyte plants in biochar-amended coastal soil: Roles of nutrient availability and rhizosphere microbial modulation. *Plant. Cell Environ.* **2018**, *41*, 517–532. [CrossRef] [PubMed]

220. Laird, D.A.; Novak, J.M.; Collins, H.P.; Ippolito, J.A.; Karlen, D.L.; Lentz, R.D.; Sistani, K.R.; Spokas, K.; Van Pelt, R.S. Multi-year and multi-location soil quality and crop biomass yield responses to hardwood fast pyrolysis biochar. *Geoderma* **2017**, *289*, 46–53. [CrossRef]

221. Knapp, S.; van der Heijden, M.G.A. A global meta-analysis of yield stability in organic and conservation agriculture. *Nat. Commun.* **2018**, *9*, 1–9. [CrossRef]

222. Biederman, L.A.; Harpole, W.S. Biochar and its effects on plant productivity and nutrient cycling: A meta-analysis. *GCB Bioenergy* **2013**, *5*, 202–214. [CrossRef]

223. Amoakwah, E.; Arthur, E.; Frimpong, K.A.; Parikh, S.J.; Islam, R. Soil organic carbon storage and quality are impacted by corn cob biochar application on a tropical sandy loam. *J. Soils Sediments* **2020**, *20*, 1960–1969. [CrossRef]

224. Fryda, L.; Visser, R.; Schmidt, J. Biochar Replaces Peat in Horticulture: Environmental Impact Assessment of Combined Biochar & Bioenergy Production. *Detritus* **2019**, *5*, 132–149. [CrossRef]

225. Thi Thu Hien, T.; Shinogi, Y.; Taniguchi, T.; Hien, T.T.T.; Shinogi, Y.; Taniguchi, T. The Different Expressions of Draft Cherry Tomato Growth, Yield, Quality under Bamboo and Rice Husk Biochars Application to Clay Loamy Soil. *Agric. Sci.* **2017**, *08*, 934–948. [CrossRef]

226. Woldetsadik, D.; Drechsel, P.; Marschner, B.; Itanna, F.; Gebrekidan, H. Effect of biochar derived from faecal matter on yield and nutrient content of lettuce (*Lactuca sativa*) in two contrasting soils. *Environ. Syst. Res.* **2018**, *6*, 2. [CrossRef]

227. Liu, B.; Cai, Z.; Zhang, Y.; Liu, G.; Luo, X.; Zheng, H. Comparison of efficacies of peanut shell biochar and biochar-based compost on two leafy vegetable productivity in an infertile land. *Chemosphere* **2019**, *224*, 151–161. [CrossRef]

228. Velez, T.I.; Moonilall, N.I.; Reed, S.; Jayachandran, K.; Scinto, L.J. Impact of melaleuca quinquenervia biochar on phaseolus vulgaris growth, soil nutrients, and microbial gas flux. *J. Environ. Qual.* **2018**, *47*, 1487–1495. [CrossRef]

229. Alotaibi, K.D.; Schoenau, J.J. Addition of Biochar to a Sandy Desert Soil: Effect on Crop Growth, Water Retention and Selected Properties. *Agronomy* **2019**, *9*, 327. [CrossRef]

230. Nzediegwu, C.; Prasher, S.; Elsayed, E.; Dhiman, J.; Mawof, A.; Patel, R. Effect of Biochar on the Yield of Potatoes Cultivated Under Wastewater Irrigation for Two Seasons. *J. Soil Sci. Plant Nutr.* **2019**, *19*, 865–877. [CrossRef]

231. Griffin, D.E.; Wang, D.; Parikh, S.J.; Scow, K.M. Short-lived effects of walnut shell biochar on soils and crop yields in a long-term field experiment. *Agric. Ecosyst. Environ.* **2017**, *236*, 21–29. [CrossRef]

232. Li, S.; Shangguan, Z. Positive effects of apple branch biochar on wheat yield only appear at a low application rate, regardless of nitrogen and water conditions. *J. Soils Sediments* **2018**, *18*, 3235–3243. [CrossRef]

233. Faloye, O.T.; Alatise, M.O.; Ajayi, A.E.; Ewulo, B.S. Effects of biochar and inorganic fertiliser applications on growth, yield and water use efficiency of maize under deficit irrigation. *Agric. Water Manag.* **2019**, *217*, 165–178. [CrossRef]

234. Kang, S.-W.; Kim, S.-H.; Park, J.-H.; Seo, D.-C.; Ok, Y.S.; Cho, J.-S. Effect of biochar derived from barley straw on soil physicochemical properties, crop growth, and nitrous oxide emission in an upland field in South Korea. *Environ. Sci. Pollut. Res.* **2018**, *25*, 25813–25821. [CrossRef]

235. Munda, S.; Bhaduri, D.; Mohanty, S.; Chatterjee, D.; Tripathi, R.; Shahid, M.; Kumar, U.; Bhattacharyya, P.; Kumar, A.; Adak, T.; et al. Dynamics of soil organic carbon mineralization and C fractions in paddy soil on application of rice husk biochar. *Biomass Bioenergy* **2018**, *115*, 1–9. [CrossRef]

236. Faria, W.M.; de Figueiredo, C.C.; Coser, T.R.; Vale, A.T.; Schneider, B.G. Is sewage sludge biochar capable of replacing inorganic fertilizers for corn production? Evidence from a two-year field experiment. *Arch. Agron. Soil Sci.* **2018**, *64*, 505–519. [CrossRef]

237. Sarma, B.; Borkotoki, B.; Narzari, R.; Kataki, R.; Gogoi, N. Organic amendments: Effect on carbon mineralization and crop productivity in acidic soil. *J. Clean. Prod.* **2017**, *152*, 157–166. [CrossRef]

238. Takaragawa, H.; Yabuta, S.; Watanabe, K.; Kawamitsu, Y. Effects of Application of Bagasse- and Sunflower Residue-derived Biochar to Soil on Growth and Yield of Oilseed Sunflower. *Trop. Agric. Dev.* **2017**, *61*, 32–39. [CrossRef]

239. Hood-Nowotny, R.; Watzinger, A.; Wawra, A.; Soja, G. The Impact of Biochar Incorporation on Inorganic Nitrogen Fertilizer Plant Uptake; An Opportunity for Carbon Sequestration in Temperate Agriculture. *Geosciences* **2018**, *8*, 420. [CrossRef]

240. Boersma, M.; Wrobel-Tobiszewska, A.; Murphy, L.; Eyles, A. Impact of biochar application on the productivity of a temperate vegetable cropping system. *N. Z. J. Crop Hortic. Sci.* **2017**, *45*, 277–288. [CrossRef]

241. Artiola, J.F.; Rasmussen, C.; Freitas, R. Effects of a Biochar-Amended Alkaline Soil on the Growth of Romaine Lettuce and Bermudagrass. *Soil Sci.* **2012**, *177*, 561–570. [CrossRef]

242. Azeem, M.; Hayat, R.; Hussain, Q.; Ahmed, M.; Pan, G.; Ibrahim Tahir, M.; Imran, M.; Irfan, M.; Mehmood-ul-Hassan. Biochar improves soil quality and N_2-fixation and reduces net ecosystem CO_2 exchange in a dryland legume-cereal cropping system. *Soil Tillage Res.* **2019**, *186*, 172–182. [CrossRef]

243. Rezaie, N.; Razzaghi, F.; Sepaskhah, A.R. Different Levels of Irrigation Water Salinity and Biochar Influence on Faba Bean Yield, Water Productivity, and Ions Uptake. *Commun. Soil Sci. Plant Anal.* **2019**, *50*, 611–626. [CrossRef]

244. El-Naggar, A.; El-Naggar, A.H.; Shaheen, S.M.; Sarkar, B.; Chang, S.X.; Tsang, D.C.W.; Rinklebe, J.; Ok, Y.S. Biochar composition-dependent impacts on soil nutrient release, carbon mineralization, and potential environmental risk: A review. *J. Environ. Manag.* **2019**, *241*, 458–467. [CrossRef]

245. Lima, I.M.; White, P.M., Jr. Sugarcane bagasse and leaf residue biochars as soil amendment for increased sugar and cane yields. *Int. Sugar J.* **2017**, *119*, 382–390.

246. Jay, C.N.; Fitzgerald, J.D.; Hipps, N.A.; Atkinson, C.J. Why short-term biochar application has no yield benefits: Evidence from three field-grown crops. *Soil Use Manag.* **2015**, *31*, 241–250. [CrossRef]

247. Ogbe, V.B.; Jayeoba, O.J.; Amana, S.M. Effect of Rice Husk as an Amendment On The Physico-Chemical Properties of Sandy-Loam Soil In Lafia, Southern-Guinea Savannah, Nigeria. *Prod. Agric. Technol.* **2015**, *11*, 44–55. [CrossRef]

248. Farrell, M.; Macdonald, L.M.; Butler, G.; Chirino-Valle, I.; Condron, L.M. Biochar and fertiliser applications influence phosphorus fractionation and wheat yield. *Biol. Fertil. Soils* **2014**, *50*, 169–178. [CrossRef]

249. Marra, R.; Vinale, F.; Cesarano, G.; Lombardi, N.; d'Errico, G.; Crasto, A.; Mazzei, P.; Piccolo, A.; Incerti, G.; Woo, S.L.; et al. Biochars from olive mill waste have contrasting effects on plants, fungi and phytoparasitic nematodes. *PLoS ONE* **2018**, *13*, e0198728. [CrossRef] [PubMed]

250. Intani, K.; Latif, S.; Islam, M.; Müller, J. Phytotoxicity of Corncob Biochar before and after Heat Treatment and Washing. *Sustainability* **2018**, *11*, 30. [CrossRef]

251. Paetsch, L.; Mueller, C.W.; Kögel-Knabner, I.; Von Lützow, M.; Girardin, C.; Rumpel, C. Effect of in-situ aged and fresh biochar on soil hydraulic conditions and microbial C use under drought conditions. *Sci. Rep.* **2018**, *8*. [CrossRef] [PubMed]

252. Hersh, B.; Mirkouei, A.; Sessions, J.; Rezaie, B.; You, Y. A review and future directions on enhancing sustainability benefits across food-energy-water systems: The potential role of biochar-derived products. *AIMS Environ. Sci.* **2019**, *6*, 379–416. [CrossRef]

253. Wang, F.; Liu, Y.; Li, X.; Liu, F. Effects of Biochar Amendment, CO_2 Elevation and Drought on Leaf Gas Exchange, Biomass Production and Water Use Efficiency in Maize. *Pak. J. Bot.* **2018**, *50*, 1347–1353.

254. Rajapaksha, A.U.; Chen, S.S.; Tsang, D.C.W.; Zhang, M.; Vithanage, M.; Mandal, S.; Gao, B.; Bolan, N.S.; Ok, Y.S. Engineered/designer biochar for contaminant removal/immobilization from soil and water: Potential and implication of biochar modification. *Chemosphere* **2016**, *148*, 276–291. [CrossRef]

255. Ok, Y.S.; Tsang, D.C.W.; Bolan, N.; Novak, J.M. *Biochar from Biomass and Waste: Fundamentals and Applications*; Elsevier: Amsterdam, The Netherlands, 2018.

256. Novak, J.M.; Johnson, M.G.; Spokas, K.A. Concentration and Release of Phosphorus and Potassium From Lignocellulosic- and Manure-Based Biochars for Fertilizer Reuse. *Front. Sustain. Food Syst.* **2018**, *2*, 54. [CrossRef]

257. Al-Wabel, M.I.; Hussain, Q.; Usman, A.R.A.; Ahmad, M.; Abduljabbar, A.; Sallam, A.S.; Ok, Y.S. Impact of biochar properties on soil conditions and agricultural sustainability: A review. *Land Degrad. Dev.* **2018**, *29*, 2124–2161. [CrossRef]
258. Nair, A.; Lang, K.; Snyder, D. *Impact of Biochar and Fertility Management on Potato Production*; Iowa State University Research and Demostration Farms Progress Reports; 2018. Available online: https://lib.dr.iastate.edu/farmprogressreports/vol2017/iss1/71/ (accessed on 21 March 2020).
259. Youssef, M.; Al-Easily, I.A.S.; Nawar, D.A.S. Impact of Biochar Addition on Productivity and Tubers Quality of Some Potato Cultivars Under Sandy Soil Conditions. *Egypt. J. Hortic.* **2017**, *44*, 199–217. [CrossRef]
260. Gámiz, B.; Velarde, P.; Spokas, K.A.; Celis, R.; Cox, L. Changes in sorption and bioavailability of herbicides in soil amended with fresh and aged biochar. *Geoderma* **2019**, *337*, 341–349. [CrossRef]

 *horticulturae*

Article

Effects of Foliar Application of Gibberellic Acid on the Salt Tolerance of Tomato and Sweet Pepper Transplants

Alessandro Miceli *, Filippo Vetrano * and Alessandra Moncada

Dipartimento Scienze Agrarie, Alimentari e Forestali, Università di Palermo, Viale delle Scienze 4, 90128 Palermo, Italy; alessandra.moncada@unipa.it

* Correspondence: alessandro.miceli@unipa.it (A.M.); filippo.vetrano@unipa.it (F.V.); Tel.: +39-091-23862219 (A.M.); +39-091-23862229 (F.V.)

Received: 1 November 2020; Accepted: 24 November 2020; Published: 28 November 2020

Abstract: Seed germination and early seedling growth are the plant growth stages most sensitive to salt stress. Thus, the availability of poor-quality brackish water can be a big limiting factor for the nursery vegetable industry. The exogenous supplementation of gibberellic acid (GA_3) may promote growth and vigor and counterbalance salt stress in mature plants. This study aimed to test exogenous supplementation through foliar spray of 10^{-5} M GA_3 for increasing salt tolerance of tomato and sweet pepper seedlings irrigated with increasing salinity (0, 25, and 50 mM NaCl during nursery growth. Tomato and sweet pepper seedlings suffered negative effects of salinity on plant height, biomass, shoot/root ratio, leaf number, leaf area, relative water content, and stomatal conductance. The foliar application of GA_3 had a growth-promoting effect on the unstressed tomato and pepper seedlings and was successful in increasing salinity tolerance of tomato seedlings up to 25 mM NaCl and up to 50 mM NaCl in sweet pepper seedlings. This treatment could represent a sustainable strategy to use saline water in vegetable nurseries limiting its negative effect on seedling quality and production time.

Keywords: *Solanum lycopersicum*; *Capsicum annuum*; seedlings; vegetable nursery; transplant production; salinity; abiotic stress; plant growth regulators; GA_3

---

1. Introduction

Greenhouse production of container-grown transplants has become a typical practice for vegetable crops in many vegetable production areas of the world [1]. Some of the advantages of using transplants are the possibility of anticipating planting, thus reducing field occupation time, the production of seedlings with higher and more consistent quality, the control of plant spacing in the field [2], the reduction of cost against direct seeding when using hybrid seeds or grafted plants, and concentration of crop maturity [2–4]. The main goal of transplant production is to produce in a short time strong, vigorous, and compact seedlings that establish and grow fast when transplanted in the field so that they can reach high yield and quality [5]. Containerized vegetable transplant production is an extremely intensive agricultural practice as seedlings need to be watered and fertilized frequently, even daily, due to the limited volume of substrate explored by the young roots that can retain only small amounts of water and nutrients. Water availability and quality can be limiting factors for the nursery vegetable industry. Vegetable growers are facing more and more frequently the availability of poor quality water due to high salt content. This is an increasingly important problem worldwide as it could limit the growth of sensitive plants such as young vegetable seedlings and result in reductions of crop yield. Many Mediterranean regions are characterized by high salinity groundwater due to the intensive draw of irrigation water that increases seawater infiltration [6]. The use of this water for growing vegetable transplants can adversely affect the growth and vigor of vegetable seedlings. The plant growth stages most sensitive to salt stress are generally seed germination and early seedling growth because plant sensitivity to salt stress usually decreases with plant ontogeny [7].

It is known that salinity activates stress response mechanisms which control the osmotic and ionic re-equilibrium, the detoxification of reactive oxygen species, and the modulation of cell growth or cell division [8]. These mechanisms may be mainly mediated by plant hormones as shown by the modifications of the endogenous phytohormone concentration recorded during seed germination and plant growth under salt stress [9]. Thus, the adverse effects of salinity could be mitigated by the exogenous application of plant growth regulators (PGR) such as auxins, gibberellins (GAs), or cytokinins [10–13]. Gibberellins are phytohormones produced by plants and fungi that act at different levels in plant metabolism ending in the modification of plant physiology and morphology. Exogenous supplementation with gibberellic acid (GA_3) promoted seedling and plant growth, improved post-harvest life and enhanced tolerance to abiotic stress (e.g., drought, heat, salinity) of many crops [13–22]. Therefore, the object of this study was to evaluate the exogenous supplementation through foliar sprays of 10^{-5} M GA_3 to increase salt tolerance of tomato and sweet pepper seedlings during nursery growth.

2. Materials and Methods

2.1. Plant Materials and Transplant Production

The effects of salt stress and gibberellic acid (GA_3) treatment on transplant production were evaluated in a nursery trial carried out during autumn 2018 in a greenhouse situated at the Department of Agricultural, Food, and Forest Sciences (SAAF—University of Palermo, Palermo, Italy) (38°6′28″ N 13°21′3″ E; altitude 49 m above sea level).

Seeds of *Solanum lycopersicum* 'Marmande' (Vilmorin, La Ménitré, France) and *Capsicum annuum* 'Dolce di Spagna' were sown into 12 polystyrene trays (104 cells each) for each species filled with a commercial substrate (SER CA-V7 Special semine, Vigorplant Italia srl, Fombio, Italy, containing 800 g m^{-3} of a mineral fertilizer NPK 12-11-18). After sowing (1 October 2018), the trays were kept in a dark room at a temperature ranging from 22 °C to 24 °C until the first emergence was observed and were then moved onto benches in the greenhouse for seedling growth. Plantlet emergence occurred 5 and 14 days after sowing for tomato and pepper, respectively. Three days after emergence, the plantlets were thinned to one per cell. When the plantlets had fully expanded cotyledons and the first true leaf (11th BBCH growth stage [23]) (5 and 11 days after emergence for tomato and pepper, respectively), half trays were treated by spraying plantlets with 10^{-5} M GA_3 (Gibrelex, Biolchim, Bologna, Italy). The gibberellic acid concentration was chosen according to previous experiments based on the effects of GA_3 levels on tomato plants [8,14,18,22,24,25].

Salt treatments started at the same growth stage for tomato and pepper (11th BBCH growth stage [23]) and were applied with an ebb and flow sub-irrigation system using water with one of three concentrations of NaCl: 0 mM (Electrical conductivity—EC 0.68 mS cm^{-1}), 25 mM (EC 3.14 mS cm^{-1}) or 50 mM (EC 5.57 mS cm^{-1}). Seedlings were sub-irrigated according to their need until they were ready for transplanting (twice a week on average).

Leaf stomatal conductance was measured one week before seedlings were ready for transplanting (17 and 25 days after emergence for tomato and pepper, respectively) using a diffusion porometer (AP4, Delta-T Devices Ltd., Cambridge, UK) on two young unshaded leaves of 20 seedlings for each species and each replicate.

The seedlings were considered ready for transplanting when they reached the 14–15th BBCH growth stage [23] (29 and 46 days from sowing for tomato and pepper, respectively). At this stage, four replicated samples of 30 seedlings for each species and each treatment were randomly selected and analyzed to evaluate their morphological characteristics (seedling height, stem diameter, and leaf number). Leaf color of each seedling was measured on the upper part of 2 randomly selected leaves, using a colorimeter (CR-400, Minolta corporation, Ltd., Osaka, Japan) that measured L* (lightness), a* (positive values for reddish colors and negative values for greenish colors) and b* (positive values for yellowish colors and negative values for bluish colors). These components were used to calculate Hue angle (h°) and Chroma (C*) as $h° = 180° + \arctan(b^*/a^*)$ [26] and $C^* = (a^{*2} + b^{*2})^{1/2}$. Then, the seedlings were separated into leaves, stem, and roots, weighed, and dried to a constant weight at 85 °C to determine the fresh and dry biomass and the shoot/root ratio for both fresh and dry weight. Before drying, the leaves were scanned at 350 dpi (Epson Perfection 4180 Photo, Seiko Epson Corp., Suwa, Japan) to obtain digital images that were analyzed with the ImageJ 1.52a software (National Institutes Health, Bethesda, MD, USA) to measure the leaf area. The specific leaf area (SLA $cm^2\ g^{-1}$ DW) was estimated as the leaf area/leaf dry weight.

To evaluate the water status at the end of seedling growth the relative water content (RWC) was determined for each species and each treatment. Ten leaves for each replicate were weighed immediately (FW) after sampling the seedlings, placed in distilled water for 4 h and then their turgid weight (TW) was calculated. The turgid leaves were dried in an oven at 80 °C for 24 h and weighed to obtain their dry weight (DW). Relative water content was calculated as $RWC = (FW − DW)/(TW − DW) \times 100$.

2.2. Statistics and Principal Component Analysis

The experimental design consisted of four replicated samples of 30 seedlings each for every combination of GA_3 treatment and NaCl level, randomly assigned in four blocks. A two-way ANOVA was performed to evaluate the effects of GA_3 treatments and NaCl levels on tomato and pepper seedlings. The mean values were compared by the least significant differences (LSD) test at $p \leq 0.05$ to identify the significant differences among treatments and the significant interactions between factors.

A principal component analysis on morphophysiological parameters of tomato and pepper seedlings was performed (SPSS version 13.0; SPSS Inc., Chicago, IL, USA) to investigate the main parameters that were most efficient in differentiating between NaCl levels and GA_3 treatments. The input matrix for the analysis consisted of the seedlings morphophysiological parameters (height, stem diameter, total, shoot and root fresh and dry weight, shoot/root ratio of fresh and dry weights, dry matter percentage, leaf number, total leaf area, SLA, stomatal conductance, RWC, L*, chroma, and hue angle). The number of principal components (PCs) was calculated by retaining only the factors with eigenvalues higher than 1.0. The correlations between the variables of the input data set were studied through the plot of the PCs. Furthermore, the initial variables were projected into the subspace defined by the first and second PCs, and correlated variables were determined.

3. Results

During the experiment, the average temperature outside the greenhouse ranged between 10.1 ± 0.3 °C (night) and 31.4 ± 0.6 °C (day), and the average net solar radiation at noon was 449 W·m^{-2}, with a day length that ranged between 8 and 9 h. Inside the greenhouse, the air temperature was on average 23.5 ± 0.4 °C and ranged between 36.8 (day) and 11.9 °C (night) (Figure 1), whereas the relative humidity was 85.2 ± 1.4% and ranged between 59.9% and 100%; the light intensity at noon was 39063 ± 2451 lux and ranged from 58728 to 1286 lux as a function of the cloudiness.

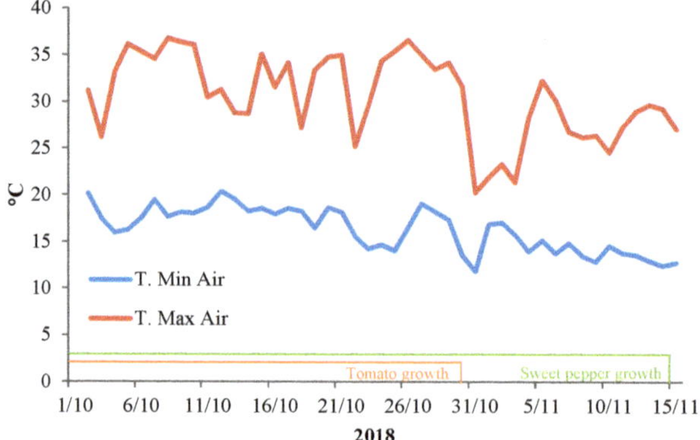

Figure 1. Daily average maximum and minimum temperatures of the air inside the greenhouse during nursery trials.

3.1. Morphophysiological Parameters of Tomato Seedlings

The emergence (at least 50% of plantlets emerged) of tomato plantlets occurred 5 days after sowing and seedlings were ready for transplanting (4–5 true leaves; 14–15th BBCH growth stage [23]) after 29 d from sowing.

The height of tomato seedlings was significantly affected by the experimental factors. After ten days from the beginning of salt stress and GA_3 spray treatment, the seedlings had less height with increasing NaCl concentration but were not influenced by GA_3, whereas at the end of the experiment their height was affected by salt stress and was modified by the GA_3 treatment (Figure 2).

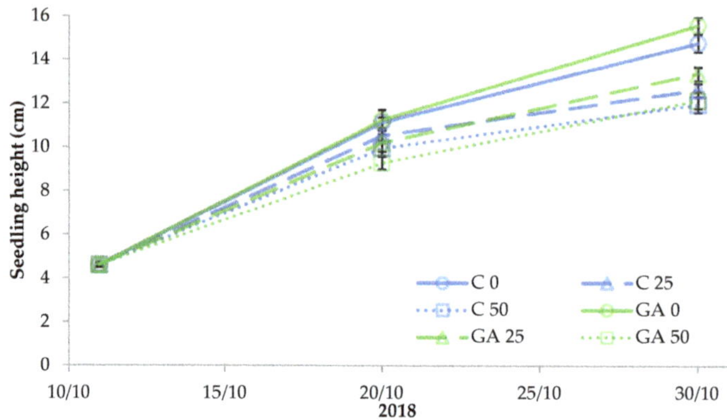

Figure 2. Effect of the seedling treatments (C, untreated control, GA, gibberellic acid spray 10^{-5} M) and salt stress (0, 25, and 50 mM NaCl) on tomato seedling growth (the least significant difference (LSD) value at $p < 0.05$ for 20/10 = 0.37 and for 30/10 = 0.73).

The tallest seedlings were those treated with GA_3 and no salt stress (15.6 cm on average) followed by the unstressed control seedlings (14.8 cm on average); the height significantly decreased under salt stress in control seedlings irrespective of NaCl concentration whereas it gradually dropped in the GA_3-treated seedlings as salt stress increased (Figure 2).

Similar to seedling height, the stem of control seedlings had a lower diameter under both levels of salt stress, while the GA_3-treated seedlings had a significantly lower stem diameter, with the highest NaCl concentration in the irrigation water (Figure 3).

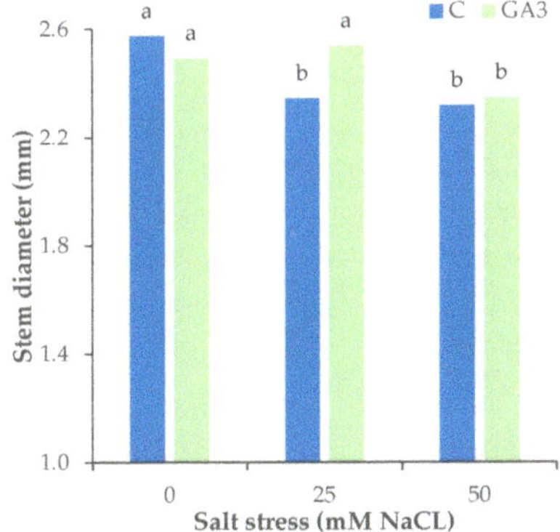

Figure 3. Effect of the seedling treatments (C, untreated control, GA3, gibberellic acid spray 10^{-5} M) and salt stress (0, 25, and 50 mM NaCl in the irrigation water) on the stem diameter of tomato seedlings. Bars with different letters are significantly different at $p < 0.05$ according to the LSD test.

The fresh weight (FW) of tomato seedlings was affected by the interaction between the GA_3 treatment and the salt stress (Table 1, Figure 4). The total fresh biomass was 2.28 g in the unstressed control seedlings and was reduced by 25.0 and 30.9% with 25 and 50 mM NaCl, respectively. GA_3-treated seedlings had a higher total fresh weight under no salt stress (2.64 g with 0 mM NaCl) or moderate stress (1.95 g with 25 mM NaCl) even if they exhibited a reduction (−26.1%) close to those of control seedlings. The root fresh weight was almost constant in control seedlings even under salt stress conditions whereas it was significantly reduced by salt stress in GA_3-treated seedlings. The fresh biomass of the stem and the leaves of the GA_3-treated seedlings were higher than control with 0 and 25 mM NaCl and dropped to comparable values with the highest NaCl concentration (50 mM). The differences in the response to salt stress between control and GA_3-treated seedlings for root and shoot fresh biomass accumulation resulted in some changes in the biomass partitioning. The shoot/root ratio was higher in GA_3-treated seedlings than the control with 0 and 25 mM NaCl whereas it dropped below 4.0 with 50 mM NaCl irrespective of GA_3 treatment (Table 1).

Table 1. Effects of gibberellic acid treatment (C, not treated; GA3, 10^{-5} M GA_3) and salt stress (0, 25, and 50 mM NaCl in the irrigation water) on morphological parameters of tomato seedlings.

Source of Variance		Seedling Fresh Weight (g FW)				Seedling Dry Weight (mg DW)					Dry Matter (%)
		Total	Roots	Stem	Leaves	Shoot/Root	Total	Roots	Stem	Leaves	Shoot/Root
Treatment											
C		[z] 1.86	0.39	0.86	0.61	3.8	191.8	24.2	74.3	93.3	7.0	11.9
GA3		2.06	0.36	0.94	0.77	4.7	210.7	23.9	81.0	105.8	7.8	11.7
NaCl (mM)											
0		2.46	0.40	1.19	0.87	5.1	235.9	26.3 a	94.4	115.3	8.2	10.5
25		1.83	0.37	0.82	0.64	4.0	197.1	24.0 ab	74.6	98.5	7.2	12.3
50		1.58	0.35	0.68	0.55	3.6	170.8	21.9 b	64.0	84.9	6.8	12.6
Treatment × NaCl											
C	0	2.28 b	0.40 ab	1.14 b	0.75 b	4.8b	236.3 a	26.5	94.0 a	115.8 a	8.3a	11.4 b
	25	1.71 d	0.40 ab	0.74 c	0.57 c	3.3c	173.0 b	23.3	64.3 b	85.5 b	6.5b	11.9 ab
	50	1.58 d	0.37 b	0.69 c	0.52 c	3.2c	166.3 b	22.8	64.8 b	78.8 b	6.3b	12.4 ab
GA3	0	2.64 a	0.41 a	1.24 a	0.99 a	5.4a	235.5 a	26.0	94.8 a	114.8 a	8.2a	9.6 c
	25	1.95 c	0.34 bc	0.91 b	0.71 b	4.7b	221.3 a	24.8	85.0 a	111.5 a	7.9a	12.7 a
	50	1.58 d	0.32 c	0.67 c	0.59 c	3.9c	175.3 b	21.0	63.3 b	91.0 b	7.2ab	12.8 a
Significance [x]											
Treatment		ns	ns	ns	**	**	**	ns	*	***	*	ns
NaCl		***	***	***	***	**	***	*	***	***	**	***
Treatment × NaCl		**	***	***	**	*	**	ns	**	**	*	**

[z] Each value is the mean of 4 replicated samples of 30 seedlings each. For each factor, values in a column followed by different letters are significantly different, according to the LSD test. [x] Significance: ns = not significant; * significant at $p < 0.05$; ** significant at $p < 0.01$; *** significant at $p < 0.001$.

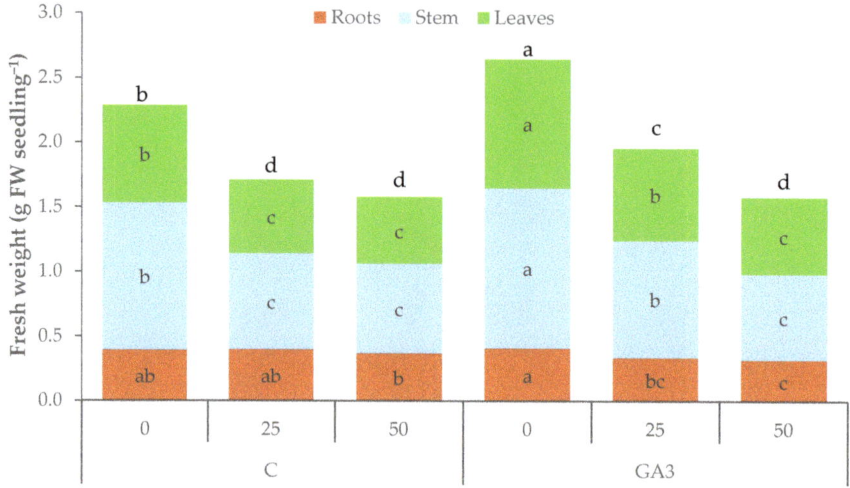

Figure 4. Effect of the seedling treatments (C, untreated control, GA3, gibberellic acid spray 10^{-5} M) and salt stress (0, 25, and 50 mM NaCl in the irrigation water) on the total, root, stem, and leaf fresh weight of tomato seedlings. Bbars of the same color with different letters are significantly different at $p < 0.05$ according to the LSD test.

The total dry weight (DW) of the untreated seedlings showed the same trend as that of fresh biomass (Table 1, Figure 5). Control seedlings exhibted a reduction of 26.8% and 29.6% of total dry weight with 25 and 50 mM NaCl, respectively. GA$_3$-treated seedlings were more tolerant to moderate salt stress (25 mM NaCl), with no significant reduction of total dry weight (−6.1%) but lowered their dry biomass to values comparable to the untreated seedlings with the highest NaCl concentration (175.3 mg DW seedling^{-1}). The GA$_3$ treatment had no significant effect on the accumulation of dry biomass in the roots, whereas increasing salt stress up to 50 mM NaCl significantly reduced root dry weight. The dry weights of the stem (94.0 and 94.8 mg DW seedling^{-1} for control and GA$_3$, respectively) and the leaves (15.8 and 114.8 mg DW seedling^{-1} for control and GA$_3$, respectively) were not affected by the GA$_3$ treatment with 0 mM NaCl. The dry biomass of the shoot (stem and leaves) of the untreated seedlings was significantly reduced under the 25 mM NaCl salt stress, with no further significant decrease at the highest NaCl concentration (−31.4 ± 0.3% and −29.0 ± 2.9% for stem and leaves, respectively). GA$_3$ spray limited the reduction of dry biomass in the shoot of the seedlings fed with 25 mM NaCl in the irrigation water; therefore, the dry weights of stems and leaves were similar to those of the unstressed seedlings. The GA$_3$ treatment was not as effective at the highest NaCl level (−33.2% and −20.7% for stem and leaves, respectively) (Table 1). The dry biomass partitioning was influenced by salt stress that caused a reduction of the shoot/root ratio but this reduction was milder in the GA$_3$-treated seedlings that showed a shoot/root ratio higher than control seedlings under the stress condition (Table 1).

The GA$_3$-treated seedlings had higher water content than controls with no salt stress as shown by the lower dry matter percentage. Salt stress had no effect on this parameter in the control seedlings while the increase was higher and significant in the GA$_3$-treated seedlings (Table 1).

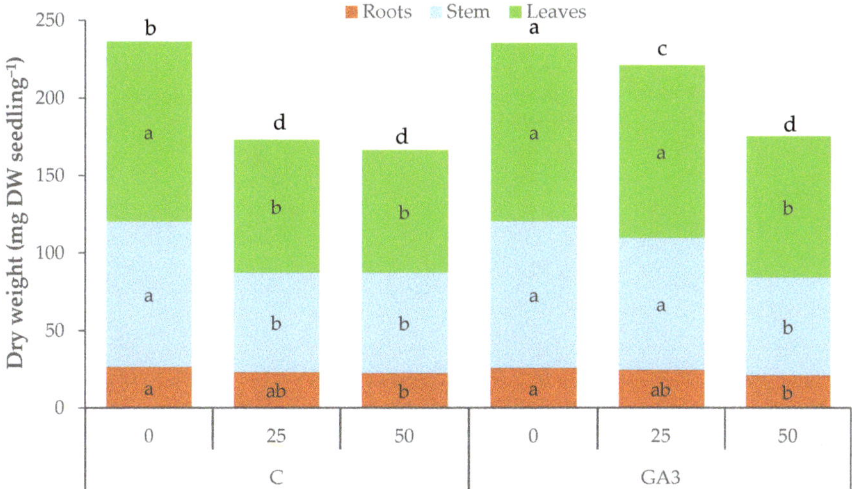

Figure 5. Effect of the seedling treatments (C, untreated control, GA3, gibberellic acid spray 10^{-5} M) and salt stress (0, 25, and 50 mM NaCl in the irrigation water) on the total, root, stem, and leaf dry weight of tomato seedlings. Bars of the same color with different letters are significantly different at $p < 0.05$ according to the LSD test).

The tomato seedlings treated with GA$_3$ recorded the highest leaf number (5.44 leaves seedling^{-1}), as they had 18% more leaves than the control seedlings (4.61 leaves seedling^{-1} on average) (Table 2). Seedling leafiness significantly decreased by increasing the concentration of NaCl in the irrigation water from 0 to 50 mM (from 5.27 to 4.75 leaves seedling^{-1}). The GA$_3$ treatment significantly affected the average leaf area that was 18.0% wider in GA$_3$-treated seedlings (10.17 cm^2 leaf^{-1}) than control (8.62 cm^2 leaf^{-1}) under no salt stress. Both NaCl level (25 and 50 mM) caused a similar reduction ($-26.3 \pm 0.8\%$) to the average leaf area of control seedlings, whereas the GA$_3$-treated seedlings reduced by 18.6% the leaf wideness with the moderate salt stress (8.28 cm^2 leaf^{-1} with 25 mM NaCl) resulting in a leaf area no different from that of the untreated and unstressed seedlings (control seedlings with 0 NaCl). Watering the GA$_3$-treated seedlings with the highest NaCl level (50 mM) further reduced their leaf area with no significant difference between treated and untreated seedlings (6.47 cm^2 leaf^{-1} on average). A similar effect of the GA$_3$ treatment and the salt stress was found for the total leaf area of the seedlings that was highest in the unstressed seedlings sprayed with GA$_3$ (58.47 cm^2 seedling^{-1}, +41.7% than control). Raising to 25 mM NaCl the salt concentration in the irrigation water, the seedlings sprayed with GA$_3$ had a total leaf area (44.17 cm^2 seedling^{-1}) higher than the untreated seedlings (29.47 cm^2 seedling^{-1}) and not statistically different from the values of leaf area recorded for the control seedlings under no salt stress (41.25 cm^2 seedling^{-1}). The increase to 50 mM NaCl of the salt concentration in the irrigation water did not cause further significant reduction of total leaf area in control seedlings while it negatively affected treated seedlings that had values not statistically different from those not treated (Table 2, Figure 6).

Table 2. Effects of the gibberellic acid treatment (C, not treated; GA3, GA_3-treated seedlings) and salt stress (0, 25, and 50 mM NaCl in the irrigation water) on leaf characteristics of tomato seedlings.

Source of Variance		Number of Leaves	Leaf Area		SLA (cm^2 g DW^{-1})	Stomatal Conductance (mmol m^2 s^{-1})	RWC(%)	L *	Chroma	Hue°
			(cm^2 Seedling^{-1})	(cm^2 Leaf^{-1})						
Treatment										
C		[z] 4.61 b	32.90	7.11	353.4	188.1	86.8 a	51.7	49.0 b	120.5 b
GA3		5.44 a	45.41	8.37	425.3	219.5	80.0 b	52.1	50.9 a	121.3 a
NaCl (mM)										
0		5.27 a	49.86	9.39	435.1	371.8	87.5 a	50.3 a	48.4 b	121.2 a
25		5.06 ab	36.82	7.35	370.9	171.6	83.8 a	52.1 ab	50.7 a	121.3 a
50		4.75 b	30.80	6.47	362.1	67.8	78.8 b	53.2 b	50.7 a	120.1 b
Treatment × NaCl										
C	0	4.79	41.25 b	8.62 b	360.3 b	369.9 a	90.0	50.1	47.6	120.9
	25	4.60	29.47 c	6.42 c	345.0 b	123.0 c	90.1	52.8	50.2	121.0
	50	4.45	27.99 c	6.29 c	354.9 b	71.3 d	80.2	52.3	49.1	119.5
GA3	0	5.75	58.47 a	10.17 a	509.9 a	373.7 a	85.0	50.6	49.2	121.5
	25	5.53	44.17 b	8.28 b	396.7 b	220.3 b	77.6	51.5	51.2	121.6
	50	5.05	33.60 c	6.65 c	369.3 b	64.4 d	77.3	54.1	52.2	120.7
Significance [x]										
Treatment		***	***	*	***	ns	***	ns	**	*
NaCl		**	***	***	**	***	**	**	**	**
Treatment × NaCl		ns	***	***	**	*	ns	ns	ns	ns

[z] Each value is the mean of 4 replicated samples of 30 seedlings each. For each factor, values in a column followed by different letters are significantly different, according to the LSD test. [x] Significance: ns = not significant; * significant at $p < 0.05$; ** significant at $p < 0.01$; *** significant at $p < 0.001$.

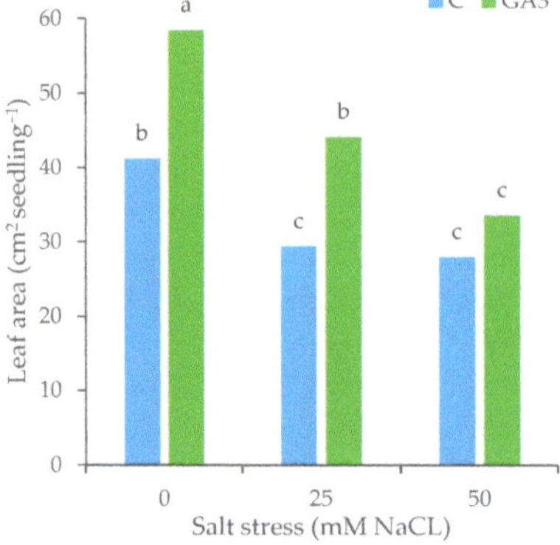

Figure 6. Effect of the seedling treatments (C, untreated control, GA3, gibberellic acid spray 10^{-5} M) and the salt stress (0, 25, and 50 mM NaCl in the irrigation water) on the total leaf area of tomato seedlings. Bars with different letters are significantly different at $p < 0.05$ according to the LSD test.

The specific leaf area (SLA) is a parameter related to dry biomass distribution in the leaves and leaf thickness (Table 2). This index showed a significant increase only under no stress conditions in the treated seedlings compared to the control seedlings (+41.5%) (Table 2).

The seedlings did not differ in stomatal conductance due to GA_3 treatment when not salt-stressed (371.8 mmol m^2 s^{-1} ± 20.9) or with the highest salt stress level (67.8 ± 3.9 mmol m^2 s^{-1}), whereas they had a different response to the moderate salt stress that determined a reduction by 66.7% in control seedlings (123.0 mmol m^2 s^{-1}) and by 41.0% in GA_3-treated seedlings (220.3 mmol m^2 s^{-1}) (Table 2).

The relative water content (RWC) was higher in the untreated seedlings and was negatively affected only by the highest NaCl level (Table 2).

The physiological and nutritional status of the plants can influence the pigment content of the leaf, thus leaf color can be an index of seedling health (Table 2). The supply of brackish water for irrigation caused a linear increase of color lightness (L*) with a significantly lighter color with the highest NaCl concentration. The color vividness of the leaves was evaluated by calculating the Chroma and exhibited a slight but significant increase by GA_3 treatment (+3.9 ± 1.3%) and salt stress (+4.7 ± 0.7%) (Table 2). Moreover, the green component of color was less intense in the control seedlings compared to the

GA$_3$-treated seedlings and in the seedlings that received the highest level of NaCl with the irrigation water, as suggested by the reduction of the hue angle.

3.2. Morphophysiological Parameters of Sweet Pepper Seedlings

Pepper plantlets emerged 14 days after sowing and seedlings were ready for transplanting (3–4 true leaves; 13–14th BBCH growth stage [23]) after 46 d from sowing.

The height of pepper seedlings was significantly affected by the experimental treatments. After ten days from the beginning of salt stress and GA$_3$ spray treatment, the control seedlings showed a reduction of plant height with increasing NaCl concentration, whereas GA$_3$-treated seedlings suffered height reduction only with 50 mM NaCl. At the end of the experiment, the seedlings sprayed with GA$_3$ (8.1 ± 0.2 cm on average) were significantly taller than the control (+27.3 ± 2.9%). Both treated and untreated pepper seedlings exhibited a linear decrease in their height as increasing salt stress from 0 to 50 mM NaCl (7.6 ± 0.7 and 6.8 ± 0.9 cm, respectively) (Figure 7).

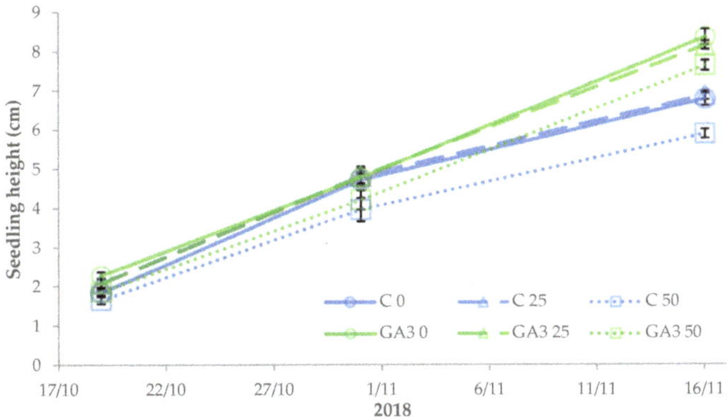

Figure 7. Effect of the seedling treatments (C, untreated control, GA, gibberellic acid spray 10^{-5} M) and salt stress (0, 25, and 50 mM NaCl) on sweet pepper seedling growth (the LSD value at $p < 0.05$ for 30/10 = 0.44 and for 16/11 = 0.80).

The diameter of the seedling stem was 2.0 ± 0.03 mm and was not influenced by the experimental treatments.

The fresh weight (FW) of pepper seedlings was affected by the GA$_3$ treatment applied to seedlings and by salt stress (Table 3, Figure 8). The mean total fresh biomass was 0.97 g in the unstressed control seedlings and was increased by 18.2 ± 1.8% due to GA$_3$ spray irrespective of salt stress. The use of brackish irrigation water negatively affected the total fresh biomass produced by pepper seedlings that significantly dropped from 1.15 g to 0.95 g with 0 and 50 mM NaCl, respectively (Table 3, Figure 8). The fresh weight of the roots of pepper seedlings showed different variations due to salt stress according to seedling treatments. Control seedlings recorded the highest root fresh biomass with 0 and 25 mM NaCl, and it dropped significantly only with the highest salt stress level, whereas GA$_3$-treated seedlings showed a reduction of root fresh weight even with the moderate salt stress (25 mM NaCl) (Table 3). The fresh biomass of the stem and the leaves significantly increased by 24.4% and 40.7%, respectively, in the unstressed GA$_3$-treated seedlings compared to the control seedlings. The salinity of the irrigation water had a lower effect on shoot fresh weight as it negatively affected only stem fresh weight in the seedlings watered with the higher NaCl level (Table 3, Figure 8).

Table 3. Effects of the gibberellic acid treatment (C, not treated; GA3, GA$_3$-treated seedlings) and salt stress (0, 25, and 50 mM NaCl in the irrigation water) on morphological parameters of sweet pepper seedlings.

Source of Variance		Seedling Fresh Weight (g FW)				Seedling Dry Weight (mg DW)					Dry Matter (%)
	Total	Roots	Stem	Leaves	Shoot/Root	Total	Roots	Stem	Leaves	Shoot/Root	
Treatment											
C	[z] 0.97 b	0.29	0.41 b	0.27 b	2.4 b	79.3 b	19.8	30.5 b	29.0 b	3.0	8.8 b
GA3	1.14 a	0.25	0.51 a	0.38 a	3.7 a	93.5 a	12.4	38.4 a	42.7 a	7.6	9.1 a
NaCl (mM)											
0	1.15 a	0.31	0.50 a	0.34	2.8 b	96.3 a	21.0	38.3 a	37.0	3.6	8.9
25	1.06 ab	0.28	0.45 a	0.33	3.0 ab	84.6 b	15.0	32.6 b	37.0	5.5	8.8
50	0.95 b	0.22	0.42 b	0.31	3.3 a	78.2 b	12.3	32.4 b	33.5	6.8	9.0
Treatment × NaCl											
C 0	1.06	0.32 a	0.46	0.29	2.3	87.3	22.0 a	34.3	31.0	3.0 d	8.8
25	0.98	0.32 a	0.39	0.27	2.1	77.2	20.0 a	28.2	29.0	2.9 d	8.7
50	0.86	0.23 b	0.38	0.25	2.8	73.2	17.3 a	28.9	27.0	3.2 d	8.9
GA3 0	1.24	0.29 a	0.55	0.39	3.2	105.3	20.0 a	42.3	43.0	4.3 c	9.0
25	1.14	0.23 b	0.52	0.39	3.9	92.1	10.0 b	37.1	45.0	8.2 b	9.0
50	1.04	0.21 b	0.46	0.37	3.9	83.2	7.3 b	35.9	40.0	10.3 a	9.2
Significance [x]											
Treatment	***	***	***	***	***	**	***	***	***	***	**
NaCl	*	***	*	ns	*	**	***	**	ns	***	ns
Treatment × NaCl	ns	**	ns	ns	ns	ns	**	ns	ns	***	ns

[z] Each value is the mean of 4 replicated samples of 30 seedlings each. For each factor, values in a column followed by different letters are significantly different, according to the LSD test. [x] Significance: ns = not significant; * significant at $p < 0.05$; ** significant at $p < 0.01$; *** significant at $p < 0.001$.

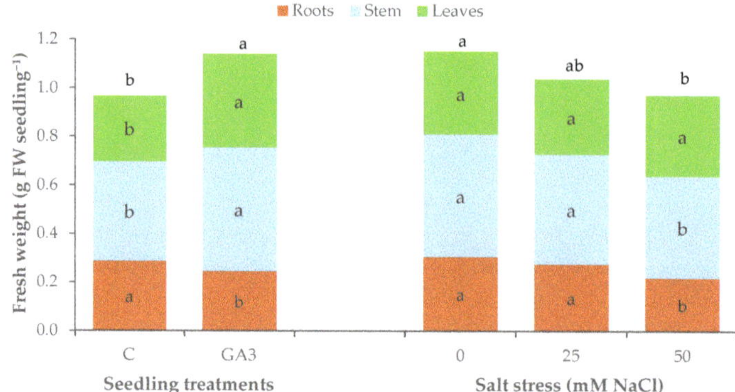

Figure 8. Effect of the seedling treatments (C, untreated control, GA3, gibberellic acid spray 10^{-5} M) and salt stress (0, 25, and 50 mM NaCl in the irrigation water) on the total, root, stem, and leaf fresh weight of sweet pepper seedlings (within each experimental factor. Bars of the same color with different letters are significantly different at $p < 0.05$ according to the LSD test.

The experimental treatments modified the fresh biomass partitioning as shown by the shoot/root ratio (Table 3). The treatment with GA$_3$ increased the portion of biomass accumulated in the shoot reaching a shoot/root ratio of 3.7 compared to 2.4 for the control seedlings. The reduction of root fresh weight due to salt stress was higher than those of the shoots determining an increasing trend of the shoot/root ratio when irrigation water salinity increased.

The effects of the experimental treatments on the dry weight (DW) of the seedlings were similar to those reported for the fresh biomass (Table 3, Figure 9). Control seedlings accumulated a total of 79.3 ± 4.3 mg DW seedling^{-1}; the treatment with GA$_3$ significantly increased the total dry weight up to 93.5 ± 6.4 mg DW seedling^{-1} (+17.8 ± 2.1%). The dry weight of the roots was 19.8 ± 1.4 mg DW seedling^{-1} in the untreated seedlings that were not affected by salt stress. GA$_3$-treated unstressed seedlings had a similar root dry weight but significantly reduced the dry weight of this part under salt stress (−56.7 ± 6.7%). The accumulation of dry matter was almost equal between stem and leaves. Both plant parts were positively influenced by GA$_3$ with an increase of 26.2 ± 2.5% for the stem and 47.3 ± 4.8% for the leaves. The stems also showed a greater effect of salt stress as regards the dry biomass accumulated as it dropped even with moderate salt stress ($-15.2 \pm 1.1\%$ for 25 and 50 mM NaCl). The shoot/root ratio of the dry weights recorded the lowest values in the untreated seedlings irrespective of salt stress (3.0 on average). Unstressed GA_3-seedlings had a shoot/root DW ratio of 4.3 that significantly increased up to 8.2 and 10.3 with 25 and 50 mM NaCl, respectively.

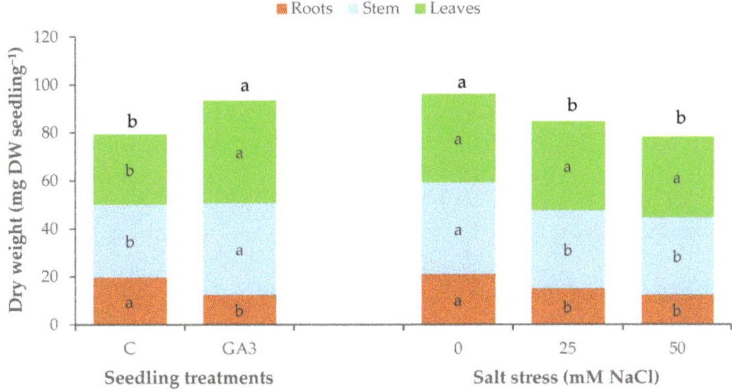

Figure 9. Effect of the seedling treatments (C, untreated control, GA3, gibberellic acid spray 10^{-5} M) and salt stress (0, 25, and 50 mM NaCl in the irrigation water) on the total, root, stem, and leaf dry weight of sweet pepper seedlings (within each experimental factor. Bars of the same color with different letters are significantly different at $p < 0.05$ according to the LSD test.

The dry matter percentage of the pepper seedling shoots was slightly but significantly influenced by GA_3 that determined a higher dry matter percentage in the treated seedlings (9.1%) compared to the untreated seedlings (8.8%) (Table 3).

The pepper seedlings treated with GA_3 had the highest leaf number (4.44 ± 0.17 leaves seedling^{-1}), with 18.4% more leaves than the untreated seedlings (3.75 ± 0.09 leaves seedling^{-1}) (Table 4). The leaf number significantly decreased in the presence of NaCl in the irrigation water ($-8.6 \pm 1.3\%$ for 25 and 50 mM NaCl) compared to 4.35 ± 0.26 leaves seedling^{-1} for the unstressed seedlings. The GA_3 treatment significantly affected the average leaf area; it was $26.6 \pm 3.4\%$ wider in treated seedlings (4.51 ± 0.41 cm^2 leaf^{-1}) than control (3.57 ± 0.32 cm^2 leaf^{-1}) (Table 4). The effect of the GA_3 treatment on the total leaf area of the seedlings was even greater as it increased this parameter from 13.44 ± 1.22 to 20.10 ± 1.97 cm^2 seedling^{-1} ($+49.9 \pm 4.8\%$). The irrigation water salinity caused a linear reduction in the average leaf area that ranged on average from 4.41 ± 0.31 cm^2 leaf^{-1}, in the unstressed plants to 3.69 ± 0.28 cm^2 leaf^{-1} with 50 mM NaCl ($-16.6 \pm 2.9\%$) (Table 4). Similarly, the total leaf area was reduced by $24.7 \pm 1.5\%$ when the salt concentration in the irrigation water increased from 0 to 50 mM (19.37 ± 1.9 and 14.64 ± 1.3 cm^2 seedling^{-1}, respectively).

As with tomato seedlings, the specific leaf area (SLA) showed a significant increase due to GA_3 only under no salt stress (542.4 cm^2 g^{-1} DW, $+9.1\%$ compared to the control seedlings) but a reduction of SLA was recorded for both seedling treatments when the NaCl level was increased up to 50 mM (395.0 ± 2.1 cm^2 g^{-1} DW) (Table 4).

The seedlings did not differ in stomatal conductance due to seedling treatments even if GA_3-treated seedlings had a slightly higher stomatal conductance (Table 4). The highest salt stress caused a reduction of $23.3 \pm 2.1\%$ compared to the stomatal conductance values with 0 and 25 mM NaCl.

The relative water content (RWC) of unstressed pepper seedlings was $93.1 \pm 0.3\%$ and was significantly lower in the salt-stressed seedlings ($88.8 \pm 0.8\%$ with 25 and 50 mM NaCl) (Table 4).

The leaf color of GA_3-treated pepper seedlings was lighter, more vivid, and less greenish than the untreated seedlings as shown by the changes in the values of L*, Chroma, and Hue angle. Leaf color was also affected by the highest NaCl concentration mainly as regards L* and Chroma modifications, resulting in darker and less vivid color (Table 4).

Table 4. Effects of the gibberellic acid treatment (C, not treated; GA3, GA$_3$-treated seedlings) and salt stress (0, 25, and 50 mM NaCl in the irrigation water) on leaf characteristics of sweet pepper seedlings.

Source of Variance		Number of Leaves	Leaf Area		SLA (cm^2 g DW^{-1})	Stomatal Conductance (mmol m^2 s^{-1})	RWC(%)	L *	Chroma	Hue°
			(cm^2 Seedling^{-1})	(cm^2 Leaf^{-1})						
Treatment										
C		[z] 3.75 b	13.44 b	3.57 b	463.4	49.8	90.8	50.5 b	58.7 b	120.3 a
GA3		4.44 a	20.10 a	4.51 a	472.3	54.0	89.6	54.8 a	61.8 a	119.0 b
NaCl (mM)										
0		4.35 a	19.37 a	4.41 a	519.7	58.5 a	93.1 a	53.2 a	60.8 a	119.7 ab
25		4.01 b	16.29 ab	4.03 ab	488.7	54.0 a	89.5 b	53.6 a	60.8 a	119.1 b
50		3.93 b	14.64 b	3.69 b	395.0	43.1 b	88.1 b	51.0 b	59.0 b	120.1 a
Treatment × NaCl										
C	0	3.93	15.41	3.93	497.1 b	55.7	93.2	50.6	59.1	120.8
	25	3.73	13.52	3.63	500.8 b	52.6	89.8	51.2	59.1	119.9
	50	3.60	11.38	3.16	392.3 c	40.7	89.5	49.6	57.7	120.2
GA3	0	4.78	23.32	4.88	542.4 a	61.3	93.0	55.9	62.6	118.7
	25	4.30	19.07	4.43	476.7 b	55.3	89.2	55.9	62.5	118.3
	50	4.25	17.90	4.21	397.8 c	45.5	86.6	52.5	60.3	120.0
Significance [x]										
Treatment		***	***	***	ns	ns	ns	***	***	***
NaCl		**	**	**	***	**	***	***	***	*
Treatment × NaCl		ns	ns	ns	**	ns	ns	ns	ns	ns

[z] Each value is the mean of 4 replicated samples of 30 seedlings each. For each factor, values in a column followed by different letters are significantly different, according to the LSD test. [x] Significance: ns = not significant; * significant at $p < 0.05$; ** significant at $p < 0.01$; *** significant at $p < 0.001$.

3.3. Principal Components Analysis

The principal components analysis revealed four principal components (PCs) with eigenvalues higher than 1.00 (Table 5), representing 69.23%, 11.04%, 9.23%, and 5.11% of the total variance, respectively. Thus, the combination of four PCs could represent the initial twenty-two variables, explaining 94.61% of the total variance.

Table 5. Correlation of variables to the factors of the principal components analysis (PCA) based on factor loadings.

Variable	PC1	PC2	PC3	PC4
Height	**0.984**	0.089	−0.022	0.087
Stem diameter	**0.934**	−0.201	−0.113	0.101
Total fresh weight	**0.976**	0.003	0.203	0.002
Root fresh weight	**0.795**	−0.443	0.052	0.294
Stem fresh weight	**0.962**	0.010	0.240	−0.060
Leaf fresh weight	**0.970**	0.122	0.178	−0.021
Shoot/Root FW	**0.810**	0.466	0.204	−0.259
Total dry weight	**0.995**	−0.003	−0.025	0.056
Root dry weight	**0.704**	**−0.609**	0.050	0.335
Stem dry weight	**0.994**	0.020	0.035	0.006
Leaf dry weight	**0.989**	0.084	−0.079	0.042
Shoot/Root DW	**0.530**	**0.720**	−0.080	−0.364
Dry matter %	**0.684**	0.075	−0.667	0.255
Leaf number	**0.855**	0.353	0.174	0.180
Plant area	**0.962**	0.114	0.202	0.028
Leaf area	**0.986**	0.074	0.137	0.027
SLA	−0.346	−0.005	**0.839**	0.254
Stomatal conductance	**0.869**	−0.114	0.381	−0.231
RWC	**−0.520**	−0.321	**0.573**	−0.073
L*	−0.256	**0.726**	0.042	**0.599**
Chroma	**−0.890**	0.260	0.314	0.025
Hue°	**0.672**	−0.442	−0.118	−0.342

Values in bold within the same factor indicate the variable with the largest correlation.

PC1 was mainly related to height, stem diameter, total, root, stem, and leaf fresh weight (FW), shoot/root (S/R) FW, total, root, stem, and leaf dry weight (DW), dry matter percentage, leaf number, plant and leaf area, stomatal conductance, relative water content (RWC), chroma and hue angle; PC2 was related to the root DW, shoot/root (S/R) DW and L*; PC3 was related to the specific leaf area

(SLA) and RWC; PC4 was related to L* (Table 5). Projecting the original variables on the plane of the two main PCs could ilustrate such relationships, as displayed in the plot of loadings (Figure 10a).

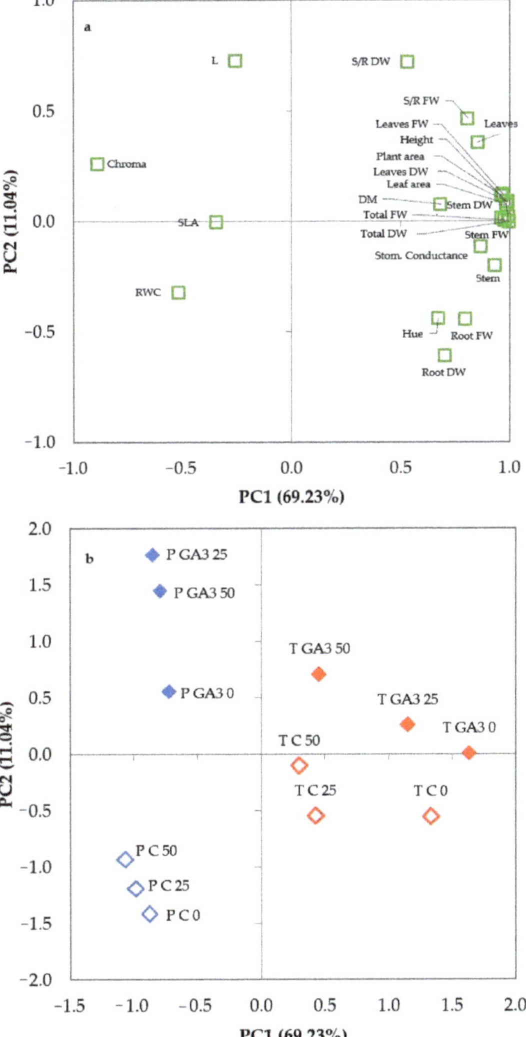

Figure 10. Plots of (**a**) loadings (morphophysiological characteristics of tomato and sweet pepper seedlings) and (**b**) scores (trials) formed by the two principal components from the Principal Component Analysis (PCA). T C: untreated tomato seedlings; T GA3; tomato seedlings spayed with 10^{-5} M of gibberellic acid (GA$_3$); P C: untreated pepper seedlings; P GA3: pepper seedlings spayed with 10^{-5} M GA$_3$; 0, 25, and 50: concentration of NaCl (mM) in the irrigation water.

The differentiation of the NaCl levels supplied with the irrigation water to tomato and pepper seedlings treated or not treated with GA$_3$ are shown in the plot of scores (Figure 10b), where two main clusters (tomato and pepper seedlings) and two sub-clusters for each main cluster (GA$_3$ treated or untreated seedlings) could be visibly distinguished. The scores of tomato seedlings were located in the positive part of the PC1 axis, whereas pepper seedlings were located in the negative part of the PC1 axis; the tomato and pepper seedlings sprayed with GA$_3$ were located in the positive part of the PC2 axis, whereas the untreated seedlings were located in the negative part of the PC2 axis. Thus, GA$_3$ treated or untreated tomato and pepper seedlings were located in a different quadrant of the plane: GA$_3$-treated tomato seedlings were in the first quadrant (+; +), control tomato seedlings were in the fourth quadrant (+; –), GA$_3$-treated pepper seedlings were in the second quadrant (–; +), and control pepper seedlings were in the third quadrant (–; –). The response of tomato and pepper seedlings differed in their sensitivity to increased salt stress (from 0 to 50 mM NaCl) as a function of the GA$_3$ treatment, as the scores of control tomato seedlings mainly reduced their PC1 values, the scores of GA$_3$-treated tomato seedlings mainly reduced their PC1 values but also increased their PC2 values, the scores of control pepper seedlings mainly increased their PC2 values but also reduced their PC1 values, and the scores of GA$_3$-treated pepper seedlings mainly increased their PC2 value. Combining the data from the plot of loadings and scores, it can be concluded that the concentration of NaCl in the irrigation water influenced the tomato and pepper seedlings in different ways according to the GA$_3$ treatment (Figure 10a,b).

4. Discussion

Irrigation water of poor quality due to salinity may represent a big issue for vegetable crops as saline conditions result in lowered growth, yield, and quality and could have negative economic effects for vegetable growers [13,27,28]. These effects are even worse for the vegetable nursery industry that has to face the higher sensitivity to salt stress of vegetable seedlings compared to mature plants [7]. The goal of this important sector is to produce well-developed and vigorous transplants [25] which could be hard or even impossible to reach when brackish water is the only available source for watering the seedlings. Growth limitation due to salinity can reduce the size of transplants which is linked to establishment success, the growth rate, and the size at harvest [29,30] and thus also reduces the commercial success of these products.

In a previous study [13], we found that salt tolerance of leaf lettuce and rocket grown in a floating system increased when GA$_3$ was added to the nutrient solution. In this experiment, we tested the possibility of increasing the salinity tolerance threshold of tomato and pepper seedlings by spraying a solution of 10^{-5} M GA$_3$ on the seedlings at the beginning of salt stress.

The reductions in the biomass of untreated tomato and pepper seedlings watered with saline water were indicative of growth limitations even with differences in the salinity tolerance threshold. The negative effects of salinity on biomass were also noted on other morphological and physiological traits such as plant height, shoot/root ratio, leaf number, leaf area, relative water content, and stomatal conductance. Similar effects were also found in many other vegetable crops grown under salt stress [13,28,31,32]. Seedling height exhibited a significant reduction of over 14% in both non-GA$_3$-treated tomato and pepper seedlings, but the latter species suffered this reduction only at the highest salt stress. The salinity of the irrigation water also affected the total fresh and dry biomass of the untreated seedlings of both species, but even for these traits, the reduction of these parameters was higher in tomato (–30.9% and –29.6% with 50 mM NaCl for total fresh and dry weight, respectively) than pepper (–19.5% and –16.1% with 50 mM NaCl for total fresh and dry weight, respectively) and progressively increased with salt stress level in pepper seedlings (–7.5% and –11.5% with 25 mM NaCl for total fresh and dry weight, respectively), whereas it reached a significantly lower value at an intermediate salt stress level in tomato seedlings (–25.0% and –26.8% with 25 mM NaCl for total fresh and dry weight, respectively), confirming that vegetables have different tolerance thresholds to salinity and that the magnitude of the negative effects can increase with different slopes with increases of salinity level [33,34]. The high electrical conductivity (EC) of the soil or the irrigation water may restrict water availability to the plants as it increases soil osmotic potential resulting in reduced water uptake and partial dehydration of the cell cytoplasm. This may lead to plasmolysis and could affect cell metabolism and macromolecule function, ending up with a slowdown or even cessation of growth [35]. The growth reduction recorded in response to salt stress could also be ascribed to a modification of nutrient uptake determined by a negative effect on nutrient availability [28,33]. Salinity can also affect nutrient translocation from the roots to the shoot, determining changes in biomass partitioning and plant morphology as shown by the variations in shoot/root ratios and the reduction of leaf number and leaf expansion in tomato and pepper seedlings. This latter effect could be related to the decrease of the relative water content of the leaves in response to increased salt stress that is generally due to lower water availability under stress conditions or to root disorders that impede sufficient water uptake to compensate for the water lost by transpiration [36,37].

The foliar application of gibberellic acid had a growth-promoting effect on the unstressed tomato and pepper seedlings. It affected some growth parameters similarly in tomato and pepper whereas others were enhanced to a greater extent in pepper than tomato. The GA_3-treated seedlings increased by about 16% their total fresh weight thanks to an increase of 34.4% of the leaf fresh weight compared to the untreated seedlings; the increase of leaf fresh weight could be explained by the higher number of leaves (+20.9%) and the greater expansion of single leaf and total plant area (+21.2% and +46.5% respectively). The exogenous supplementation of gibberellic acid at a very low concentration can improve the morpho-physiological and yield characteristics of many vegetable crops [14,18,20,21,38,39] and has been already successfully applied to tomato transplant production [22]. Endogenous gibberellins are important plant hormones that act in regulating plant growth and development by promoting division and elongation of the cells [40,41]. Many other mechanisms are triggered by gibberellins and contribute to the improvement of biomass accumulation in vegetative parts [42,43] (stimulating the synthesis of DNA, RNA, and protein, the multiplication of ribose and polyribosomes, and improve enzyme activities and membrane permeability [44–49]). The plants supplemented with exogenous GA_3 may significantly modify their morphological traits and direct the biomass allocation toward leaves and stem, increasing the shoot/root ratio [20,50], as found in the present work with tomato and pepper seedlings. These changes could significantly enhance transplant quality as it has been shown that the increase in shoot weight is related to seedling vigor and plant earliness [51].

The use of gibberellic acid was also effective in modifying the response of seedlings to salt stress even if to a different extent for tomato and pepper. Foliar treatment with GA_3 delayed the onset of salt stress symptoms and limited growth reduction of tomato seedlings at the intermediate salinity level, resulting in fresh and dry weight of the shoot similar to those of the unstressed untreated tomato seedlings. The effect of the exogenous GA_3 was more evident on pepper seedlings that maintained comparable or even higher values than unstressed untreated seedlings in many morphological characteristics. Similar effects of low doses of exogenous GA_3 supplementation were also found in lettuce and rocket grown in an hydroponic floating system with brackish nutrient solutions [13]. Plant hormones are involved in the response of vegetables to salt stress and mediate the activation of tolerance mechanisms [52] through complex crosstalk between hormones signaling pathways [53]. Decreased growth under salt stress of many species has been linked to a reduced level of bioactive endogenous gibberellins (GAs) caused by the increase of the two stress hormones ethylene and abscisic acid (ABA). It is known that ABA and GAs reciprocally influence each other's metabolism. The increase of ABA levels in seedlings may reduce bioactive GAs levels, thus reducing cell elongation and consequently stem elongation, leaf expansion, and root growth [54]. In contrast, gibberellins may promote the catabolism of ABA [55], so the exogenous supplementation of gibberellic acid might mitigate the negative responses mediated by ABA of plants to salt stress. This may be due to the GA-mediated activation of important metabolic pathways (ribose and polyribosome multiplication; DNA, RNA, and protein synthesis) [56–59] that could influence the repartitioning of the internal resources toward the aerial part of the plants thus increasing the biomass of seedling shoot. Moreover, exogenous GA_3 may increase membrane permeability [60,61] leading to improved uptake and utilization of water and mineral nutrients [15,62,63] and enhanced transport of photosynthates [64–66], which end up increasing biomass accumulation. This could explain the higher tolerance of the GA_3-treated seedlings to salinity compared to the untreated seedlings. In our experiment, tomato and pepper seedlings sprayed with GA_3 reacted to moderate (tomato) or even high (pepper) salt stress increasing significantly the accumulation of assimilates in the leaves and the stems compared to untreated seedlings, confirming that gibberellins may noticeably change biomass partitioning thus increasing the shoot/root ratio [13,20,22,50]. High dry weights of stems and shoots in vegetable transplants is an index of seedling vigor and has been related to higher establishment success in the field and to reduced time from transplanting to production, thus promoting plant earliness [22,51,67,68]. The increase in NaCl level decreased the physiological age of the seedlings as resulted from the reduction of leaf number, whereas GA$_3$ was confirmed to promote tomato seedling growth rate [22] and was effective in increasing the physiological age of tomato and pepper seedlings under salt stress conditions. As well as leaf number, gibberellic acid also increased the average leaf area under moderate (tomato) or even high (pepper) salt stress, resulting in a wider total leaf area compared to the untreated seedlings, despite the reduction of the relative water content under salt stress. This parameter is related to the cell turgor and its reduction can negatively affect leaf expansion. The higher biomass accumulation of GA$_3$-treated seedlings under salt stress compared to control seedlings could be ascribed to the wider photosynthetic area. A supply of exogenous GA$_3$ can stimulate nitrogen redistribution in plants and improve N utilization in leaves and stems by promoting the translocation of assimilates towards shoot apices and young leaves, thus determining the production of more leaves and the increase of physiological age [19]. This latter parameter could have a strong effect on early growth and production after plant establishment [22,68].

The quality of vegetable transplants is closely related to the morphological characteristics of the seedlings and the biomass accumulated, thus the adverse effects of salt stress on plant morphology and leaf development could negatively affect the agronomic and economic value of transplants. The foliar treatment with GA$_3$ was effective in counteracting the negative impact of salinity on the growth and development of tomato and pepper seedlings but to a different extent in these two species as shown by the results of the principal component analysis. Tomato and pepper are considered moderately sensitive to salinity but have different tolerance thresholds and critical salinity values for yield loss [34]. The effect of salinity may vary as a function of species or even of varieties within a species, and can be widely different according to development stages and environmental or agronomic factors [7,28]. The principal components analysis (PCA) highlighted the different responses of the two species to salinity. Tomato seedlings showed similar harmful effects when exposed to moderate and high salt stress whereas the pepper negative response was more gradual and slightly increased at every NaCl level. The tested species also differed for the effects determined by GA$_3$ treatment. Under no salt stress, pepper seedlings showed a higher growth-promoting effect from the GA$_3$ foliar spray than tomato. The application of GA$_3$ counterbalanced the response to salinity of pepper and tomato seedlings but it acted to a different extent and probably on distinct plant adaptation systems, as GA$_3$ almost completely overrode the negative effects of all NaCl levels in pepper seedlings, while tomato seedlings showed an increased salt tolerance thanks to GA$_3$ only up to 25 mM NaCl. These results confirmed that the effects of exogenous GA$_3$ supply may vary according to species and environmental conditions [13,20,21,69].

5. Conclusions

Salinity of irrigation water negatively affected the growth of tomato and sweet pepper seedlings during nursery transplant production by affecting biomass accumulation, leaf number, leaf area, and delaying the physiological age. Tomato seedlings were more sensitive to salinity and suffered significant negative effects even with moderate salt stress (25 mM NaCl), whereas sweet pepper seedlings suffered progressive growth reduction that was significant only with the highest salt stress (50 mM NaCl). The spray application of 10^{-5} M GA$_3$ exerted a growth-promoting effect on unstressed seedlings and was successful in increasing salinity tolerance of tomato seedlings up to 25 mM NaCl and up to 50 mM NaCl in sweet pepper seedlings. This treatment could represent a sustainable strategy for use with saline water in vegetable nurseries, limiting negative effects on seedling quality and production time, but this needs to be further validated on other vegetable species.

Author Contributions: Conceptualization, A.M. (Alessandro Miceli), F.V. and A.M. (Alessandra Moncada); Data curation, A.M. (Alessandro Miceli) and F.V.; Formal analysis, A.M. (Alessandro Miceli) and A.M. (Alessandra Moncada); Investigation, A.M. (Alessandro Miceli), F.V. and A.M. (Alessandra Moncada); Methodology, A.M. (Alessandro Miceli), F.V. and A.M. (Alessandra Moncada); Supervision, A.M. (Alessandro Miceli) and A.M. (Alessandra Moncada); Validation, A.M. (Alessandro Miceli), F.V. and A.M. (Alessandra Moncada); Writing—original draft, A.M. (Alessandro Miceli), F.V. and A.M. (Alessandra Moncada); Writing—review & editing, A.M. (Alessandro Miceli), F.V. and A.M. (Alessandra Moncada). All authors have read and agreed to the published version of the manuscript.

Funding: This research received no external funding.

Acknowledgments: The authors would like to thank Marina Di Simone, Monia Gnoffo, and Roberta Lo Nardo for their precious help and support.

Conflicts of Interest: The authors declare no conflict of interest.

References

1. Nicola, S.; Cantliffe, D.J. Increasing cell size and reducing medium compression enhance lettuce transplant quality and field production. *HortScience* **1996**, *31*, 184–189. [CrossRef]
2. Russo, V.M. Organic vegetable transplant production. *HortScience* **2005**, *40*, 623–628. [CrossRef]
3. Caracciolo, G.; Moncada, A.; Prinzivalli, C.; D'Anna, F. Effects of planting dates on strawberry plug plant performance in Sicily. *Acta Hortic.* **2009**, *842*, 155–158. [CrossRef]
4. Swiader, J.M.; Ware, G.W.; McCollum, J.P. *Producing Vegetable Crops: Teacher's Manual*; Interstate Publishers: Crete, IL, USA, 1992; ISBN 0813429048.
5. Kloepper, J.W.; Reddy, M.S.; Rodriguez-Kabana, R.; Kenney, D.S.; Kokalis-Burelle, N.; Martinez-Ochoa, N.; Vavrina, C.S. Application for rhizobacteria in transplant production and yield enhancement. *Acta Hortic.* **2004**, *631*, 217–229. [CrossRef]
6. Mariani, L.; Ferrante, A. Agronomic Management for Enhancing Plant Tolerance to Abiotic Stresses—Drought, Salinity, Hypoxia, and Lodging. *Horticulturae* **2017**, *3*, 52. [CrossRef]
7. Foolad, M.R. Recent Advances in Genetics of Salt Tolerance in Tomato. *Plant Cell. Tissue Organ Cult.* **2004**, *76*, 101–119. [CrossRef]
8. Maggio, A.; Barbieri, G.; Raimondi, G.; de Pascale, S. Contrasting Effects of GA 3 Treatments on Tomato Plants Exposed to Increasing Salinity. *J. Plant Growth Regul.* **2010**, *29*, 63–72. [CrossRef]
9. Wang, Y.; Mopper, S.; Hasenstein, K.H. Effects of salinity on endogenous ABA, IAA, JA, and SA in Iris hexagona. *J. Chem. Ecol.* **2001**, *27*, 327–342. [CrossRef]
10. Egamberdieva, D. Alleviation of salt stress by plant growth regulators and IAA producing bacteria in wheat. *Acta Physiol. Plant.* **2009**, *31*, 861–864. [CrossRef]
11. Khan, M.A.; Gul, B.; Weber, D.J. Action of plant growth regulators and salinity on seed germination of Ceratoides lanata. *Can. J. Bot.* **2004**, *82*, 37–42. [CrossRef]
12. Afzal, I.; Basra, S.A.; Iqbal, A. The effects of seed soaking with plant growth regulators on seedling vigor of wheat under salinity stress. *J. Stress Physiol. Biochem.* **2005**, *1*, 6–14.
13. Vetrano, F.; Moncada, A.; Miceli, A. Use of Gibberellic Acid to Increase the Salt Tolerance of Leaf Lettuce and Rocket Grown in a Floating System. *Agronomy* **2020**, *10*, 505. [CrossRef]
14. Khan, M.M.A.; Gautam, C.; Mohammad, F.; Siddiqui, M.H.; Naeem, M.; Khan, M.N. Effect of gibberellic acid spray on performance of tomato. *Turk. J. Biol.* **2006**, *30*, 11–16.
15. Khan, N.A.; Ansari, H.R. Effect of gibberellic acid spray during ontogeny of mustard on growth, nutrient uptake and yield characteristics. *J. Agron. Crop Sci.* **1998**, *181*, 61–63. [CrossRef]
16. Pal, P.; Yadav, K.; Kumar, K.; Singh, N. Effect of gibberellic acid and potassium foliar sprays on productivity and physiological and biochemical parameters of parthenocarpic cucumber cv.'Seven Star F1'. *J. Hortic. Res.* **2016**, *24*, 93–100. [CrossRef]
17. Shah, S.H. Effects of Salt Stress on Mustard As Affected By Gibberellic Acid Application. *Gen. Appl. Plant Physiol.* **2007**, *33*, 97–106.
18. Gelmesa, D.; Abebie, B.; Desalegn, L. Effects of gibberellic acid and 2, 4-dichlorophenoxyacetic acid spray on fruit yield and quality of tomato (Lycopersicon esculentum Mill.). *J. Plant Breed. Crop Sci.* **2010**, *2*, 316–324.

19. Khan, N.A.; Mir, R.; Khan, M.; Javid, S. Effects of gibberellic acid spray on nitrogen yield efficiency of mustard grown with different nitrogen levels. *Plant Growth Regul.* **2002**, *38*, 243–247. [CrossRef]
20. Miceli, A.; Moncada, A.; Sabatino, L.; Vetrano, F. Effect of Gibberellic Acid on Growth, Yield, and Quality of Leaf Lettuce and Rocket Grown in a Floating System. *Agronomy* **2019**, *9*, 382. [CrossRef]
21. Miceli, A.; Vetrano, F.; Sabatino, L.; D'Anna, F.; Moncada, A. Influence of preharvest gibberellic acid treatments on postharvest quality of minimally processed leaf lettuce and rocket. *Horticulturae* **2019**, *5*, 63. [CrossRef]
22. Moncada, A.; Vetrano, F.; Esposito, A.; Miceli, A. Fertigation Management and Growth-Promoting Treatments Affect Tomato Transplant Production and Plant Growth after Transplant. *Agronomy* **2020**, *10*, 1504. [CrossRef]
23. Feller, C.; Bleiholder, H.; Buhr, L.; Hack, H.; Hess, M.; Klose, R.; Meier, U.; Stauss, R.; van den Boom, T.; Weber, E. Phanologische Entwicklungsstadien von Gemusepflanzen II. Fruchtgemuse und Hulsenfruchte. *Nachr. Dtsch. Pflanzenschutzd.* **1995**, *47*, 217–232.
24. Kazemi, M. Effect of gibberellic acid and potassium nitrate spray on vegetative growth and reproductive characteristics of tomato. *J. Biol. Environ. Sci.* **2014**, *8*, 1–9.
25. Choudhury, S.; Islam, N.; Ali, M. Growth and Yield of Summer Tomato as Influenced by Plant Growth Regulators. *Int. J. Sustain. Agric.* **2013**, *5*, 25–28. [CrossRef]
26. McGuire, R.G. Reporting of objective color measurements. *HortScience* **1992**, *27*, 1254–1255. [CrossRef]
27. Yamaguchi, T.; Blumwald, E. Developing salt-tolerant crop plants: Challenges and opportunities. *Trends Plant Sci.* **2005**, *10*, 615–620. [CrossRef]
28. Shannon, M.C.; Grieve, C.M. Tolerance of vegetable crops to salinity. *Sci. Hortic.* **1998**, *78*, 5–38. [CrossRef]
29. Gama, P.B.S.; Inanaga, S.; Tanaka, K.; Nakazawa, R. Physiological response of common bean (*Phaseolus vulgaris* L.) seedlings to salinity stress. *Afr. J. Biotechnol.* **2007**, *6*, 79–88.
30. Bayuelo-Jiménez, J.S.; Debouck, D.G.; Lynch, J.P. Growth, gas exchange, water relations, and ion composition of Phaseolus species grown under saline conditions. *Field Crops Res.* **2003**, *80*, 207–222. [CrossRef]
31. Shahbaz, M.; Ashraf, M.; Al-Qurainy, F.; Harris, P.J.C. Salt Tolerance in Selected Vegetable Crops. *Crit. Rev. Plant Sci.* **2012**, *31*, 303–320. [CrossRef]
32. Moncada, A.; Vetrano, F.; Miceli, A. Alleviation of Salt Stress by Plant Growth-Promoting Bacteria in Hydroponic Leaf Lettuce. *Agronomy* **2020**, *10*, 1523. [CrossRef]
33. Grieve, C.M.; Grattan, S.R.; Maas, E. V Plant salt tolerance. *Agric. Salin. Assess. Manag.* **2012**, *2*, 405–459.
34. FAO Drainage Paper 61, Agricultural Drainage Water Management in Arid and Semi-Arid Areas, Annex 1. Crop Salt Tolerance Data. Available online: http://www.fao.org/3/y4263e/y4263e0e.htm (accessed on 10 September 2020).
35. Le Rudulier, D. Osmoregulation in rhizobia: The key role of compatible solutes. *Grain Legum.* **2005**, *42*, 18–19.
36. Srivastava, J.P.; Gupta, S.C.; Lal, P.; Muralia, R.N.; Kumar, A. Effect of salt stress on physiological and biochemical parameters of wheat. *Annu. Arid Zo.* **1988**, *27*, 197–204.
37. Shalhevet, J. Plants under salt and water stress. In *Plant Adaptation to Environmental Stress*; Fowden, L., Mansfield, T., Stoddart, J., Eds.; Chapman and Hall: London, UK, 1993; p. 133.
38. Lee, I.-J. Practical application of plant growth regulator on horticultural crops. *J. Hort. Sci* **2003**, *10*, 211–217.
39. Khan, N.A. Comparative effect of modes of gibberellic acid application on photosynthetic biomass distribution and productivity of rapeseed-mustard. *Physiol. Mol. Biol. Plants* **2003**, *9*, 141–145.
40. Achard, P.; Gusti, A.; Cheminant, S.; Alioua, M.; Dhondt, S.; Coppens, F.; Beemster, G.T.S.; Genschik, P. Gibberellin Signaling Controls Cell Proliferation Rate in Arabidopsis. *Curr. Biol.* **2009**, *19*, 1188–1193. [CrossRef]
41. Ubeda-Tomás, S.; Federici, F.; Casimiro, I.; Beemster, G.T.S.; Bhalerao, R.; Swarup, R.; Doerner, P.; Haseloff, J.; Bennett, M.J. Gibberellin Signaling in the Endodermis Controls Arabidopsis Root Meristem Size. *Curr. Biol.* **2009**, *19*, 1194–1199. [CrossRef]
42. Emongor, V.E. Effect of benzyladenine and gibberellins on growth, yield and yield components of common bean (*Phaseolus vulgaris*). *UNISWA Res. J. Agric. Sci. Technol* **2002**, *6*, 65–72. [CrossRef]
43. Gupta, V.N.; Datta, S.K. Influence of gibberellic acid (GA3) on growth and flowering in chrysanthemum (Chrysanthemummorifolium, Ramat) cv. Jayanti. *Indian J. Plant Physiol.* **2001**, *6*, 420–422.
44. Huttly, A.K.; Phillips, A.L. Gibberellin-regulated plant genes. *Physiol. Plant.* **1995**, *95*, 310–317. [CrossRef]

45. Van Huizen, R.; Ozga, J.A.; Reinecke, D.M. Influence of auxin and gibberellin on in vivo protein synthesis during early pea fruit growth. *Plant Physiol.* **1996**, *112*, 53–59. [CrossRef]

46. Cohn, N.S.; Zhang, L.; Mitchell, J.P.; Vierheller, C.-Z.J. Gibberellin-stimulated changes in abundance of two mRNAs in the developing shoot of dwarf peas (*Pisum sativum* L.). *Int. J. Plant Sci.* **1994**, *155*, 498–505. [CrossRef]

47. Shiri, Y.; Solouki, M.; Ebrahimie, E.; Emamjomeh, A.; Zahiri, J. Gibberellin causes wide transcriptional modifications in the early stage of grape cluster development. *Genomics* **2020**, *112*, 820–830. [CrossRef]

48. Siddiqui, M.H.; Khan, M.N.; Mohammad, F.; Khan, M.M.A. Role of nitrogen and gibberellin (GA3) in the regulation of enzyme activities and in osmoprotectant accumulation in Brassica juncea L. under salt stress. *J. Agron. Crop Sci.* **2008**, *194*, 214–224. [CrossRef]

49. De Freitas, S.T.; Jiang, C.-Z.; Mitcham, E.J. Mechanisms involved in calcium deficiency development in tomato fruit in response to gibberellins. *J. Plant Growth Regul.* **2012**, *31*, 221–234. [CrossRef]

50. Hedden, P.; Thomas, S.G. Gibberellin biosynthesis and its regulation. *Biochem. J.* **2012**, *444*, 11–25. [CrossRef]

51. Liptay, A.; Nicholls, S. Nitrogen Supply during Greenhouse Transplant Production Affects Subsequent Tomato Root Growth in the Field. *J. Am. Soc. Hortic. Sci.* **1993**, *118*, 339–342. [CrossRef]

52. Javid, M.G.; Sorooshzadeh, A.; Moradi, F.; Ali, S.; Modarres, M. Review article The role of phytohormones in alleviating salt stress in crop plants. *Annu. Rev. Plant Biol.* **2011**, *5*, 726–734.

53. Yuan, K.; Rashotte, A.M.; Wysocka-Diller, J.W. ABA and GA signaling pathways interact and regulate seed germination and seedling development under salt stress. *Acta Physiol. Plant.* **2011**, *33*, 261–271. [CrossRef]

54. Sun, T. Gibberellin Metabolism, Perception and Signaling Pathways in Arabidopsis. *Arab. Book* **2008**, *6*, e0103. [CrossRef] [PubMed]

55. Gonai, T.; Kawahara, S.; Tougou, M.; Satoh, S.; Hashiba, T.; Hirai, N.; Kawaide, H.; Kamiya, Y.; Yoshioka, T. Abscisic acid in the thermoinhibition of lettuce seed germination and enhancement of its catabolism by gibberellin. *J. Exp. Bot.* **2004**, *55*, 111–118. [CrossRef] [PubMed]

56. Evins, W.H.; Varner, J.E. Hormonal control of polyribosome formation in barley aleurone layers. *Plant Physiol.* **1972**, *49*, 348–352. [CrossRef] [PubMed]

57. Broughton, W.J.; McComb, A.J. Changes in the pattern of enzyme development in gibberellin-treated pea internodes. *Ann. Bot.* **1971**, *35*, 213–228. [CrossRef]

58. Johri, M.; Varner, J.E. Enhancement of RNA synthesis in isolated pea nuclei by gibberellic acid. *Proc. Natl. Acad. Sci. USA* **1968**, *59*, 269. [CrossRef]

59. Roth-Bejerano, N.; Lips, S.H. Hormonal regulation of nitrate reductase activity in leaves. *New Phytol.* **1970**, *69*, 165–169. [CrossRef]

60. Wood, A.; Paleg, L.G. The influence of gibberellic acid on the permeability of model membrane systems. *Plant Physiol.* **1972**, *50*, 103–108. [CrossRef]

61. Wood, A.; Paleg, L.G. Alteration of liposomal membrane fluidity by gibberellic acid. *Funct. Plant Biol.* **1974**, *1*, 31–40. [CrossRef]

62. Al-Wakeel, S.A.M.; Dadoura, S.S.; Hamed, A.A. Interactive effects of water stress and gibberellic acid on mineral composition of fenugreek plant. *Egypt. J. Physiol. Sci.* **1994**, *18*, 269–272.

63. Ansari, H. Effect of Some Phytohormones and NPK on Growth and Metabolism of Mustard. Ph.D. Thesis, Aligarh Muslim University, Aligarh, India, 1996.

64. Daie, J.; Watts, M.; Aloni, B.; Wyse, R.E. In vitro and in vivo modification of sugar transport and translocation in celery by phytohormones. *Plant Sci.* **1986**, *46*, 35–41. [CrossRef]

65. Estruch, J.J.; Peretó, J.G.; Vercher, Y.; Beltrán, J.P. Sucrose loading in isolated veins of Pisum sativum: Regulation by abscisic acid, gibberellic acid, and cell turgor. *Plant Physiol.* **1989**, *91*, 259–265. [CrossRef] [PubMed]

66. Mulligan, D.R.; Patrick, J.W. Gibberellic-acid-promoted transport of assimilates in stems of *Phaseolus vulgaris* L. *Planta* **1979**, *145*, 233–238. [CrossRef] [PubMed]

67. Soundy, P.; Cantliffe, D.J.; Hochmuth, G.J.; Stoffella, P.J. Management of nitrogen and irrigation in lettuce transplant production affects transplant root and shoot development and subsequent crop yields. *HortScience* **2005**, *40*, 607–610. [CrossRef]

68. Vetrano, F.; Miceli, C.; Angileri, V.; Frangipane, B.; Moncada, A.; Miceli, A. Effect of Bacterial Inoculum and Fertigation Management on Nursery and Field Production of Lettuce Plants. *Agronomy* **2020**, *10*, 1477. [CrossRef]

69. Cleland, R.E. Introduction: Nature, occurrence and functioning of plant hormones. In *Biochemistry and Molecular Biology of Plant Hormones*; Hooykaas, P.J.J., Hall, M.A., Libbenga, K.R., Eds.; Elsevier: Amsterdam, The Netherlands, 1999; Volume 33, pp. 3–22.

 *horticulturae*

Article

An Appraisal of Biodegradable Mulch Films with Respect to Strawberry Crop Performance and Fruit Quality

Maria Giordano, Ciro Gianmaria Amoroso, Christophe El-Nakhel, Youssef Rouphael, Stefania De Pascale * and Chiara Cirillo *

Department of Agricultural Sciences, University of Naples Federico II, 80055 Portici, Italy; maria.giordano@unina.it (M.G.); cirogianmaria.amoroso@unina.it (C.G.A.); christophe.elnakhel@unina.it (C.E.-N.); youssef.rouphael@unina.it (Y.R.)
* Correspondence: depascal@unina.it (S.D.P.); chiara.cirillo@unina.it (C.C.)

Received: 29 July 2020; Accepted: 19 August 2020; Published: 21 August 2020

Abstract: *Fragaria × ananassa* is a fruit grown all over the world, appreciated for its organoleptic and nutraceutical properties. Together with other berry fruits, it is rich in bioactive molecules that make it a beneficial fruit for human health. However, strawberry cultivation is influenced by pre- and post-harvest factors. Being a small plant, its fruit comes into direct contact with the soil and, as such, can quickly decompose. To reduce this inconvenience, farmers have used different strategies to mulch the soil, and the most useful method is polyethylene mulch films that are not biodegradable. The focus on environmentally sustainable agriculture can be represented by a transition to biodegradable mulch films. In our study, ten biodegradable mulch films were used to understand their effectiveness in covering the soil during the cultivation cycle of strawberry cv. Rociera. Polyethylene film was considered the control. The best yield and the highest number of fruits with greatest size and quality were obtained on polyethylene, BioFlex® (P2), Bio 6, and Bio 7 films. On BioFlex® (P2) and Bio 3 biodegradable films, strawberries showed a higher calcium and magnesium content, respectively. These results may encourage growers toward the use of eco-sustainable agricultural practices, such as biodegradable mulch films.

Keywords: sustainable agriculture; marketable production; antioxidant molecules; mineral content; strawberry; weed biomass

---

1. Introduction

Strawberry belongs to the *Rosaceae* family, *Fragaria* genus, with wild and cultivated species of great agronomic importance all over the world, incorporating a wide range of ecological spread, from temperate to tropical and subtropical regions, and from sea level to high altitudes [1]. Nowadays, all cultivated varieties derive from *Fragaria × ananassa*, developed in 1700 in Europe from a hybridization of two natural hybrid species from America, *Fragaria chilosensis* and *Fragaria virginiana*. Currently, there are 11 species of *Fragaria* found in their native habitats in Europe and North and South America. The fruit of all species are dry achenes on the surface of a fleshy receptacle. The differences among the various species, or within the same species, can be seen at a morphological level, in the different color and shape of the achenes, and different morphology of the stolons and leaves [2]. Worldwide, strawberry consumption is calculated in millions of tons [3,4]. In 2018, Italy and Spain dedicated an area of 4717 ha and 7032 ha to strawberry cultivation, respectively, resulting in a production of 119,223 and 344,679 tons, respectively [5]. Strawberries are mainly consumed as a fresh product, but its organoleptic characteristics make it ideal also for processed products, such as liqueurs, syrups, jam, ice cream and essences.

Berries are a rich source of bioactive compounds with potential beneficial effects on human health [6] and have thus been extensively studied [7]. The high content of polyphenols, flavonoids, and anthocyanins makes strawberry a functional food, hence a fruit with beneficial effects on human health [8,9]. Indeed, Giampieri et al. [10] showed that extracts of transgenic lines of three strawberry cultivars contained high levels of anthocyanins, which exerted cytotoxic effects on HepG2, a human hepatic cancer cell line. The antioxidant compounds present in berries, including strawberry, reduce oxidative DNA damage and can up- or down-regulate different genes involved in the inhibition of carcinogen activation, inhibition of oncogenes and metastasis. Additionally, it can reduce the risk of neurodegenerative pathologies including Lou Gehrig's, Parkinson's, Alzheimer's, and Huntington's diseases [11]. Moreover, the protective action of anthocyanins was demonstrated by Basu et al. [12], reducing non-fatal myocardial infarction when strawberries and blueberries were consumed more than three times a week. In addition, strawberry consumption reduced blood glucose levels, obesity, inflammatory responses [13], and apoptosis in breast cancer cells [14,15]. The consumption of berry fruit, including strawberries, may mitigate the risk of type 2 diabetes (T2D). Calvano et al. [16] and Davis et al. [17] showed that dietary supplementation of freeze-dried strawberries reduced postprandial hyperglycemia and hyperinsulinemia in overweight and obese adults. The same authors showed that adult consumption of freeze-dried strawberries (50 g day^{-1}) for eight weeks, significantly decreased total and LDL cholesterol. Strawberry extracts are potent antimicrobial agents, as in the case of *Vibrio cholerae* [18]. They are also rich in fiber, folate and manganese [8]. Furthermore, strawberry can fight cellular aging, because its polyphenols are activators of AMP-activated protein kinase (AMPK; [19,20]), molecules involved in the biogenesis of mitochondria in eukaryotic cells, and in the cellular response against ROS-induced oxidative stress damage. Giampieri et al. [20] demonstrated that the consumption of strawberries significantly increased the expression of the AMPK cascade genes in older rats. Nevertheless, high-sensitivity TNF-α (hs-TNF-α) and soluble tumor necrosis factor receptor (sTNF-R2), two biomarkers of inflammation and lipid peroxidation, were significantly decreased in men and woman with knee osteoarthritis, when administered a dose of 50 g of strawberry powder for 12 weeks [21].

The nutraceutical and sensorial properties of strawberry can vary based on cultivar, climatic and geographical conditions, ripening stage, and conservation methods [8,22,23]. A limitation of strawberry is the quick deterioration of fruit after harvest [24]. A solution adopted in strawberry production is to keep fruit clean, dry and free from soil contact to prevent triggering the decomposition by using low density polyethylene films (LDPE) [1,25]. These films have been widely adopted for various crops including strawberry, because their use speeds up crop cycles, assures high and good quality yields, and blocks weed growth [1,3,26,27].

LDPE can be transparent, colored, or dark. Transparent films pass 50% of ultraviolet (UV) rays (220–380 nm), black ones restrain UV rays but are penetrated by short wavelengths (780–2500 nm) and longer wavelength (>2500 nm) infrared rays [3]. Polyethylene films are very resistant to degradation, thanks to their high molecular weight and their hydrophobicity. Unfortunately, these characteristics render polyethylene a slowly degrading product for possibly hundreds of years [3,25]. Due to climatic conditions and chemical products used in agriculture, plastic films can be used for one or at most two cultivation cycles and then need to be removed. Their removal from the soil at the end of a crop cycle is only partial, since they can be torn apart by harvest machinery and, consequently may be dispersed throughout the harvested area. Normally, these films should be recovered from the soil, disposed of in landfills, and recycled or even incinerated appropriately. Plastic waste disposal represents a high cost for farmers, who prefer to burn them directly in the field, thus endangering their health and the environment. Moreover, mulch films increase the serious problem of plastic dispersal in the environment that the world is now facing [3,28,29]. The use of plastic products in agriculture is estimated to be in the millions of tons per year, and at least 10% of the total amount originates from mulch films [25,28,30].

Biodegradable films can represent an excellent solution to the adverse environmental impact derived from plastic use in agriculture. They have proven to be a good alternative to polyethylene films and bare soil in short cultivation cycles as stated by Cozzolino et al. [31] and in the case of strawberry cultivation as noted by Morra et al. [32]. The main practical and ecological advantages of these films are that they can be both left on the field and buried in the soil to be degraded by microorganisms. Indeed, fungi, bacteria, and algae can transform the residues of these films into carbon dioxide, methane, water, and biomass [28,29]. The biodegradable films currently in use are mainly based on starch and cellulose, polyhydroxybutyrate/valerate copolymers, and polylactic acid polymers; molecules that are susceptible to UV- and visible light-facilitated photooxidation or thermoxidized at high temperatures [25,28,33,34]. A limitation of biodegradable films is that, similar to PE films, they are subjected to weathering and chemical substances used on crops, as well as to soil microorganism attack. Therefore, they risk incomplete soil coverage for the entire crop cycle. Moreover, Nestby and Guéry [35], in a work conducted over three years on three strawberry cultivars on mulch, showed that the marketable yield could be reduced up to 58% due to a significant mildew infestation. According to European standard values [36], PE films (dark, transparent, or thermal) used in agriculture must have particular features, such as the tensile strength at break or the tensile elongation at break of at least 16 MPa and 180–250%, respectively. Conversely, as stated by Scarascia-Mugnozza et al. [28], biodegradable films have both tensile stress and tensile elongation at break lower than those of PE set by the European Standard. However, their mechanical properties fall within the range required to ensure an adequate soil coverage during the entire strawberry growing cycle. Moreover, on the basis of ecotoxicological tests, the authors demonstrated the absence of soil ecotoxicity at the end of the crop cycle after burying the material.

To balance biodegradation and physical-mechanical properties of a mulch film, it would be essential to obtain a biodegradable film, which endures until the end of the crop cycle. Furthermore, biodegradable films should yield a satisfactory amount of a high-quality final product, together with adequate weed control, similar to PE films. According to Andrade et al. [37], biodegradable mulch films, formed by starch, allowed adequate ground cover and weed suppression during a strawberry autumn-winter cycle.

Based on the aforementioned, the aim of our work was to compare different mulch films in order to determine their ability to cover the soil during the cultivation of strawberry, control weed infestation, and distinguish their effect on berry production, quantity, and quality as compared to black PE mulch film.

2. Materials and Methods

2.1. Experimental Site, Plant Material and Growth Conditions

The experiment was conducted in a greenhouse at the experimental farm of ADESVA Technological Center, located in Lepe, Huelva-Spain (lat. 37°15′ N; long 7°12′ W, 48 a.s.l.). Site temperatures varied between a maximum of 30 °C and a minimum of 11 °C, generally in July and January, respectively. The soil had a sandy loam texture (66% sand, 19% silt, and 15% clay; pH 7.2, organic matter 1.9%, available P_2O_5 of 20 mg kg^{-1}, exchangeable K of 3270 mg kg^{-1}). Strawberry cultivar Rociera distributed by Fresas Nuevos Materiales (Huelva, Spain), was used in the experiment. It is highly resistant to *Botrytis* and mildew, with uniform coloring and round-shaped fruit, long stems that facilitate harvest, and produces a high percentage of first category fruit.

On 7 August 2017 the greenhouse was divided into four macrotunnels (6.6 m wide) covered by plastic material, with an area of 264 m^2/macrotunnel. On 13th August, in each macrotunnel, five soil beds were prepared (40 cm wide, 40 m long) 50 cm apart. In each bed, two furrows 25 cm apart were then created to accommodate two parallel lines of plants. On 20th September, the soil was disinfected with TELOPIC C35 (1,3-dichloropropene 80.3% p/v (equivalent to 60.8% p/p) + chloropicrin 44.0% p/v (equivalent to 33.3% p/p; p/v: weight/volume, w/w: weight/weight), at the rate of two irrigations per day, for ten minutes each irrigation, for a total of 300 kg ha^{-1}. From the vegetative period of October/December until the productive period (from January to May), the following fertilizing units were distributed: total nitrogen (234.15 kg ha^{-1}), phosphoric anhydride (75.67 kg ha^{-1}), potassium oxide (216.28 kg ha^{-1}), total calcium (229.15 kg ha^{-1}), magnesium oxide (48.61 kg ha^{-1}), potassium chloride (52.47 kg ha^{-1}). A drip irrigation system was adopted with 1–2 irrigations of 10–25 min daily, for a total of 4684.63 m^3 ha^{-1} of water.

2.2. Mulch Installation, Transplanting, Experimental Design

On 18 October 2017, each soil bed was covered by mulch films. Eleven different films were used, and their thickness and black pigment percentages are shown in Table 1:

Table 1. Mulch film types, thickness and black pigment percentage.

Mulch film	Thickness (µm)	Black Pigment (%)
Polyethylene (P1)	35	9
BioFlex® 1130 (P2)	20	9
BioFlex® 1821 (P3)	20	9
Bio M 16 F54 (Bio 1)	20	9
Bio M 17 F53 (Bio 2)	25	9
Bio M 4b F28 (Bio 3)	20	9
Bio M 5b F28 (Bio 4)	20	9
Bio M 5b F28 (Bio 5)	25	9
Bio M 17 F53 (Bio 6)	40	5
Bio M 4b F28 (Bio 7)	40	5
Bio M 5b F28 (Bio 8)	25	5

Black pigment percentage added to the mixture.

P1 is a black low-density polyethylene (LDPE) plastic film. BioFlex® 1130 (P2) and BioFlex® 1821 (P3) (FKuR Kunststoff GmbH, Germany) are commercial biodegradable films, formed from polylactic acid (PLA)/copolyester blends, obtained from a cornstarch fermentation process. They completely degrade in the soil and are compostable according to EN 13432 certification. Furthermore, BioFlex® 1130 (P2) has a minimum biobased carbon content of 10% (calculated according to ASTM D6866 certification), tear resistance of 100/110 N / mm (according to ASTM D 1922 certification), Spencer impact test of 420 N / mm (according to ASTM D 3420 certification). BioFlex® 1821 (P3) has a minimum biobased carbon content of 10% (calculated according to ISO 16620), tear resistance of 100/180 N/mm (according to ASTM D 1922 certification), Spencer impact test of 240 N/mm (according to ASTM D 3420 certification). Biodegradable films (Bio 1–8) are based on starch, but the manufacturer has not revealed its exact concentration. The black color of all films was obtained by adding from 5 to 9% black pigment to the mixture during film preparation.

The experiment involved a completely randomized design, in which the treatments were 11 different films with three repetitions, resulting in 33 plots in total, randomly distributed in the four macrotunnels representing the greenhouse. In three macrotunnels, only the three central beds, out of total five prepared, were taken into consideration for the experiment, resulting in nine different mulch films per macrotunnel (three mulch treatments per bed). The fourth macrotunnel contained six plots with the remaining film treatment repetitions distributed on two central beds. This partition of the film treatments replicated each treatment three times and distributed the repetitions among the different macrotunnels. Each experimental plot included 25 representative plants placed in double rows. On 18th October (1st day after transplant: 1 DAT) the plants were transplanted in each bed at a plant density of 65,000 plants ha^{-1}.

2.3. Environmental Control

The weather data were monitored thanks to a weather control station set up under one of the four greenhouses. It consisted of a data logger (Watermark 900M), responsible for collecting and sending, via GPRS, the data of the different sensors in use. December, January, and February were the coldest months with mean temperatures of 11.5 °C, 11.0 °C, and 10.7 °C, respectively. May and October were the hottest, with mean temperatures of 19.1 °C and 17.4 ° C, respectively. Relative humidity showed an increasing trend from October, reaching the highest value of 75.7% in January. Similar values were also found in March and April. Solar radiation exhibited a decreasing trend from October to January, with the lowest values in January (9.4 MJ m^{-2} per day). From January to May, however, the trend reversed, and the highest values were found in May with 26.0 MJ m^{-2} per day.

2.4. Productive Parameters, Yield and Quality Measurements

Twice a week, or three times a week in March-April (maximum production period), the fruit was collected from all the 25 plants forming each plot, for a total of 28 harvests expressed in one cumulative value. Moldy fruit was also collected and weighed. The production was assessed by weighing the fruit and dividing them into two main categories, expressed in gram per m^2. The first category (Cat. I) included fruit weighing more than 15 g, with no defects in shape or color. The second category (Cat. II) included fruit weighing less than 15 g and with few defects in shape and color. Marketable yield was determined by the sum of the total weight of the Cat. I and Cat. II fruit. As for the fruit average weight determination, once a month, or twice a month in March and April, Cat. I fruits were collected by plot and weighed, then divided by their count.

Fruit flesh firmness was also determined on two opposite sides at the equatorial zone, of five previously peeled fruit with similar color sampled from each plot, using a digital penetrometer (FT 327, Effegi, Milan, Italy) equipped with a 'star' plunger for strawberries (model 53207, Turoni, Forlì, Italy) and expressed as N. Measurements were performed three times throughout the cycle, at the beginning of the harvest season (February; 128 DAT) in the middle (March; 153 DAT) and at the end of the season (May; 202 DAT).

On the same five fruits used for the flesh firmness measurement, the fruit soluble solids content (° Brix) was evaluated using an Atago digital refractometer (model: PR-32) at a temperature of 20 °C.

2.5. Weed Biomass

Weed biomass was assessed in two steps: during the cropping season, collected only from the central part of each bed in April (178 DAT), and at the end of the season in May (202 DAT) when the plants were removed. The weeds were collected from the central part of each bed and also from the sides and above the beds. Weeds were cut and weighed, and the weed biomass was expressed as grams fresh weight m^{-2}.

2.6. Quality Parameters

At the end of the last harvest, 100 g of fruits were collected per plot. They were frozen at −80 °C and subsequently freeze-dried to determine total phenol content and hydrophilic (HAA) and ABTS antioxidant activities (ABTS-AA), following use of the Folin–Ciocalteau [38], DMPD [39] and ABTS [40] methods, respectively, as previously described in Rouphael et al. [41]. Results were expressed as mg gallic acid equivalents per 100 g^{-1} dry weight (dw), mmol Trolox per 100 g^{-1} dw, and mmol ascorbic acid equivalents per 100 g^{-1} dw, respectively.

Total protein content was estimated by determining first the total nitrogen content of fruit by the Kjeldahl method [42], and then by multiplying the total N value by a factor of 6.25 following official method 976.05 of the AOAC (Association of Official Analytical Chemists [43]). Data were expressed as g per 100 g^{-1} dw.

2.7. Malic and Citric Acid Content and Mineral Profile

Malic and citric acid content, and minerals were determined using an ion chromatography model ICS-3000, (Dionex, Sunnyvale, CA, USA). Two different columns were used: an IonPac CS12A (4 × 250 mm) analytical column for cation determination (K, Ca, Mg, Na); and a IonPac AS11-HC analytical column (4 × 250 mm) for anion (nitrate, P, SO_4) and malic and citric acid determination [41]. Data were expressed as g kg^{-1} dw.

2.8. Statistical Analysis

The Shapiro–Wilk and Kolmororov–Smirnov procedures were performed to verify that the data had a normal distribution, and the Levene, O'Brien and Bartlet tests were conducted to verify the homogeneity of variances. The experimental data were subjected to analysis of variance (ANOVA) using the SPSS 20 software package (SPSS Inc., Chicago, IL). In particular, the cumulative yield for each category was calculated by adding the fruit yield at each harvest. Duncan's multiple range test was performed for mean comparisons on each of the significant ($p \leq 0.05$) variables measured.

3. Results

3.1. Marketable Production and Mulching Performance

The highest marketable yield was observed when P1 film was used (6995 g m^{-2}; Table 2), though it was not significantly different from the biodegradable films P2, Bio 6, and Bio 7 that produced slightly less on average (−6.8%). Conversely, Bio 1, 2, and 4 showed the lowest marketable yield, with a 25% reduction compared to the P1 film. P3, Bio 3, 5, and 8 resulted in intermediate yield values. The marketable yield was divided into two categories: strawberries with a weight greater than 15 g and without defects in shape or color that belong to the first category, and strawberries with a weight less than 15 g and with slight defects that were attributed to the second category. The highest yield of Cat.I fruits was on the beds covered with LDPE film (P1; 5729 g m^{-2}), P2, Bio 6, and Bio 7. Production of Cat. II fruit ranged between 966 g m^{-2} (Bio 2) and 1422 g m^{-2} (Bio 5), but no significant differences between the mulch treatments were noticed (Table 2). P1 film resulted in a minor amount of moldy fruit but was not significantly different from most of the other treatments, except for Bio 3, Bio 4, and Bio 5, which registered the greatest quantity of moldy fruit (~167 % more than P1). Mulch treatments did not affect fruit mean weight, while they had a significant effect on fruit number. The highest fruit number was harvested in P1 treatment (250 no. m^{-2}), even though it was not significantly different from P2, P3, Bio 5, Bio 6, and Bio 7.

The resistance to degradation of mulch films can be indirectly deduced by the number of weeds they allowed to grow. LDPE P1 film did not allow the weeds to emerge throughout the crop cycle by maintaining the full coverage of the beds. Among the biodegradable films, Bio 6 and 7 had the lowest weed biomass with 220 and 225 g m^{-2}, respectively, though not significantly different from most mulch treatments except for Bio 1 with 660 g m^{-2}.

3.2. Quality Analysis

The mulch films had no significant effects on fruit soluble solids content or flesh firmness (Table 3). Similarly, there was no significant effect on total protein, total phenols, ABTS antioxidant activity, or malic and citric acid content (Table 4). On the contrary, differences were evident in hydrophilic antioxidant activity levels among the treatments, with the highest values obtained on P2 film (12.80 mmol ascorbic acid eq. 100 g^{-1} dw), which was not significantly different from P1, Bio 2, Bio 3 and Bio 4, while the lowest HAA value was on Bio 8 (10.61 mmol of ascorbic acid eq. 100 g^{-1} dw), which was not significantly different from most mulching treatments, except for P1, P2, and Bio 3 (Table 4).

Table 2. Effects of different mulch films on strawberry yield, yield components, and weed biomass of strawberry.

Mulch Film	Cat. I Fruits [z] (g m^{-2})	Cat. II Fruits [y] (g m^{-2})	Moldy Fruits (g m^{-2})	Marketable Yield (g m^{-2})	Fruit Average Weight (g)	Fruit Number (no. m^{-2})	Weed Biomass (g m^{-2})
Polyethylene	5729 ± 43 a [x]	1265 ± 116	393 ± 128 c	6995 ± 109 a	27.98 ± 1.0	250 ± 5.8 a	0 ± 0 d
BioFlex 1	5238 ± 221 abc	1246 ± 28	620 ± 174 bc	6484 ± 239 ab	27.64 ± 1.3	236 ± 14 abc	337 ± 97 abc
BioFlex 2	4756 ± 109 bcde	1247 ± 232	654 ± 22 bc	6003 ± 326 bc	26.7 ± 1.5	226 ± 17 abcd	468 ± 95 ab
Bio 1	4253 ± 175 ef	981 ± 41	801 ± 153 abc	5234 ± 213 c	29.62 ± 2.0	178 ± 14 de	660 ± 52 a
Bio 2	4331 ± 98 ef	966 ± 155	600 ± 123 bc	5298 ± 194 c	27.67 ± 0.6	192 ± 11 cde	292 ± 32 abc
Bio 3	4410 ± 392 def	1410 ± 134	1181 ± 201 a	5820 ± 424 bc	29.28 ± 2.0	201 ± 20 bcde	588 ± 266 ab
Bio 4	3965 ± 156 f	1249 ± 164	1045 ± 74 ab	5214 ± 261 c	29.91 ± 1.4	174 ± 1.5 e	382 ± 81 abc
Bio 5	4594 ± 23 cdef	1422 ± 156	928 ± 132 ab	6016 ± 150 bc	29.15 ± 3.1	211 ± 21 abcde	280 ± 61 abc
Bio 6	5336 ± 166 ab	1231 ± 43	821 ± 151 abc	6568 ± 152 ab	26.92 ± 1.5	246 ± 19 ab	220 ± 108 bc
Bio 7	5083 ± 204 abcd	1419 ± 99	727 ± 54 bc	6502 ± 195 ab	29.91 ± 0.7	217 ± 3.5 abcde	225 ± 66 bc
Bio 8	4722 ± 359 bcde	1084 ± 120	672 ± 156 bc	5805 ± 433 bc	29.72 ± 3.2	198 ± 18 bcde	377 ± 176 abc
Significance	*** [w]	ns	*	***	ns	*	**

[z] Fruits weighing more than 15 g, with no defects in shape or color; [y] Fruits weighing less than 15 g and with few defects in shape and color. Marketable yield = Cat. I + Cat. II fruits. [x] All data are expressed as mean ± standard error, n = 3. Different letters within each column indicate significant differences according to Duncan's multiple-range test ($p \leq 0.05$). [w] ns,*,**, *** Non significant or significant at $p \leq 0.05$, 0.01, and 0.001, respectively.

Table 3. Effects of different mulch films on total soluble solids (°Brix) and flesh firmness (N) of greenhouse strawberry fruit harvested at 128, 153, and 202 days after transplant (DAT).

Mulch Film.	Total Soluble Solids (°Brix)			Flesh Firmness (N)		
	128 DAT	153 DAT	202 DAT	128 DAT	153 DAT	202 DAT
Polyethylene	12.57 ± 0.22 [z]	10.53 ± 0.44	9.82 ± 0.50	2.32 ± 0.19	2.39 ± 0.16	2.19 ± 0.13
BioFlex 1	11.90 ± 0.56	10.11 ± 0.61	10.31 ± 1.25	2.40 ± 0.14	2.33 ± 0.19	2.26 ± 0.10
BioFlex 2	10.37 ± 0.63	10.99 ± 0.14	10.08 ± 0.92	2.48 ± 0.19	2.48 ± 0.22	2.25 ± 0.05
Bio 1	11.49 ± 0.74	10.53 ± 0.58	9.34 ± 0.44	2.25 ± 0.14	2.55 ± 0.06	2.20 ± 0.09
Bio 2	12.86 ± 0.30	11.04 ± 0.23	9.91 ± 0.66	2.55 ± 0.24	2.57 ± 0.10	2.37 ± 0.12
Bio 3	10.83 ± 0.08	10.52 ± 0.57	9.61 ± 0.48	2.51 ± 0.16	2.40 ± 0.15	2.15 ± 0.06
Bio 4	12.72 ± 1.36	10.79 ± 0.27	10.18 ± 0.35	2.64 ± 0.08	2.50 ± 0.13	2.26 ± 0.09
Bio 5	10.76 ± 0.24	10.91 ± 0.44	10.51 ± 0.31	2.46 ± 0.13	2.42 ± 0.19	2.30 ± 0.03
Bio 6	12.79 ± 0.50	11.19 ± 0.23	9.71 ± 0.37	2.36 ± 0.17	2.50 ± 0.17	2.29 ± 0.07
Bio 7	11.34 ± 0.55	10.60 ± 0.56	9.24 ± 0.41	2.58 ± 0.07	2.26 ± 0.07	2.17 ± 0.04
Bio 8	11.48 ± 1.26	11.88 ± 0.54	9.69 ± 0.26	2.59 ± 0.17	2.44 ± 0.09	2.24 ± 0.04
Significance	ns [y]	ns	ns	ns	ns	ns

[z] All data are expressed as mean ± standard error, n = 3. Absence of letters within each column indicate no significant differences according to Duncan's multiple-range test ($p \leq 0.05$). [y] ns Non-significant.

Table 4. Effects of different mulch films on total protein, antioxidant activities, total phenols, and malic and citric acids of greenhouse strawberry.

Mulch Film	Total Proteins (g 100 g^{-1} DW)	LAA (mmol Trolox 100 g^{-1} dw)	HAA (mmol ascorbic ac. eq. 100g^{-1} dw)	Total Phenols (mg gallic ac. eq. 100g^{-1} dw)	Malic Acid (g kg^{-1} dw)	CITRIC ACID (g kg^{-1} dw)
Polyethylene	7.86 ± 0.41 [z]	52.33 ± 0.79	12.33 ± 0.30 ab	7.22 ± 0.51	19.95 ± 0.51	72.76 ± 4.81
BioFlex 1	7.60 ± 0.48	54.00 ± 2.31	12.80 ± 0.46 a	7.50 ± 0.46	21.55 ± 0.92	74.28 ± 2.75
BioFlex 2	6.85 ± 0.28	57.74 ± 2.15	11.20 ± 0.42 bc	7.46 ± 0.19	19.59 ± 1.97	73.40 ± 3.61
Bio 1	6.98 ± 0.12	54.71 ± 2.90	11.44 ± 0.55 bc	7.31 ± 0.43	18.16 ± 1.22	69.57 ± 0.51
Bio 2	6.53 ± 0.06	56.32 ± 0.46	11.97 ± 0.23 abc	7.41 ± 0.12	17.73 ± 0.51	71.40 ± 1.76
Bio 3	7.81 ± 0.67	55.27 ± 2.67	12.19 ± 0.36 ab	7.00 ± 0.36	19.79 ± 0.99	71.51 ± 1.38
Bio 4	7.98 ± 0.28	57.18 ± 0.51	11.47 ± 0.13 abc	7.01 ± 0.21	19.35 ± 1.96	74.22 ± 2.02
Bio 5	7.58 ± 0.53	53.70 ± 0.73	11.31 ± 0.18 bc	7.29 ± 0.29	18.16 ± 0.10	68.50 ± 2.97
Bio 6	7.39 ± 0.16	55.14 ± 1.14	11.37 ± 0.42 bc	6.84 ± 0.23	18.76 ± 1.10	72.85 ± 1.44
Bio 7	7.91 ± 0.59	54.84 ± 1.66	11.15 ± 0.08 bc	7.19 ± 0.25	17.47 ± 1.06	69.56 ± 0.75
Bio 8	7.90 ± 0.36	55.08 ± 1.90	10.61 ± 0.82 c	6.96 ± 0.36	18.81 ± 0.54	74.46 ± 0.81
Significance	ns [y]	ns	*	ns	ns	ns

[z] All data are expressed as mean ± standard error, n = 3. Different letters within each column indicate significant differences according to Duncan's multiple-range test ($p \leq 0.05$).
[y] ns,* Non-significant or significant at $p \leq 0.05$, respectively.

3.3. Mineral Profile

Phosphorus, potassium, sulfate, and sodium content were not influenced by mulch film treatments. Significant differences among mulches were found for nitrate, calcium, and magnesium content (Table 5). The lowest nitrate content was recorded with strawberries grown on Bio1 and Bio 2 films, with 1.34 and 1.35 g kg^{-1} dw, respectively. These results were not significantly different from most of the other film treatments, except for P2, Bio 3, and Bio 7, which had higher accumulations of nitrate in the fruits. Moreover, fruits grown in P1 and P2 film treatments had the highest calcium content (~1.74 g kg^{-1} dw). As for magnesium, Bio 3 and Bio 4 showed the highest content of around 1.58 g kg^{-1} dw, while Bio 2 exhibited the lowest fruit content (1.16 g kg^{-1} dw; Table 5).

Table 5. Effects of different mulch films on mineral profile of greenhouse strawberry.

Mulch Film	NO3 (g kg^{-1} dw)	P (g kg^{-1} dw)	K (g kg^{-1} dw)	Ca (g kg^{-1} dw)	Mg (g kg^{-1} dw)	SO4 (g kg^{-1} dw)	Na (g kg^{-1} dw)
Polyethylene	2.38 ± 0.37 abc [z]	4.45 ± 0.23	16.11 ± 0.94	1.68 ± 0.07 ab	1.35 ± 0.06 de	2.33 ± 0.21	0.25 ± 0.05
BioFlex 1	2.71 ± 0.36 ab	4.52 ± 0.45	17.05 ± 0.69	1.79 ± 0.09 a	1.49 ± 0.05 bc	2.12 ± 0.38	0.61 ± 0.12
BioFlex 2	1.71 ± 0.05 bc	4.11 ± 0.20	15.52 ± 0.19	1.51 ± 0.05 c	1.34 ± 0.01 de	1.95 ± 0.46	0.29 ± 0.01
Bio 1	1.34 ± 0.10 c	3.95 ± 0.16	14.95 ± 0.43	1.41 ± 0.01 c	1.26 ± 0.01 ef	1.39 ± 0.03	0.37 ± 0.07
Bio 2	1.35 ± 0.13 c	3.75 ± 0.25	14.68 ± 0.87	1.41 ± 0.06 c	1.16 ± 0.05 f	1.61 ± 0.50	0.23 ± 0.06
Bio 3	2.74 ± 0.67 ab	4.66 ± 0.13	17.13 ± 0.69	1.38 ± 0.05 c	1.62 ± 0.04 a	1.57 ± 0.12	0.43 ± 0.13
Bio 4	2.24 ± 0.37 abc	5.08 ± 0.32	16.92 ± 0.98	1.40 ± 0.04 c	1.54 ± 0.06 ab	1.58 ± 0.17	0.59 ± 0.12
Bio 5	2.07 ± 0.28 abc	4.47 ± 0.51	16.22 ± 0.58	1.48 ± 0.05 c	1.49 ± 0.05 bc	1.71 ± 0.24	0.38 ± 0.04
Bio 6	2.04 ± 0.20 abc	4.66 ± 0.21	15.77 ± 0.17	1.43 ± 0.05 c	1.38 ± 0.05 cde	1.60 ± 0.22	0.27 ± 0.04
Bio 7	2.83 ± 0.28 a	4.89 ± 0.47	17.53 ± 0.65	1.53 ± 0.06 bc	1.39 ± 0.01 cd	1.48 ± 0.20	0.44 ± 0.29
Bio 8	2.23 ± 0.14 abc	4.60 ± 0.26	17.18 ± 0.36	1.51 ± 0.01 c	1.34 ± 0.02 de	1.68 ± 0.07	0.24 ± 0.06
Significance	* [y]	ns	ns	***	***	ns	ns

[z] All data are expressed as mean ± standard error, n = 3. Different letters within each column indicate significant differences according to Duncan's multiple-range test ($p \leq 0.05$). [y] ns, *, *** Non-significant or significant at $p \leq 0.05$ and 0.001, respectively.

4. Discussion

Mulching is a common practice in horticulture that embodies a prime necessity because fruit in contact with the soil tends to perish in short notice, such as strawberry. Among the different ways of soil mulching in strawberry cultivation, PE film is the most widely used all over the world. In our study, strawberry grown on PE film had the highest marketable yield and the highest number of fruits belonging to the first category, encompassing berries weighing greater than 15 g and exempt of any defects. These results are comparable to some biodegradable films such as P2, Bio 6, and Bio7. Other biodegradable films (Bio 1, 2, and 4) led to a reduction in production of up to 25% compared to PE film. The different yields obtained among the mulch films may reflect the specific effect they have on the soil hydrothermal regime. Generally, PE film generates higher soil temperatures, since it can absorb more solar radiation, and retains the heat between the soil and the film. Furthermore, PE film is made of high molecular weight molecules and is hydrophobic [3,30], therefore acting as a waterproof barrier that slows down or blocks the evaporation of water from the soil. Thus, it maintains a constant soil moisture, and allows steam to condense and infiltrate into the soil. By measuring the water vapor permeability of PE films, Bilck et al. [3] found that it was up to 250 times lower than the biodegradable films they compared. The better soil hydrothermal conditions achieved under PE films, compared to other types of mulch films, are considered responsible for better flowering and fruiting of strawberry plants [1]. Biodegradable films have been found to be more porous and, therefore, less efficient in reducing evaporation [24]. Many authors have confirmed the greater growth of strawberry plants when mulched with black PE compared to other types of mulch [1,44–46]. Others such as Scarascia-Mugnozza et al. [28] and Filippi et al. [47], reported the opposite results.

Both the different chemical composition and the physical characteristics, such as thickness and color, favor the best performance of PE film, compared to the other films. The PE film in our work had a thickness of 35 µm and was black in color. Bio 6 and Bio 7 films were more transparent than PE but were 40 µm thick, which might have led to their comparable performance to the control P1 and with less weed biomass. Moreover, P2 also performed comparably to P1, which could be attributed to the fact that has a high Spencer impact test making it different from P3 which is characterized by a lower one. The other biodegradable films were black as well but were thinner than PE. These characteristics probably allowed the PE film to remain intact until the end of the crop cycle, and to completely block weed growth. The biodegradable films remained intact until March / April, the middle of the production season. Then, they showed a degradation that increased in the following months, especially with Bio 1, 6, and 7 films. However, Bio 6 and Bio 7 films resulted in a lower weed total biomass than the other biodegradable films (with 220 and 225 g m^{-2} of weeds), which may be attributed to their thickness. The films that resulted in the greatest weed biomass were Bio 1, Bio 3, and BioFlex® (P3), films that were characterized by a low thickness. Gupta and Acharya [48], Tarara [49], and Singh et al. [44] attributed the best yield of strawberries cultivated on PE film to the complete suppression of weeds. In our study, the microbial degradation of the biodegradable films may have been accelerated by water supplied through irrigation, as it was also shown in the work of Costa et al. [29]. Varying growth of weeds in the presence of four types of biodegradable mulch films was also shown in the cultivation of tomato plants in the work of Cowan et al. [50]. A good mulching effect with biodegradable films (Mater Bi) was shown by Filippi et al. [47] in melon crop.

In support to our results, infrared spectroscopy analysis carried out by Scarascia-Mugnozza et al. [28] on biodegradable starch-based films for the cultivation of strawberry plants showed that 59 days after contact with the soil, the peak corresponding to starch -OH groups were very small. Moreover, an electron microscopic analysis of the films showed holes corresponding to starch particle disappearance or reduction by soil microorganisms. By analyzing mechanical properties of biodegradable films, such as elongation at break and stress at break, after 124 days of strawberry cultivation, the above-mentioned authors also found that these properties decreased significantly with biodegradable films. Similarly, Bilck et al. [3], by analyzing various mechanical parameters of starch biodegradable and PE films such as strength, elasticity, and rigidity, found that PE films remained flexible and without breaks at the end of the strawberry plant cultivation cycle. Instead, the biodegradable films showed a 50% reduction in their tensile strength and a 120% reduction in their elasticity. Their stiffness increased by 50%, and there was a reduction in deformation, showing them to be more fragile than PE films.

Strawberry, like other berry fruits (raspberries, blackberries, cranberries, and blueberries), is rich in nutraceutical molecules, such as ascorbic acid, phenols, anthocyanins, and flavonoids, which content can vary with the cultivar choice and fruit ripening stage [9,51]. These molecules have strong antioxidant power, capable of preventing or blocking already triggered radical reactions. Another important characteristic of strawberry is its flavor, which is due to the percentage of TSS, mainly glucose, and then fructose and sucrose [9]. According to Singh et al. [1], the best microclimate, together with weed suppression with a PE film, may lead to a better quality strawberry fruit in terms of a higher TSS and ascorbic acid content, and reduced acidity, compared to fruit mulched with other material (clear PE mulch or with straw). However, in our study, neither TSS nor the firmness of the fruit was influenced by the type of mulch film. This TSS data is supported by the results obtained using various mulch films by Costa et al. [29]. As well, the results concerning total protein, phenols, ABTS antioxidant activity, and malic and citric acid content were not influenced by the mulch treatment. Differences were found only in the hydrophilic antioxidant activity. This latter increased in the presence of biodegradable films used in the cultivation of two strawberry cultivars [32]. A higher content of antioxidants may be related to a condition of abiotic or biotic stress, following climatic conditions or pathogenic attacks or diseases.

The effect of the different mulch films on fruit mineral content in our study was evident for nitrate, calcium, and magnesium content. Both calcium and magnesium are essential nutrients for plant growth and development. Their accumulation in strawberries increases their nutraceutical properties. The different composition of nitrate, calcium, and magnesium may depend on different penetration and proliferation of roots in the soil, which can influence the absorption of these nutrients. Kumar and Dey [24], for example, found that strawberry root length, volume, and weight were higher in the presence of both PE and biodegradable film, compared to bare soil. The authors found higher N, P, and K absorption in strawberries grown with PE and biodegradable films, compared to plants grown in bare soil.

5. Conclusions

Currently, there is a great effort to reduce the use of plastic, due to the pollution it is causing. Agriculture is becoming more aware of this dramatic problem. Therefore, adoption of environmentally sustainable practices is increasing, such as the replacement of classic polyethylene films used for soil mulch with biodegradable films that offer similar results. Our study shed light on some biodegradable films, such as BioFlex® (P2), Bio 6, and Bio 7, that were capable of maintaining fruit production comparable to that of black polyethylene film. These films also made it possible to have a greater total number of fruits, including Cat.I fruits, and an overall high marketable yield and greater weed suppression (especially for Bio 6 and Bio 7) than the other biodegradable films. Such results represent sustainable alternatives to polyethylene film use, encouraging farmers towards the adoption of biodegradable films, where production and quality are maintained and pollution is reduced. Nonetheless, more in-depth focus on the thickness of biodegradable films could improve the performance of this green practice.

Author Contributions: Conceptualization, Y.R. and C.C.; methodology, C.G.A.; software, C.G.A.; validation, M.G., C.G.A. and C.E.-N.; formal analysis, C.G.A.; investigation, C.G.A.; resources, S.D.P.; data curation, M.G. and C.G.A.; writing—original draft preparation, M.G.; writing—review and editing, M.G., C.G.A., C.E.-N., Y.R., S.D.P. and C.C.; visualization, Y.R. and C.C.; supervision, Y.R., S.D.P. and C.C.; project administration, Y.R. and C.C.; funding acquisition, S.D.P. All authors have read and agreed to the published version of the manuscript.

Funding: This research received no external funding.

Acknowledgments: The authors would like to thank Magdalena Torres (Head of Adesva Agronomy Department), Adesva Technological Center of Agro-industry, José Lopez Medina (University of Huelva). We would like as well to thank all the partners of Biomulch Project and Antonio Pannico for the statistical analysis support.

Conflicts of Interest: The authors declare no conflict of interest.

References

1. Singh, R.; Sharma, R.R.; Jain, R.K. Planting time and mulching influenced vegetative and reproductive traits in strawberry (*Fragaria* × *ananassa* Duch.) in India. *Fruits* **2005**, *60*, 395–403. [CrossRef]
2. Liston, A.; Cronn, R.; Ashman, T.L. *Fragaria*: A genus with deep historical roots and ripe for evolutionary and ecological insights. *Am. J. Bot.* **2014**, *101*, 1686–1699. [CrossRef]
3. Bilck, A.P.; Grossmann, M.V.E.; Yamashita, F. Biodegradable mulch films for strawberry production. *Polym. Test.* **2010**, *29*, 471–476. [CrossRef]
4. Whitaker, V.M.; Knapp, S.J.; Hardigan, M.A.; Edger, P.P.; Slovin, J.P.; Bassil, N.V.; Hytönen, T.; Mackenzie, K.K.; Lee, S.; Jung, S.; et al. A roadmap for research in octoploid strawberry. *Hortic. Res.* **2020**, *7*, 33. [CrossRef] [PubMed]
5. FAO. Available online: http://www.fao.org/faostat/en/#data/QC (accessed on 15 June 2020).
6. Cianciosi, D.; Jesús Simal-Gándara, J.; Forbes-Hernández, T.J. The importance of berries in the human diet. *Mediterr. J. Nutr. Metab.* **2019**. [CrossRef]
7. Yeung, A.W.K.; Tzvetkov, N.T.; Zengin, G.; Wang, D.; Xu, S.; Mitrovic, G.; Brncic, M.; Dall'Acqua, S.; Pirgozliev, V.; Kijjoa, A.; et al. The berries on the top. *J. Berry Res.* **2019**. [CrossRef]
8. D'Urso, G.; Piacente, S.; Pizza, C.; Montoro, P. Metabolomics of healthy berry fruits. *Curr. Med. Chem.* **2018**, *25*, 4888–4902. [CrossRef]
9. Minutti-López Sierra, P.; Gallardo-Velázquez, T.; Osorio-Revilla, G.; Meza-Márquez, G. Chemical composition and antioxidant capacity in strawberry cultivars (*Fragaria* × *ananassa* Duch.) by FT-MIR spectroscopy and chemometrics. *Cyta J. Food* **2019**, *17*, 724–732. [CrossRef]

10. Giampieri, F.; Gasparrini, M.; Forbes-Hernandez, T.J.; Mazzoni, L.; Capocasa, F.; Sabbadini, S.; Alvarez-Suarez, J.M.; Afrin, S.; Rosati, C.; Pandolfini, T.; et al. Overexpression of the anthocyanidin synthase gene in strawberry enhances antioxidant capacity and cytotoxic effects on human hepatic cancer cells. *J. Agric. Food Chem.* **2018**, *66*, 581–592. [CrossRef]

11. Mazzoni, L.; Perez-Lopez, P.; Giampieri, F.; Alvarez-Suarez, J.M.; Gasparrini, M.; Forbes-Hernandez, T.J.; Quiles, J.L.; Mezzetti, B.; Battino, M. The genetic aspects of berries: From field to health. *J. Sci. Food Agric.* **2016**, *96*, 365–371. [CrossRef]

12. Basu, A. Role of berry bioactive compounds on lipids and lipoproteins in diabetes and metabolic syndrome. *Nutrients* **2019**, *11*, 1983. [CrossRef] [PubMed]

13. Parelman, M.A.; Storms, D.H.; Kirschke, C.P.; Huang, L.; Zunino, S.J. Dietary strawberry powder reduces blood glucose concentrations in obese and lean C57BL/6 mice, and selectively lowers plasma C-reactive protein in lean mice. *Br. J. Nutr.* **2012**, *108*, 1789–1799. [CrossRef] [PubMed]

14. Pinto, M.S.; de Carvalho, J.E.; Lajolo, F.M.; Genovese, M.I.; Shetty, K. Evaluation of antiproliferative, anti-type 2 diabetes, and antihypertension potentials of ellagitannins from strawberries (*Fragaria* × *ananassa* Duch.) using in vitro models. *J. Med. Food* **2010**, *13*, 1027–1035. [CrossRef] [PubMed]

15. Somasagara, R.R.; Hegde, M.; Chiruvella, K.K.; Musini, A.; Choudhary, B.; Raghavan, S.C. Extracts of strawberry fruits induce intrinsic pathway of apoptosis in breast cancer cells and inhibits tumor progression in mice. *PLoS ONE* **2012**, *7*, e47021. [CrossRef] [PubMed]

16. Calvano, A.; Izuora, K.; Oh, E.C.; Ebersole, J.L.; Lyonse, T.J.; Basu, A. Dietary berries, insulin resistance and type 2 diabetes: An overview of human feeding trials. *Food Funct.* **2019**, *10*, 6227–6243. [CrossRef] [PubMed]

17. Davis, D.W.; Navalta, J.W.; McGinnis, G.R.; Serafica, R.; Izuora, K.; Basu, A. Effects of acute dietary polyphenols and post-meal physical activity on postprandial metabolism in adults with features of the metabolic syndrome. *Nutrients* **2020**, *12*, 1120. [CrossRef]

18. Khalifa, H.O.; Kamimoto, M.; Shimamoto, T.; Shimamoto, T. Antimicrobial effects of blueberry, raspeberry, and strawberry aqueous extracts and their effects on virulence gene expression in *Vibrio cholera*. *Phytother. Res.* **2015**, *29*, 1791–1797. [CrossRef]

19. Forbes-Hernandez, T.J.; Giampieri, F.; Gasparrini, M.; Afrin, S.; Mazzoni, L.; Cordero, M.D.; Mezzetti, M.; Quiles, J.L.; Battino, M. Lipid accumulation in HepG2 cells is attenuated by strawberry extract through AMPK activation. *Nutrients* **2017**, *9*, 621. [CrossRef]

20. Giampieri, F.; Alvarez-Suarez, J.M.; Cordero, M.D.; Gasparrini, M.; Forbes-Hernandez, T.J.; Afrin, S.; Santos-Buelga, C.; González-Paramás, A.M.; Astolfi, P.; Corrado Rubini, C.; et al. Strawberry consumption improves aging-associated impairments, mitochondrial biogenesis and functionality through the AMP-activated protein kinase signaling cascade. *Food Chem.* **2017**, *234*, 464–471. [CrossRef]

21. Basu, A.; Kurien, B.T.; Tran, H.; Byrd, B.A.; Maher, J.; Schell, J.; Masek, E.; Barrett, J.R.; Lyons, T.J.; Bettse, N.M.; et al. Strawberries decrease circulating levels of tumor necrosis factor and lipid peroxides in obese adults with knee osteoarthritis. *Food Funct.* **2018**, *9*, 6218–6226. [CrossRef]

22. Laugale, V.; Strautina, S.; Krasnova, I.; Seglina, D.; Kampuss, K. The influence of cultivation system on biochemical content of strawberry fruits. *J. Hortic. Res.* **2014**, *22*, 85–92. [CrossRef]

23. Diel, M.D.; Pinheiro, M.V.M.; Thiesen, L.A.; Altíssimo, B.S.; Holz, E.; Schmidt, D. Cultivation of strawberry in substrate: Productivity and fruit quality are affected by the cultivar origin and substrates. *Ciência Agrotecnol.* **2018**, *42*, 229–239. [CrossRef]

24. Kumar, S.; Dey, P. Effects of different mulches and irrigation methods on root growth, nutrient uptake, water-use efficiency and yield of strawberry. *Sci. Hortic.* **2011**, *127*, 318–324. [CrossRef]

25. Adhikaria, R.; Bristowb, K.L.; Caseya, P.S.; Freischmidta, G.; John, W.; Hornbucklec, J.W.; Adhikarie, B. Preformed and sprayable polymeric mulch film to improve agricultural water use efficiency. *Agric. Water Manage.* **2016**, *169*, 1–13. [CrossRef]

26. De Vetter, L.W.; Zhang, H.; Ghimire, S.; Watkinson, S.; Miles, C.A. Plastic biodegradable mulches reduce weeds and promote crop growth in day-neutral strawberry in western Washington. *Hortscience* **2017**, *52*, 1700–1706. [CrossRef]

27. Scaringelli, M.A.; Giannoccaro, G.; Prosperi, M.; Lopolito, A. Adoption of biodegradable mulching films in agriculture: Is there a negative prejudice towards materials derived from organic wastes? *Ital. J. Agron.* **2016**, *11*, 92–99. [CrossRef]

28. Scarascia-Mugnozza, G.; Schettini, E.; Vox, G.; Malinconico, M.; Immirzi, B.; Pagliara, S. Mechanical properties decay and morphological behaviour of biodegradable films for agricultural mulching in real scale experiment. *Polym. Degrad. Stab.* **2006**, *91*, 2801–2808. [CrossRef]

29. Costa, R.; Saraiva, A.; Carvalho, L.; Duarte, E. The use of biodegradable mulch films on strawberry crop in Portugal. *Sci. Hortic.* **2014**, *173*, 65–70. [CrossRef]

30. Ao, L.; Qin, L.; Kang, H.; Zhou, Z.; Su, H. Preparation, properties and field application of biodegradable and phosphorus-release films based on fermentation residue. *Int. Biodeterior. Biodegrad.* **2013**, *82*, 134–140. [CrossRef]

31. Cozzolino, E.; Giordano, M.; Fiorentino, N.; El-Nakhel, C.; Pannico, A.; Mola, I.D.; Mori, M.; Kyriacou, M.C.; Colla, G.; Rouphael, Y. Appraisal of biodegradable mulching films and vegetal-derived biostimulant application as eco-sustainable practices for enhancing lettuce crop performance and nutritive value. *Agronomy* **2020**, *10*, 427. [CrossRef]

32. Morra, L.; Bilotto, M.; Cerrato, D.; Coppola, R.; Leone, V.; Mignoli, E.; Pasquariello, M.S.; Petriccione, M.; Cozzolino, E. The Mater-Bi ® biodegradable film for strawberry (*Fragaria × ananassa* Duch.) mulching: Effects on fruit yield and quality. *Ital. J. Agron.* **2016**, *11*, 203. [CrossRef]

33. Sivan, A. New perspectives in plastic biodegradation. *Curr. Opin. Biotechnol.* **2011**, *22*, 422–426. [CrossRef] [PubMed]

34. Scott, G. Photo-biodegradation of plastics. A systems approach to plastic waste and litter. *Degrad. Mater.* **2018**, 143–178. [CrossRef]

35. Nestby, R.; Guéry, S. Balanced fertigation and improved sustainability of June bearing strawberry cultivated three years in open polytunnel. *J. Berry Res.* **2017**, *7*, 203–216. [CrossRef]

36. Comite' Europe'en de Normalisation (C.E.N.). *EN-13655 Plastics: Mulching Thermoplastic Films for Use in Agriculture and Horticulture*; Comite' Europe'en de Normalisation (C.E.N.): Brussels, Belgium, 2002.

37. Andrade, S.C.; Palha, M.d.G.; Duarte, E. Biodegradable mulch films performance for autumn-winter strawberry production. *J. Berry Res.* **2014**, *4*, 193–202. [CrossRef]

38. Singleton, V.L.; Orthofer, R.; Lamuela-Raventós, R.M. Analysis of Total phenols and other oxidation substrates and antioxidants by means of Folin-Ciocalteu reagent. In *Methods in Enzymology*; Academic Press: Cambridge, MA, USA, 1999; Volume 299, pp. 152–178. [CrossRef]

39. Fogliano, V.; Verde, V.; Randazzo, G.; Ritieni, A. Method for measuring antioxidant activity and its application to monitoring the antioxidant capacity of wines. *J. Agric. Food Chem.* **1999**, *47*, 1035–1040. [CrossRef]

40. Pellegrini, N.; Re, R.; Yang, M.; Rice-Evans, C. Screening of dietary carotenoids and carotenoid-rich fruit extracts for antioxidant activities applying 2,20-azinobis(3-ethylenebenzothiazoline-6-sulfonic acid radical cation decolorization assay. *Methods Enzymol.* **1999**, *299*, 379–384. [CrossRef]

41. Rouphael, Y.; De Pascale, S.; Colla, G.; Ritieni, A.; Cardarelli, M. Phenolic compositions, antioxidant activity and mineral profile in two seed-propagated artichoke cultivars as affected by microbial inoculants and planting time. *Food Chem.* **2017**, *234*, 10–19. [CrossRef]

42. Bremner, J.M. Total nitrogen. In *Methods of Soil Analysis*; Black, C.A., Evans, D.D., White, D.D., Ensminger, E., Clark, F.E., Eds.; American Society of Agronomy: Madison, WI, USA, 1965; pp. 149–1178.

43. Association of Official Analytical Chemists. *AOAC Official Method of Analysis*, 17th ed.; Association of Official Analytical Chemists: Washington, DC, USA, 2000.

44. Deb, P.D.; Sangma, D.K.; Prasad, B.V.G.; Bhowmick, N.; Dey, K. Effect of different mulches on vegetative growth of strawberry (cv. Tioga) under red and lateritic zone of west Bengal. *Int. J. Basic Appl. Biol.* **2014**, *2*, 77–80.

45. Sharma, R.R.; Sharma, V.P. Mulch influences plant growth, albinism and fruit quality in strawberry (*Fragaria × ananassa* Duch). *Fruits* **2003**, *58*, 221–227. [CrossRef]

46. Badiyala, S.D.; Aggarwal, G.C. Note on effect of mulches on strawberry production. *Indian J. Agric. Res.* **1981**, *51*, 832–834.

47. Filippi, F.; Magnani, G.; Guerrini, S.; Ranghino, F. Agronomic evaluation of green biodegradable mulch on melon crop. *Ital. J. Agron.* **2011**, *6*, e18. [CrossRef]

48. Gupta, R.; Acharya, C.L. Effect of mulch induced hydrothermal regime on root growth, water-use efficiency, yield and quality of strawberry. *J. Indian Soc. Soil Sci.* **1993**, *41*, 17–25.

49. Tarara, J.M. Microclimate modification with plastic mulch. *HortScience* **2000**, *35*, 169–180. [CrossRef]

50. Cowan, J.S.; Miles, C.A.; Andrews, P.K.; Inglis, D.A. Biodegradablemulch performed comparably to polyethylene in high tunnel tomato (*Solanum lycopersicum* L.) production. *Sci. Food Agric.* **2014**, *94*, 1854–1864. [CrossRef] [PubMed]
51. Gramza-Michałowska, A.; Bueschke, M.; Kulczyński, B.; Gliszczyńska-Świgło, A.; Kmiecik, D.; Bilska, A.; Purłan, M.; Wałęsa, L.; Ostrowski, M.; Filipczuk, M.; et al. Phenolic compounds and multivariate analysis of antiradical properties of red fruits. *J. Food Meas. Charact.* **2019**, *13*, 1739–1747. [CrossRef]

 *horticulturae*

Article

Plant Extract Treatments Induce Resistance to Bacterial Spot by Tomato Plants for a Sustainable System

Kamal A. M. Abo-Elyousr [1,2,*], Najeeb M. Almasoudi [1], Ahmed W. M. Abdelmagid [2,3], Sergio R. Roberto [4,*] and Khamis Youssef [5,*]

[1] Department of Arid Land Agriculture, King Abdulaziz University, Jeddah 80208, Saudi Arabia; nalmasoudi@kau.edu.sa
[2] Department of Plant Pathology, University of Assiut, Faculty of Agriculture, Assiut 71526, Egypt; Ahmed.Abdelmagid@umanitoba.ca
[3] Plant Science Department, Manitoba University, Winnipeg, MB R3T 2N2, Canada
[4] Department of Agronomy, Agricultural Research Center, Londrina State University, Londrina, PR 86057-970, Brazil
[5] Agricultural Research Center, Plant Pathology Research Institute, 9 Gamaa St., Giza 12619, Egypt
* Correspondence: kaaboelyousr@agr.au.edu.eg (K.A.M.A.-E.); sroberto@uel.br (S.R.R.); youssefeladawy@yahoo.com or youssefeladawy@arc.sci.eg (K.Y.)

Received: 27 April 2020; Accepted: 19 June 2020; Published: 22 June 2020

Abstract: The aim of this study is to assess the effect of extracts of *Nerium oleander*, *Eucalyptus chamadulonsis* and *Citrullus colocynthis* against bacterial spot disease of tomato and to investigate the induction of resistance by tomato (*Solanum lycopersicum*) in order to promote a sustainable management system. The antibacterial activity of aqueous and ethanol plant extracts was tested against *Xanthomonas axonopodis* pv. *vesicatoria*, isolate PHYXV3, in vitro and in vivo. The highest antibacterial activity in vitro was obtained with *C. colocynthis*, *N. oleander* and *E. chamadulonsis*, respectively. In vivo, ethanol extracts of *N. oleander* and *E. chamadulonsis* were more effective than aqueous extracts in reducing pathogen populations on tomato leaves. Under greenhouse conditions, application of the plant extracts at 15% (v/v) to tomato plants significantly reduced disease severity and increased the shoot weight of 'Super Marmande' tomato. In most cases, plant extracts significantly increased total phenol and salicylic acid content of tomato plants compared to either healthy or infected ones. In addition, *C. colocynthis* and *E. chamadulonsis* extracts significantly increased peroxidase activity while only *E. chamadulonsis* increased polyphenol oxidase after infection with the causal agent. The results indicated that the plant extracts showed promising antibacterial activity and could be considered an effective tool in integrated management programs for a sustainable system of tomato bacterial spot control.

Keywords: Induced resistance; polyphenol oxidase; peroxidase; plant extract; bacterial spot

---

1. Introduction

Worldwide tomato (*Solanum lycopersicum*) production is 182,256,458 MT with a harvested area around 4762,457 ha [1]. Tomato crops are facing many bacterial diseases including bacterial spot caused by *Xanthomonas axonopodis* pv. *vesicatoria* [2,3]. The application of copper bactericides, alone or in combination with other pesticides and antibiotics (especially streptomycin), is the primary tool to control such bacterial diseases [4]. However, several limitations have restricted their use including bactericide resistance, market demands related to residues and worries of human health [5]. Thus, safe alternative control methods need to be tested and developed [6–8]. Among the alternative means, many studies have proven the antimicrobial activity of various plant species against bacterial spot diseases [9,10]. In addition, several studies have screened extracts of various higher plant leaves, fruit and seeds for controlling phytopathogenic bacteria [11,12]. Basically, higher plants are considered to be one of the major sources of bioactive, economically significant natural compounds, pesticides and pharmaceuticals [13].

Numerous studies were mentioned that plant extracts such as *Cesalpinia coriaria* and essential oils have shown satisfactory antifungal and antibacterial properties in vivo [14,15]. Recently, many higher plant products have attracted the attention of researchers seeking phytochemicals to evaluate for their antimicrobial properties [16–18].

Leaf extracts of several species, e.g., *Eucalyptus globules*, *Datura stramonium*, *Ocimum* spp., *Salix* spp., *Rosmarinus officinalis*, *Cydonia oblonga* and *Foeniculum vulgare*, have been used successfully to control diverse plant diseases [12,19,20]. Application of some chemicals and plant extracts have induced resistance in plants against many causal pathogens including bacteria, fungi, viruses and nematodes [21,22]. In addition, plant extracts have been found to induce a defense response in infected plants [23]. Geetha and Shetty [24] mentioned that the mode of action of plant extracts against bacterial pathogens may enhance natural host defense mechanisms by increasing the activity of some antioxidant enzymes, e.g., peroxidases, polyphenol oxidase or the accumulation of phenolic compounds [25]. This may directly affect the survival of the pathogens or act indirectly on plant metabolism [26]. In addition, polyphenol oxidases (PPOs) catalyze the oxidation of several phenols to o-quinones [27].

The present study aimed to assess the effectiveness of extracts of three plant species against *X. axonopodis* pv. *vesicatoria* in vitro and the suppression of bacterial spot disease of tomato plants under greenhouse conditions. Furthermore, the potential change of total phenols, salicylic acid content, peroxidase and polyphenoloxidase activity in response to application of the extracts was investigated.

2. Materials and Methods

2.1. Seeds, Growth of Seedlings and Bacterial Isolates

'Super Marmande' tomato seeds used in this study were kindly provided by the Vegetable Dept. Faculty of Agriculture, Assiut University, Egypt. Pots of 30 cm diameter and containing sand (3 kg/pot) were used to grow two seedlings. The pots were kept on a greenhouse bench at 30 ± 5 °C and 68%–80% RH, irrigated as required and fertilized with 30 mL of 0.01g of NPK formulation (12:4:6). *Xanthomonas campestris* pv. *vesicatoria* (Doidge) Dye isolate PHYXV3 was provided from the collection of the Department of Plant Pathology, Faculty of Agriculture, Assiut University, Egypt. This isolate was identified, and its pathogenicity was confirmed according to Abo-Elyousr and El-Hendawy [3].

2.2. Preparation of Leaf Extracts

Methods described by Abo-Elyousr and Asran [12] were followed to prepare the aqueous and 70% ethanol extracts. Briefly, 10 g of fresh leaves of *Nerium oleander*, *Eucalyptus chamadulonsis* and *Citrullus colocynthis* were collected from a private farm belonging to Assiut Governorate and washed several times with distilled water. They were crushed in 100 mL of sterile water or ethanol (1:10 w/v) using a mortar and pestle and filtered through double-layered cheesecloth. The ethanolic filtrate was concentrated by exposition to 60 °C in a water bath for 30 min to evaporate ethanol and kept in dark glass bottles. The final extracts were diluted with distilled water to reach a 10% and 15% v/v concentration. These concentrations were sterilized by 0.2 m disposable syringe filters and kept in bottles in the dark until used [28].

2.3. Effect of Plant Extracts on Pathogen Growth In Vitro

To test the effect of three plant extracts on the growth of the bacterial pathogen, the paper disc diffusion method was used [29]. Sterilized filter paper discs (5 mm) were placed in Petri plates containing nutrient sucrose agar (NSA) medium and inoculated with 100 µL of the bacterial suspension of PHYXV3 (5×10^6 CFU mL^{-1}). Sterilized filter paper discs were moistened with 5 µL of each concentration (10% and 15%) of plant extract. As check treatments, water and ethanol served as controls. After 48 h of incubation at 27 °C, the inhibition zones around the paper disc were measured. Four plates were used for each treatment, as replicates, and the whole trial was repeated twice.

2.4. Effect of Plant Extracts on Disease Severity and Pathogen Population

2.4.1. Preparation of Inoculum and Inoculation Methods

Inoculum of PHYXV3 was prepared from cultures shaken for 48 h and incubated at 28 ± 2 °C for 48 h. The cultures were centrifuged at 6000× g at room temperature for 20 min and were then suspended in sterile distilled water. The bacterial concentration was photometrically adjusted using a spectrophotometer (Spectronic® 20 Genesys, Schutt Labortechnik, Cambridge CB5 8HY, UK) to 5×10^6 CFU mL^{-1}.

Four-week-old tomato seedlings were inoculated with PHYXV3 by spraying each plant with 20 mL of the pathogen solution using small hand sprayer. This volume was enough to cover plant leaves with the pathogen solution. Inoculated plants were covered by polyethylene bags for 48 h at 25–27 °C in the greenhouse to maintain the humidity and to prevent inoculation desiccation and inoculum establishment on the plant. Two days after inoculation, tomato seedlings were treated with a 15% v/v concentration of each plant extract and were kept under greenhouse conditions. Fourteen days after inoculation, disease severity was recorded based on the scale of Abbasi et al. [30]. Four replicates (two seedlings each) were used per each treatment and greenhouse experiments were repeated twice.

2.4.2. Determination of Pathogen Population on Tomato Leaves

The number of colonies of PHYXV3 (5×10^6 CFU mL^{-1}) was estimated one week after the time the tomato plants were treated with the plant extracts. About 5 mm of tomato leaf was excised and homogenized in 1 mL of sterile distilled water. Several dilutions were made and then 0.1 mL of each dilution was spread on plates containing nutrient sucrose agar (NSA) medium. Petri dishes were incubated for 72 h at 26 °C, and then recovered colonies were counted. Four replicates were used from each treatment and the results were expressed as CFU/g [3].

2.4.3. Determination of Fresh and Dry Weight

At the end of experiment (90 days), the seedlings from each treatment were removed, washed with distilled water to eliminate the soil, blotted with tissue paper, and then dried at 60 °C for 72 h when shoot dry weight was recorded.

2.5. Determination of Total Phenols and Salicylic Acid Contents

2.5.1. Preparation of Samples

Eight leaves were collected from eight seedlings per treatment. One gram of tomato leaves was ground in liquid nitrogen and homogenized in 10 mL of 80% methanol, and the homogenate was centrifuged at 10,000× g for 30 min at 4 °C. The pellet was discarded after addition of ascorbic acid (0.02 g·mL^{-1}). A rotary evaporator was used to evaporate the supernatant at 65 °C and the process was repeated three times, each for 5 min. The residues were dissolved in 5 mL of 80% methanol. For each treatment four replicates were used [31].

2.5.2. Total Phenol Content

The methods described by Sahin et al. [32] were followed to determine the total phenol content in leaves. Total phenol content was determined by spectrophotometer (Spectronic® 20 Genesys, Schutt Labortechnik, Cambridge CB5 8HY, UK) at 767 nm as mg·g^{-1} plant fresh weight using gallic acid as standard. Total phenol content was expressed as mg of gallic acid per g plant material.

2.5.3. Salicylic Acid Content

Salicylic acid (SA) content was estimated using a method modified from Datet al [33]. A 500 µL sample of leaf homogenate was mixed with 250 µL of 10-N HCl and 1 mL of methanol. Samples were incubated in a water bath at 80 °C for 2 h. One mL of methanol was added to the mixture and each sample was neutralized with 4–5 drops of 1-M NaHCO$_3$. The optical density (OD) was measured at 254 nm, and SA content was calculated and expressed as µg of SA per g plant material.

2.6. Enzymatic Activities

To determine peroxidase (PO) and polyphenol oxidase (PPO) activities, 1 g of fresh tomato leaf tissue was homogenized in liquid nitrogen with 10 mL of 0.1-M sodium acetate buffer (pH 5.2) and the mixture was centrifuged at 1000× g for 30 min at 4°C. For each treatment, four replicates were used. Using Bradford reagent, the total protein content of the supernatant was determined [34].

2.6.1. Peroxidase Activity (PO)

Activity of peroxidase was determined using guaiacol as a substrate according to the method described by Putter [35]. The reaction mixture was composed of 0.2 mL supernatant, 1 mL 0.1-M sodium acetate buffer (pH 5.2), 0.2 mL 1% w/v guaiacol and 0.2 mL of 1% H$_2$O$_2$. The mixture was incubated for 5 min at 25 °C then measured at 436 nm by spectrophotometer (Spectronic® 20 Genesys, Schutt Labortechnik, Cambridge CB5 8HY, UK). Peroxidase activity was calculated as the change in absorbance units (Au) and expressed as change in Au per mg protein. Extraction buffer served as a blank reference.

2.6.2. Polyphenol Oxidase (PPO) Activity

Polyphenol oxidase activity was determined according to the method described by Batra and Kuhn [36]. The reaction mixture was 0.5 mL of supernatant, 2 mL 50-mM Sorensen phosphate buffer (pH 6.5) and 0.5 mL of the substrate pyrocatechol (10%, Sigma Aldrich, Missouri, USA). The reaction mixture was incubated in water bath for 2 h at 37 °C and measured at 410 nm. The activity of PPO was calculated as OD at 410 nm and expressed as OD·mg protein^{-1}.

2.7. Statistical Analysis

All experiments were repeated twice, and percentage data were arcsine transformed before analyses to normalize variance. Data were subjected to one-way analysis of variance (ANOVA) using Statistica Software Ver. 6.0 (Stat Soft, Inc., Tulsa, OK, USA). Fisher's protected least significant difference was used at $p \leq 0.05$ to distinguish the differences among various treatments [37]. When applicable, error bars are shown.

3. Results and Discussion

3.1. Effect of Plant Extracts on Pathogen Growth In Vitro

The plant extracts had variable effect on the growth of the pathogen in vitro (Figure 1). The greatest reduction was obtained with both extracts of *C. colocynthis* at 15% (ethanol (0.81 mm) and water (0.65)) followed by the ethanol extract of *N. oleander* (ethanol (0.59 mm) and water (0.50)), while the least reduction was obtained from *E. chamadulonsis* (ethanol (0.44 mm) and water (0.51)). Overall, the inhibition of pathogen growth increased as the concentration increased from 10% to 15%. Thus, 15% was used for further experiments. The results herein agree with those reported by Abo-Elyousr and Asran [12], in which garlic (*Allium sativum*) extract had a strong antibacterial activity against bacterial wilt in vitro followed by *Datura* spp and then *N. oleander*. Our previous study reported that gas chromatography mass spectrometry (GC-MS) analysis of the aqueous extract of *C. colocynthis* contained 37 compounds and their derivatives, including imidazole. In addition, Rahman and

Gray [38] mentioned that dimericcarbazole was the most effective compound against *Escherichia coli* and *Proteus vulgaris*, both Gram-negative bacteria.

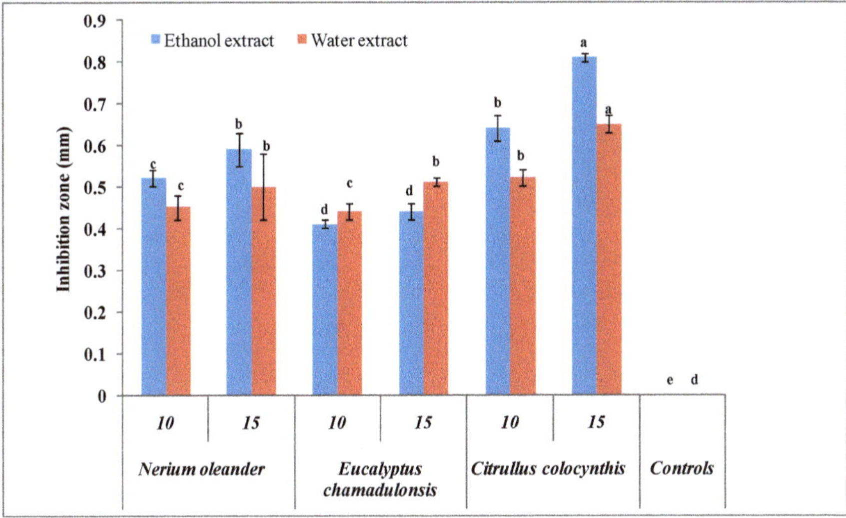

Figure 1. Antibacterial activity of 10% and 15% concentrations of plant extracts of *Nerium oleander*, *Eucalyptus chamadulonsis* and *Citrullus colocynthis* against *Xanthomonas axonopodis* pv. *vesicatoria* PHYXV3 in vitro after 48 h of incubation at 27 °C. Four replicate plates were used for each treatment and the experiment was repeated twice. Columns with the same letters are not significantly different according to Fisher's protected least significant difference at $p \leq 0.05$.

3.2. Determination of Pathogen Population on the Tomato Leaves

Application of the plant extracts at the 15% concentration reduced the number of bacteria in tomato leaves compared to the infected control (Table 1). The greatest reduction was achieved by *C. colocynthis* [ethanol (3.0 CFU/g) and water (4 CFU/g)] followed by ethanol extract of *N. oleander* (5.0 CFU/g) and *E. chamadulonsis* (5.1 CFU/g), with no difference between the latter two. The lowest reduction was with water extracts of *E. chamadulonsis* (7.7 CFU/g) and *Nerium oleander* (6.2 CFU/g), respectively, which did not differ. The results assumed that the application of the plant extracts reduced the number of the bacterial pathogens by toxicity within the cytoplasm to the pathogen. These results are in agreement with Draz et al. [39] who revealed the bioactivity of certain components of plant extracts.

Table 1. Effect of 15% concentrations of water and ethanol plant extracts of *Nerium oleander*, *Eucalyptus chamadulonsis* and *Citrullus colocynthis* on the population of *Xanthomonas axonopodis* pv. *vesicatoria* PHYXV3 in 'Super Marmande' tomato leaves after seven days of inoculation.

Plant Extracts	Method of Extract	CFU/g [z]
Nerium oleander	Water extract	6.2 ± 0.15 b [y]
	Ethanol extract	5.0 ± 0.72 c
Eucalyptus chamadulonsis	Water extract	7.7 ± 0.38 b
	Ethanol extract	5.1 ± 0.23 c
Citrullus colocynthis	Water extract	4.0 ± 0.15 d
	Ethanol extract	3.0 ± 0.45 d
Controls	Infected	9.7 ± 0.53 a
	Healthy	0 e

[z] Bacterial population was recorded one week after inoculation with the isolate PHYXV3 and is expressed as log 10^8 colony forming units (CFU) g^{-1} leaf tissue. Four replicates were used from each treatment. [y] Values in the column followed by the same letters are not significantly different according to Fisher's protected least significant difference at $p \leq 0.05$ level.

3.3. Effect of Plant Extracts on Disease Severity and Dry Weight of Shoots

Treatment with extracts of *C. colocynthis* gave the greatest reduction of disease severity, 20.0% and 21.2% for ethanol and water extracts, respectively, with no significant difference between them, followed by *N. oleander* and *E. chamadulonsis*, respectively (Table 2). Furthermore, *C. colocynthis* gave the greatest shoot dry weight (44.7 and 43.0 for ethanol and water, respectively), while the other treatments also increased shoot weight compared to the infected control, but without statistical difference between them. These results agree with Hassan et al. [40] who mentioned that the application of certain plant extracts reduced wilt diseases of potato plants and increased the yield s of tubers compared to an infected control. This may be due to the lower numbers of the pathogen in treated plants and the induction of some antioxidant enzymes that can reduce pathogens in the tissues [29]. Similarly, Drazet et al. [39] used five plant extracts (pomegranate (*Punica granatum*), acalypha (*Acalypha wilkesiana*), henna (*Lawsonia inermis*), lantana (*Lantana camara*) and chinaberry (*Melia azedarach*) to induce resistance in wheat (*Triticum aestivum*) against rust disease and all plant extracts reduced disease severity and increased yield components.

Table 2. Disease severity (%) and shoot dry weight of 'Super Marmande' tomato treated with extracts of *Nerium oleander*, *Eucalyptus chamadulonsis* and *Citrullus colocynthis* after inoculation with *Xanthomonas axonopodis* pv. *vesicatoria* PHYXV3.

Plant Extracts	Method of Extract	Disease Severity (%)	Shoot Weight (gm)
Nerium oleander	Water extract	29.13 ± 0.10 [z] c [y]	25.2 ± 0.30 c
	Ethanol extract	30.00 ± 1.51 c	25.5 ± 0.38 c
Eucalyptus chamadulonsis	Water extract	39.90 ± 0.68 b	26.8 ± 0.60 c
	Ethanol extract	33.13 ± 0.10 cd	24.2 ± 0.30 c
Citrullus colocynthis	Water extract	21.23 ± 0.17 e	43.0 ± 1.51 ab
	Ethanol extract	20.00 ± 0.76 e	44.7 ± 0.45 ab
Controls	Infected	45.23 ± 0.98 a	13.2 ± 0.30 d
	Healthy	0 f	45.2 ± 0.91 a

[z] Mean ± SE of 4 replicates (two seedlings each) for each treatment. [y] Values in the columns followed by the same letters are not significantly different according to Fisher's protected least significant difference at $p \leq 0.05$ level.

3.4. Effect of Plant Extracts on Total Phenol and Salicylic Acid Contents

Overall, total phenol content was significantly higher in treated versus infected or healthy control plants with the exception of the ethanol extract of *C. colocynthis*. In the present study, treatment with the plant extracts increased the accumulation of phenolic substances in response to pathogen infection (Figure 2). Values of phenol content ranged from 3.13 to 3.57 mg gallic acid g^{-1} for the water extraction method and from 2.81 to 3.74 mg gallic acid g^{-1} for the ethanol extraction method for the three species. Accumulation of phenolic compounds at the infection site was correlated with the suppression of pathogen development since these compounds are toxic to phytopathogenic bacteria. Resistance maybe also improved by increasing the pH of the plant cell cytoplasm due to an increase in phenolic acid substances which prevents the development of the pathogen [41].

Salicylic acid (SA) content significantly increased with application of *C. colocynthis* aqueous (6.4 µg salicylic acid g^{-1}) and ethanol (6.37 µg salicylic acid g^{-1}) extracts after 7 days of treatment, though there was no significance difference between them (Figure 3). The results showed statistically significant increases in SA levels for all treatments versus control. However, it is also clear that ethanolic and aqueous extracts showed similar SA levels for all treatments within species and that *C. colocynthis* extracts induced the largest SA increases among the treatments. De Meyer et al. [42] suggested that SA accumulation is important for expression of several modes of plant disease resistance. In addition, SA mediates plant defense against pathogens, accumulating in both infected and distal leaves in response to pathogen infection [29]. The enhanced SA contents are a prerequisite for expression of systemic acquired resistance against *R. solanacreaum* in potato (*Solanum tuberosum*) plants [43,44], and these results may also indicate that applied plant extracts induce pathogen resistance in plants either via the activation of a signaling pathway that is dependent on SA or via the activation of a novel signaling cascade that is not dependent on SA, jasmonic acid or ethylene signaling.

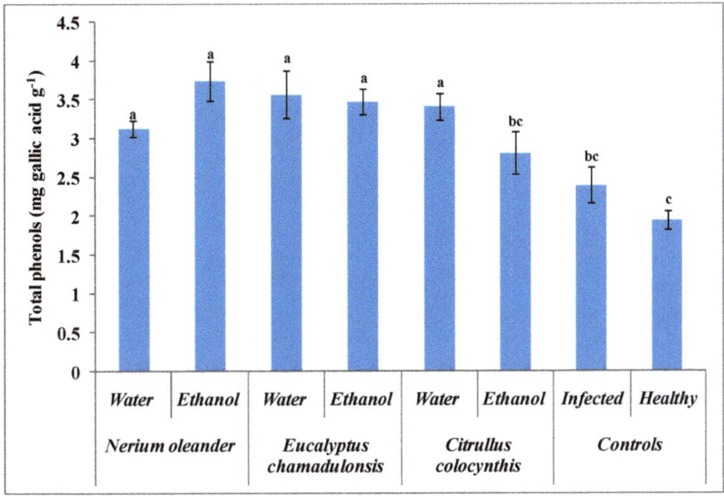

Figure 2. Effect of 15% concentrations of plant extracts of *Nerium oleander*, *Eucalyptus chamadulonsis* and *Citrullus colocynthis* on total phenols of 'Super Marmande' tomato after 7 days of treatment with the extracts and inoculated with *Xanthomonas axonopodis* pv. *vesicatoria* PHYXV3. Four replicates were used for each treatment. Columns with the same letters are not significantly different according to Fisher's protected least significant difference at $p \leq 0.05$.

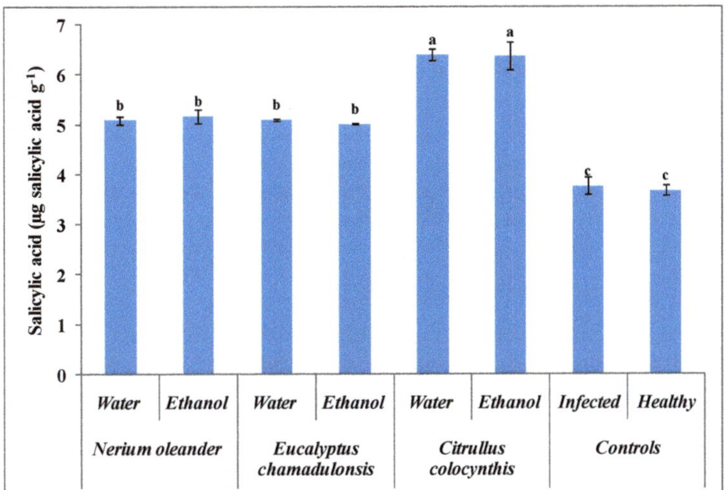

Figure 3. Effect of 15% concentrations of plant extracts of *Nerium oleander*, *Eucalyptus chamadulonsis* and *Citrullus colocynthis* on salicylic content of 'Super Marmande' tomato after 7 days of treatment with the extracts and inoculated with *Xanthomonas axonopodis* pv. *vesicatoria* PHYXV3. Four replicates were used for each treatment. Columns with the same letters are not significantly different according to Fisher's protected least significant difference at $p \leq 0.05$.

3.5. Effect of Plant Extracts on Peroxidase (PO)

Peroxidase (PO) activity significantly increased in infected plants treated with extracts of all species except *Nerium oleander* (Figure 4). *C. colocynthis* extract caused the most increase in PO activity followed by *E. chamadulonsis*. Values for *C. colocynthis* were 6.9 and 6.25 Au per mg protein for water and ethanol extraction, respectively. The results agree with prior studies [45]. The results suggested that the plant extracts promoted an increase in defense– related peroxidase enzyme activity. Several investigators have reported that enhanced peroxidase activity was associated with plant defense against fungal, bacterial and viral pathogens [46].

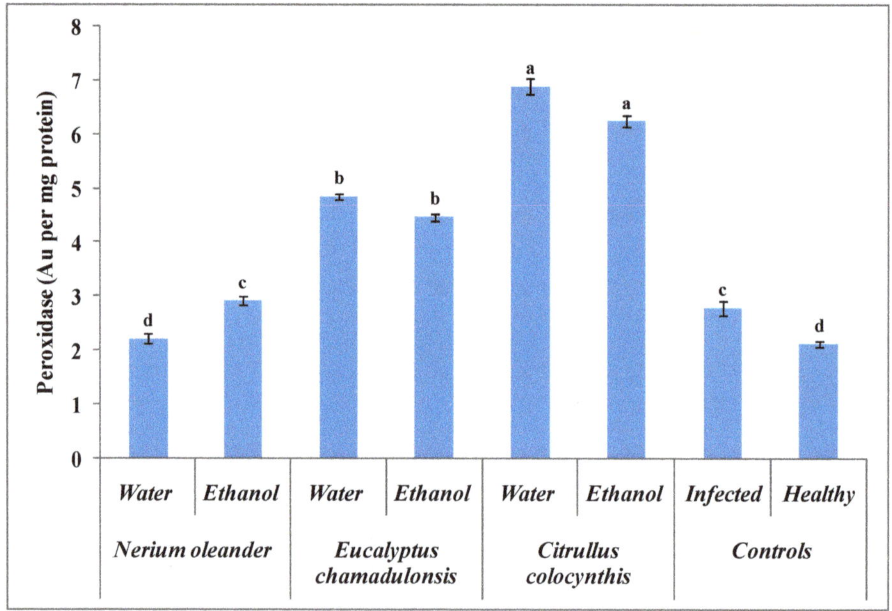

Figure 4. Effect of 15% concentrations of plant extracts of *Nerium oleander*, *Eucalyptus chamadulonsis* and *Citrullus colocynthis* on peroxidase activity as absorbance units (Au) per mg protein of 'Super Marmande' tomato after 7 days of treatment with the extracts and inoculated with *Xanthomonas axonopodis* pv. *vesicatoria* PHYXV3. Four replicates were used for each treatment. Columns with the same letters are not significantly different according to Fisher's protected least significant difference at $p \leq 0.05$.

3.6. Effect of Plant Extract on Polyphenol Oxidase (PPO)

In most cases, PPO activity showed no significant difference between infected tomato plants versus extract-treated plants using *N. oleander* and *C. colocynthis* extracts (Figure 5). The highest PPO activity was obtained following applications of *E. chamadulonsis* extracts 7 days post-application (1.25 and 1.6 nm mg^{-1} protein for water and ethanol, respectively). Such results agree with other investigations [40,46]. The importance of PPO activity in disease resistance probably stems from its property to oxidize phenolic compounds to quinines, which are frequently more toxic to pathogens than the original phenol [46]. In addition, Draz et al. [39] mentioned that biochemical analyses proved a significant increase in the plant contents of total phenolics and oxidative enzymes activities (PPO and POX) after treated with certain plant extracts.

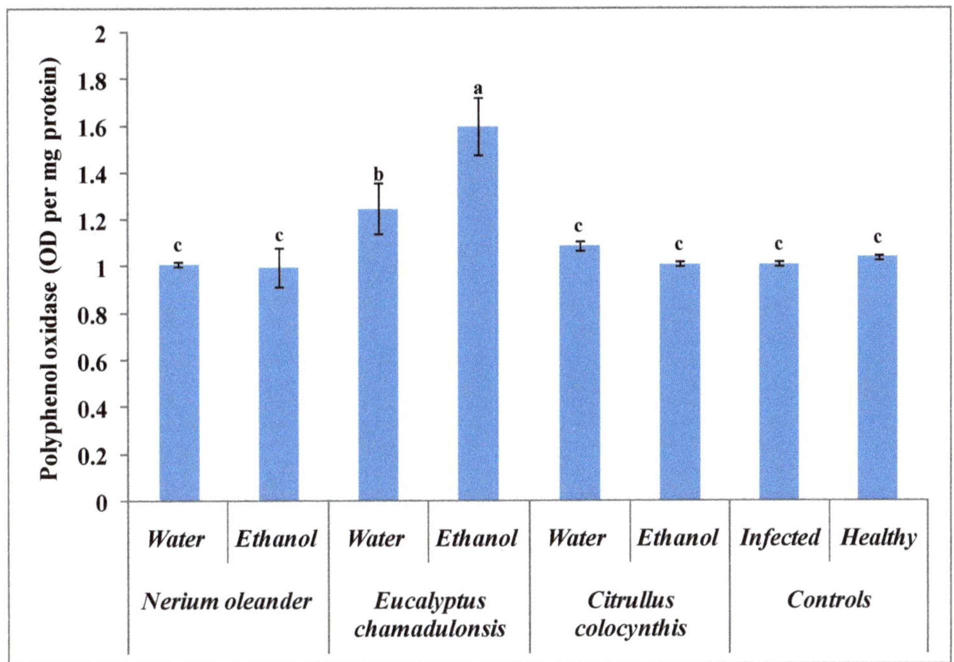

Figure 5. Effect of 15% concentrations of plant extracts of *Nerium oleander*, *Eucalyptus chamadulonsis* and *Citrullus colocynthis* on polyphenol oxidase activity as optical density (OD) per mg protein of 'Super Marmande' tomato after 7 days of treatment with the extracts and inoculated with *Xanthomonas axonopodis* pv. *vesicatoria* PHYXV3. Four replicates were used for each treatment. Columns with the same letters are not significantly different according to Fisher's protected least significant difference at $p \leq 0.05$.

4. Conclusions

The antibacterial activity of aqueous and ethanol extracts of *N. oleander*, *E. chamadulonsis* and *C. colocynthis* plants were tested against *Xanthomonas axonopodis* pv. *vesicatoria* in vitro and inoculated plants in vivo. In in vitro tests, the highest antibacterial activity was achieved by extracts of *C. colocynthis*, followed by *N. oleander* and then *E. chamadulonsis*. Application of the extracts to tomato plants significantly reduced disease severity and increased shoot weight of 'Super Marmande' tomato plants. Overall, plant extracts significantly increased total phenol and salicylic acid content treated plants compared to healthy or infected ones. In addition, *C. colocynthis* and *E. chamadulonsis* significantly increased peroxidase activity while only *E. chamadulonsis* increased polyphenol oxidase. Thus, alternative means for controlling plant disease—such as plant extracts—may be able to replace or reduce the use of bactericides. Leaf extracts of these species showed promising antibacterial activity against the causal agent of tomato bacterial spot and could become a part of an integrated pest management program for controlling the disease.

Author Contributions: K.A.M.A.-E. conceived the research idea. N.M.A. and A.W.M.A. helped to collect the data. S.R.R. and K.Y. analyzed the data and wrote the study. All authors have read and agreed to the published version of the manuscript.

Funding: This research received no external funding.

References

1. FAOSTAT. Available online: http://www.fao.org/faostat/en/#data/QC (accessed on 25 March 2020).
2. El-Hendawy, H.H.; Osman, M.E.; Sorour, N.M. Biological control of bacterial spot of tomato caused by *Xanthomonas campestris* pv. *vesicatoria* by *Rahnella aquatilis*. *Microbiol. Res.* **2005**, *160*, 343–352. [CrossRef] [PubMed]
3. Abo-Elyousr, K.A.M.; El-Hendawy, H.H. Integration of *Pseudomonas fluorescens* and acibenzolar-S-methyl to control bacterial spot disease of tomato. *Crop Prot.* **2008**, *27*, 1118–1124. [CrossRef]
4. Vallad, G.E.; Pernezny, K.L.; Balogh, B.; Wen, A.; Figueiredo, J.F.L.; Jones, J.B.; Momol, T.; Muchovej, R.; Havranek, N.; Abdallah, N.; et al. Comparison of kasugamycin to traditional bactericides for the management of bacterial spot on tomato. *Hortscience* **2010**, *45*, 1834–1840. [CrossRef]
5. Janse, J. Prevention and control of bacterial pathogens and diseases. In *Phytobacteriol Principles Practice*; CABI Publishing: Wallingford, UK, 2005; p. 360.
6. Youssef, K.; Hashim, A.F.; Margarita, R.; Alghuthaymi, M.A.; Abd-Elsalam, K.A. Fungicidal efficacy of chemically-produced copper nanoparticles against *Penicillium digitatum* and *Fusarium solani* on citrus fruit. *Philipp. Agric. Sci.* **2017**, *100*, 69–78.
7. Hussien, A.; Ahmed, Y.; Al-Essawy, A.; Youssef, K. Evaluation of different salt-amended electrolysed water to control postharvest moulds of citrus. *Trop. Plant Pathol.* **2018**, *43*, 10–20. [CrossRef]
8. Youssef, K.; de Oliveira, A.G.; Tischer, C.A.; Hussain, I.; Roberto, S.R. Synergistic effect of a novel chitosan/silica nanocomposites-based formulation against gray mold of table grapes and its possible mode of action. *Int. J. Biol. Macromol.* **2019**, *141*, 247–258. [CrossRef] [PubMed]
9. Kagale, S.T.; Marimuthu, T.; Thaynmanavan, P.; Nandakumar, P.; Samiyappan, R. Antimicrobial activity and induction of systemic resistance in rice by leaf extract of *Datura metel* against *Rhizoctonia solani* and *Xanthomona soryzae* pv. oryzae. *Physiol. Mol. Plant Pathol.* **2004**, *65*, 91–100. [CrossRef]
10. Balestra, G.M.; Heydari, A.D.; Ceccarelli, E.O.; Quattrucci, A. Antibacterial effect of *Allium sativum* and *Ficus carica* extracts on tomato bacterial pathogens. *Crop Prot.* **2009**, *28*, 807–811. [CrossRef]
11. Hassan, M.A.E.; Bereika, M.F.F.; Abo-Elnaga, H.I.G.; Sallam, M.A.A. Management of potato bacterial wilt using plant extracts, essential oils, antagonistic bacteria and resistance chemical inducers. *Assiut J. Agric. Sci.* **2008**, *39*, 141–159.
12. Abo-Elyousr, K.A.M.; Asran, M.R. Antibacterial activity of certain plant extracts against bacterial wilt of tomato. *Arch. Phytopathol. Plant Prot.* **2009**, *42*, 573–578. [CrossRef]
13. Hostettmann, K.; Wolfender, J.L. The search for biologically active secondary metabolites. *Pestic. Sci.* **1997**, *51*, 471–482. [CrossRef]
14. Mohana, D.C.; Raveesha, K.A. Anti-bacterial activity of *Caesalpinia coriaria* (Jacq.) Willd. against plantpathogenic *Xanthomonas pathovars*: An eco-friendly approach. *J. Agric. Technol.* **2006**, *2*, 317–327.
15. Bagy, H.M.M.; Abo-Elyousr, K.A.M. Antibacterial activity of some essential oils on bacterial spot disease of tomato plant caused by *Xanthomonas axonopodis* pv. vesicatoria. *Int. J. Phytopathol.* **2019**, *8*, 53–61. [CrossRef]
16. Joseph, B.M.; Darand, A.; Kumar, V. Bioefficacy of plant extracts to control *Fusarium solani* f.sp. *melangenae* incitant of brinjal wilt. *Glob. J. Biotechnol. Biochem.* **2008**, *3*, 56–59.
17. Khan, R.; Barira, I.; Mohd, A.; Shazi, S.; Anis, A.; Manazir, A.; Mashiatullah, S.; Asad, U.K. Antimicrobial activity of five herbal extracts against multidrug resistant (MDR) strains of bacteria and fungus of clinical origin. *Molecules* **2009**, *14*, 586–597. [CrossRef]
18. Salem, E.A.; Youssef, K.; Sanzani, S.M. Evaluation of alternative means to control postharvest Rhizopus rot of peaches. *Sci. Hortic.* **2016**, *198*, 86–90. [CrossRef]
19. Gurjar, S.M.; Shahid, A.; Masood, A.; Kangabam, S.S. Efficacy of plant extracts in plant disease management. *Agric. Sci.* **2012**, *3*, 425–433. [CrossRef]
20. Zahid, N.Z.; Abbasi, N.A.; Hafiz, A.I.; Hussain, A.; Ahmad, Z. Antifungal activity of localfennel (*Foeniculum vulgare* Mill) extracts to growth responses of some soil diseases. *Afr. J. Microbiol. Res.* **2012**, *6*, 46–51.
21. Kuc, J. Induced immunity to plant diseases. *Bioscience* **1982**, *32*, 854–860.
22. Fallanaj, F.; Sanzani, S.M.; Youssef, K.; Zavanella, C.; Salerno, M.G.; Ippolito, A. A new perspective in controlling postharvest citrus rots: The use of electrolyzed water. *Acta Hortic.* **2015**, *1065*, 1599–1606. [CrossRef]

23. Srivastava, S.; Singh, V.P.; Kumar, R.; Srivastava, M.; Sinha, A.; Simon, S. In vitro evaluation of Carbendazim 50% WP, antagonists and botanicals against *Fusarium oxysporum* f.sp. *psidii* associated with rhizosphere soil of guava. *Asian J. Plant Pathol.* **2011**, *5*, 46–53. [CrossRef]

24. Geetha, H.M.; Shetty, H.S. Induction of resistance in pearl millet against mildew disease caused by *Sclerospora graminicola* using benzothiadiazole, calcium chloride and hydrogen peroxide—A comparative evaluation. *Crop Prot.* **2002**, *21*, 601–610. [CrossRef]

25. Hassan, M.E.M.; Abd-El-Rahman, S.S.; El-Abbasi, I.H.; Mikhail, M.S. Change in peroxidase activity due to resistance induced against faba bean chocolate spot disease. *Egypt. J. Phytopathol.* **2007**, *35*, 35–48.

26. Khan, W.; Prithiviraj, B.; Smith, D.L. Phytosynthetic response of corn and soybean to foliar application of salicylates. *Plant Physiol.* **2003**, *160*, 485–493. [CrossRef]

27. Oliveira, C.M.; Ferreira, A.C.S.; de Freitas, V.; Silva, A.M. Oxidation mechanisms occurring in wines. *Food Res. Int.* **2011**, *44*, 1115–1126. [CrossRef]

28. Abd-Rabboh, M.S.; El Shennawy, M.Z. Effect of some plant extracts on sugar beet powdery mildew. *Egypt. J. Phytopathol.* **2016**, *44*, 49–56.

29. Hassan, M.A.E.; Abo-Elyours, K.A.M. Activation of tomato plant defence responses against bacterial wilt caused by *Ralstonia solanacearum* using DL-3-aminobutyricacid (BABA). *Eur. J. Plant Pathol.* **2013**, *146*, 145–157. [CrossRef]

30. Abbasi, P.A.; Al-Dahmani, J.; Sahin, F.; Hoitink, H.A.J.; Miller, S.A. Effect of compost amendments on disease severity and yield of tomato in conventional and organic production systems. *Plant Dis.* **2002**, *86*, 156–161. [CrossRef]

31. Rapp, A.; Ziegler, A. Bestimmungder Phenolcarbonsaurein Rebblattern Weintraubeund Weinmittels Polamyid-Dunnschicht Chromatographie. *Vitis* **1973**, *12*, 226–236.

32. Sahin, F.; Gulluce, M.; Daferera, D.; Sokmen, A.; Sokmen, M.; Polissiou, M.; Agar, G.; Ozer, H. Biological activities of the essential oils and methanol extract of *Origanum vulgares* sp. *vulgare* in the Eastern Anatolia region of Turkey. *Food Control* **2004**, *15*, 549–557. [CrossRef]

33. Dat, J.F.; Foyer, C.H.; Scott, I.M. Changes in salicylic acid and antioxidants during induced thermo tolerance in mustard seedling. *Plant Physiol.* **1998**, *118*, 1455–1461. [CrossRef] [PubMed]

34. Bradford, M. A rapid and sensitive method for the quantitation of Microgram quantities of protein utilizing the principle of protein dye binding. *Anal. Biochem.* **1976**, *72*, 248–250. [CrossRef]

35. Putter, J. Peroxidase. In *Methoden der Enzymatischen Analyses*; Bergmeyer, H.U., Ed.; Verlag Chemie: Weinheim, Germany, 1974; p. 725.

36. Batra, G.K.; Kuhn, C.W. Polyphenoloxidase and peroxidase activities associated with acquired resistance and it inhibition by 2-thiouracilin virus infected soybean. *Physiol. Plant Pathol.* **1975**, *5*, 239–248. [CrossRef]

37. Hashim, A.F.; Youssef, K.; Abd-Elsalam, K.A. Ecofriendly nanomaterials for controlling gray mold of table grapes and maintaining postharvest quality. *Eur. J. Plant Pathol.* **2019**, *154*, 377–388. [CrossRef]

38. Rahman, M.M.; Gray, A.I. A benzoisofuranone derivative and carbazole alkaloids from *Murraya koenigii* and their antimicrobial activity. *Phytochemstry* **2005**, *66*, 1601–1606. [CrossRef]

39. Draz, I.S.; Amal, A.E.; Abdelnaser, A.E.; Hassan, M.E.; Abdel-Wahab, A.I. Application of plant extracts as inducers to challenge leaf rust of wheat. *Egypt. J. Biol. Pest Control* **2019**, *29*, 6. [CrossRef]

40. Hassan, M.A.E.; Bereika, M.F.F.; Abo-Elnaga, H.I.G.; Sallam, M.A.A. Direct Antimicrobial Activity and Induction of systemic resistance in potato plants against bacterial wilt disease by plant extracts. *Plant Pathol. J.* **2009**, *24*, 352–360. [CrossRef]

41. Abo-Elyousr, K.A.M.; Hussein, M.A.M.; Allam, A.; Hassan, M. Enhanced onion resistance against stemphylium leaf blight disease, caused by *Stemphylium vesicarium*, by di-potassium phosphate and benzothiadiazole treatments. *Plant Pathol. J.* **2008**, *24*, 171–177.

42. De Meyer, G.; Capieau, K.; Audenaert, K.; Buchala, A.; Métraux, J.P.; Hofte, M. Nanogram amounts of salicylic acid produced by the *Rhizobacterium pseudomonas aeruginosa* 7NSK2 activate the systemic acquired resistance pathway in Bean. *Mol. Plant Microbe Interact.* **1999**, *12*, 450–458. [CrossRef]

43. Zimmerli, L.; Jakab, G.; Metraux, J.P.; Mauch-Mani, B. Potentiation of pathogen—Specific defense mechanisms in Arabidopsis by β-Aminobutyric acid. *Proc. Natl. Acad. Sci. USA* **2000**, *97*, 12920–12925. [CrossRef]

44. Zimmerli, L.; Metraux, J.P.; Mauch-Mani, B. Beta–Aminobutyric acid-induced protection of Arabidopsis against the necrotrophic fungus *Botrytis cinerea*. *Plant Physiol.* **2001**, *126*, 517–523. [CrossRef] [PubMed]

45. Seleim, M.A.A.; Abo-Elyousr, K.A.M.; Mohamed, A.A.; Al-Marzoky, H.A. Peroxidase and polyphenoloxidase activities as biochemical markers for biocontrol efficacy in the control of tomato bacterial wilt. *J. Plant. Physiol. Pathol.* **2014**, *2*, 2–8. [CrossRef]
46. Safdarpour, F.; Khodakaramain, G. Endophytic bacteria suppress bacterial wilt of tomato caused by *Ralstonia solanacearum* and Activate defense–related metabolites. *Biol. J. Microorg.* **2018**, *6*, 39–52.

horticulturae

Article

Tomato Breeding for Sustainable Crop Systems: High Levels of Zingiberene Providing Resistance to Multiple Arthropods

João Ronaldo Freitas de Oliveira [1], Juliano Tadeu Vilela de Resende [2], Renato Barros de Lima Filho [1], Sergio Ruffo Roberto [2,*], Paulo Roberto da Silva [3], Caroline Rech [1] and Cristiane Nardi [1]

[1] Department of Agronomy, Midwestern State University, Simeão Camargo Varela de Sá 03, Vila Carli, Guarapuava, Paraná ZIP 85040-080, Brazil; joaoroliveira@yahoo.com.br (J.R.F.d.O.); delimafilho.renato@yahoo.com (R.B.d.L.F.); caroline_rech@outlook.com (C.R.); cnardi@alumni.usp.br (C.N.)

[2] Agricultural Research Center, Londrina State University, Celso Garcia Cid Road, km 380, P.O. Box 10.011, Londrina ZIP 86057-970, Brazil; jvresende@uel.br

[3] Genetics and Plant Molecular Biology Laboratory, Midwestern State University, Simeão Camargo Varela de Sá 03, Vila Carli, Guarapuava, Paraná ZIP 85040-080, Brazil; prsilva@unicentro.br

* Correspondence: sroberto@uel.br

Received: 13 April 2020; Accepted: 29 May 2020; Published: 5 June 2020

Abstract: In sustainable cropping systems, the management of herbivorous arthropods is a challenge for the high performance of the tomato crop. One way to reduce the damage caused by these pests is the use of resistant cultivars within a sustainable integrated management system. The host selection of *Tetranychus urticae*, *Bemisia tabaci*, and *Tuta absoluta* was evaluated, characterizing their preference among the tomato genotypes RVTZ2011-79-503-143, RVTZ2011-79-335-164, RVTZ2011-79-185-250 (high zingiberene content—HZC), and RVTZ2011-79-117-273 (low zingiberene content—LZC). Such genotypes were selected in the F_2BC_2 generation (the F2 generation of the 2th backcross towards *Solanum lycopersicum* after the inicial interspecific cross *S. lycopersicum* × *S. habrochaites* var. *hirsutum*), resulting from crossing *Solanum habrochaites* var. *hirsutum* PI-127826 (HZC and resistant to mites) and the commercial cv. Redenção (*S. lycopersicum*) (LZC and susceptible to mites). In choice and no-choice bioassays by *T. urticae*, and in choice bioassays by *B. tabaci* and *T. absoluta*, arthropods preferred to stay and oviposit in an LZC genotype. In contrast, genotypes with HZC showed repellency to pests and induced a non-preference for oviposition. The F_2BC_2 genotypes selected for HZC are considered sources of resistance genes to these pests for tomato breeding programs, and therefore have excellent potential for sustainable cropping systems. These results represent an advance in obtaining tomato genetic materials which can be used in sustainable production systems with less loss from pests.

Keywords: genetic resistance; natural allelochemicals; organic production; plant defense

1. Introduction

The adoption of genetically resistant plant materials is one of the main strategies for pest management in sustainable tomato crops, in which the use of pesticides should be reduced or mitigated due to their adverse effects on the environment and human health [1–3]. A large number of biochemical and morphological characteristics have been related to tomato resistance to several key and secondary crop pests. Biochemical factors such as allelochemicals, proteins, and several metabolites are often associated with the morphological structures of species like trichomes, mesoderm, and parenchyma, among many other [4–6]. The introgression of genes from wild species has been shown to be the main alternative for tomato breeding programs, as there are several records of resistance in relation to arthropod pest species in this crop [7–12].

As a way to identify and measure the resistance of tomato genotypes to pest arthropods, biological and behavioral bioassays have been used to characterize cultivars that have resistance mechanisms of the antibiosis or antixenosis type to herbivores in comparison to susceptibility patterns [13]. Thus, it is also possible to identify the mechanisms and degrees of resistance of seedlings to the investigated pest arthropods. In tomato, the resistance mechanism of wild species is associated with the presence and exudation of allelochemicals present mainly in glandular trichomes of stems, leaves, and fruit [14–18]. Zingiberene is an allelochemical with deterrent and repellent actions against pest arthropods, and was found in high concentrations in *Solanum habrochaites* var. *hirsutum* PI-127826 in the glandular trichomes of types IV and VI [19–21]. The inheritance of the content of this allelochemical shows a high heritability, with values above 80%; it is controlled by two genes with incomplete dominance, which allows genetic gains regarding resistance when selecting seedlings with high zingiberene content [22–25].

Recently, the tomato genotype access PI-127826 was been used to increase the degrees of resistance in commercial varieties of *Solanum lycopersicum* though interspecific crosses [20,25–27] in order to obtain isogenic lines for hybrid production in conventional and sustainable production systems. However, the reactions triggered by the allelochemicals on the pests have been varied and complex, resulting in different levels of response in relation to the attack. Therefore, it is necessary to understand the effect of zingiberene and its stereoisomers on the behavior and biology of arthropod pests to determine the level of resistance of the genotypes. In addition, for a tomato breeding program, it is essential to characterize the resistance of generations and confirm the presence of the desirable characteristics of the parents which can be employed in future crossings.

Among the pest arthropods that occur in tomato, *Tuta absoluta* (Meyrick) (Lepidoptera: Gelechiidae), *Bemisia tabaci* (Gennadius) (Hemiptera: Aleyrodidae) and *Tetranychus urticae* (Koch) (Acari: Tetranychidae) are cosmopolitan and recurrent in crops grown in diverse production systems. For these arthropods, the deleterious effect of zingiberene has been evidenced in genotypes related to *S. habrochaites* var. *hirsutum*. However, despite all the studies on this subject, obtaining commercial cultivars with acceptable levels of yield and resistance is still an objective yet to be achieved [6,28–32].

In order to characterize resistant tomato genotypes to be used in breeding programs, we investigated the degrees of resistance of F_2BC_2 genotypes with high levels of zingiberene which were obtained from interspecific crossing between *S. habrochaites* var. *hirsutum* and *S. lycopersicum*. The resistance degree of the genotypes to *T. urticae*, *B. tabaci*, and *T. absoluta* were verified upon host selection bioassays.

2. Material and Methods

2.1. Location

The bioassays were carried out at the Midwestern Parana State University, Guarapuava, Paraná, Brazil (25°23′00″ S, 51°29′38.50″ W, elevation 1024 m.a.s.l.). The local climate is classified as humid subtropical mesothermal (Cfb), with moderate summers, winters with frost incidence, and an average annual rainfall of 1711 mm.

2.2. Tomato Genotypes and Growing Seedlings

The tomato genotypes selected by Zanin et al. [25] from a F_2BC_1 population for high levels of zingiberene (RVTZ 2011-079-117, RVTZ 2011-079-185, RVTZ 2011-079-335, and RVTZ 2011-079-503) were used as parents and pollen sources for crosses with *Solanum lycopersicum* cv. Redenção, thus obtaining the F_1RC_2 generation which, after self-fertilization, originated the F_2BC_2 population.

The selection of contrasting genotypes occurred in 400 seedlings of the F_2BC_2 population. The levels of zingiberene were quantified in young and expanded leaflets according to the methodology proposed by de Freitas et al. [33]. Therefore, six leaf disks (1 cm in diameter) were sampled from each genotype, placed in tubes containing 2 mL of hexane, and then vortexed for 30 s. After that, the leaf disks were removed and the absorbance of the solution was measured using a spectrophotometer (Cary series, UV-Vis Spectrophotometer) at a wavelength of 270 nm. The selection of genotypes was carried out based on the absorbance values, which were above 0.300 nm for the thee high-zingiberene genotypes and below 0.150 nm for the low-zingiberene genotype.

As low and high allelochemical standards, 50 tomato seedlings of cv. Redenção and 50 seedlings of the wild genotype *S. habrochaites* var. *hirsutum*, respectively, were used.

The F_2BC_2 genotypes selected indirectly for laboratory resistance based on zingiberene content were RVTZ 2011-79-503-143, RVTZ 2011-79-503-164, RVTZ 2011-79-503-250 (high zingiberene content—HZC) and, to use as a susceptibility standard in bioassays, RVTZ 2011-79-117-273 (low zingiberene content—LZC) (Table 1). The genotypes were cloned from axillary shoots; rooted; transplanted into polyethylene pots (5 L) containing a mixture of commercial substrate and soil (1:1), nitrogen, phosphorus and potassium chemical fertilizer (4:14:8), and acidity correction according to crop demand.

Table 1. Absorbancy at 270 nm (means ± SD) indicating the zingiberene content of tomato genotypes (*Solanum lycopersicum* cv. Redenção, *Solanum habrochaites* var. *hirsutum*, and genotypes of F_2BC_2).

Genotype	Absorbancy (270 nm)	Classification
S. lycopersicum cv. Redenção	0.032 ± 0.01	Susceptible standard (SS)
RVTZ 2011-79-117-273	0.100 ± 0.05	Low zingiberene content (LZC)
RVTZ 2011-79-335-164	0.322 ± 0.04	High zingiberene content (HZC)
RVTZ 2011-79-185-250	0.330 ± 0.07	High zingiberene content (HZC)
RVTZ 2011-79-503-143	0.809 ± 0.18	High zingiberene content (HZC)
S. habrochaites var. *hirsutum*	0.946 ± 0.19	Resistant standard (RS)

2.3. Arthropod Rearing

Adults of *T. urticae* were collected and kept on common bean seedlings (*Phaseolus vulgaris*), under controlled conditions (25 ± 2 °C, 70% ± 4% relative humidity—RH, 12 h photophase), with daily irrigation. The age-controlled breeding consisted of 10 adult mites (3 males and 7 females) kept on individual bean leaflets, arranged on a sponge moistened with distilled water, and packed in plastic trays (200 × 300 mm) with the bottom facing up. The mites were kept on the leaflets for 24 h and then the adults were removed, leaving only the eggs.

Individual *B. tabaci* and *T. absoluta* were collected in greenhouses from tomatoes in commercial production. For the rearing of *B. tabaci*, individuals were kept and multiplied on common bean seedlings, grown in pots, and packed in cages in a greenhouse at 25 ± 3 °C with daily irrigation. Rearing for age control was carried out, keeping bean seedlings with *B. tabaci* nymphs in cages made of voile fabric (100 × 100 cm) in a protected environment free from infestation. From five to eight days after emergence, the adults were used for bioassays. To rear *T. absoluta*, the individuals were kept on tomato seedlings cv. Santa Cruz, grown in pots, and placed in voile cages (100 × 100 cm) in a greenhouse at 25 ± 3 °C. The eggs deposited on the leaves were removed at intervals of three days and transferred to similar cages containing seedlings free from infestation.

2.4. Host Selection Bioassays

2.4.1. Tetranychus urticae

Choice and no-choice bioassays were carried out in Petri dish arenas (6 cm in diameter) and covered with a sponge and a cotton layer both moistened in distilled water. In each choice test dish, two leaf discs (3 cm in diameter) were placed under the cotton layer with the abaxial region facing upwards. Two genotypes were placed in each dish, and all of them were tested in combination two by two in order to evaluate the preference of the mites (Figure 1a). The discs were connected to each other by a transparent coverslip (18 × 18 mm), in the center of which four adult females of *T. urticae* were released, allowing free choice between the discs of the different tomato genotypes. The dishes were kept for 24 h in an air-conditioned chamber (25 ± 1 °C, 70% ± 4% RH, 12 h photophase). Then, the live mites present on each leaf disc were counted, in addition to the eggs deposited on the leaf surface.

Figure 1. Material used in the host selection bioassays. (**A**) Arena for *Tetranychus urticae* with two tomato genotypes connected by a transparent coverslip where the mites were released. (**B**) Tomato leaflet wrapped in cotton wool and plastic tube. (**C**) Tomato genotypes (6 treatments) arranged equidistant presenting the design of the *Bemisia tabaci* and *Tuta absoluta* bioassays. Each leaflet wrapped in cotton wool was introduced into a plastic tube containing water in order to maintain the humidity.

The experiment was designed with six treatments (genotypes) that were combined in pairs (total of 14 combinations) with ten repetitions per pair, and the arenas were distributed in a completely randomized design.

For the no-choice test, a leaf disc (3 cm in diameter) of each genotype was inserted in each arena, on which six adult females were transferred with the aid of a stereoscopic microscope (Nikon SMZ745T, Japan). The females were kept on the leaf discs for 24 h, and then the live mites and eggs deposited on the abaxial face of each disc were counted. The experiment was carried out in a completely randomized design, with six treatments (genotypes) and ten repetitions. The dishes were kept in an air-conditioned chamber (25 ± 1 °C, 70% ± 4% RH, 12 h photophase).

2.4.2. *Bemisia tabaci*

To evaluate the host selection behavior for the oviposition of *B. tabaci*, a choice test was adopted using circular plastic arenas (29.0 cm in diameter × 10.0 cm in height) with the cover lined with voile fabric to allow air circulation. From six equidistant holes in the side walls of the arena, tomato leaflets were inserted in a horizontal position, with the petiole kept externally and the leaf area internally within the arena. In order to keep the leaflet moist, the petiole was wrapped in cotton wool before being inserted into the plastic tube (0.3 cm in diameter × 10.0 cm in length) filled with water, whose function was to keep the cotton and petiole moist (Figure 1B,C). Thus, each arena contained a leaflet of each tomato genotype, comprising six leaflets, which were available for the choice of the insects.

In each arena, 30 individuals of *B. tabaci* were inserted and released with the aid of a suction device (transparent plastic tube 0.3 cm in diameter) from a hole made in the bottom of the container. After 24 h, the individuals and eggs present on the abaxial face of each leaflet were counted in order to calculate the attractive and preference index.

This bioassay was carried out in a completely randomized design with 6 treatments (genotypes) and 20 repetitions, and the arenas were kept in a greenhouse with temperature control (25 ± 3 °C).

2.4.3. *Tuta absoluta*

The behavior and deposition of the *T. absoluta* were evaluated in choice bioassays, using arenas similar to those described above for *B. tabaci*, allowing the free choice of insects in relation to the genotypes. However, for *T. absoluta* the arenas were made with a rectangular EVA (ethylene vinyl acetate) plate arranged in a conical shape forming a tube (25.0 cm in diameter × 40.0 cm in height), with side holes arranged in an equidistant way, which allowed the coupling of the leaflets in the horizontal position. The leaflets were inserted in the arena with the leaf area on the inside and the petiole wrapped in cotton and a plastic tube on the outside (Figure 1B,C). In turn, the upper and lower openings of the arena were lined with voile fabric to prevent the escape of insects and to allow air to circulate.

In the center of each arena, 12 females were released (10 to 12 days old), which remained for 72 h until the number of individuals and the number of eggs deposited on the adaxial and abaxial surfaces of the leaflets of each genotype were evaluated.

The bioassay was carried out in a completely randomized design, with 6 treatments (genotypes) and 10 repetitions. The insects were kept in air-conditioned rooms (25 ± 1 °C, 70% ± 4% RH, photophase 12 h).

2.5. Data Analysis

The number of adults and eggs of *T. urticae*, *B. tabaci*, and *T. absoluta* present in each treatment in the choice and no-choice tests were submitted to a normality analysis (Bartlet, $p \leq 0.05$) and analysis of variance (ANOVA, $p \leq 0.05$), and the means were compared by the Tukey test ($\alpha \leq 0.05$). The non-normal data were transformed by $(x + 0.5)^{1/2}$. These analyses were performed using the Statistica 7.0 software [34].

The data referring to the choice bioassays for *T. urticae* were analyzed using the Chi-square test ($p \leq 0.05$) and Kruskal–Wallis ($p \leq 0.05$), respectively, to test the hypothesis of equality between the observed and expected frequencies from individuals in each treatment.

In the choice bioassays for *B. tabaci* and *T. absoluta*, the attractiveness index was estimated as AI = 2G/(G + S), where AI = attractiveness index; G = number of insects attracted to the evaluated genotype; and S = number of insects attracted to the susceptible standard (*S. lycopersicum* cv. Redenção). The AI values varied between zero and two, in which AI = 1 indicates a similar attraction between the evaluated genotypes (repellent test plant) and the standard (attractive test plant); AI < 1 corresponds to less attraction (greater repellency) to the genotype; and AI > 1 indicates a greater attraction to the evaluated genotype in relation to the standard. This index is an adaptation of the formula mentioned by Lin et al. [35] and used by Baldin et al. [36].

The oviposition preference index (OPI) was also calculated as OPI = [(T − S)/(T + S)] × 100, where T = number of eggs counted in the evaluated treatment and S = number of eggs counted in the standard genotype (*S. lycopersicum* cv. Redenção). This index ranges from +100 (high preference) to −100 (non-preference or oviposition inhibition). The classification of genotypes was performed by comparing the average of eggs from the treatments with the average of the *S. lycopersicum* cv. Redenção [37].

3. Results

3.1. Tetranychus Urticae Host Selection

The combinations of genotypes offered to *T. urticae* influenced the choice of females, which showed a rejection of the F_2BC_2 genotypes with HZC (RVTZ 2011-79-503-143, RVTZ 2011-79-503-164, and RVTZ 2011-79-503-250) when combined with the standard susceptibility tomato cv. Redenção (Chi-square, GL = 1, $p < 0.05$) (Figure 2). Although it was with less intensity, this response pattern was repeated when this cultivar was compared to the low-zingiberene genotype (RVTZ 2011-79-503-273) (Chi-square, GL = 1, $p = 0.016$).

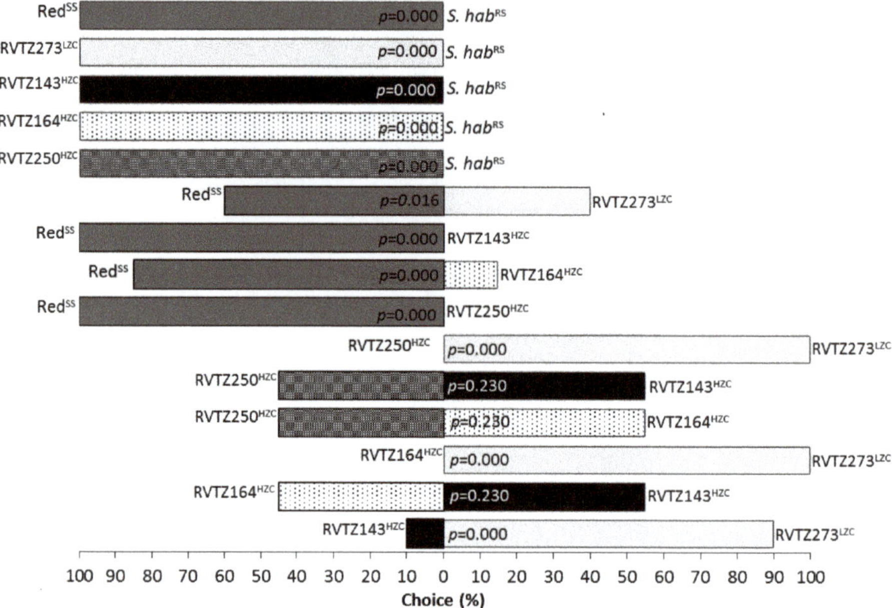

Figure 2. Frequency of choice of *Tetranychus urticae* in relation to tomato F_2BC_2 genotypes (n = 10). Red = *Solanum lycopersicum* cv. Redenção; PI-127826 = *Solanum habrochaites* var. *hirsutum* PI-127826; 273 = RVTZ 2011-79-117-273 (with low zingiberene content); 143 = RVTZ 2011-79-503-143; 164 = RVTZ 2011-79-503-164; and 250 = RVTZ 2011-79-503-250 (with high zingiberene content). p-value indicates the statistical significance according to the Chi-square test. SS = Susceptible Standard genotype; RS = Resistant standard genotype; HZC = high zingiberene content; LZC = low zingiberene content.

Less attractiveness compared to the standard resistance genotype *S. habrochaites* var. *hirsutum* PI-127826 was observed in all the evaluated combinations, even when compared to the genotypes selected for HZC in the F_2BC_2 generation (Chi-square, GL = 1, p < 0.05). *T. urticae* females did not differentiate between the genotypes with HZC, presenting a similar frequency of choice regardless of the combined genotypes (Chi-square, GL = 1, p > 0.05).

In no-choice bioassays, the number of live mites was influenced by the presence of zingiberene. *S. lycopersicum* cv. Redenção leaflets provided more than 90% survival after 24 h of exposure of mites, with an average of 5.75 living individuals. The LZC genotype RVTZ 2011-79-117-273 did not differ from the standard susceptible genotype (Tukey, GL = 5, α = 0.912). In turn, HZC triggered acari mortality, as occurred with the resistant standard genotype (Tukey, GL = 5, α < 0.0001). The average number of eggs deposited in the leaflets was higher in the susceptible cultivar and in the low allelochemical genotype, differing from the high-content genotypes (Tukey, GL = 5, α < 0.0001). These genotypes did not differ among themselves (Tukey, GL = 5, α > 0.1) (Table 2).

3.2. Bemisia tabaci Host Selection

S. lycopersicum cv. Redenção was the most attractive for permanence and oviposition, followed by RVTZ 2011-79-117-273 (LZC) (Tukey, GL = 5, p = 0.01). *S. habrochaites* var. *hirsutum* PI-127826 and other genotypes selected for their high levels of zingiberene triggered the same insect responses (RVTZ 2011-79-503-143, RVTZ 2011-79-335-164, RVTZ 2011-79-185-250) (Tukey, GL = 5, α > 0.9) (Table 3).

Table 2. Number (± standard error) of live adults and eggs of *Tetranychus urticae* mites on the abaxial face of leaf discs of tomato genotypes in no-choice oviposition bioassays (*n* = 10).

Genotypes	Number of Individuals	Number of Eggs
Solanum lycopersicum cv. Redenção	5.75 ± 0.12 a	46.6 ± 1.80 a
RVTZ 2011-79-117-273 [LZC]	4.25 ± 0.14 ab	27.2 ± 1.20 b
RVTZ 2011-79-503-143 [HZC]	3.25 ± 0.15 b	3.1 ± 0.60 c
RVTZ 2011-79-335-164 [HZC]	3.50 ± 0.14 b	3.9 ± 0.53 c
RVTZ 2011-79-185-250 [HZC]	3.25 ± 0.30 b	0.7 ± 0.14 c
Solanum habrochaites var. *hirsutum*	3.75 ± 0.12 b	1.8 ± 0.18 c
F-value	81.9	328
ANOVA *p*-value	0.03	0.0001

[LZC] low zingiberene content; [HZC] high zingiberene content. Means followed by the same letter in a column do not differ statistically from each other by the Tukey test (GL = 5, $\alpha \leq 0.05$). For analysis, the average number of eggs was transformed by $(x + 0.5)^{1/2}$.

Table 3. Number (± standard error) of *Bemisia tabaci* live adults and eggs on the abaxial face of leaflets from tomato genotypes in the choice bioassay (*n* = 10).

Genotypes	Number of Individuals	Number of Eggs
Solanum lycopersicum cv. Redenção	13.00 ± 0.90 a	47.10 ± 3.70 a
RVTZ 2011-79-117-273 [LZC]	12.90 ± 1.0 b	35.30 ± 3.50 b
RVTZ 2011-79-503-143 [HZC]	0.70 ± 0.20 c	1.45 ± 0.50 c
RVTZ 2011-79-335-164 [HZC]	1.20 ± 0.30 c	1.80 ± 0.40 c
RVTZ 2011-79-185-250 [HZC]	1.00 ± 0.15 c	1.75 ± 0.35 c
Solanum habrochaites var. *hirsutum*	1.20 ± 0.30 c	2.50 ± 0.40 c
F-value	683.8	177.8
ANOVA *p*-value	0.0001	0.0001

[LZC] low zingiberene content; [HZC] high zingiberene content. Means followed by the same letter in each column do not differ statistically from each other by the Tukey test ($\alpha \leq 0.05$). For analysis, the average number of eggs was transformed by $(x + 0.5)^{1/2}$.

According to the values of the attractiveness index calculated after 24 h of infestation, all the genotypes with a HZC were classified as repellent to *B. tabaci* when compared to the susceptible standard genotype *S. lycopersicum* cv. Redenção (Figure 3a). In addition, the preference index for oviposition calculated in the choice bioassay classified all the materials as non-preferred to *B. tabaci* oviposition when compared to the susceptible *S. lycopersicum* cv. Redenção, even for the low zingiberene genotype (Figure 3b).

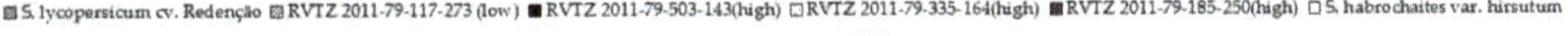

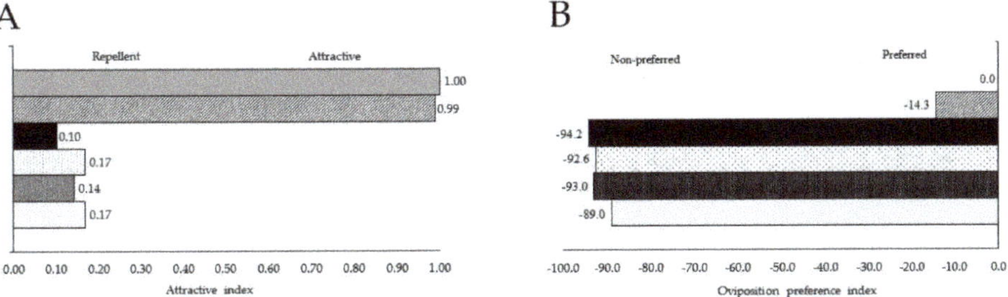

Figure 3. *Bemisia tabaci* responses to tomato genotypes in choice bioassays (*n* = 10). (**A**) Attractiveness index of genotypes for adults of *B. tabaci*. (**B**) Oviposition preference index for *B. tabaci* in relation to genotypes. The data regarding the number of individuals were transformed by $(x + 0.5)^{1/2}$.

3.3. Tuta Absoluta Host Selection

T. absoluta females preferred to remain in the LZC genotypes. Nevertheless, they did not differ from RVTZ 2011-79-335-164, which was selected for the high content of this compound (Tukey, GL = 5, α > 0.1) (Table 4). However, even in the genotype with a HZC, *T. absoluta* females oviposited compared to the susceptible standard and that with a LZC (Tukey, GL = 5, α < 0.05) (Table 4). Therefore, the number of eggs deposited was significantly higher in the standard susceptibility cultivar than in the other genotypes (Tukey, GL = 5, α < 0.05) (Table 4).

Table 4. Number (± standard error) of *Tuta absoluta* adults and eggs on the adaxial and abaxial sides of leaflets of tomato genotypes in a choice bioassay ($n = 10$).

Genotypes	Number of Adults	Number of Eggs
Solanum lycopersicum cv. Redenção	1.4 ± 0.15 a	47.0 ± 1.40 a
RVTZ 2011-79-117-273 [LZC]	1.4 ± 0.15 a	37.0 ± 1.30 b
RVTZ 2011-79-503-143 [HZC]	0.5 ± 0.16 b	4.7 ± 0.10 c
RVTZ 2011-79-335-164 [HZC]	1.0 ± 0.2 ab	8.2 ± 0.30 c
RVTZ 2011-79-185-250 [HZC]	0.6 ± 0.15 b	9.1 ± 0.20 c
Solanum habrochaites var. *hirsutum*	0.5 ± 0.16 b	3.6 ± 0.09 c
F-value	6.1	111.6
ANOVA *p*-value	0.0001	0.0001

[LZC] low zingiberene content; [HZC] high zingiberene content. Means followed by the same letter in each column do not differ statistically from each other by the Tukey test ($\alpha \leq 0.05$). For analysis, the average number of eggs was transformed by $(x + 0.5)^{1/2}$.

The calculated indices demonstrated that the LZC and susceptible genotypes were more attractive to *T. absoluta* than the genotypes with a HZC (Figure 4a). In contrast, the HZC genotypes presented less attractiveness (RVTZ 2011-79-335-164 and RVTZ 2011-79-185-250) or repellency (RVTZ 2011-79-503-143) to *T. absoluta* (Figure 4a). Furthermore, the selected HZC genotypes suppressed *T. absoluta* oviposition, triggering more than 60% of the oviposition index when compared with the standard genotype (*S. lycopersicum* cv. Redenção) (Figure 4b).

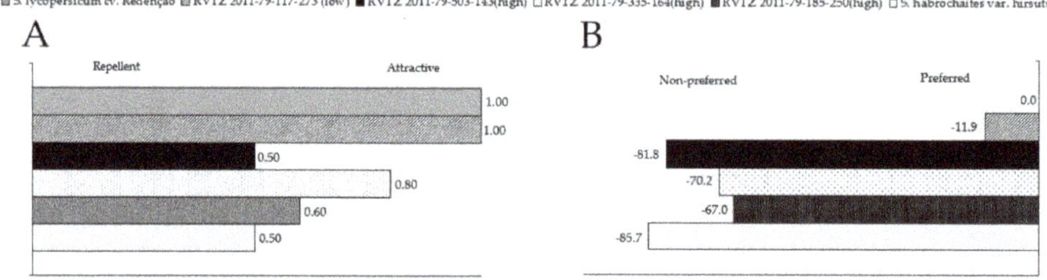

Figure 4. *Tuta absoluta* responses to tomato genotypes in choice bioassays. (**A**) Attractiveness index of genotypes for adults of *T. absoluta*. (**B**) Oviposition preference index of *T. absoluta* in relation to genotypes. The data regarding the number of individuals were transformed by $(x + 0.5)^{1/2}$.

4. Discussion

Genetics and plant breeding are two main sciences in the service of man that contribute in various fields of activities, especially food production. Genetic improvement considers the relationship of physical, chemical, biological, economic, and social components to the environment, aiming at the best possible adjustment and resulting in the sustainability of a productive system. Thus, genetic improvement is essential for plants to adapt sustainably to the growing environment, providing greater productivity and less possible damage to the system balance.

Tomato production is characterized by low sustainability, largely due to the high number of applications for pest and disease control. In this context, breeding aimed at the greater sustainability of the systems has sought to develop genotypes more tolerant to abiotic and biotic stresses. The use of wild tomato genotypes as a source of resistance to pests has been used on a recurring basis, aiming to seek adjustment to the sustainable production system.

The resistance in these cases is related to the presence of allelochemicals that are exuded by the tomato leaf trichomes, including zingiberene, a volatile sesquiterpene with a proven effect in pest control. The best way to introduce resistance in commercial cultivars is though interspecific crossings. In the present work, resistance was introduced though backcrosses and, as shown, the results were promising.

Glandular trichomes are specialized structures found on the surface of about 30% of all vascular plants and are responsible for a significant portion of a plant's secondary chemistry. Glandular trichomes are an important source of allelochemicals, which provide the plant with protection against herbivores and pathogens. The storage compartment of glandular trichomes is usually located at the tip of the morphological structure and is part of the glandular cell, or cells that are metabolically active. This knowledge now helps classic breeding programs, as well as targeted genetic engineering, with the aim of optimizing the density and physiology of trichomes for the purpose of biocides to improve crop protection [38,39]

Glandular trichomes and zingiberene contribute to increased tomato resistance against herbivore attack [39]. The ability to accumulate zingiberene in glandular trichomes is being transferred to commercial cultivars through crossbreeding, contributing to the increased resistance of plants to mites and whiteflies [40]. The presence of zingiberene in tomato genotypes induced a non-preference for the oviposition of arthropod pests. Such evidence demonstrates that F_2BC_2 seedlings selected for high zingiberene content have desirable degrees of resistance to arthropods, and that selection for this characteristic has been successful in terms of the biological response of the target organisms. These genotypes, in some cases, showed similar degrees of resistance to the standard *S. habrochaites* var. *hirsutum* PI-127826, which was expected due to the high heritability (81.9%) of zingiberene content character in plants [22,25].

Normally, the genetic inheritance of a concentration of allelochemicals associated with the presence of glandular trichomes on tomato species is mediated by at most two larger genes, and is associated with smaller genes with an additive effect. The heritability in a broad sense was estimated at above 50%, considered as moderate to high [41–43]. Heterozygous genotypes for zingiberene showed levels of allelochemicals intermediate to those of their parents of high and low content, indicating incomplete dominant gene action for the content of allelochemicals [7].

The concentration of this allelochemical is controlled by the two larger gene loci, with incomplete dominance in the sense of the lowest content in populations of interspecific crosses of *S. lycopersicum* with *S. habrochaites* var. *hirsutum* [22]. Other studies related to the presence of zingiberene reported that genetic inheritance was controlled by a single dominant gene of *S. habrochaites* var. *hirsutum* in the generation F_2 and in the BC progeny of *S. lycopersicum* with *S. habrochaites* var. *hirsutum*, and also as a single recessive gene in the F_2 generation and in the BC progeny of *S. habrochaites* var. *hirsutum* × *S. habrochaites* var. *glabratum* [33,44].

This fact, together with the high heritability of the character, may reflect satisfactory genetic gains in backcrossing over generations, using indirect selection for resistance to arthropod pests based on the zingiberene content [25]. The most important aspect of the genetic breeding is hereditary variability and, particularly, the genetic component, which has a strong influence on the response to selection. The information generated from the character association studies appeared as an important indicator for a selection program. The present study also contributed to the suitability of several characters for indirect selection, because the selection for one or more characteristics resulted in a correlated response in several other studies [45].

The two larger gene loci are located very close to each other on chomosome 8 (ShZIS and ShSBS) [40]. However, it is assumed that only one of the gene loci is primarily responsible for the synthesis of zingiberene [44]. Evidence suggested that this locus must contain ShZIS synthase, which produces 7-epizingiberene, a zingiberene sesquiterpene stereomer, in glandular trichomes from Z-Z-farnesyl-diphosphate (zFPP) in plastids, which promotes resistance to various tomato pests.

It is also assumed that ShZIS is allelic to ShSBS and linked to zFPS. Gonzales-Vigil et al. [46] suggested that ShZIS originated before ShSBS, with the mutagenesis of only three amino acids at the active site of a synthesizing enzyme, predominantly zingiberene, elucidating the probable pathway for the synthesis of this sesquiterpene. Thus, indirect selection for resistance facilitated the process of genetic improvement, as it allows the use of tools that minimize the difficulties of selecting in large populations of plants, as in the case of segregating populations [7].

The environment can interfere in expression of zingiberene content, and thus low heritability can be great challenge for breeding programs. In-depth knowledge of quantitative genetic parameters and, in particular, genetic variation and heritability, help to overcome these challenges. Although the expression of the characters is influenced by the environment, the high heritability described in the literature implies success in transfer to progeny through crosses [47].

The high heritability of the traits is the fundamental basis for the genetic improvement of this particular trait. In addition, the nature and magnitude of genetic variation that governs inheritance of yield and yield components are essential for breeding. In general, heritability is the reason for variation due to differences between genotypes and the total phenotypic variation of a trait in a population [48]. Heritability estimates, along with genetic advancement, are more accurate in predicting genetic gain under selection than heritability alone. In addition, the relationship between the yield and the characteristics attributable to yield is of high importance for the direct and indirect selection of characteristics [49].

Resistance depends on the relationship of the target insect to the levels of zingiberene. The bioassays provided important results, inferring that the adults of *T. urticae* (Figure 2) and *B. tabaci* (Table 3) rejected the genotypes with high zingiberene content. For these two arthropods, the number of eggs deposited on leaflets of the susceptible standard treatment (*S. lycopersicum* cv. Redenção) and the low zingiberene genotype was higher compared to the high content genotypes and *S. habrochaites* var. *hirsutum*.

The results of this work corroborated the research of others who evaluated the selection of genotypes from different generations resulting from interspecific crosses between *S. habrochaites* var. *hirsutum* and *S. lycopersicum*. Interspecific hybrids with a high content of zingiberene, derived from a diallel and demonstrated by means of GCA (General Combining Ability) and SCA (Specific Combining Ability), have a high potential for use in improving tomato pest resistance [50]. Studies reported that genes responsible for high levels of zingiberene were passed to segregating generations through crosses, maintaining the resistance of plants in advanced populations with commercial backgrounds [20,40,51].

The high number of eggs deposited by *T. urticae* (Table 2), *B. tabaci* (Table 3), and *T. absoluta* (Table 4) on the leaflets of the RVTZ genotype 2011-79-117-273 demonstrated that, although it has a low zingiberene content, it is highly susceptible to the evaluated pests. Studies with other tomato lines have also shown a positive correlation between zingiberene content and resistance to these pest arthropods [7,52,53]. Genotypes with a high zingiberene content (ZGB-703 and ZGB-704) had a lower oviposition and a lower number of nymphs of whitefly when compared with genotypes with a low acyl-sugar (AS) and zingibrene content (Santa Clara, TOM-695, TOM-556, and TOM-584) [54].

Bleeker et al. [46] observed a severe reduction in the fertility of two of the main tomato pests, *B. tabaci* and *T. urticae*, mediated by a zingiberene stereoisomer, 7-epizingiberene, also present in the PI-127826 genotype. Bleeker et al. [40] found that the survival rate and reproductive success of *B. tabaci* adults in F_2 genotypes from the cross between PI-127826 and *S. lycopersicum* were severely compromised, although the levels of 7-epizingiberene on the F_2-7 and F_2-71 leaf surface were lower than that observed in wild tomato PI-127826.

For *T. absoluta*, although adults remained on the leaflets of genotypes with high levels of zingiberene, their rejection was verified by the reduced number of eggs deposited compared to the susceptibility standard and the low allelochemical content genotype (Table 4). Homozygous lines with a high zingiberene content, simple heterozygous genotypes for zingiberene, and controls with a low zingiberene content ('Débora Max' and 'TOM-684') and a high zingiberene content (PI-127826) were subjected to infestation with adult pinworm from South America. High zingiberene genotypes showed significantly lower egg counts than the low ones [7]. The damage by *T. absoluta* was greater in the low zingiberene controls, as in the present study.

Silva et al. [55] found that volatile compounds such as sesquiterpenes negatively affected the preference (antixenosis) or survival (antibiosis) of herbivorous insects in tomatoes by modulating the flight behavior and oviposition of females of *T. absoluta* during mating. In addition, they observed that females landed more frequently and oviposited in the most susceptible genotype. However, this proportions of preference for feeding and oviposition were not substantially higher in the resistant genotypes when compared to susceptible ones.

Al-Bayati [56], when evaluating nine tomato hybrids (BC_3F_3 and BC_3F_4) selected based on the variation of the type IV trichome density and zingiberene concentration of an interspecific cross with *S. habrochaites*, concluded that the resistance to the spider mite had been successfully transferred to hybrids through backcrossing and indirect selection. There was a significant negative correlation between almost all the behavioral and biological responses of the mites with the density of type IV trichomes and the zingiberene content. Therefore, these results are in agreement with those obtained in the present study for the relationship between zingiberene and pests.

The results of this work showed that the genotypes with a high zingiberene content affected the behavior and development of arthropods. Thus, it confirms that the use of *S. habrochaites* var. *hirsutum* PI-127826 enabled the introgression of genes to increase the degrees of resistance in the F_2BC_2 genotypes. Finally, the results suggested that the genotypes RVTZ 2011-79-503-143, RVTZ 2011-79-335-164, and RVTZ 2011-79-185-250 can be used as potential sources of genes for resistance to *T. urticae*, *B. tabaci*, and *T. absoluta* in tomato breeding programs. The results also showed that the genetic improvement of this species, mediated by allelochemicals present in wild species, is an excellent alternative for obtaining cultivars with better adaptation to sustainable production systems.

Author Contributions: Methodology, validation, and original draft preparation, J.R.F.d.O.; project administration and funding acquisition, J.T.V.d.R.; methodology and validation, R.B.d.L.F. and C.R.; review and editing, S.R.R. and P.R.d.S.; formal analyses, review and editing, and funding acquisition, C.N. All authors have read and agreed to the published version of the manuscript.

Funding: This research was supported by the National Institute of Science and Technology–Semiochemicals in Agriculture (FAPESP and CNPq–Grants #2014/50871-0 and #465511/2014-7, respectively). Support Program for Centers of Excellence (Pronex-FA/CNPq (PI) #01/2018). J.R.F.d.O. was sponsored partially by the Coordination for the Improvement of Higher Education Personnel (CAPES Finance Code #001).

Acknowledgments: The authors thank the anonymous reviewers for their helpful comments.

Conflicts of Interest: The authors declare no conflict of interest.

References

1. Fantke, P.; Jolliet, O. Life cycle human health impacts of 875 pesticides. *Int. J. Life Cycle Assess.* **2016**, *21*, 722–733. [CrossRef]
2. Pang, N.; Fan, X.; Fantke, P.; Zhao, S.; Hu, J. Dynamics and dietary risk assessment of thiamethoxam in wheat, lettuce and tomato using field experiments and computational simulation. *Environ. Pollut.* **2020**, *256*, 113285. [CrossRef] [PubMed]
3. Michelotto, M.D.; Carrega, W.C.; Pirotta, M.Z.; de Godoy, I.J.; Lourencao, A.L.; Martins, A.L.M. Control of thips ('*Enneothips flavens*' Moulton.) with synthetic and biological insecticides in different peanut genotypes. *Aust. J. Crop Sci.* **2019**, *13*, 1074. [CrossRef]

4. Seki, K. Leaf-morphology-assisted selection for resistance to two-spotted spider mite *Tetranychus urticae* Koch (Acari: Tetranychidae) in carnations (*Dianthus caryophyllus* L.). *Pest Manag. Sci.* **2016**, *72*, 1926–1933. [CrossRef] [PubMed]

5. Stout, M.J.; Kurabchew, H.; Leite, G.L.D. Host-Plant Resistance in Tomato. In *Sustainable Management of Arthropod Pests of Tomato*; Elsevier: Amsterdam, The Netherlands, 2018; pp. 217–236.

6. Dawood, M.H.; Snyder, J.C. The alcohol and epoxy alcohol of zingiberene, produced in trichomes of wild tomato, are more repellent to spider mites than zingiberene. *Front. Plant Sci.* **2020**, *11*, 35. [CrossRef] [PubMed]

7. Maluf, W.R.; de Fátima Silva, V.; das Graças Cardoso, M.; Gomes, L.A.A.; Neto, Á.C.G.; Maciel, G.M.; Nízio, D.A.C. Resistance to the South American tomato pinworm *Tuta absoluta* in high acylsugar and/or high zingiberene tomato genotypes. *Euphytica* **2010**, *176*, 113–123. [CrossRef]

8. Dias, D.; Resende, J.; Marodin, J.; Matos, R.; Lustosa, I.; Resende, N. Acyl sugars and whitefly (*Bemisia tabaci*) resistance in segregating populations of tomato genotypes. *Genet. Mol. Res.* **2016**, *15*, 1–11. [CrossRef] [PubMed]

9. Lucini, T.; Faria, M.V.; Rohde, C.; Resende, J.T.V.; de Oliveira, J.R.F. Acylsugar and the role of trichomes in tomato genotypes resistance to *Tetranychus urticae*. *Arthropod-Plant Interact.* **2015**, *9*, 45–53. [CrossRef]

10. Goiana, E.S.; Dias-Pini, N.S.; Muniz, C.R.; Soares, A.A.; Alves, J.C.; Vidal-Neto, F.C.; Bezerra Da Silva, C.S. Dwarf-cashew resistance to whitefly (*Aleurodicus cocois*) linked to morphological and histochemical characteristics of leaves. *Pest Manag. Sci.* **2020**, *76*, 464–471. [CrossRef]

11. Cherif, A.; Verheggen, F. A review of *Tuta absoluta* (Lepidoptera: Gelechiidae) host plants and their impact on management strategies. *Biotechnol. Agron. Société Environ.* **2019**, *23*, 270–278.

12. Silva, A.A.D.; Carvalho, R.D.C.; Andrade, M.C.; Zeist, A.R.; Resende, J.T.V.D.; Maluf, W.R. Glandular trichomes that mediate resistance to green peach aphid in tomato genotypes from the cross between *S. galapagense* and *S. lycopersicum*. *Acta Sci. Agron.* **2019**, *41*, e42704. [CrossRef]

13. Follett, P.A. Insect-plant interactions: Host selection, herbivory, and plant resistance–an introduction. *Entomol. Exp. Appl.* **2017**, *162*, 1–3. [CrossRef]

14. Lucini, T.; Resende, J.; Oliveira, J.; Scabeni, C.; Zeist, A.; Resende, N. Repellent effects of various cherry tomato accessions on the two-spotted spider mite *Tetranychus urticae* Koch (Acari: Tetranychidae). *Genet. Mol. Res.* **2016**, *15*, 10.4328. [CrossRef] [PubMed]

15. Sohabi, F.; Nooryazdan, H.; Gharati, B.; Saeidi, Z. Plant resistance to the moth *Tuta absoluta* (Meyrick) (Lepidoptera: Gelechiidae) in tomato cultivars. *Neotrop. Entomol.* **2017**, *46*, 203–209. [CrossRef] [PubMed]

16. Savi, P.; De Moraes, G.; Junior, A.B.; Melville, C.; Carvalho, R.; Lourenção, A.; Andrade, D. Impact of leaflet trichomes on settlement and oviposition of *Tetranychus evansi* (Acari: Tetranychidae) in African and South American tomatoes. *Syst. Appl. Acarol.* **2019**, *24*, 2559–2576.

17. Yao, Q.; Peng, Z.; Tong, H.; Yang, F.; Xing, G.; Wang, L.; Zheng, J.; Zhang, Y.; Su, Q. Tomato plant flavonoids increase whitefly resistance and reduce spread of tomato yellow leaf curl virus. *J. Econ. Entomol.* **2019**, *112*, 2790–2796. [CrossRef]

18. Ali, A.; Rakha, M.; Shaheen, F.A.; Srinivasan, R. Resistance of certain wild tomato (*Solanum* spp.) accessions to *Helicoverpa armigera* (Hübner) (Lepidoptera: Noctuidae) based on choice and no-choice bioassays. *Fla. Entomol.* **2019**, *102*, 544–548. [CrossRef]

19. Rakha, M.; Hanson, P.; Ramasamy, S. Identification of resistance to *Bemisia tabaci* Genn. in closely related wild relatives of cultivated tomato based on trichome type analysis and choice and no-choice assays. *Genet. Resour. Crop Evol.* **2017**, *64*, 247–260. [CrossRef]

20. Oliveira, J.; Resende, J.; Maluf, W.; Lucini, T.; Lima Filho, R.; Lima, I.; Nardi, C. Trichomes and allelochemicals in tomato genotypes have antagonistic effects upon behavior and biology of *Tetranychus urticae*. *Front. Plant Sci.* **2018**, *9*, 1132. [CrossRef]

21. Tissier, A. Glandular trichomes: What comes after expressed sequence tags? *Plant J.* **2012**, *70*, 51–68. [CrossRef]

22. Lima, I.; Resende, J.; Oliveira, J.; Faria, M.; Resende, N.; Lima-Filho, R. Indirect selection of industrial tomato genotypes rich in zingiberene and resistant to *Tuta absoluta* Meyrick. *Genet. Mol. Res.* **2015**, *14*, 15081–15089. [CrossRef]

23. Zhou, F.; Pichersky, E. More is better: The diversity of terpene metabolism in plants. *Curr. Opin. Plant Biol.* **2020**, *55*, 1–10. [CrossRef]

24. Andrade, M.C.; Da Silva, A.A.; Neiva, I.P.; Oliveira, I.R.C.; De Castro, E.M.; Francis, D.M.; Maluf, W.R. Inheritance of type IV glandular trichome density and its association with whitefly resistance from *Solanum galapagense* accession LA1401. *Euphytica* **2017**, *213*, 52. [CrossRef]

25. Zanin, D.S.; Resende, J.T.; Zeist, A.R.; Oliveira, J.R.; Henschel, J.M.; Lima Filho, R.B. Selection of processing tomato genotypes resistant to two spotted spider mite. *Hortic. Bras.* **2018**, *36*, 271–275. [CrossRef]

26. Quesada-Ocampo, L.; Vargas, A.; Naegele, R.; Francis, D.; Hausbeck, M. Resistance to crown and root rot caused by *Phytophthora capsici* in a tomato advanced backcross of *Solanum habrochaites* and *Solanum lycopersicum*. *Plant Dis.* **2016**, *100*, 829–835. [CrossRef] [PubMed]

27. Resende, J.; da Silva, A.; Gabriel, A.; Zeist, A.; Favaro, R.; Nascimento, D.; Zeist, R.; Camargo, C. Resistance to *Helicoverpa armigera* mediated by zingiberene and glandular trichomes in tomatoes for industrial processing. *Genet. Mol. Res.* **2018**, *17*. [CrossRef]

28. Carter, C.D.; Snyder, J.C. Mite responses in relation to trichomes of *Lycopersicon esculentum x L. hirsutum* F 2 hybrids. *Euphytica* **1985**, *34*, 177–185. [CrossRef]

29. Zeist, A.R.; da Silva, A.A.; de Resende, J.T.V.; Maluf, W.R.; Gabriel, A.; Zanin, D.S.; Guerra, E.P. Tomato breeding for insect-pest resistance. In *Recent Advances in Tomato Breeding and Production*; Intech Open: London, UK, 2018.

30. Heinz, K.M.; Zalom, F.G. Variation in trichome-based resistance to *Bemisia argentifolii* (Homoptera: Aleyrodidae) oviposition on tomato. *J. Econ. Entomol.* **1995**, *88*, 1494–1502. [CrossRef]

31. Nombela, G.; Beitia, F.; Muñiz, M. Variation in tomato host response to *Bemisia tabaci* (Hemiptera: Aleyrodidae) in relation to acyl sugar content and presence of the nematode and potato aphid resistance gene Mi. *Bull. Entomol. Res.* **2000**, *90*, 161–167. [CrossRef]

32. Leite, G.; Picanço, M.; Guedes, R.; Zanuncio, J. Role of plant age in the resistance of *Lycopersicon hirsutum f. glabratum* to the tomato leafminer *Tuta absoluta* (Lepidoptera: Gelechiidae). *Sci. Hortic.* **2001**, *89*, 103–113. [CrossRef]

33. De Freitas, J.A.; Maluf, W.R.; das Graças Cardoso, M.; de Oliveira, A.C.B. Seleção de plantas de tomateiro visando à resistência à artrópodes-praga mediada por zingibereno. *Acta Sci. Agron.* **2000**, *22*, 919–923. [CrossRef]

34. Statsoft, I. *STATISTICA (Data Analysis Software System)*; Version 11; Science Open: Berlin, Germany, 2012.

35. Lin, H.; Kogan, M.; Fischer, D. Induced resistance in soybean to the Mexican bean beetle (Coleoptera: Coccinellidae): Comparisons of inducing factors. *Environ. Entomol.* **1990**, *19*, 1852–1857. [CrossRef]

36. Baldin, E.; Vendramim, J.; Lourenção, A. Resistência de genótipos de tomateiro à mosca-branca *Bemisia tabaci* (Gennadius) biótipo B (Hemiptera: Aleyrodidae). *Neotrop. Entomol.* **2005**, *34*, 435–441. [CrossRef]

37. Baldin, E.; Toscano, L.; Lima, A.; Lara, F.; Boiça, A., Jr. Preferência para oviposição de *Bemisia tabaci* biótipo B por genótipos de *Cucurbita moschata* e *Cucurbita maxima*. *Boletín Sanid. Vegetal. Plagas* **2000**, *26*, 409–413.

38. Sperotto, R.A.; Grbic, V.; Pappas, M.; Leiss, K.; Kant, M.; Wilson, C.R.; Santamaria, M.E.; Gao, Y. Plant responses to phytophagous mites/thips and search for resistance. *Front. Plant Sci.* **2019**, *10*, 866. [CrossRef]

39. Glas, J.J.; Schimmel, B.C.; Alba, J.M.; Escobar-Bravo, R.; Schuurink, R.C.; Kant, M.R. Plant glandular trichomes as targets for breeding or engineering of resistance to herbivores. *Int. J. Mol. Sci.* **2012**, *13*, 17077–17103. [CrossRef]

40. Bleeker, P.M.; Mirabella, R.; Diergaarde, P.J.; VanDoorn, A.; Tissier, A.; Kant, M.R.; Prins, M.; de Vos, M.; Haring, M.A.; Schuurink, R.C. Improved herbivore resistance in cultivated tomato with the sesquiterpene biosynthetic pathway from a wild relative. *Proc. Natl. Acad. Sci. USA* **2012**, *109*, 20124–20129. [CrossRef]

41. Saeidi, Z.; Mallik, B.; Kulkarni, R. Inheritance of glandular trichomes and two-spotted spider mite resistance in cross *Lycopersicon esculentum* "Nandi"and L. pennellii "LA2963". *Euphytica* **2007**, *154*, 231–238. [CrossRef]

42. Alba, J.M.; Montserrat, M.; Fernández-Muñoz, R. Resistance to the two-spotted spider mite (*Tetranychus urticae*) by acylsucroses of wild tomato (*Solanum pimpinellifolium*) trichomes studied in a recombinant inbred line population. *Exp. Appl. Acarol.* **2009**, *47*, 35–47. [CrossRef]

43. Rodríguez-López, M.; Garzo, E.; Bonani, J.; Fereres, A.; Fernández-Muñoz, R.; Moriones, E. Whitefly resistance traits derived from the wild tomato *Solanum pimpinellifolium* affect the preference and feeding behavior of *Bemisia tabaci* and reduce the spread of Tomato yellow leaf curl virus. *Phytopathology* **2011**, *101*, 1191–1201. [CrossRef]

44. Rahimi, F.; Carter, C. Inheritance of zingiberene in Lycopersicon. *Theor. Appl. Genet.* **1993**, *87*, 593–597. [CrossRef] [PubMed]

45. Dutta, P.; Hazari, S.; Karak, C.; Talukdar, S. Study on genetic variability of different tomato (*Solanum lycopersicum*) cultivars grown under open field condition. *IJCS* **2018**, *6*, 1706–1709.

46. Gonzales-Vigil, E.; Hufnagel, D.E.; Kim, J.; Last, R.L.; Barry, C.S. Evolution of TPS20-related terpene synthases influences chemical diversity in the glandular trichomes of the wild tomato relative *Solanum habrochaites*. *Plant J.* **2012**, *71*, 921–935. [CrossRef]

47. Zörb, C.; Piepho, H.-P.; Zikeli, S.; Horneburg, B. Heritability and variability of quality parameters of tomatoes in outdoor production. *Research* **2020**, *2020*, 6707529. [CrossRef]

48. Adhikari, B.; Joshi, B.; Shestha, J.; Bhatta, N. Genetic variability, heritability, genetic advance and trait association study for yield and yield components in advanced breeding lines of wheat. *Nepal. J. Agric. Sci.* **2018**, *17*, 229.

49. Aditya, J.; Bhartiya, A. Genetic variability, correlation and path analysis for quantitative characters in rainfed upland rice of *Uttarakhand hills*. *J. Rice Res.* **2013**, *6*, 24–34.

50. Fernández-Muñoz, R.; Salinas, M.; Álvarez, M.; Cuartero, J.s. Inheritance of resistance to two-spotted spider mite and glandular leaf trichomes in wild tomato *Lycopersicon pimpinellifolium* (Jusl.) Mill. *J. Am. Soc. Hortic. Sci.* **2003**, *128*, 188–195. [CrossRef]

51. Lima, I.P.; Resende, J.T.; Oliveira, J.R.; Faria, M.V.; Dias, D.M.; Resende, N.C. Selection of tomato genotypes for processing with high zingiberene content, resistant to pests. *Hortic. Bras.* **2016**, *34*, 387–391. [CrossRef]

52. Neiva, I.P.; Andrade Júnior, V.C.d.; Maluf, W.R.; Oliveira, C.M.; Maciel, G.M. Role of allelochemicals and trichome density in the resistance of tomato to whitefly. *Ciência Agrotecnol.* **2013**, *37*, 61–67. [CrossRef]

53. Oliveira, C.M.d.; Andrade Júnior, V.C.d.; Maluf, W.R.; Neiva, I.P.; Maciel, G.M. Resistance of tomato strains to the moth *Tuta absoluta* imparted by allelochemicals and trichome density. *Ciência Agrotecnol.* **2012**, *36*, 45–52. [CrossRef]

54. Neiva, I.P.; Silva, A.A.d.; Resende, J.F.; Carvalho, R.d.C.; Oliveira, A.M.S.d.; Maluf, W.R. Tomato genotype resistance to whitefly mediated by allelochemicals and Mi gene. *Chil. J. Agric. Res.* **2019**, *79*, 124–130. [CrossRef]

55. Silva, D.B.; Weldegergis, B.T.; Van Loon, J.J.; Bueno, V.H. Qualitative and quantitative differences in herbivore-induced plant volatile blends from tomato plants infested by either *Tuta absoluta* or *Bemisia tabaci*. *J. Chem. Ecol.* **2017**, *43*, 53–65. [CrossRef] [PubMed]

56. AL-Bayati, A.S. *Breeding for Tomato Resistance to Spider Mite Tetranychus urticae Koch (Acari: Tetranychidae)*; University of Kentucky: Lexington, KY, USA, 2019.

 *horticulturae*

Article

A Strategic Approach to Value Chain Upgrading—Adopting Innovations and Their Impacts on Farm Households in Tanzania

Jesse Steffens *, Kathleen Brüssow and Ulrike Grote

Institute for Environmental Economics and World Trade, Leibniz University Hannover, Königsworther Platz 1, 30167 Hannover, Germany; bruessow@iuw.uni-hannover.de (K.B.); grote@iuw.uni-hannover.de (U.G.)
* Correspondence: jessesteffens@web.de

Received: 26 March 2020; Accepted: 22 May 2020; Published: 2 June 2020

Abstract: The level of agricultural productivity in Sub-Saharan Africa remains far below the global average. This is partly due to the scarce use of production- and process-enhancing technologies. This study aims to explore the driving forces and effects of adopting innovative agricultural technologies in food value chains (FVC). These enhancing FVC technologies are referred to as upgrading strategies (UPS) and are designed to improve specific aspects of crop production, postharvest processing, market interaction, and consumption. Based on cross-sectional data collected from 820 Tanzanian farm households, this study utilized the adaptive lasso to analyse the determinants of UPS. To measure the impact of their adoption on well-being, this study applied the propensity score matching approach (PSM). Results from the adaptive lasso suggested that access to credit, experience of environmental shocks and social capital were the main drivers of UPS adoption. In contrast, the engagement in off-farm wage employment impeded adoption. The results from the PSM suggested that UPS adoption has a positive and significant impact on well-being among sampled households, especially with respect to their total value of durable goods and commercialization. The paper suggests that the promotion of social capital and access to financial capital is pivotal in enhancing the adoption of innovative UPS in the farming sector.

Keywords: value chain analysis; innovations; adaptive lasso; propensity score matching; Tanzania

1. Introduction

Over the last decades, many technological improvements were promoted to increase productivity in the agricultural sector in Sub-Saharan Africa (SSA). Nevertheless, the average agricultural growth rate is still well below the targeted 6% as declared by the Comprehensive Africa Agriculture Development Program [1]. This is also the case for Tanzania. Here, about 70% of the economic value is derived from agriculture and most of the population lives in rural areas and their main source of livelihood is linked to food value chains (FVCs) [2]. FVCs link participants and activities that bring an agricultural product from production at the farm gate to final consumption, with value being added at each stage [3]. Nonetheless, huge portions of what rural farmers produce is consumed within the households, which point at short subsistence-oriented FVCs [4]. Due to the absence of agricultural technologies and sustainable storage facilities, estimated output losses amount to 30% and more throughout FVCs [5,6]. Adopting innovations as improvements in agriculture are necessary, particularly in terms of production at farm level as well as yields and cultivation intensity, in order to promote FVCs. Nonetheless, farmers face multiple constraints such as reliance on rainfall, low soil fertility and weak market systems. Most of these factors diminish yields and efficient trading, hindering farmers from sustaining their basic needs and increasing their income [7,8]. To encounter these challenges, many development interventions and strategies are contentious or have already been implemented [8–10]. Therefore, it is pivotal that innovative upgrading strategies (UPS) alter the diverse obstacles that rural farmers are facing to stimulate FVCs most effectively. For the purpose of this study, UPS are defined as sets of good practices and agricultural technologies used for securing food along the value chain at local and regional levels [11]. They may, by their nature, target improving efficiency, agricultural output and livelihoods by introducing machinery at their location [12]. Another target of UPS is to smoothen temporal food availability, enhance stored grain quality and increase poor farmers' incomes through increased opportunities for market interaction [13]. Succinctly, UPS should stimulate value addition and simultaneously address food security, poverty reduction and income stability through the effective sustainable management of resources [11,14]. Additionally, UPS should fit into existing local and regional FVCs, must consider the local relational household context and be jointly developed with local stakeholders [15]. This study focuses on three specific UPS, namely, a maize-sheller, a millet-thresher and storage superbags. These three devices enhance either production and/or processing stages within the FVCs.

The driving forces of rural farmers decisions about whether to adopt a certain innovation or strategy are very closely linked to the "innovation-diffusion theory" by Rogers [16]. According to the theory, a few farmers are initially willing to try an innovation. As these few early adopters "spread the word" more and more people become aware of the innovation and over time, the innovation spreads. Finally, the more risk averse and poorer farmers adopt the innovation. The assumption that the adoption of innovations is influenced by social interaction and the perceived need for change is reinforced by empirical findings [16,17]. Thus, most of the constraints related to adoption are lack of credit, lack of access to information and markets, unfavourable geographical areas and poor infrastructure, risk aversion and social capital of farmers [16,18–24].

Numerous empirical studies also examined the relationship between demographic and socioeconomic factors and adoption behaviour [18,24]. Younger farmers have been found to more likely adopt a new agricultural technology than older ones. However, older farmers possess more physical capital and are more experienced in adopting UPS, hence, the impact of age on technology adoption is ambiguous [9]. Better educated households are more likely to adopt new technologies and are more likely to benefit from their social network [23]. Labour input is used as a proxy and is associated with the nucleus size of a household [25]. The larger the household size, the more labour is available for agricultural production and the higher the likelihood of adoption. In addition, the farm size increases the adoption of innovations [26,27].

Access to credit helps rural farmers access inputs and labour. A higher asset score is associated with a higher probability of adopting technologies [28]. The existence of off-farm income acts as a strategy to overcome the capital and credit constraints related to intense capital-related technologies faced by rural households [29]. According to Ellis and Freeman [30], off-farm income may substitute for borrowed capital in rural households where capital markets or credit facilities are dysfunctional, hence, increasing the likelihood of adoption. Other studies report that it may reduce the labour input to on-farm activities, therefore reducing the likelihood of adoption [31].

Social capital in the form of social groups in a cooperative enhances trust, as a result improving idea exchange and spreading the exchange of information. Farmers who can learn about the benefits of a particular innovation share this information within the group and spread it within their social network. Therefore, collective processing and the production of crop cultivation enhance the probability of adopting an UPS [32,33]. According to Barrett [22], farmers engage more in effective FVCs if they are provided with postharvest handling activities such as storage. This implies that the likelihood of adopting an enhancing market-oriented storing technology increases if households decide to store crops for selling. The distance from homestead to markets is seen as a path-leading driver for technology adoption. The closer farm land is to main roads or market centres, the more farmers benefit from transportation facilities, hence increasing the likelihood of adoption [34].

Adoption decisions are also influenced by the household's perception of land security [35]. Results of a study conducted in Ethiopia by Teklewold and Köhlin [36] show that a high degree of risk aversion decreases the probability of adopting soil conservation practices. Cavatassi et al. [37] argue that unexpected climatic disasters such as droughts or floods may drive farmers to avoid adopting any UPS. Farmers who have been most vulnerable to extreme weather events are less likely to use process-enhancing fertilizer since the plot will be affected on an interim basis. In this context, a climate-related shock may additionally lead to an income loss. Following the framework of Grothmann and Patt [38], farmers that experience climate-related shocks in a higher frequency or severity have an increased likelihood of adopting several strategies. This implies that the farmers either respond precautiously with long-term strategies that might involve some monetary investment such as an UPS or they respond reactively.

There are only a few impact studies analysing performance enhancing machinery and optimized market storage in SSA regions and in Tanzania in particular. Those that do are predominantly ex ante impact studies [10]. The ex-ante impact assessment studies conducted in Tanzania showed a positive impact of UPS, resulting in higher income and market participation measured by the household commercialization index (HCI) [10]. A study conducted in Nigeria revealed that farmers who adopted UPS machinery devices for improved processing activities realized beneficial outcomes [12]. These beneficial outcomes ranged from increased efficiency in the process of shelling, lowering labour input of shelling and reducing wastage of grains produced, to creation of employment for the youth. UPS in the form of improved bags for market-oriented storage has proven successful in Tanzania, Mozambique, Ghana and Malawi. The study showed that higher prices, at around 50%, were obtained for grains and maize [13,39].

Studies conducted in Tanzania and Ethiopia showed a positive impact of improved processing technologies on consumption expenditures for durable goods [23,40]. Thus, the increase in consumption expenditure on durable goods serves as a proxy for their well-being and indicates that if the value of durable goods increases, it shows a rise in overall well-being of rural farmers. Shiferaw et al. [41] investigated the role of process-enhancing pigeon pea varieties by using the augmented double-hurdle model. Their results suggested that household income improved by up to 80% for those who used the agricultural technology. Furthermore, their disease-induced yield losses decreased by about 50% for local varieties and about 5% for the new varieties. In summary, postharvest loss decreased significantly and therefore, the rural households were able to achieve a higher FVC output and increase their income levels.

The objectives of this paper are twofold: First, what are the determinants of the adoption of these upgrading strategies? Second, how do the upgrading strategies impact the well-being of Tanzanian households? Distinct from the huge number of studies that already exist on determinants of agricultural technologies [15,16,18–20,25,26], this study utilized the adaptive lasso to contribute to a more precise analysis of the determinants. To estimate the impact of upgrading strategies for rural farmers, propensity score matching was applied to control for hidden and self-selection bias.

2. Data and Methodology

2.1. Study Area and Data Collection

The United Republic of Tanzania had a human development index of 0.528 in 2019, which ranks the least developed East African Country on position 159 out of 189 in the world [42]. Tanzania has a diversified landscape, which results in highly variable local conditions [43]. This is also true for the study area (Figure 1).

Figure 1. Location of Tanzania and the sample sites. Study sites are located in Kilosa district in Morogoro region and Chamwino district in Dodoma region (dashed area). Source: Own production using ArcGIS, Trans-Sec (2014).

Annex Table A1 illustrates key distinctions and economic and geographic characteristics of the two study sites. The first one, Dodoma depends on rain-fed farming, especially of millet and sorghum, which further contributes to low agricultural productivity [44]. Noteworthy is that Dodoma focuses on small-scale livestock keeping [45]. The main roads are poorly maintained [46] and the villages are often isolated from nearby markets and cities which hinder farmers from participating in trade. Almost every area in Tanzania was able to lower their level of poverty, but in Dodoma, it continuously increased, while simultaneously commercialization remained subsistence oriented [47]. The second region, Morogoro, is less dry and more diverse in terms of its food system [48]. The sample villages are more closely located to the large handling centre Dar es Salaam and the coast, which is favourable for rural farmers' trade. Nonetheless, the high dependence on agriculture and low level of commercialization in both regions are associated with low income levels.

The households for the survey were selected from the Kilosa district in Morogoro and the Chamwino district in Dodoma. The survey was conducted in 2016 and covered 820 households. In the two-step sampling procedure, six villages with homogeneous attributes were first chosen from the two heterogeneous regions. These villages include Ilakala, Changarawe, and Nyali in Kilosa and Ilolo, Idifu, and Ndebwe in Chamwino. The criteria for the selection process of the six villages are based on several comparable but differing socioeconomic and agroecological conditions [11]. In the second step, 150 households per village were randomly drawn based on household lists and proportionately to their subvillage size [11].

Before the survey, the maize-sheller, millet-thresher, and superbags were introduced to the farmers in the treatment villages. Additionally, farmers were provided with workshops and trainings for the respective innovations. Farmers could choose whether to adopt an UPS or not. Ndebwe and Nyali represent control villages without any interventions. While the maize-sheller was only introduced in Morogoro region, the farmers of Dodoma region only had the choice for the millet-thresher. The machinery devices intend to upgrade the FVC by reducing postharvest losses and increase the quality of the grains. Farmers use the machine once per year after harvest. Both the maize-sheller and millet-thresher are capital intense and require five to six people to operate [12]. To receive access to one of the machines, the farmers had to group up and develop a financial plan. They share the cost of the purchase (maize-sheller about 5,100,000 T.Sh and millet-thresher about 3,600,000 T.Sh), maintenance, and transport of the machine. The superbags increase the quality of grains and enable farmers to obtain competitive prices during lean seasons, as they can store their grain for longer time periods. The price of the bags range from 4000 to 10,000 T.Sh [13]. The household questionnaire covered several topics, including basic sociodemographic characteristics of household members, the perception of environmental shocks and changes in climatic conditions, such as precipitation rates. This is of specific interest since most Tanzanian farmers depend on rain-fed agriculture. The diverse income-generating activities, such as agricultural production, crop cultivation on farm size, off-farm employment, livestock earnings, and returns on capital assets, were surveyed as well. Of particular interest in this study are the value of crop production and the sales value of cultivated crops in determining the household commercialization index. To cover basic infrastructure—such as access to financial credit facilities, distances to markets for each village, and availability of extension services—a village questionnaire was additionally developed.

2.2. Methodology

2.2.1. Adaptive Lasso to Identify Determinants of Adoption

To identify factors that are associated with the decision to adopt an UPS, a three-step procedure is used [49,50]. In the first step, the variables that most likely influence the adoption decision need to be identified. In the second step, the adaptive lasso is then used to determine the factors of adoption, and in the third step, logistic regression models are applied. The binary logistic regression is applied due to the dichotomous nature of the dependent variable (adopter and nonadopter), while the multinomial logistic regression is applied allowing households to adopt more than one UPS. For the purpose of this study, an adopter was defined as a household that uses one of the three presented UPS. All other households that did not adopt the innovation were nonadopters.

For the first step, a summarizing list of the variables can be found in Annex Table A2. Nonetheless, it is unclear if every variable influences the decision to adopt an UPS in Tanzania. One of the major obstacles in microstudies with cross-sectional data is that most of the influencing factors for adoption of agricultural technologies are based on a specific regional context. However, often researchers use variables that have common acceptance in literature, even if it is not appropriate for each context and microstudy area [25]. Therefore, it seems reasonable to only use a subset of variables based on previous studies [51]. Indeed, including all 20 variables, which incorporate household-demographics, assets, social capital factors, climatic as well as geographic variables, and specific characteristics such as risk behaviour, awareness, and the perceived tenure status, would reduce the possibility of omitted variable bias. Nevertheless, in this case, the variance of the estimates would be high, which means that for different samples, the estimates will vary strongly with the result of inaccurate predictions. Therefore, as a remedy, this study uses a statistical model selection procedure, the adaptive lasso [51] given by:

$$(\hat{\beta}_{0,AL}, \hat{\vec{\beta}}_{AL})^{logistic} = \underset{\beta_0,\, \vec{\beta}}{\text{argmin}} \sum_{i=1}^{N} -y_i(\beta_0 + \vec{x}_i\vec{\beta}) + \log\left(1 + \exp\left(\beta_0 + \vec{x}_i\vec{\beta}\right)\right) + \lambda \sum_{j=1}^{J} \hat{w}_j\left|\beta_j\right| \tag{1}$$

This procedure is an extension of the lasso by Tibshirani [52], where $\vec{x}_i = (x_{i,1},...,x_{i,J})'$ is the J linearly independent predictors, β_0 is the intercept, and $\vec{\beta} = (\beta_1,...,\beta_J)$ is the parameter vector.

The important part of this equation is the regulation parameter $\lambda \geq 0$. It controls the amount of shrinkage applied to the estimates and is chosen using k-fold-cross-validation. If the regulation parameter is exactly $\lambda = 0$, lasso nests the standard ordinary least squares estimation. When λ increases, the coefficients continuously shrink towards zero, with the result that for very high λ, some coefficients are exactly zero. Variable selection and parameter estimation are executed simultaneously, meaning an increase in squared bias is thereby traded in for a larger decrease in variance of the estimates. As this paper considers a dichotomous classification, the dependent variable is binary. When applying a linear model to this problem, the probability of $y_i = 1$ given, the values $\vec{x}_i$ are estimated. To ensure that the estimated probabilities of the dependent variable are in the interval [0,1], the logistic regression model can be used [18].

In addition to the ordinary lasso, it has weights $\vec{w} = 1/|\vec{\beta}_j|^y$ assigned to the coefficients. The weights are calculated by determining $\vec{\beta}$ for the full set of explanatory variables using logistic regression. The adaptive lasso has consistency in variable selection for $J > 2$ and asymptotic normality of the estimates, meaning the adaptive lasso fulfils the oracle property [51]. The results of the computation of the adaptive lasso is done in R.

As a third step, binary and multinomial logistic regressions are applied to the subset of variables. After identifying factors that influence the adoption decision with the adaptive lasso, the logistic regression detects the magnitude and direction of the factors. It can be represented as follows [53,54]:

$$E(y_i|\vec{x}_i) = \Pr(y_i = r|\vec{x}_i) = \frac{1}{1 + \exp\left(-\beta_0 - \vec{x}_i\vec{\beta}\right)}, \; r = 0, 1, 2, 3, \; i = 1, 2, ..., 820 \tag{2}$$

The binary model describes the probability of whether adoption has taken place $y_i = 1$ or the alternative 1-Pr for nonadoption $y_i = 0$. Since three UPS are used in this study, the multinomial logistic regression holds that r = {1, 2, 3}, where r = 1 corresponds to the probability that household i adopts the first UPS, r = 2 the second UPS, r = 3 the third UPS, and r = 0 corresponds to the case for no adoption.

The coefficients in the logistic regression model are estimated using the maximum likelihood estimation method. Furthermore, to determine the magnitude, direction, and likelihood, the marginal effects are calculated using the delta method [55]. Additionally, to test if the two independent samples correspond to the same distribution, the Wilcoxon (Mann–Whitney test) rank-sum test is used [56].

2.2.2. Propensity Score Matching to Measure Impacts on Well-being of Rural Households

Estimating the impact of UPS adoption on the well-being of rural farmers using observational data is not an easy task—because of the necessity to identify the counterfactual situation had they not adopted the UPS. This is due to the fact that the farmers are not randomly distributed across the two groups (adopters and nonadopters); rather they are systematically selected by developing agencies based on similar characteristics [23,40]. To overcome the selection bias in the results, this study uses propensity score matching (PSM) [57]. PSM is a common method used when a small treated group needs to be compared to a large control group as it is the case in this study [58,59]. This study uses observational cross-sectional data, where the surveyed households declared if they adopted a certain UPS or not. It is impossible that the same household is observed with and without the adoption of a certain UPS at the same time.

The basic idea of PSM in this study is to build up groups of explanatory variables. Each group possesses relatively similar characteristics with the only difference being the adoption decision. Therefore, the outcome for each household that received the treatment is compared to a similar household that did not received the treatment. To appraise the similarity of characteristics, the propensity score $p(\vec{x}_i)$ is applied as a balancing score, describing the probability of being an adopter given the observed characteristics of $\vec{x}_i$ [58]. Applying the adaptive lasso causes another beneficial effect. The variable selection model reduces the used variables, making it easier to build up groups with characteristics similar to those of the explanatory variables. Eliminating irrelevant variables is only useful when there is a clear census on the unrelated outcome [60].

After computing propensity scores, the average treatment effect on the treated based on propensity score matching can be estimated as follows [61]:

$$\text{ATT}^{\text{PSM}} = E\{E[Y1|A = 1, p(\vec{x}_i)] - E[Y0|A = 0, p(\vec{x}_i)]|A = 1\} \tag{3}$$

where (Y1) is an outcome variable in the form of an income indicator for a specific household which is compared to a similar household outcome (Y0). For A = 1, the household received the treatment meaning that it adopted an UPS, while for A = 0, it did not.

This study uses three different matching algorithms to calculate similar propensity scores following Caliendo and Kopeinig [57]. These include Nearest Neighbour Matching (NNM), Kernel Based Matching (KBM), and Radius Caliper Matching (RCM). For NNM, the five nearest neighbours of household adopters vs. nonadopters were matched with the most similar propensity scores. While simultaneously increasing the variance of the matches and reducing bias, the matching will be executed with replacements, meaning that nonadopters can be used more than once [57,62]. If the distance between households becomes too large, it will likely result in bad matches. To circumvent this risk, RCM is suggested as an altered approach. RCM only includes control units within the given propensity score caliper of 0.01 [63]. KBM utilizes weighted averages to compose the counterfactual outcome. Higher weights are allocated to those with a propensity score close to the treated observations and vice versa. Figure 2 displays propensity score distribution and common support areas. Lastly, Rosenbaum boundaries were calculated as a robustness check. The boundaries identify hidden biases caused by possible unobservable factor heterogeneity. The hidden bias adjusts for the chance to receive the treatment by a factor $\Gamma \geq 1$ and misstates the implication about the ATT [64].

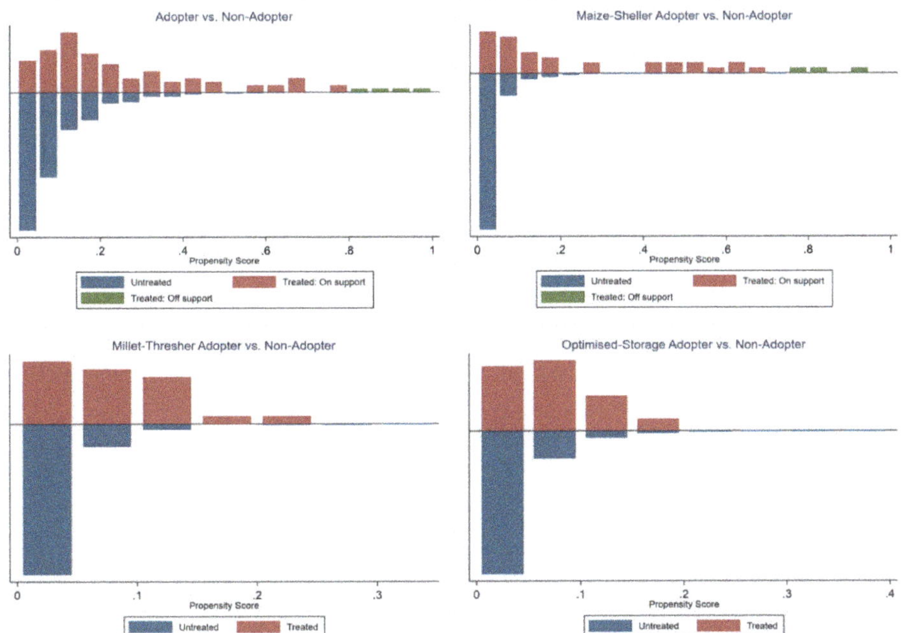

Figure 2. Propensity score distribution and common support for treated and untreated groups. Source: Own calculation based on Trans-SEC data (2014).

The impact variables of the three UPS and their expected impact on the well-being of rural farmers are presented in Table 1 based on previous literature presented in Section 1.

Table 1. Well-being indicators.

Variable	Description	Exp. Direction	Source
Total annual net income	Total available net income of household (PPP US $ 2010)	+	Graef et al.; Shiferaw [11,41]
Income from crops	Income generated from crops (PPP US $ 2010)	+	Kassie et al. [65]
Postharvest loss	Loss after harvest of crops and grains in %	−	Bokusheva et al.; Tefera [66,67]
Household Commercia-lization Index (HCI)	1 = fully commercialized 0 = fully subsistence-oriented	+	Carletto et al. [68]
Total value of durable goods	accounts for goods with durability <1 year	+	Amare et al.; Asfaw et al. [23,40]

3. Results and Discussion

3.1. Characteristics of Adoption

As illustrated in Annex Table A3, 91 households adopted the maize-sheller, millet-thresher, or storage superbags, representing approximately 11% of the whole population in the sample, while 729 did not adopt. About one-third of the adopters and about half of the 729 nonadopters are located in the Dodoma region. Although sociodemographic characteristics such as age, gender, or household size do not differ significantly between adopters and nonadopters, education and household assets such as asset score, farm size, and off-farm wage employment differ significantly. Adopters also differ most strongly from nonadopters with respect to social capital factors such as being a member in agricultural cooperatives. In addition, adopters have on average a higher awareness of changing soil fertility, they experienced more environmental shocks, and suffered from higher income losses due to shock in comparison to nonadopters.

Annex Table A4 reports differences between adopters and nonadopters regarding well-being indicators. Strong significant differences exist for the value of durable goods. Adopters in the Morogoro region have a significantly higher value of durable goods (75.26), in comparison to the nonadopters (25.67). In the Dodoma region, the adopters' value of durable goods is below average (25.43) for the adopters with a value of (13.98) as well for the nonadopters accounting for a value of (18.45). Furthermore, the HCI of the adopters is slightly higher than of the nonadopters in Morogoro but a lot higher than the average of the two sample sites. This indicates that FVCs in Morogoro are more developed than in Dodoma and that the component market interaction in the FVC are utilized with a higher density. In contrast, the descriptive results of the HCI in Dodoma indicate a rather subsistence-oriented agriculture. The total income from crop production in the two different regions does not seem to differ significantly between the adopters and nonadopters. Results indicate that Dodoma has lower income from crop production for adopters (372 PPP US $) and for nonadopters (342 PPP US $) than Morogoro and, furthermore, the incomes are below the average of the sample (414 PPP US $). Although the total annual income per HH does not differ significantly between adopters and nonadopters in Dodoma, the adopters in Morogoro are slightly better off than the nonadopters.

3.2. Determinants of Adoption

To analyse the determinants of the decision to adopt an UPS, the logistic regression with all variables, the adaptive lasso and the logistic regression with only a few selected variables are applied. The regression results are shown in Table 2. When comparing the three methods, it can be seen that nine out of eleven variables, which the adaptive lasso indicated to be relevant, are significant. Furthermore, the coefficients corresponding to these variables are similar for all three methods. The interpretation of the adaptive lasso coefficients is the same as in the logistic regression. The standard deviation of the parameters is, however, much smaller when using only the relevant variables. This emphasizes that the adaptive lasso works as intended. In the following, only the variables, which the adaptive lasso indicated to be relevant, are discussed.

With regard to off-farm wage employment, households that are additionally involved in off-farm activities outside the farming sector are significantly less likely (−0.997) to adopt an UPS. These results are in contrast to the study of Ellis and Freeman [30] who reported that the off-farm income could act as an income buffer to diminish the constraint of obtaining high capital-intense agricultural technologies. Nevertheless, the results are reasonable since farmers who additionally generate income from off-farm activities are less dependent on agriculture and hence adopt less agriculture-intense UPS [31].

The logistic regression also indicates that a household shows a significantly lower probability of adoption (−0.921), if there is a higher awareness about changing soil fertility. This is not necessarily surprising because changing soil fertility can mean that it may change to the better or to the worse. If the soil quality changes to the worse, then the yields and, accordingly, the output are likely to decrease and the household might not want to invest into processing machines or storage bags. This result is expected to be different if the UPS would be related to any soil fertility-enhancing investments as in the case of Lee [32] and Afolami et al. [69]. Then, the probability of adoption would likely increase in order to halt the deterioration and degradation of the soil. If the above result, however, relates to the less likely event that the soil fertility changes to the better, the household is more likely to invest also in processing and storage facilities in anticipation of higher yields. Further research is needed to verify these assumptions.

Contrary to expectations, the results show, on the 5% significance level, that the experience of environmental shocks is positively correlated with the decision to adopt. In an earlier study about improved processing methods in Ethiopia, Cavatassi et al. [37] found a negative relation between experiencing an environmental shock and adoption. However, the results indicate that farmers use adoption as an ex ante coping strategy to mitigate the climatic risks. In Tanzania, environmental shocks such as droughts or floods occur frequently. Especially, the semiarid Dodoma region is prone to droughts. Therefore, the adoption of UPS, especially of the storage superbags, helps rural farmers to cope because they can store their crops for a longer time period. Additionally, farmers are able to sustain their families with food, in case of flooding or storms that destroy their harvest.

As expected, being part of a microcredit group facilitates the likelihood (0.863) of adopting an UPS in our case study. On the one hand, this shows that social capital facilitates the adoption, especially since several investors are needed to make first, the financial plan and then, to buy any of the two processing machines as a group. Furthermore, the idea of a microcredit group is to enhance the access of poor farmers to financial capital. Being member in such a group may thus open the access to credit for such an investment. This is in line with findings from several studies, as access to credit helps rural dwellers overcome the constraint barrier of capital-intense resources, such as agricultural inputs and technologies of greater costs in the form of machinery devices [18,25,70]. At the same time, it has to be noted that a much higher proportion of adopters are members in microcredit groups but that the adoption rate is generally very low. The descriptive results of Annex Table A3 underline the importance of accessing credit in facilitating the adoption decision. Most households stated they did not adopt due to high costs.

Table 2. Factors that influence the adoption decision of upgrading strategies (UPS) in Tanzania—estimates from logistic regression and the adaptive lasso.

Adoption Variables	Logit Regression (All Variables)		Logit Regression (Adaptive Lasso)		Adaptive Lasso Computation
			N = 91		
	Coef	m.e	Coef	m.e	$\overrightarrow{\beta}_{AL}$
HH head is male	0.067	0.052	–	–	–
	(0.342)	(0.272)			
Age of HH head in years	0.01	0.007	–	–	–
	(0.008)	(0.006)			
Education years of schooling HH head	−0.019	−0.001	–	–	–
	(0.044)	(0.035)			
Household size	−0.013	−0.001	–	–	–
	(0.055)	(0.004)			
Livestock keeping	−0.427	−0.033	−0.404	−0.0323	−0.3677549
	(0.319)	(0.025)	(0.311)	(0.024)	
Off-farm wage employment	−0.948 ***	−0.075 ***	−0.997 ***	−0.0798 ***	−0.9905171
	(0.29)	(0.023)	(0.277)	(0.022)	
Farm size	−0.015	−0.012	–	–	–
	(0.091)	(0.007)			
Perceived land security	0.176	0.013	0.194 *	0.01552 *	0.1453635
	(0.113)	(0.008)	(0.11)	(0.008)	
Awareness	−0.875 ***	−0.069 ***	−0.921 ***	−0.073 ***	−0.9037974
	(0.32)	(0.025)	(0.322)	(0.025)	
Asset Score	0	0	–	–	–
	(0.001)	(0)			
Microcredit group	0.756 **	0.060 **	0.863 **	0.069 **	0.8305872
	(0.372)	(0.029)	(0.358)	(0.028)	
Store for selling	0.533	0.042	0.525	0.042	0.4643561
	(0.434)	(0.034)	(0.414)	(0.033)	
Collective processing	2.850 ***	0.226 ***	2.764 ***	0.221 ***	2.7129524
	(0.447)	(0.033)	(0.437)	(0.032)	
Collective production	0.712 *	0.056 *	0.695 *	0.055 *	0.6982264
	(0.372)	(0.029)	(0.36)	(0.028)	
Income loss due to shock	0.001	0.001	–	–	–
	(0.001)	(0.001)			
Experienced environmental shock	0.977 **	0.077 **	0.865 **	0.069 **	0.8168528
	(0.387)	(0.031)	(0.384)	(0.03)	
Member in an agricultural organization	0.894 ***	0.071 ***	0.988 ***	0.079 ***	0.9699722
	(0.271)	(0.022)	(0.259)	(0.021)	
Prepared to take risk	0.047	0.003	–	–	–
	(0.05)	(0.003)			
Distance to next market	0.012	0	–	–	–
	(0.011)	(0)			
Located in Morogoro	0.328	0.026	0.460 *	0.036 *	0.4145265
	(0.309)	(0.024)	(0.268)	(0.021)	
Constant	−4.308 ***	–	−3.495 ***	–	−3.3054075
	(1.009)		(0.698)		
Pseudo R^2	0.205		0.198		
Wald Chi squared (20;11;36)	102.54 ***		93.69 ***		
Prob > Chi^2	0		0		
Log pseudolikelihood	−227		−229		
N	820		820		

Standard error in parentheses; marginal effects are displayed for logistic regression; $\overrightarrow{\beta}_{AL}$ applied for the logistic regression and the multinomial logistic regression; *p*-value *p*-values: * $p < 0.10$, ** $p < 0.05$, *** $p < 0.01$.

Our results further show that also other social network factors, such as being a member of an agricultural organization or doing collective processing activities for postharvest handling, both facilitate the adoption of UPS at the 1% significance level. Based on the innovation-diffusion theory, it has been shown that information spreads more easily and faster in such networks because the farmers constantly exchange new information and constraints to the technology itself diminishes. Furthermore, as mentioned above, social capital is likely to trigger such group investment as needed in the case of a processing machine because there is trust among the members and because it might be easier to get access to the needed investment. Additionally, collective processing activities enhance the learning process of postharvest handling. These findings are in line with those from Lee [32] and Kassie et al. [33] and also correspond to the earlier mentioned descriptive results in Annex Table A2, which highlight that adopters are significantly more engaged in social networks than nonadopters

The results of our analysis also show that living in the Morogoro region favours the likelihood of households to adopt an UPS (0.460). Living in Morogoro as compared to Dodoma increases the probability for adoption by approximately 4%. This may indicate that the region around Morogoro offers higher diversity in terms of agroecological conditions and cultivation, higher productivity on average, and better access to markets. Indeed, these factors have been identified as the most important barriers for adoption as pointed out by a study in Nigeria [12]. Therefore, when a household is located at the Morogoro region, these barriers are on average lower than in Dodoma, and therefore, our farm households located in the Morogoro region are more likely to adopt an UPS than those from the Dodoma region.

The perceived land security status significantly favours the probability to adopt an UPS but only at the 10% level. This is feasible because the most frequent answers of the households on plot characteristics reveal that most of the plots are government owned and since Tanzania's development is relatively low in terms of political stability, the government plays a crucial role in everyday life. The district of Kilosa has been suffering, which is augmented by bloody land conflicts between pastoralists and crop farmers. Farmers that perceive their land as more secure are more willing to try out new agricultural technologies, thus, knowing that they can realize returns from their investments on several farm activities [9].

When running the multinomial logistic regression to further analyse the determinants of the three individual UPS, some further interesting insights are revealed Annex Table A5. First, the same variables turn significant as in the binary case but not for every UPS alike. Then, the social capital variables such as being a member in an agricultural organization or being involved in collective processing strongly favour the likelihood of adopting the millet-thresher and maize-sheller. According to the study conducted by Isham [21] and Barrett [22], not only is the exchange of workers beneficial inside collective cooperatives but also the share and flow of information is higher than in noncooperatives. This enables farmers to take part in higher quality FVCs, which positively correlates with the likelihood of adoption. The nature of the collective activity seems to be dependent on the specific UPS, because for storage superbags adopters only, collective production is significant at the 1% level. This indicates that each UPS is different in usage and one needs to consider regional and local farmer's needs [32]. As expected, the higher experience of an income loss due to a shock increases the probability to adopt the superbags (0.002). This result support the hypothesis that the farmers act either proactively with long-term strategies to avoid losing more fractions of their anyhow subsistence-oriented FVC, or act only reactively, which yields to the same result of preventing upcoming loss [38].

3.3. Upgrading Strategies to Improve the Agriculture Value Chains

To assess the impact of UPS adoption, the households that adopted are compared to nonadopters based on indicators covering the well-being of rural households. The assessment is performed through the average treatment effect on the treated (ATT) based on PSM. As a robustness check, the ATT^{PSM} is estimated using three matching algorithms, including NNM, RCM, and KBM. For the case of multinomial adoption, the common support condition, kernel density plots are used to assess the probability of receiving each treatment level for all observations. The kernel density plot Annex Figure A1 suggests sufficient overlap among the treatment levels, despite a slightly left-skewed distribution for the treatment millet-thresher and storage superbags.

Assessing the impact of UPS adoption, the results for the ATT in Table 3 demonstrate that the adopter of the three presented UPS have a significant higher well-being than nonadopters. Table 3 provides evidence that the calculated ATT of NNM adopters are slightly better off in terms of value of durable goods (18.25), with a significance level of 10%. If the value of durable goods increases, more expenditure on the consumption of durable goods is spent, which further indicates a better well-being of the rural households. This would suggest that the additional income generated out of the increased output by adopting UPS is rather spent on primary needs to sustain sufficient nutrition. Therefore, according to Maslow (1943), primary needs need to be satisfied first before realizing higher overall improvement. The well-being impact results on consumption expenditures on durable goods match the studies conducted by Afolami et al. [69], Amare et al. [23], as well as Asfaw et al. [40]. This finding is highly important because UPS are not only process-enhancing innovations but they also consider the whole improvement sequence of an agricultural FVC.

Table 3. Propensity score matching: the impact of UPS adoption on well-being in Tanzania.

	Nearest Neighbour		Radius		Kernel		Γ
	ATT	S.E	ATT	S.E	ATT	S.E	
Adopter vs. Nonadopter	62.26	384.24	201.66	323.80	125.43	337.8	
Total annual net income per HH (PPP $2010)							
Total net income from crop production per HH (PPP $2010)	−126.95	120.11	12.61	113.27	−95.18	110.79	
Total value of durable goods per HH	18.25 *	9.49	10.91	8.64	9.62	10.17	–
HCI per HH	0.03	0.044	0.056	0.038	0.02	0.033	
% of postharvest loss	0.008	0.017	−0.004	0.016	0.13	0.012	
Maize-Sheller vs. Nonadopter	66.36	575.22	73.90	722.71	−42.12	703.54	
Total annual net income per HH (PPP $2010)							
Total net income from crop production per HH (PPP $2010)	−132.53	230.63	−81.67	254.46	−119.19	290.41	
Total value of durable goods per HH	35.09 *	21.08	34.29 *	18.67	26.67	26.87	3.6
HCI per HH	0.11 *	0.067	0.14 **	0.071	0.13 **	0.062	3.5
% of postharvest loss	0.048	0.035	0.053	0.034	0.05	0.034	
Millet-Thresher vs. Nonadopter	−51.93	796.24	328.32	571.59	546.73	505.55	
Total annual net income per HH (PPP $2010)							
Total net income from crop production per HH (PPP $2010)	−14.57	185.48	51.84	176.65	−8.46	154.08	
Total value of durable goods per HH	−28.02	19.31	−18.24	12.57	−17.84	13.99	
HCI per HH	−0.18 **	0.077	−0.14 **	0.073	−0.16 **	0.060	3.4
% of postharvest loss	−0.005	0.018	−0.006	0.011	−0.011	0.010	
Optimized Storage vs. Nonadopter	−577.55	380.65	−193.17	246.32	−164.6	233.82	
Total annual net income per HH (PPP $2010)							
Total net income from crop production per HH (PPP $2010)	−184.72	182.12	−120.05	166.13	−96.00	141.14	
Total value of durable goods per HH	4.37	13.88	4.67	13.73	8.67	11.47	
HCI per HH	0.06	0.076	0.05	0.058	0.05	0.065	
% of postharvest loss	−0.016	0.16	−0.02 *	0.10	−0.016 *	0.01	4.4

ATT: average treatment effect on the treated; * $p < 0.1$, ** $p < 0.05$, *** $p < 0.01$ when compared to nonadopting farmers; S.E.: bootstrapped standard errors; Γ: Rosenbaum bounds (critical level for hidden bias).

Taking a closer look at each UPS separately gives a clearer picture of the dynamics and impact of adopting UPS in Tanzania. For all three matching algorithms, the adoption of the maize-sheller has a positive and significant ATT for the HCI at the 5% level for RCM as well as KM and 10% for NNM. Adopters of the machinery device maize-sheller improve the process of shelling, therefore, increasing the amount produced for maize with the same amount of input. Following this, the UPS triggers the FVC process that further allows selling more at markets and achieving higher prices for the crops. The higher HCI indicates that more of the grains shelled by the adoption of a maize-sheller is traded at local spot markets or handling centres. The Morogoro region is especially favoured by the geographical proximity next to Dar es Salam. Then again, the increase of the HCI might lead to more economic surplus, resulting in higher income, which is further spent on durable goods, again increasing the well-being of rural households. This is confirmed by the statistical significance for the adoption of the maize-sheller on the value of durable goods for NNM and RCM. The results are consistent with a study conducted by Carletto et al. [68], who measured the degree of commercialization. The study concludes the positive linkage between increased commercialization and improved nutritional status, which reflects the results of the adoption of maize-sheller in this study [68]. Nonetheless, the results of the ATT for NNM at the 5% level for the HCI (-0.18) suggest that for the millet-thresher adopters in Dodoma region, especially, market infrastructures are dysfunctional, meaning the UPS cannot embrace its full potential.

Regarding storage superbags adopters, the ATT shows a significantly negative impact for RCM and KM at the 10% level. Adopters of this UPS have approximately 2% lower postharvest losses. As expected, the results suggest that the superbags decrease the postharvest losses due to the nature of the innovation [67,71]. The superbags enhance the quality of the grains because they are not affected by insects or pesticides. Accordingly, the outcome of the FVC is increased in terms of higher income generated when selling the grains at local spot markets. [12]. Nonetheless, even if the postharvest losses decrease, it does not reflect an increase in higher income or better well-being. The result itself does not show a clear relation to the crops produced. If the farmers do not produce sufficient staple crops, the reduction is not that high to achieve the overall goal of improved well-being.

Regarding the influence of hidden bias represented by Γ, Table 3 lists the Rosenbaum bounds for all significant results. The results concerning total value of durable goods are very robust against hidden bias, since even a three-fold increase of hidden bias does not affect their significance. To continue, the same holds true for the HCI index per household as well as percentage loss of postharvest handling.

4. Summary and Conclusions

This study evaluated the determinants of adopting an upgrading strategy and their potential impacts on household well-being measured by different indicators in rural Tanzania. The analysis used cross-sectional data collected from 820 farm households in Tanzania. In order to investigate factors that influenced the adoption decision, the adaptive lasso was utilized. The causal impact of UPS adoption was then estimated by using the average treatment effect on the treated with propensity score matching. This allowed the estimation of a more accurate effect of UPS adoption on well-being of the households by controlling for the selection bias.

Results of the adaptive lasso highlighted the importance of social capital variables for the likelihood of adopting UPS. Particularly, collective processing and production as well as membership in an agriculture-related organization act as the main factors circumventing constraints to UPS adoption. This suggests that the promotion of social capital is pivotal in enhancing the adoption of innovations and technologies in the farming sector. However, further research on a more detailed differentiation of social network factors is necessary for developing a clearer understanding of how the information exchange of farmers can be used more effectively. In addition, access to credit has been found to be of great importance for promoting technology adoption. Poor farmers without adequate collateral tend to be excluded from formal financial services due to high transaction costs and incomplete information. Thus, financial institutions hesitate to offer them services. Consequently, poor farmers may not be able to invest in new technologies and agricultural productivity-enhancing activities. With respect to the factor "Awareness" (−0.921) about changing soil fertility, further research is needed to better understand the direction of the changes. If farmers are aware of a declining soil fertility, then it is understandable that they have no interest in investing into processing technologies as they are likely to expect declining yields in the future. At the same time, the government should offer UPS to improve soil fertility. Such an innovation is then more likely to be taken up by the farmers.

The impact estimation of the PSM revealed that the adoption of UPS has a significant positive impact on household well-being in rural Tanzania. Results confirmed that the three UPS have a positive impact on the value of durable goods, although the results are not consistent for each UPS separately. This demonstrates the importance of differentiating each UPS individually. The results generally highlighted the potential role of UPS in enhancing rural household welfare, as indicated by the HCI, which should eventually result in higher income. This would translate into higher food security, lower poverty levels and greater ability to withstand environmental risks.

Despite the comprehensiveness of the data, it cannot account for time-varying influences. Therefore, panel data is needed. Furthermore, as already mentioned, another limitation to this study is the limited adoption rate by participants throughout the questionnaire. Therefore, the results of the determinants as well as for the impact need to be approached with caution. Moreover, the definition of an adopter in the present study could lead to the misinterpretation of the impact of individual UPS, because households were able to adopt multiple UPS in addition to the three examined ones.

Nonetheless, the beneficial outcomes of adopting UPS raise the question why such a high proportion of rural households in Tanzania did not adopt UPS. Furthermore, it poses the question why households that adopted the improved machinery did not simultaneously adopt the improved bags to benefit from synergy effects. Overall, the analysis of the determinants of adoption identified lack of access to credit, absence of social networks and lack of information resulting in insufficient awareness as major key constraints to adoption of UPS. The results indicated possible policy interventions which enhance the adoption of UPS. Policy makers could create incentives for rural farmers to establish agricultural cooperatives, both financially and institutionally, to strengthen social capital and access to physical capital. Policy should also focus on the integration of rural farmers who have been unable to build sufficient social network links to increase agricultural productivity and welfare among them. Additionally, the government could improve infrastructure to make financial services more accessible or facilitate microcredit schemes to poor rural farmers. Extension services could promote awareness campaigns on UPS to improve soil fertility, combined with tailored information services on local farms' soil characteristics, enabling a policy mix that facilitates and accelerates adoption. Overall, a strategic approach of upgrading FVCs is indispensable for an effective and efficient improvement of rural farmer's livelihood in Tanzania.

Author Contributions: Conceptualization, J.S., K.B. and U.G.; methodology, J.S. and K.B.; formal analysis, J.S.; investigation, J.S. and K.B.; resources, J.S., K.B. and U.G.; data curation, J.S., K.B. and U.G.; writing—original draft preparation, J.S.; writing—review and editing, J.S., U.G. and K.B.; visualization, J.S.; supervision, K.B. and U.G.; project administration, K.B. and funding acquisition, U.G. All authors have read and agreed to the published version of the manuscript.

Funding: This research received no external funding.

Acknowledgments: This publication is a product of the project "Innovating Strategies to Safeguard Food Security using Technology and Knowledge Transfer: A People-centered Approach (TransSEC)" (http://www.trans-sec.org/) funded by the German Federal Ministry of Education and Research (BMBF) and cofinanced by the German Federal Ministry of Economic Cooperation and Development (BMZ). The views expressed are those of the authors and may not under any circumstances be regarded as stating an official position of the BMBF and BMZ.

Conflicts of Interest: The authors declare that they have no conflict of interest.

Appendix A

Table A1. Basic characteristics of the two sample sites.

Category	Chamwino—Dodoma Region	Kilosa—Morogoro Region
Crop system	Based on sorghum and millet	Based on maize, sorghum, legumes, rice, and horticulture
Commercialization	Subsistence	Subsistence to semi-commercial
Poverty	GDP per capita 690,000 T.Sh,	GDP per capita 1,000,000 T.Sh.
Highland	Flat plains and small hills	Flat plains, highlands and more divers dry alluvial valleys
Livestock	Highly dependent on livestock	Partly dependent on livestock
Climate	Semiarid (350–500 mm)	Predominantly subhumid (600–800 mm)
Markets	Bad infrastructure and weak market access	Medium infrastructure and weak market access
Productivity	Low to medium	Low to high
Land pressure	Medium and high	High

Source: Own compilation based on [10,13].

Table A2. Summary of variables used in this study based on literature.

Variables	Description	Exp. Direction	Source
Household Demographics			
gender	1 = If household head is male	+/=/−	Seymour et al.; Doss and Morris [24,72]
Age	Age of household head in years	+/=/−	Kassie et al.; Feder et al. [9,18]
Education	Years of schooling of household head	+	Amare et al.; Feder et al. [18,23]
Household size	Number of nucleus household members	+	Adeyele et al.; Doss [12,25]
Household Assets			
Asset score	Value of assets in USD (ratio)	+	Teklewold; Morris [28,59]
Livestock	1 = Household keeps livestock	−	Barrett [22]
Farm size	Size of agricultural land owned by household (ha)	+	Uaiene et al.; Feder et al. [18,26]
Off-farm wage-employment	1 = Household has off-farm employment activities	+/−	Ellis and Freeman; Goodwin and Mishra [30,31]
Household Social Capital			
Microcredit group	1 = Household head is part of a microcredit group	+	Abdulai and Huffmann [70]
Member in agricultural organization	1 = Household head is member of any agricultural organization	+	Isham; Kassie et al. [21,33]
Store for selling	1 = Household stores for selling	+	Tefera [67]
Collective processing	1 = Household does collective processing	+	Lee [32]
Collective production	1 = Household does collective production	+	Lee [32]

Table A2. *Cont.*

Variables	Description	Exp. Direction	Source
Household Specific Characteristics			
Awareness	1 = Household head is aware of changing soil fertility (better or worse)	+	Afolami et al. [69]
Prepared to take risk	0 = Household head is absolutely risk averse 10 = HH head is willing to take risk	+	Teklewold and Köhlin [36]
Perceived land security	Perceived tenure status of land security (0 = not secure) − (3 = very secure)	+	Kassie et al. [9]
Household Climate Change			
Experienced environmental shock	1 = Household experienced environmental shock	−	Cavatassi [37]
Income loss due to shock	Average on household income loss due to environmental shock	−	Grothmann and Patt [38]
Geographics			
Distance to next market	Distance from homestead to next market (km)	−	Mwangi and Kariuki; Idrisa et al. [34,73]
Located in Morogoro	1 = Household located in Morogoro	+	URT [2]

Source: Own consideration based on previous studies.

Table A3. Summary statistics of rural farmer's adoption scheme in Tanzania.

Variable	Pooled Sample			By Subsample		
	Total Sample	Adopter	Nonadopter	Maize-Sheller	Millet-Thresher	Storage Superbags
	N = 820	N = 91	N = 729	N = 37 (1)	N = 23 (2)	N = 31 (3)
Household Demographics						
Gender (1 = HH head is male)	0.76 (0.42)	0.79 (0.4)	0.76 (0.42)	0.91 ** (0.28)	0.74 (0.45)	0.67 (0.47)
Age (HH head in years)	51.15 (16.55)	51.22 (16.2)	51.14 (16.6)	47.4 (13.47)	55.6 (11.87)	52.51 (20.76)
Education (HH head years schooling)	4.55 (3.44)	5.22 *** (3.27)	4.46 (3.45)	6.27 *** (2.7)	4.47 (3.19)	4.51 (3.67)
Household size (member)	5.25 (2.35)	5.27 (2.51)	5.24 (2.32)	5.46 (1.79)	5.13 (2,00)	5.16 (3.48)
Household Assets						
Asset score (PPP US $ 2010)	58.87 (125.86)	74.4 *** (98.66)	56.93 (129.77)	97.33 *** (114)	82.27 ** (119.19)	41.18 (36.78)
Livestock (1 = HH owns livestock)	0.8 (0.39)	0.75 (0.43)	0.81 (0.39)	0.76 (0.43)	0.96 (0.21)	0.61 *** (0.49)
Farm size (ha)	2.21 (1.71)	2.65 *** (1.58)	2.16 (1.72)	2.94 *** (1.52)	2.69 *** (1.4)	2.26 (1.74)
Off-farm wage employment (1 = yes)	0.42 (0.49)	0.24 *** (0.43)	0.44 (0.49)	0.27 ** (0.45)	0.08 *** (0.28)	0.32 (0.47)

Table A3. *Cont.*

Variable	Pooled Sample			By Subsample		
	Total Sample	Adopter	Nonadopter	Maize-Sheller	Millet-Thresher	Storage Superbags
	N = 820	N = 91	N = 729	N = 37 (1)	N = 23 (2)	N = 31 (3)
Household Social Capital						
Access to credit (1 = yes)	0.09	0.17 **	0.08	0.24 ***	0.13	0.13
	(0.29)	(0.38)	(0.28)	(0.44)	(0.34)	(0.34)
Member in organization (1 = yes)	0.37	0.59 ***	0.34	0.70 ***	0.65 ***	0.42
	(0.48)	(0.49)	(0.47)	(0.46)	(0.49)	(0.5)
Storing (1 = HH does store for selling)	0.89	0.92	0.89	0.89	1 *	0.9
	(0.3)	(0.26)	(0.3)	(0.31)	(0)	(0.3)
Collective processing (1 = HH does collective processing)	0.04	0.20 ***	0.01	0.38 ***	0.17 ***	0.03
	(0.19)	(0.4)	(0.13)	(0.49)	(0.38)	(0.18)
Collective production (1 = HH does collective production)	0.1	0.14	0.09	0.1	0.04	0.25 ***
	(0.29)	(0.35)	(0.29)	(0.31)	(0.21)	(0.44)
Household Specific Characteristics						
Awareness (1 = yes)	0.45	0.30 ***	0.47	0.20 ***	0.52	0.25 ***
	(0.43)	(0.38)	(0.43)	(0.33)	(0.43)	(0.35)
Risk attitude HH head (0 = fully risk averse) (10 = fully prepared to take risk)	5.56	6.21 *	5.48	6.86 ***	6.34	5.35
	(2.73)	(2.56)	(2.74)	(2.2)	(2.51)	(2.82)
Perceived land security (0 = not secure at all) (3 = very secure)	1.87	1.89	1.86	1.74	2.05	1.96
	(1.11)	(1.02)	(1.13)	(0.98)	(1.19)	(0.92)
Household Climate Effect						
Environmental shock (1 = yes)	0.47	0.56 *	0.46	0.63 **	0.41	0.59 **
	(0.34)	(0.34)	(0.33)	(0.33)	(0.38)	(0.31)
Income loss due to shock (PPP US $ 2010)	708.1	106.5 ***	663.5	1371.17 ***	860.79 *	853.44
	(971.92)	(1318.6)	(910.91)	(1572.94)	(978.13)	(1162.52)
Geographics						
Distance to market (km)	9.55	12.27	9.21	13.24 **	6.56	15.35
	(11.31)	(14.77)	(10.77)	(17.2)	(2)	(16.14)
Region (1 = Morogoro)	0.48	0.66 ***	0.46	1 ***	0.00 ***	0.74 ***
	(0.5)	(0.47)	(0.49)	(0)	(0)	(0.44)

Mean values (with standard deviation in parentheses) across schemes tested for statistically significant differences compared to nonadopting farmers using Mann–Whitney test; * $p < 0.1$, ** $p < 0.05$, *** $p < 0.01$//Max VIF 1.45.

Table A4. Well-being indicators of rural farmers in Tanzania.

Variable	Semi-arid Dodoma Region		Semi-humid Morogoro Region		Total Sample
	Adopter	Nonadopter	Adopter	Nonadopter	N = 820
	N = 31	N = 390	N = 60	N = 339	
HH well-being indicators					
Total annual income per HH (PPP US $ 2010)	1657.12	1411.68	1764.47 *	1311.49	1405.35
	(1575.77)	(1966.44)	(2335.7)	(2508.95)	(2221.26)
Total income from crop production per HH (PPP US $ 2010)	372.55	342.45	447.39	496.15	414.81
	(386.27)	(496.41)	(992.87)	(771.02)	(666.51)
Total value of durable goods per HH	13.98	18,45	75.26 ***	25.67	25.43
	(11.53)	(35.19)	(107.08)	(41.37)	(48.33)
Percentage of postharvest loss	0.016	0.021	0.057 ***	0.031	0.028
	(0.31)	(0.06)	(0.13)	(0.08)	(0.07)
HCI per HH	0.13	0.17	0.50 *	0.44	0.3
	(0.16)	(0.22)	(0.29)	(0.28)	(0.29)

Mean values (with standard deviation in parentheses) across schemes tested for statistically significant differences compared to nonadopting farmers using Mann–Whitney test; * $p < 0.1$, ** $p < 0.05$, *** $p < 0.01$.

Table A5. Factors that influence the adoption decision of UPS in Tanzania—estimates from multinomial logistic regression and the adaptive lasso.

Adoption Variables Base = 0	Multinomial Logistic Regression (Adaptive Lasso)					
	Maize-Sheller		Millet-Thresher		Storage Superbags	
	1		2		3	
	Adopter N = 37		Adopter N = 23		Adopter N = 31	
	Coef	$\vec{\beta}_{AL}$	Coef	$\vec{\beta}_{AL}$	Coef	$\vec{\beta}_{AL}$
HH head is male	1.524 (0.936)	1.38	−0.106 (0.515)	−0.012	−0.6 (0.395)	−0.468
Age of HH head in years	− −	−	− −	−	− −	−
Education years of schooling HH head	− −	−	− −	−	− −	−
Household size	0.068 (0.069)	0.042	−0.169 (0.106)	−0.125	0.041 (0.083)	0
Livestock keeping	− −	−	− −	−	− −	−
Off-farm wage employment	− −	−	− −	−	− −	−
Farm size (ha)	− −	−	− −	−	− −	−
Perceived land security	− −	−	− −	−	− −	−
Awareness	− −	−	− −	−	− −	−
Asset score	− −	−	− −	−	− −	−
Microcredit group	− −	−	− −	−	− −	−
Store for selling	− −	−	− −	−	− −	−
Collective processing	3.384 *** (0.576)	1.555	2.478 *** (0.75)	0.641	0.382 (0.848)	−0.641
Collective production	0.318 (0.672)	0.153	−0.469 (1.067)	−0.436	1.204 *** (0.428)	1.022
Income loss due to shock	0.002 (0.001)	0.008	0.002 * (0.001)	0.002	0.005 (0.002)	−0.002
Experienced environmental shock	1.393 * (0.768) *	0.692	0.093 (0.766)	−0.445	1.184 ** (0.571)	0.445
Member in any agricultural organization	0.997 ** (0.462)	0.356	1.208 ** (0.491)	0.63	0.274 (0.384)	−0.356
Prepared to take risk	0.084 (0.084)	0.034	0.154 (0.099)	0.065	−0.019 (0.071)	−0.039
Distance to next market	- -		- -		- -	
Located in Morogoro	17.60 *** (0.351)	2.42	−18.04 *** (0.35)	−3.329	1.246 *** (0.438)	0.425
Constant	−24.07 *** (2.281)	−4.793	−3.712 *** (1.218)	4.938	−4.593 *** (0.894)	−0.138
Pseudo R^2	0.264					
Wald Chi squared (20;11;36)	14,990.16 ***					
Prob > Chi^2	0.000					
Log pseudolikelihood	−282.75					
N	820					

Standard error in parentheses; marginal effects are displayed for Logistic regression; $\vec{\beta}_{AL}$ applied for the logistic regression and the multinomial logistic regression; *p*-values: * $p < 0.10$, ** $p < 0.05$, *** $p < 0.01$.

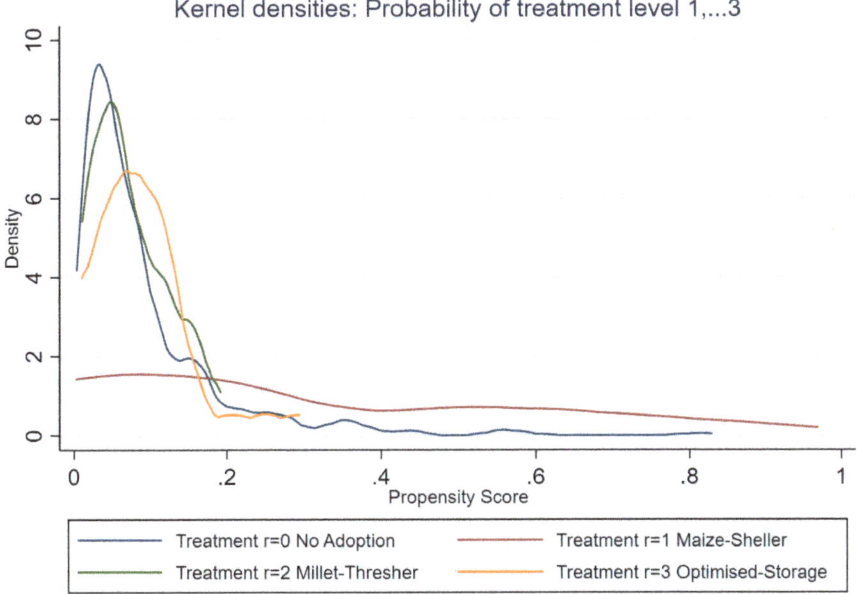

Figure A1. Kernel densities of the probability of treatment level r = 1, . . . ,3. Note: Matching on Nearest Neighbour Matching (NNM) for upgrading strategies (UPS) [1,2,3] and binary case [0,1].

References

1. African Union Inaugural Biennial Review Report of the African Union Commission on the Implementation of the Malabo Declaration on Accelerated Agricultural Growth and Transformation for Shared prosperity and Improved Livelihoods. Available online: http://www.donorplatform.org/news-caadp/au-summit-1st-biennial-review-on-the-status-of-agriculture-in-africa-triggers-unique-momentum-249.html (accessed on 19 August 2019).
2. United Nations Development Programme (UNDP); United Republic of Tanzania (URT). *Tanzania Human Development Report 2017. Social Policy in the Context of Economic Transformation*; Economic and Social Research Foundation: Dar es Salaam, Tanzania, 2017; Available online: http://hdr.undp.org/sites/default/files/thdr2017launch.pdf (accessed on 5 May 2019).
3. FAO. Addressing marketing and processing constraints that inhibit agrifood exports. A guide for policy analysts and planners. In *Food and Agriculture Organization*; Agricultural Services Bulletin 60; FAO: Rome, Italy, 2005.
4. Neven, D. Developing Sustainable Food Value Chains. In *Guiding Principles*; Food and Agriculture Organization of the United Nations (FAO): Rome, Italy, 2014; pp. 86–89.
5. Malabo Montpellier Panel. Mechanized: Transforming Africa's Agriculture Value Chains. 2018. Available online: www.mamopanel.org/media/uploads/files/MaMo_2018_Mechanisiert_web.pdf (accessed on 20 August 2019).
6. High Level Panel of Experts on Food Security and Nutrition of the Committee on World Food Security (HLPE). Food Losses and Waste in the Context of Sustainable Food Systems. Rome, Italy. 2014. Available online: http://www.fao.org/3/a-i3901e.pdf (accessed on 20 August 2019).
7. Manda, J.; Alene, A.D.; Gardebroek, C.; Kassie, M.; Tembo, G. Adoption and Impacts of Sustainable Agricultural Practices on Maize Yields and Incomes: Evidence from Rural Zambia. *J. Agric. Econ.* **2016**, *67*, 130–153. [CrossRef]
8. Mnimbo, T.S.; Lyimo-Macha, J.; Urassa, J.K.; Mahoo, H.F.; Tumbo, S.D.; Graef, F. Influence of gender on roles, choices of crop types and value chain upgrading strategies in semi-arid and sub-humid Tanzania. *Food Secur.* **2017**, *9*, 1173–1187. [CrossRef]

9. Kassie, M.; Teklewold, H.; Jaleta, M.; Marenya, P.; Erenstein, O. Understanding the adoption of a portfolio of sustainable intensification practices in eastern and southern Africa. *Land Use Policy* **2015**, *42*, 400–411. [CrossRef]

10. Graef, F.; Uckert, G.; Schindler, J.; König, H.J.; Mbwana, H.A.; Fasse, A.; Mwinuka, L.; Mahoo, H.; Kaburire, L.N.; Saidia, P.; et al. Expert-based ex-ante assessments of potential social, ecological, and economic impacts of upgrading strategies for improving food security in rural Tanzania using the ScalA-FS approach. *Food Secur.* **2016**, *9*, 1255–1270. [CrossRef]

11. Graef, F.; Sieber, S.; Mutabazi, K.; Asch, F.; Biesalski, H.K.; Bitegekof, J.; Bokelmann, W.; Bruentrup, M.; Dietrich, O.; Elly, N.; et al. Framework for participatory food security research in rural food value chains. *Glob. Food Secur.* **2014**, *3*, 8–15. [CrossRef]

12. Adeyele, O.; Bako, S.; Afiemo, O.G.; Alli-Balogun, K.; Agbo, R. Role of local innovation in mechanisation of maize shelling: Evidence from Igabi, Chikun and Kajuru Local Government Areas, Kaduna State Nigeria. *J. Agric. Ext. Rural Dev.* **2015**, *7*, 170–175. [CrossRef]

13. Trans-SEC. 2014. Available online: http://project2.zalf.de/trans-sec/public/ (accessed on 15 October 2017).

14. Schindler, J.; Graef, F.; König, H.J.; Mchau, D. Developing community based food security criteria in rural Tanzania. *Food Secur.* **2017**, *9*, 1285–1298. [CrossRef]

15. Riisgaard, L.; Bolwig, S.; Matose, F.; Ponte, S.; du Toit, A.; Halberg, N. A Strategic Framework and Toolbox for Action Research with Small Producers in Value Chains. Copenhagen: DIIS. Working Paper. 2008. Available online: https://www.econstor.eu/bitstream/10419/44681/1/573597898.pdf (accessed on 20 March 2018).

16. Rogers, E.M. *Diffusion of Innovations*, 5th ed.; Free Press: New York, NY, USA, 2003.

17. Adesina, A.A.; Zinnah, M.M. Technology characteristics, farmers' perceptions and adoption decisions: A Tobit model analysis in Sierra Leone. *Agric. Econ.* **1993**, *9*, 297–311. [CrossRef]

18. Feder, G.R.; Just, R.E.; Zilberman, D. Adoption of agricultural innovations in developing countries: A survey. *Econ. Dev. Cult. Chang.* **1985**, *33*, 255–298. [CrossRef]

19. Ghadim, A.K.A.; Pannell, D.J. A conceptual framework of adoption of an agricultural innovation. *Agric. Econ.* **1999**, *21*, 145–154. [CrossRef]

20. Sunding, D.; Zilberman, D. The Agricultural Innovation Process: Research and Technology Adoption in a Changing Agricultural Sector. In *Handbook of Agricultural Economics, Volume 1A Agricultural Production*; Gardner, B.L., Rausser, G.C., Eds.; Elsevier: New York, NY, USA, 2001; Available online: https://www.sciencedirect.com/science/article/pii/S1574007201100071 (accessed on 20 March 2018).

21. Isham, J. The effect of social capital on fertilizer use. *J. Afr. Econ.* **2002**, *11*, 39–60. [CrossRef]

22. Barrett, C.B. Smallholder market participation: Concepts and evidence from eastern and southern Africa. *Food Policy* **2008**, *33*, 299–317. [CrossRef]

23. Amare, M.; Asfaw, S.; Shiferaw, B. Welfare impacts of maize-pigeonpea intensification in Tanzania. *Agric. Econ.* **2012**, *43*, 1–17. [CrossRef]

24. Seymour, G.; Doss, C.R.; Paswel, M.; Ruth, S.M.D.; Passarelli, S. Women's empowerment and the adoption of improved maize varieties: Evidence from Ethiopia, Kenya and Tanzania. In Proceedings of the Agricultural & Applied Economics Association's Annual Meeting, Boston, MA, USA, 31 July–2 August 2016; Available online: https://cgspace.cgiar.org/handle/10568/76523 (accessed on 20 March 2018).

25. Doss, C.R. Analyzing technology adoption using microstudies: Limitations, challenges, and opportunities for improvement. *Agric. Econ.* **2006**, *34*, 207–219. [CrossRef]

26. Uaiene, R.N.; Arndt, C.; Masters, W.A. "Determinants of Agricultural Technology Adoption in Mozambique." International Food policy Research Institute, Mozambique. 2009. Available online: https://tropicalsoybean.com/sites/default/files/Determinants%20of%20Agricultural%20Technology%20Adoption%20in%20Mozambique_Uaiene.pdf (accessed on 30 May 2020).

27. Pender, J.; Gebremedhin, B. Determinants of agricultural and land management practices and impacts on crop production and household income in the highlands of Tigray, Ethiopia. *J. Afr. Econ.* **2007**, *17*, 395–450. [CrossRef]

28. Morris, S.S.; Carletto, C.; Hoddinott, J.; Christiaensen, L.J.M. Validity of rapid estimates of household wealth and income for health surveys in rural Africa. *J. Epidemiol. Community Health* **2000**, *54*, 381–387. [CrossRef]

29. Reardon, T.; Stamoulis, K.; Pingali, P. Rural Nonfarm Employment in Developing Countries in an era of Globalization. *Agric. Econ.* **2007**, *37*, 173–183. [CrossRef]

30. Ellis, F.; Freeman, H.A. Rural Livelihoods and Poverty Reduction Strategies in Four African Countries. *J. Dev. Stud.* **2004**, *40*, 1–30. [CrossRef]

31. Goodwin, B.K.; Mishra, A.K. Farming efficiency and the determinants of multiple job holding by farm operators. *Am. J. Agric. Econ.* **2004**, *86*, 722–729. [CrossRef]

32. Lee, D.R. Agricultural Sustainability and Technology Adoption: Issues and Policies for Developing Countries. *Am. J. Agric. Econ.* **2005**, *87*, 1325–1334. [CrossRef]

33. Kassie, M.; Jaleta, M.; Shiferaw, B.; Mmbando, F.; Mekuria, M. Adoption of interrelated sustainable agricultural practices in smallholder systems: Evidence from rural Tanzania. *Technol. Forecast. Soc. Chang.* **2013**, *80*, 525–540. [CrossRef]

34. Idrisa, Y.L.; Ogunbameru, B.O.; Madukwe, M.C. Logit and Tobit analyses of the determinants of likelihood of adoption and extent of adoption of improved soybean seed in Borno State, Nigeria. *Greener J. Agric. Sci.* **2012**, *2*, 37–45. [CrossRef]

35. Gebremedehin, B.; Scott, M.S. Investment in soil conservation in northern Ethiopia: The role of land tenure security and public programs. *Agric. Econ.* **2003**, *29*, 69–84. [CrossRef]

36. Teklewold, H.; Köhlin, G. Risk preferences as determinants of soil conservation decisions in Ethiopia. 2010. Available online: https://www.researchgate.net/publication/46456062_Risk_Preferences_as_Determinants_of_Soil_Conservation_Decisions_in_Ethiopia (accessed on 20 March 2018).

37. Cavatassi, R.; Lipper, L.; Narloch, U. Modern variety adoption and risk management in drought prone areas: Insights from the sorghum farmers of eastern Ethiopia. *Agric. Econ.* **2011**, *42*, 279–292. [CrossRef]

38. Grothmann, T.; Patt, A. Adaptive capacity and human cognition: The process of individual adaptation to climate change. *Glob. Environ. Chang.* **2005**, *15*, 199–213. [CrossRef]

39. Guidi, D. Sustainable Agriculture Enterprise: Farming Strategies to Support Smallholder Inclusive Value Chains for Rural Poverty Alleviation. CID Research Fellow and Graduate Student Working Paper No. 53. Center for International Development at Harvard University. 2011. Available online: https://ideas.repec.org/p/cid/wpfacu/53.html (accessed on 20 March 2018).

40. Asfaw, S.; Shiferaw, B.; Simtowe, F.; Lipper, L. Impact of modern agricultural technologies on smallholder welfare: Evidence from Tanzania and Ethiopia. *Food Policy* **2012**, *37*, 283–295. [CrossRef]

41. Shiferaw, B.A.; Kebede, T.A.; You, L. Technology adoption under seed access constraints and the economic impacts of improved pigeonpea varieties in Tanzania. *Agric. Econ.* **2008**, *39*, 309–323. [CrossRef]

42. United Nations Development Programme (UNDP). Human Development Indices and Indicators 2018. 2019. Available online: http://hdr.undp.org/sites/all/themes/hdr_theme/country-notes/TZA.pdf (accessed on 3 February 2020).

43. Rowhani, P.; Lobell, D.B.; Linderman, M.; Ramankutty, N. Climate variability and crop production in Tanzania. *Agric. Forest Meteorol.* **2011**, *151*, 449–460. [CrossRef]

44. United Republic of Tanzania (URT). National Sample Census of Agriculture 2007/2008: Regional Report: Morogoro Region. Dar es Salaam: URT. 2012. Available online: https://www.ocgs.go.tz/php/ReportOCGS/Zanzibar%20Agriculture%20Sample%20Census%20Livestock%20Report%20Final%202009.pdf (accessed on 20 March 2018).

45. National Bureau of Statistics (NBS). Basic Facts and Figures on Human Settlements 2012. Tanzania Mainland. Dar es Salaam: Ministry of Finance. 2014. Available online: https://www.nbs.go.tz/index.php/en/census-surveys/environmental-statistics/75-basic-facts-and-figures-on-human-settlements-2012-tanzania-mainland (accessed on 22 March 2018).

46. Kiunsi, R.B. A Review of Traffic Congestion in Dar es Salaam City from the Physical Planning Perspective. *J. Sustain. Dev.* **2013**, *6*, 94–103. [CrossRef]

47. Minot, N.; Simler, K.; Benson, T.; Kilama, B.; Luvanda, E.; Makbel, A. Poverty and Malnutrition in Tanzania: New Approaches for Examining Trends and Spatial Patterns. Washington D.C.: International Food Policy Research Institute. 2006. Available online: http://www.repoa.or.tz/documents_storage/Research%20and%20Analysis/IFPRI%20Report.pdf (accessed on 21 March 2018).

48. Mnenwa, R.; Maliti, E. A comparative analysis of poverty incidence in farming systems of Tanzania. Special Paper 10/4, Research on Poverty Alleviation (REPOA), Dar es Salaam. 2010. Available online: http://www.repoa.or.tz/documents/10-4_web-1.pdf (accessed on 20 March 2018).

49. Schneider, U.; Wagner, M. Catching Growth Determinants with the adaptive lasso. *German Econ. Rev.* **2012**, *13*, 71–85. [CrossRef]

50. Ali, D.A.; Derick, B.; Deininger, K. Personality Traits, Technology Adoption, and Technical Efficiency: Evidence from Smallholder Rice Farms in Ghana. World Bank Policy Research Working Paper No. 7959. 2017. Available online: https://ssrn.com/abstract=2910738 (accessed on 15 November 2017).
51. Zou, H. The Adaptive Lasso and Its Oracle Properties. *J. Am. Stat. Assoc.* **2006**, *101*, 1418–1429. [CrossRef]
52. Tibshirani, R. Regression Shrinkage and Selection via the Lasso. *J. Royal Stat. Soc.* **1996**, *58*, 267–288. [CrossRef]
53. Hosmer, D.W.; Lemeshow, S. *Applied Logistic Regression*; Wiley: New York, NY, USA, 1989.
54. Chan, Y.H. Multinomial logistic regression. Biostatistics 305. *Singap. Med. J.* **2005**, *46*, 259–269.
55. Wooldridge, J.M. *Econometric Analysis of Cross Section and Panel Data*, 2nd ed.; Massachusetts Institute of Technology, MIT Press: Cambridge, MA, USA, 2010; pp. 561–585.
56. Wijnand, H.; Van De Velde, R. Mann–Whitney/Wilcoxon's nonparametric cumulative probability distribution. *Comput. Methods Programs Biomed.* **2000**, *63*, 21–28. [CrossRef]
57. Caliendo, M.; Kopeinig, S. Some practical guidance for the implementation of propensity score matching. *J. Econ. Surv.* **2008**, *22*, 31–72. [CrossRef]
58. Rosenbaum, P.R.; Rubin, R. The central role of the propensity score in observational studies for causal effects. *Biometrika* **1983**, *70*, 41–55. [CrossRef]
59. Teklewold, H.; Kassie, M.; Shiferaw, B.; Köhlin, G. Cropping system diversification, conservation tillage and modern seed adoption in Ethiopia: Impacts on household income, agrochemical use and demand for labor. *Ecol. Econ.* **2013**, *93*, 85–95. [CrossRef]
60. Bryson, A.; Dorsett, R.; Purdon, S. The Use of Propensity Score Matching in the Evaluation of Labour Market Policies. Working Paper 4, London: Department for Work and Pensions. 2002. Available online: http://eprints.lse.ac.uk/4993/ (accessed on 20 March 2018).
61. Becker, S.O.; Ichino, A. Estimation of Average Treatment Effects Based on Propensity Scores. *Stata J.* **2002**, *2*, 358–377. [CrossRef]
62. Heckman, J.; Ichimura, H.; Todd, P.E. Matching as an econometric evaluation estimator: Evidence from evaluating a job training programme. *Rev. Econ. Stud.* **1997**, *64*, 605–654. [CrossRef]
63. Dehejia, R.H.; Wahba, S. Propensity Score Matching Methods for Non-Experimental Causal Studies. *Rev. Econ. Stat.* **2002**, *84*, 151–161. [CrossRef]
64. Rosenbaum, P.R. Covariance Adjustment in Randomized Experiments and Observational Studies. *Stat. Sci.* **2002**, *17*, 286–327.
65. Kassie, M.; Shiferaw, B.; Muricho, G. Agricultural technology, crop income, and poverty alleviation in Uganda. *World Dev.* **2011**, *39*, 1784–1795. [CrossRef]
66. Bokusheva, R.; Finger, R.; Fischler, M.; Berlin, R.; Marín, Y.; Pérez, F.; Paiz, F. Factors determining the adoption and impact of a postharvest storage technology. *Food Secur.* **2012**, *4*, 279–293. [CrossRef]
67. Tefera, T.; Abass, A.B. Improved Postharvest Technologies for Promoting Food Storage, Processing, and Household Nutrition in Tanzania. International Institute of Tropical Agriculture. 2012. Available online: https://cgspace.cgiar.org/bitstream/handle/10568/24886/aresa_postharvest.pdf?sequence=1 (accessed on 20 March 2018).
68. Carletto, C.; Corral, P.; Guelfi, A. Agricultural commercialization and nutrition revisited: Empirical evidence from three African countries. *Food Policy* **2017**, *67*, 106–118. [CrossRef]
69. Afolami, C.; Obayelu, A.; Vaughan, I. Welfare impact of adoption of improved cassava varieties by rural households in South Western Nigeria. *Agric. Food Econ.* **2015**, *3*, 1–17. [CrossRef]
70. Abdulai, A.; Huffman, W.E. The Diffusion of New Agricultural Technologies: The Case of Crossbred-Cow Technology in Tanzania. *Am. J. Agric. Econ.* **2005**, *87*, 645–659. [CrossRef]
71. De Groote, H.; Hall, M.D.; Spielman, D.J.; Mugo, S.; Andam, K.; Munyua, B.G. Options for pro-poor maize seed market segmentation in Kenya. *Afr. J. Biotechnol.* **2011**, *10*, 4699–4712.
72. Doss, C.R.; Morris, M. How does gender affect the adoption of agricultural innovations? The case of improved maize technology in Ghana. *Agric. Econ.* **2001**, *25*, 27–39. [CrossRef]
73. Mwangi, M.; Kariuki, S. Factors determining adoption of new agricultural technology by smallholder farmers in developing countries. *J. Econ. Sustain. Dev.* **2015**, *6*, 208–216.

 *horticulturae*

Review

The Potential of Introduction of Asian Vegetables in Europe

Jungha Hong and Nazim S. Gruda *

Institute of Plant Sciences and Resource Conservation, Division of Horticultural Sciences, Faculty of Agriculture, University of Bonn, 53121 Bonn, Germany; jungha.hong.miru@gmail.com
* Correspondence: ngruda@uni-bonn.de

Received: 1 May 2020; Accepted: 30 June 2020; Published: 3 July 2020

Abstract: Increasing longevity, along with an aging population in Europe, has caused serious concerns about diet-related chronic diseases such as obesity, diabetes, cardiovascular diseases, and certain cancers. As recently noted during the coronavirus pandemic, regular exercise and a robust immune system complemented by adequate consumption of fruit and vegetables are recommended due to their known health benefits. Although the volume of fresh vegetable consumption in the EU is barely growing, demand for diversified, nutritious, and exotic vegetables has been increasing. Therefore, the European market for fresh Asian vegetables is expected to expand across the EU member states, and the introduction of new vegetables has enormous potential. We conducted this review to address the high number and wide range of Asian vegetable species with a commercial potential for introduction into the current European vegetable market. Many of them have not received any attention yet. Four Asian vegetables: (1) Korean ginseng sprout, (2) Korean cabbage, (3) Coastal hog fennel and (4) Japanese (Chinese or Korean) angelica tree, are further discussed. All of these vegetables possess several health benefits, are increasingly in demand, are easy to cultivate, and align with current trends of the European vegetable market, e.g., vegetables having a unique taste, higher value, are decorative and small. Introducing Asian vegetables will enhance the diversity of nutritious horticultural products in Europe, associated with all their respective consumption benefits. Future research on the Asian vegetable market within Europe is needed. In addition, experimental studies of Asian vegetables under practical conditions for their production in different European environments are required. Economic, social, and ecological aspects also ought to be considered.

Keywords: health; aging population; consumption of fruit and vegetables; diversification; market trend; Korean ginseng sprout; Ssamchoo; *Peucedanum japonicum*; *Aralia elata* (Miq.) Seem

---

1. Introduction

Health and well-being are important factors of a modern lifestyle. The health conditions of individuals and households have economic significance for both developing and developed countries, including the EU member states [1]. Additionally, an aging population is considered a significant challenge in the coming decades for the European Union [2]. Worldwide chronic diseases are increasing as leading causes of death every year, with scientific evidence linking these chronic diseases to diet [3]. These diet-related chronic diseases include obesity, diabetes, cardiovascular diseases (CVDs), and certain cancers [3–6].

Obesity is associated with various health problems including CVD, type 2 diabetes mellitus, musculoskeletal disorders such as osteoarthritis, work disability and respiratory disorders [7]. According to a 2016 report from the Organization for Economic Co-operation and Development/ European Union (OECD/EU [8], 16% of adults in the EU member states were obese in 2004, an increase from 11% in 2000. Moreover, overweight or obese children have a higher risk of poor health in adulthood [9].

Diabetes causes a higher risk of developing cardiovascular diseases such as heart attack and stroke [8]. Across EU countries, 7% of adults reported having diabetes in a self-reported interview survey in 2014. The rate of diabetes is associated with the level of education and age [8].

CVDs are the leading cause of mortality in nearly all EU countries [8]. CVDs caused almost 40% of all deaths in 2013 [8]. CVDs include a wide range of diseases, such as ischemic heart diseases (e.g., heart attack) and cerebrovascular diseases (e.g., stroke) [8].

Cancer is the second leading cause of mortality after cardiovascular diseases in EU countries [8]. 26% of all deaths were caused by cancer in 2013 [8]. Cancer mortality rates for men are higher than for women in all EU countries [8]. Considered risk factors for colorectal cancer are age, a diet high in fat, and genetic background [8].

According to WHO, the COVID-19 pandemic is caused by a novel coronavirus [10,11]. A study in China reported that approximately 50% of patients confirmed with COVID-19 had underlying chronic diseases such as CVDs (40%) and diabetes [12]. Another retrospective analysis in Wuhan showed that obesity was commonly observed in critical patients and death from COVID-19 [13]. According to this study, 88.24% of patients with a BMI (body mass index) over 25 kg/m^2 were in the non-survivor group [13]. Based on current evidence, COVID-19 is considered to be highly associated with a compromised immune system of individuals that are affected by age, gender and comorbidities (CVDs, diabetes, cancer) [14]. Understanding the significant correlation among diet, nutrition, infection, and immunity [15], a diversified, balanced, and healthy diet with macro-, micro- and phytonutrients, mainly found in vegetables, may promote healthy immune responses for prevention of chronic diseases [16].

European consumers are growing more interested in diversified, nutritious, and exotic vegetables. The cultivation and consumption of Asian vegetables are increasing in Europe. For instance, Chinese cabbage (*Brassica pekinensis*) [17–20] and pak choi (*Brassica chinensis*) [17,18,20,21], which were unknown some years ago, are the two most popular Asian vegetables in Europe. These leafy vegetables originating from China [22] are essential in northeastern Asia. However, as they started to be commercially cultivated in Europe, bolting was a significant concern for European growers [23]. Bolting is induced by vernalization which is a process of accelerated flowering by exposure to prolonged cold [24] that results in elongation of the flower stalk [25]. Vernalization lets the plants bloom and produce seeds instead of leafy heads [26], which then lead to low yield and quality loss of pak choi [27] and Chinese cabbage. As a result, many research studies of cultivation have shown how to control bolting and early flowering of Chinese cabbage and pak choi.

Mizuna [17,20], a cultivar of *Brassica rapa* var. *niposinica*, and water spinach (*Ipomoea aquatica*) [17,20,21] are other Asian vegetables introduced to Europe. They both are less well known than Chinese cabbage and pak choi in Europe. Mizuna is a Japanese vegetable [28] that is mostly cultivated in open fields but also inside greenhouses in Italy [29,30]. There is also a concern about bolting for commercial production of mizuna [28]. In contrast, to our knowledge, water spinach is consumed but is not commercially cultivated in Europe.

Currently, shiso (*Perilla frutescens* var. *crispa*), which is a Japanese herb, is cultivated in greenhouses in Germany [31]. Choy sum (*Brassica rapa* var. *parachinensis*), kai choi (*Brassica juncea*) [17–21], edamame (*Glycine hispida*), Japanese murasaki sweet potato (*Ipomoea batatas*), Okinawa sweet potato (*Ipomoea batatas* cv. 'Ayamurasaki') are produced in Portugal [32].

Carotenoids, anthocyanins, flavonoids, and other phenolic compounds in vegetables may prevent and counter many common diseases such as high blood pressure, diabetes, cancer, vision loss, heart disease, and several intestinal disorders. The link between fruit and vegetable consumption and a lower risk of mortality has been increasing [33–35]. Cardiovascular mortality may notably decrease with high consumption of fruit and vegetables [36]. It was also noticed that not only the elderly and chronic disease sufferers, but also smokers and alcohol drinkers were in the high-risk patient group during the Covid-19 pandemic. Thus, eating vegetables is suggested together with sporting activities to boost the immune system [16,37].

To our best knowledge, no published literature to date provides a comprehensive review of Asian vegetables that have the potential to be introduced in Europe. The objective of this paper is to contribute to an increasing diversity of vegetable consumption in Europe.

2. Vegetable Consumption in Europe

According to Shahbaz et al. [38], vegetables provide a variety of health benefits. They are generally low in fat and calories yet rich in vitamins and dietary fibers. Moreover, they present a valuable source of mineral nutrients, essential amino acids, antioxidants and phytochemicals such as carotenoids, anthocyanins, flavonoids, and other phenolic compounds.

Besides the nutritional attributes, vegetables also provide a diverse range of taste, aroma, texture, and color, which significantly increase the variety in food and satisfy a myriad of personal preferences [39]. Although increased daily vegetable intake has been recommended for health benefits, and the EU has made efforts to advocate it, European vegetable consumption remains below the recommended level [40–42]

2.1. '5-A-day' Campaign of Daily Vegetable & Fruits Consumption

WHO recommended daily consumption of a minimum 400 g of fruits and vegetables to prevent chronic diseases in 1990 [43]. After that, the '5-a-day' fruit and vegetable campaign was launched in several European countries such as the UK, Germany, and France [43]. The '5-a-day' fruit and vegetable campaign recommended at least five portions of fruits and vegetables a day, excluding potatoes and other starchy tubers [44]. Indeed, evidence provided by a meta-analysis showed that consumption of vegetables and fruit—up to five servings a day—was associated with a lower risk of all causes of mortality, especially cardiovascular mortality [36]. In addition, the consumption of similar amounts of vegetables appears to be significantly healthier than that of fruits [43,45] because the fructose found in fruit can be a major factor of obesity [46–48].

2.2. EU Policy to Promote Vegetable & Fruits Consumption

The EU has made efforts to promote a well-balanced diet to prevent chronic diseases. One of the policy objectives of the EU is the promotion of fruit and vegetable consumption, especially in schools and workplaces [49]. The European School Fruit Scheme, which supplies fresh fruit and vegetables to children together with educational measures, was introduced in the schools of 24 EU countries in 2009–2010 [8]. Since then, the European School Fruit Scheme has been positively showing a significant increase in fruit and vegetable intake frequency by children [8]. Caton et al. [40], reported the importance of learning to eat vegetables at an early age. The European Commission is monitoring progress in the consumption of fruit and vegetables to maintain a balance against a worsening trend of poor diets [8,50]. The WHO European Food and Nutrition Action Plan 2015-20 was designed to reduce the burden of preventable diet-related NCDs (Noncommunicable diseases) and obesity [8].

2.3. Daily Vegetable (& Fruits) Consumption of Young (Girls & Boys) and Adults in the EU

According to OECD/EU [8], over one in three girls and one in four boys ate vegetables daily in EU countries in 2013–14. The average daily vegetable consumption of adults was 51% in 28 EU countries in 2014. More women eat vegetables daily than men and older adults eat more fruits and vegetables daily than younger people do. About 12% of adults in 28 EU countries have reported a daily consumption of at least five fruits and vegetables. People with higher education levels tend to eat fruits and vegetables more often than those with lower education levels.

2.4. Daily Vegetable Consumption of European Countries—A Systematic Comparison with Asian Countries

According to the Food and Agriculture Organization of the United Nations (FAO) (1999) [51], about 75 percent of the world's food is generated from only 12 plants and five animal species. A recent study by international scientists in collaboration with the FAO identified a total of 1097 vegetable species worldwide [52] of which 392 vegetable crops are cultivated and consumed [39]. However, even in the current globalization era, there are significant differences between countries. More than 200 species are known to be produced in Asian countries whereas the maximum number of vegetable species is only 60 coming from 16 plant families in Germany. Considering the fact that approximately 81% of the consumption/capita/year is concentrated on 23 vegetable species, it is clear that the diversity of vegetables in Germany is relatively weak [53]. This situation is reflected in the average rate of daily vegetable consumption.

Data sourced by OECD Health Statistics (2017) [54] showed that the average rate of daily vegetable consumption among adults in OECD 32 member countries is 59.8%. The data of European countries are all lower than of those other countries including Korea (99.1%), Australia (99.0%), New Zealand (95.3%) and the United States (92.4%) (unfortunately, no data for China were recorded) [54].

China was the world's largest fresh vegetable producer in 2017 [55] and was the world's biggest vegetable exporter in 2018 [56]. The FAO and World Bank reported in 2017 [57] that China was likely to consume the most vegetables per capita (1033 g/day). In other Asian countries, South Korea consumed 540.14 g, Vietnam 467.25 g, Japan 249.61 g, and Thailand 109.3 g per day [57].

Among the Mediterranean countries, Croatia consumed the most vegetables per capita (699.03 g/day), followed by Portugal (418.62 g/day), Greece (410.84 g/day), and Spain (401.52 g/day) [57]. In Northwestern Europe, Belgium consumed the most vegetables (379.46 g/day), followed by Denmark (276.44 g/day), France (264.85 g/day) and Germany (261.29 g/day) [57].

Despite the current recommendations and the apparent health benefits, vegetable consumption in European countries is below the recommended level [40–42]. For instance, in Germany, the consumption per capita has increased from 186.3 g to 261.38 g in the past 30 years [53,57]. However, consumption has stagnated at approximately 94–95 kg/capita/year for many years now. The recent numbers include the comprehensive campaigns that were previously mentioned, including the 5X a day campaigns, activities for pre- and school-children, as well as information targeting adults concerning health issues.

Therefore, one possible way to increase the consumption of vegetables in EU countries could be diversification. Consumers are looking for new products, especially in the winter with reduced varieties available, which could be an excellent opportunity for new vegetables coming from Asian countries.

3. The Potential for the Introduction of New Asian Vegetables in Europe and the Market Trends

The consumption of fresh vegetables in the EU is barely increasing, and the European vegetable market is highly competitive. However, the current demand for healthy, high-quality and attractive exotic vegetables is growing steadily. Therefore, the market for fresh Asian vegetables has the potential to expand across the EU member states. But, how can we define Asian vegetables?

The meaning of "Asian vegetables" varies and the term is often used ambiguously as "Oriental vegetables" in literature. Asian vegetables can be defined as vegetables that have originated, are cultivated or are commonly consumed in Northeast Asia (China, Korea, and Japan) or in Southeast Asia (Thailand, Indonesia, Vietnam, the Philippines, and Malaysia). In this study, we will use the term Asian vegetables to mean the vegetables from both regions.

A list of Asian vegetables is provided in Tables 1–6. In the list, some well-known vegetables that can be found in other countries are not included, such as onions (*Allium cepa*), potatoes (*Solanum tuberosum*), sweet potatoes (*Ipomoea batatas*), tomatoes (*Solanum lycopersicum*), and okra (*Abelmoschus esculentus*) even though they are commonly used in Asian cuisines. However, we still included these vegetables if they are unique from their western counterparts, such as Asian green onions (*Allium fistulosum*), Japanese pumpkin (*Cucurbita maxima*), and Korean zucchini (*Cucurbita moschata*). Native Asian vegetables such as soybeans and Chinese cabbage are also included.

Table 1. Sprout vegetables.

Vegetable	Scientific Names	Used Parts	Culinary Uses	Reference	Vegetable	Scientific Names	Used Parts	Culinary Uses	References
Mung bean sprouts/Mungo bean sprouts/Small bean sprouts	*Phaseolus aureus/Vigna radiata*	sprouts	eaten raw, blanched, stir-fried	[17,18,21]	Soybean sprouts	*Glycine hispida/G. max*	sprouts	blanched, boiled	[17,18,21]

Table 2. Tree vegetables.

Vegetable	Scientific Names	Used Parts	Culinary Uses	Reference	Vegetable	Scientific Names	Used Parts	Culinary Uses	References
Aralia sprout/Dureup/Fatsia sprout/Korean angelica tree	*Aralia elata* (Miq.) Seem./*Aralia elata* var. elata	young shoots	blanched, pickled	[18,19,58]	Alangium/Trilobed-leaf alangium	*Alangium platanifolium* var. *trilobum* (Miq.) Ohwi	young leaves	eaten raw young leaves, blanched, pickled	[18,58]
Bamboo shoot	*Bambusa* spp./*Bambusa* sp./*Dendrocalamus* sp./*Phyllostachys* sp.	young shoots	blanched, boiled, pickled, stir-fried	[17,18,21]	Ginkgo nut	*Ginkgo biloba*	seeds	cooked, grilled	[18]
Burning bush spindle tree	*Euonymus alatus* (Thunb.) Siebold	young shoots	blanched	[18]	Blue Japanese oak acorn/Ring-cup oak acorn	*Quercus glauca* Thunb.	acorns	its starch is cooked and made jelly	[18,58]
Chinese cedar	*Cedrela sinensis* Juss.	young shoots	blanched, pickled, deep fried	[18,19]	Konara oak acorn	*Quercus serrata* Murray	acorns	its starch is cooked and made jelly	[18,58]
Devil's bush/Siberian-ginseng	*Eleutherococcus senticosus*	young shoots	blanched	[18,58]	Sansho pepper/Sichuan pepper	*Zanthoxylum sp.*	husks of seeds	adding in stews, foods, oiled	[20]
Matrimony-vine	*Lycium chinense*	young shoots	blanched young leaf buds	[17,18]	East Asian arrow root/Kudzu	*Dolichos japonica/Pueraria hirsute/P. lobata* (Willd.) Ohwi/*P. thunbergiana/*	roots	eaten raw, made juice, dried and boiled tea, made jelly, powdered	[17,18,58]
Sayur manis/weet leaf bush	*Sauropus androgynus*	shoot tips, young leaves	steamed, grilled, cooked in soups, stir-fried	[21]					

Table 3. Leaf and Stem Vegetables.

Vegetable	Scientific Names	Used Parts	Culinary Uses	Reference	Vegetable	Scientific Names	Used Parts	Culinary Uses	References
Asian goldenrod	*Solidago virgaurea* subsp. *asiatica* Kitam. ex H. Hara	young shoots	blanched	[18,19,58]	Komatsuna	*Brassica rapa* var. *komatsuna/*var. *perviridis*	leaves, stalks	eaten raw, stir-fried, boiled, pickled, cooked in soups	[20]
Asian royal fern/Royal fern	*Osmunda japonica* Thunb.	young shoots	boiled, stir-fried	[18,19,58]	Korean cabbage/Ssamchoo	*Brassica lee* ssp. namai cv. Ssamchoo	leaves, stalks	eaten raw	[18]
Bracken/Eastern brakenfern/Tender fern fronds	*Pteridium aquilinum* var. *latiusculum* (Desv.) Underw. ex A.Heller	young shoots	boiled, stir-fried	[17,19,58]	Mibuna greens	*Brassica rapa* var. *japonica/*var. *nipposinica*	leaves, stalks	eaten raw	[20]
Edible aster/Rough aster	*Aster scaber* Thunberg	young shoots	eaten raw	[18,19,58]	Mizuna greens/Potherb-mustard cabbage	*Brassica rapa* var. *japonica/*var. *nipposinica*	leaves, stalks	eaten raw, stir-fried, cooked in soups	[17,20]
Fulvous daylily	*Hemerocallis fulva* L.	young shoots	blanched	[18,19]	Sweet potato	*Ipomoea batatas*	leaves, stalks	blanched	[18]
Japanese spikenard/Manchurian spikenard/Mountain asparagus/Udo	*Aralia cordata* var. *continentalis* (Kitag.) Y.C.Chu	young shoots	blanched, deep fried, cooked as pancakes	[18,20]	Turnip greens	*Brassica rapa*	leaves, stalks	boiled, stir-fried, salted	[17,18]
Pea shoots	*Pisum sativum*	young shoots	stir-fried	[20]	Chinese taro/Japanese taro/Taro	*Colocasia esculenta*	stalks	boiled, stir-fried	[17,18]

Table 3. *Cont.*

Vegetable	Scientific Names	Used Parts	Culinary Uses	Reference	Vegetable	Scientific Names	Used Parts	Culinary Uses	References
Tartarian aster/Tatarinow's aster	*Aster tataricus* L. f.	young shoots	blanched	[18,19,58]	Citronella grass/Lemon grass	*Cymbopogon citratus*	stalks	cooked in curries, soups	[20]
Water shield	*Brasenia schreberi* J.F.Gmelin	young shoots	eaten raw, cooked in soups, dried, grounded	[20,58]	Sweet potato	*Ipomoea batatas*	stalks	blanched, stir-fried, salted	[17]
Hot pepper	*Capsicum annuum*	young shoots, leaves	Blanched, cooked in soups and in pancakes	[18]	Amaranthus/Chinese spinach/Edible amaranth/Pig weed	*Amaranthus gangeticus/A. mangostanus* L./*A. tricolor*	young leaves, stems	blanched, stir-fried, cooked in soups	[17–21,58]
Komarov's Russian thistle/Okahijiki	*Salsola komarovii* Iljin	young shoots, leaves	eaten raw, blanched, steamed	[20,58]	Ashitaba	*Angelica keiskei*	young leaves, stems	eaten raw, blanched, cooked in pancakes, pickled	[18,19]
Korean-mint	*Agastache rugosa*	young shoots, leaves	eaten raw young leaves, blanched, cooked in soups, as pancakes	[18]	Coastal hog fennel	*Peucedanum japonicum* Thunb.	young leaves, stems	eaten raw, pickled, blanched, deep fried	[18,19,58]
Okinawan spinach/Velvet plant	*Gynura bicolor*	young shoots, leaves	eaten raw, blanched, stir-fried, deep-fried, cooked in soups, stews	[20]	Curled mallow/Mallows	*Malva verticillata/M. verticillata* var. *crispa* L.	young leaves, stems	cooked in soups, stir-fried	[18–20]
Stringy stonecrop	*Sedum sarmentosum*	young shoots, leaves	eaten raw	[18,19]	East Asian hogweed	*Heracleum moellendorffii* Hance	young leaves, stems	eaten raw, blanched, pickled, stir-fried	[18,19,58]
Ceylon spinach/Malabar nightshade/Tsurumurasaki	*Basella alba/B. cordifolia B. rubra/*	shoots, leaves	stir-fried, cooked in soups	[17,20,21]	Field aster	*Aster yomena* (Kitam.) Honda	young leaves, stems	blanched, stir-fried, cooked in soups, deep fried	[18,58]
Castor	*Ricinus communis* L.	young leaves	blanched	[18]	Gondre/Korean thistle	*Cirsium setidens* (Dunn) Nakai	young leaves, stems	cooked in soups, pickled, dried leaves are cooked as side dishes	[18,19,58]
Mugwort	*Artemesia* spp./*A. princeps* Pampanini/*A. vulgaris*	young leaves	cooked in soups, in rice cakes, pancakes	[17,20,21]	Green garlic	*Allium sativum*	young leaves, stems	eaten raw, blanched, steamed, pickled	[18]
Chinese chives/Flat chives/Garlic chives	*Allium senescens/A. senescens* var. *minor/A. thunbergii/A. tuberosum*	leaves	eaten raw, stir-fried, pickled, cooked in pancakes	[18–21]	Japanese angelica	*Ligusticum acutilobum*	young leaves, stems	eaten raw, blanched, pickled	[18]
Perilla	*Perilla frutescens*	leaves	eaten raw, pickled, stir-fried	[17,18,20]	Japanese atractylodes/Ovate-leaf atractylodes	*Atractylodes japonica/A. ovata* (Thunb.) DC.	young leaves, stems	eaten raw, blanched, cooked in pancakes	[18,58]
Soybean	*Glycine max* (L.) Merr.	leaves	blanched, pickled, salted	[18]	Lesser solomon's seal/Solomon's seal	*Polygonatum odoratum* var. *pluriflorum* (Miq.) Ohwi	young leaves, stems	pickled	[18,58]
Asian plantain	*Plantago asiatica* L.	young leaves, stalks	blanched, cooked in soups, stir-fried	[18,58]	Riverside wormwood/Selenge wormwood	*Artemisia selengensis* Turcz. ex Besser	young leaves, stems	blanched	[18,58]
Butterbur/Giant butterbur/Sweet coltsfoot/	*Petasites japonicus* (Siebold & Zucc.) Maxim.	young leaves, stalks	blanched, pickled	[17–20,58]	Swamp cabbage/Water convolvulus/Water spinach	*Ipomoea aquatica/I. reptans*	young leaves, stems	Eaten raw, blanched, boiled, stir-fried	[17,20,21]
Chamnamul/Short-fruit pimpinella	*Pimpinella brachycarpa* (Kom.) Nakai	young leaves, stalks	blanched	[18,19,58]	Three-leaf ladybell	*Adenophora triphylla* var. *japonica* (Regel) H. Hara	young leaves, stems	eaten raw, blanched	[18,58]
Deltoid synurus	*Synurus deltoides* (Aiton) Nakai	young leaves, stalks	blanched, cooked in rice cakes	[18,19,58]	Ulleungdo aster	*Aster glehnii* F.Schmidt	young leaves, stems	blanched, deep fried	[18,58]

Table 3. *Cont.*

Vegetable	Scientific Names	Used Parts	Culinary Uses	Reference	Vegetable	Scientific Names	Used Parts	Culinary Uses	References
Fischer's ragwort/Gomchwi	*Ligularia fischeri* (Ledeb.) Turcz.	young leaves, stalks	eaten raw, cooked in soups, picked, blanched	[18,19,58]	Water celery/Water dropwort/Water parsley	*Oenanthe javanica/O. stolonifera/*	young leaves, stems	eaten raw, blanched, cooked in soups, as pancake	[17–20]
Honewort/Mitsuba	*Cryptotaenia canadensis*	young leaves, stalks	eaten raw, stir-fried, deep fired, cooked in soups	[17]	Chinese broccoli/Chinese kale	*Brassica oleracea* var. *alboglabra*	leaves, stems	boiled, steamed, stir-fried	[17,20,21]
Korean bellflower	*Campanula takesimana* Nakai	young leaves, stalks	blanched	[18,19,58]	Chrysanthemum/Chrysanthemum greens/Crown daisy/Garland chrysanthemum	*Chrysanthemum coronarium*	leaves, stems	eaten raw, cooked in stews, blanched	[17–21]
Narrow spiked Ligularia/Narrow-head ragwort	*Ligularia stenocephala* (Maxim.) Matsum. & Koidz.	young leaves, stalks	eaten raw, cooked in soups, picked, blanched	[18,19,58]	Stem lettuce	*Lactuca sativa* var. *augustana*/var. *asparagina*	leaves, stems	blanched, stir-fried	[20]
Pumpkin	*Cucurbita moschata*	young leaves, stalks	steamed	[18]	Manchurian wild rice/Water bamboo/Wild rice	*Zizania latifolia* (Griseb.) Turcz. ex Stapf	swollen stem bases	eaten raw, stir-fried	[20,58]
Rape	*Brassica napus* L.	young leaves, stalks	blanched, cooked in soups	[18,19]	Green onion/Japanese bunching onions/Oriental bunching onion/Scallions/Shallots/Spring onion/Welsh onions	*Allium fistulosum*	leaves, bulbs	eaten raw, stir-fried, salted, cooked in soups, pancakes	[17–21]
Salt sandspurry	*Spergularia marina* (L.)	young leaves, stalks	eaten raw, blanched, cooked in pancakes	[18]	Korean wild chive Wild garlic	*Allium monanthum* Maxim.	leaves, bulbs	eaten raw cooked in soups, pickled	[18,19,58]
Bok choy/Chinese cabbage/Pak choi/White cabbage/White-mustard cabbage	*Brassica campestris* var. *chinensis*/*B. rapa* var. *chinensis*	leaves, stalks	eaten raw, blanched, stir-fried	[17,18,20,21]	Korean youngia	*Youngia sonchifolia*	young leaves, roots	salted	[18,19]
Chinese cabbage	*Brassica pekinensis*/*B. pe-tsai*	leaves, stalks	eaten raw, salted, cooked in soups, as pancakes, pickled, stir-fried	[17–20]	Shepherd's purse	*Capsella bursa-pastoris*	young leaves, roots	blanched, cooked in soups	[18–20]
Chinese celery	*Apium graveolens*	leaves, stalks	eaten raw, stir-fried, cooked in soups, steamed, pickled	[20,21]	Toothed ixeridium/Toothed ixeris	*Ixeridium dentatum* (Thunb.) Tzvelev	young leaves, roots	blanched, cooked as pancakes, salted	[18,19,58]
Chinese clover	*Medicago denticulate*/*M. hispida*	leaves, stalks	stir-fried	[20]	Young radish	*Raphanus sativus* L.	young leaves, roots	salted	[18,19]
Chinese mustard/Green mustard/Leaf-mustard cabbage/Mustard cabbage/Mustard greens	*Brassica juncea*/*B. juncea* var. *rugosa*	leaves, stalks	salted, pickled, stir-fried	[17–21]	Korean ginseng sprout	*Panax ginseng* Meyer	young leaves, stems, roots	eaten raw	[59]
East Asian wildparsley/Mitsuba/Honewort	*Cryptotaenia canadensis*/*C. japonica* Hassk.	leaves, stalks	eaten raw, blanched, stir-fried, deep fired, cooked in soups	[17–20,58]	Mioga ginger	*Zingiber mioga* (Thunb.) Roscoe	flower buds	blanched, cooked in soups, pickled	[18]
Horseradish/Japanese horseradish/Wasabi	*Eutrema japonica*/*Wasabia japonica*	leaves, stalks	eaten raw, cooked in soups, picked, blanched	[18]	Banana bud	*Musa* spp.	flowers	eaten raw, blanched, stir-fried, steamed, cooked in soups, curries	[21]
Japanese radish greens/Korean radish greens	*Raphanus sativus*	leaves, stalks	dried leaves are cooked in soups, stir-fried	[18]					

Table 4. Fruit, pod and seed vegetables.

Vegetable	Scientific Names	Used Parts	Culinary Uses	Reference	Vegetable	Scientific Names	Used Parts	Culinary Uses	References
Angled gourd/Dishcloth gourd/Vegetable gourd/Luffa/Sponge gourd	*Luffa acutangula*	fruits	cooked in soups, stir-fried	[17,20,21]	Spaghetti marrow	Cucurbita pepo var. fastigata	fruits	steamed, baked, stir-fried	[21]
Balsam-pear/Bitter gourd/Bitter melon	*Momordica charantia*	fruits	steamed, stir-fried, cooked in soups, in stews	[17,19–21]	Baby corn/Miniature corn	*Zea mays* var. *rugosa*	corns with cobs	eaten raw, stir-fried	[21]
Bottle gourd/Edible bottle gourd/White-flowered gourd	*Lagenaria leucantha/L. siceraria/L. vulgaris/*	fruits	eaten raw, dried, stir-fried, cooked in soups	[17–21]	Asparagus bean/Cow pea/Long bean/Yard long bean/Yard-Long beans	*Dolichos sesquipedalis/Vigna sesquipedalis/Vigna sinensis* var. *sesquipedalis*	pods	eaten raw, stir-fried, cooked in stews	[17,20,21]
Chinese preserving melon/Wax gourd/Winter gourd	*Benincasa cerifera/B. hispida*	fruits	eaten raw, cooked in soups, in stews, pickled, steamed, stir-fried	[17,18,20,21]	Asparagus pea/Goa beans Winged bean	*Psophocarpus tetragonolobus*	pods	blanched, stir-fried	[17,21]
Eggplant	*Solanum melongena*	fruits	steamed, stir-fried, deep fried, pan fried, grilled, cooked in soups, in stews	[18,21]	Edible-podded peas	Pisum sativum var. macrocarpon	pods	stir-fried	[17]
Fiery little bird's-eye chili	*Capsicum frutescens*	fruits	eaten raw, cooked	[21]	Snow pea/Sugar pea	Pisum sativum var. saccaratum	pods	blanched, stir-fried	[21]
Finger-length chili	*Capsicum annuum* cv. group longum	fruits	eaten raw, cooked	[21]	Adzuki bean	Phaseolus angularis/Vigna angularis	beans	cooked gruel, sweet paste	[18,20]
Hairy melon/Jointed gourd	*Benincasa hispida* var. *chieh-gua*	fruits	eaten raw, steamed, cooked in soups, stir-fried	[20]	Borlotti bean/Red-streaked bean	*Phaseolus vulgaris* cv.	beans	simmered, cooked in soups, stews	[21]
Japanese pumpkin/Kabocha squash	*Cucurbita maxima/C. moschata/* some hybrids between the two	fruits	Steamed, deep fried, cooked pancakes, porridge	[18,20]	Cowpeas	*Vigna sinensis/V. unguiculate* L. Walp.	beans	grounded, cooked gruel, pancakes, jelly	[17–19]
Korean zucchini/Young pumpkin	*Cucurbita moschata*	fruits	stir-fried, cooked in soups, pancakes	[18,19]	Jwinunikong	*Rhynchosia nulubilis*	beans	cooked, roasted, boiled tea	[18]
Long eggplant	Solanum melongena var. serpentinum	fruits	steamed, stir-fried	[17,19]	Mungbean	*Vigna radiata*	beans	grounded, cooked gruel and pancakes, jelly	[18,19]
Oriental cucumber	*Cucumis sativus*	fruits	eaten raw, salted, pickled	[20]	Soya bean/Soybeans	*Glycine hispida/G. max*	beans	grounded, cooked and made tofu, pastes, sauces, sweets, side dishes	[17–20]
Oriental pickling melon	*Cucumis melo* var. *conomon*	fruits	pickled	[18]	Sword bean	*Canavalia ensiformis*	beans	cooked, boiled, steamed, roasted and boiled tea	[18,19]
Oriental pickling melon/Pickling melon	*Cucumis melo* var. *conomon*	fruits	Eaten raw, pickled, baked, stir-fried, boiled, cooked in soups	[17,20]	Twisted cluster bean	*Parkia speciosa*	beans	stir-fried	[21]
Papaya	*Carica papaya*	fruits	eaten raw, simmered, cooked in soups	[21]	Perilla seed	*Perilla frutescens*	seeds	roasted and made oil, roasted or grounded and used as seasonings	[18]
Pea eggplant	*Solanum torvum*	fruits	eaten raw, cooked in curries	[21]	Sesame seeds	*Sesame indicum*	seeds	roasted and made oil, roasted or grounded and used as seasonings	[18,20]
Snake gourd	Trichosanthes cucumerina var. anguina	fruits	stir-fried, cooked in curries	[21]	Water caltrop/Water chestnut	Trapa bicornis /T. bispinosa /T. natans	seeds	boiled	[17,20,58]

Table 5. Root vegetables.

Vegetable	Scientific Names	Used Parts	Culinary Uses	Reference	Vegetable	Scientific Names	Used Parts	Culinary Uses	Reference
Arrowhead/Chinese potato/Swamp potato	*Sagittaria sagittifolia/S. sinensis*	tubers	boiled, fried, stir-fried, grilled	[17,20]	Gegeol radish	*Raphanus sativus*	roots, leaves	salted, pickled	[18]
Chinese artichoke	*Stachys affinis/S. Sieboldii/S. tubifera*	tubers	eaten raw, blanched, stir-fried, steamed, pickled	[20]	Japanese turnips	*Brassica rapa* var. *rapifera*	roots, leaves	eaten raw, boiled, stir-fried, cooked in soups	[20]
Chinese potato/Chinese yam/Cinnamon-vine/East Asian mountain yam/Yam	*Dioscorea batatas/D. japonica* Thunb./*D. opposite*	tubers	eaten raw, steamed, cooked gruel and pancakes	[17–20,58]	Korean radish	*Raphanus sativus*	roots, leaves	eaten raw, salted, cooked in soups, dried	[18,19]
Baker's garlic/Rakkyo	*Allium chinense*	bulbs	pickled	[20]	Small radish	*Raphanus sativus* L.	roots, leaves	salted	[18,19]
Chinese taro/Japanese taro/Taro/	*Colocasia esculenta*	corms	cooked in soups, stews, grilled, steamed, simmered	[17–21]	Balloon-flower	*Platycodon grandifloras* (Jacq.) A.DC.	roots	eaten raw, boiled as tea	[18,19,58]
Gonyak	*Amorphophallus konjac*	corms	Its starch is cooked and made jelly	[18]	Bonnet bellflower/Deodeok/Lance Asiabell	*Codonopsis lanceolate* (Siebold & Zucc.) Benth. & Hook.f. ex Trautv.	roots	eaten raw, grilled, salted	[18,19,58]
Water chestnut	*Eleocharis dulcis* (Burm.f.) Trin. ex Hensch./*E. tuberosa*	corms	eaten raw, boiled, pickled, grilled	[20,21,58]	Burdock/Great burdock	*Arctium lappa/A. majus/Lappa major, L. edulis*	roots	blanched, pickled, stir-fried, deep fried, cooked in soups	[17,19,20]
Lotus roots	*Nelumbo nucifera*	rhizomes	eaten raw, steamed, stir-fried, pickled	[17–21]	Horseradish/Japanese horseradish/Wasabi/Wild wasabi	*Eutrema japonica* (Miq.) Koidz./*Wasabia japonica*	roots	grated roots are used as seasoning 'Wasabi' for Sushi	[18–20,58]
Daikon	*Raphanus sativus longipinnatus*	roots, leaves	pickled, simmered, cooked in soups, dried	[17–19,21]					

Table 6. Mushrooms.

Vegetable	Scientific Names	Used Parts	Culinary Uses	Reference	Vegetable	Scientific Names	Used Parts	Culinary Uses	References
Juda's ear	*Auricularia auricula-judae*		stir-fried	[18]	Pine mushroom	*Tricholoma matsutake* Sing.		eaten raw, grilled	[18]
Lesser rock tripe/Plated rock tripe/Smooth rock tripe	*Manna lichen*		blanched, stir-fried	[18]	Straw Mushroom	*Volvariella volvacea*		stir-fried, cooked in soups	[21]
Oak Mushroom/Shiitake Mushroom	*Lentinus edodes*		stir-fried, cooked in stews, soups, pancakes	[18,21]	Winter mushroom	*Flammulina velutipes* Sing.		stir-fried, cooked in stews, soups	[18]

Due to the limited number of publications and information in Europe, we added some commonly consumed or unique Asian vegetables to the list. Interestingly, some vegetables that had been traditionally gathered in the wild have recently been cultivated to provide more choice for consumers looking for diversity and health benefits. These vegetables are becoming more available in both online and offline markets.

Since vegetables have various common English names in the literature, we use their scientific names to avoid confusion and for better accuracy. The Asian vegetables are categorized under six groups: (i) sprout vegetables (Table 1), (ii) tree vegetables (Table 2), (iii) leaf and stem vegetables (Table 3), (iv) fruit, pod and seed vegetables (Table 4), (v) root vegetables (Table 5), and (vi) mushrooms (Table 6).

For the list of Asian vegetables in Tables 1–6, we gathered information from three published sources in English which introduce Asian vegetables, mainly from China, Japan and Southeast Asia. We also referred to information from published literature in Korean from various South Korean government ministries for more specific data. Additionally, we searched for the popularity of each species on the Internet. If the vegetable name did not appear in offline or online markets or in recipes and comments of internet users, we did not include the vegetable name in the list. Furthermore, we discuss four Asian vegetables and their future potential in the European market.

4. Four Recommended Asian Vegetables

Several Asian vegetables that were traditionally gathered wild, harvested from fields or forests, are now sold in online and offline markets in South Korea as 'Ssam' or 'Namul' vegetables. Vegetable-consuming cultures can also be found in other Asian countries. Pickled vegetables are consumed in China as 'Pao Cai' and in Japan as 'Tsukemono' and 'Asazuke'. Additionally, most Southeast Asian countries and China typically consume vegetables by using the stir-fried cooking method. The vegetables are consumed mostly in the forms of 'Kimchi', 'Ssam' and 'Namul' as we explain further below.

'Kimchi' consists of salted and fermented vegetables. It is served as a side dish in almost every meal and can be made with various kinds of vegetables such as Chinese cabbage, radish (*Raphanus sativus*), cucumber (*Cucumis sativus*), and spring onion (*Allium fistulosum*). Approximately 300 varieties of 'Kimchi' are estimated to exist in Korea [60]. 'Ssam' consists of wrapping fillings such as cooked meat, rice (*Oryza sativa*) with fresh leafy vegetables such as lettuce (*Lactuca sativa*) and perilla leaves (*Perilla frutescens*). Recently, various new leafy vegetables have been introduced for 'Ssam' and landed successfully in the vegetable market. 'Namul' are seasoned fresh vegetables prepared either blanched or raw as side dishes. Various vegetables, not only cultivated but also gathered wild, are used for 'Namul'.

Below, we present four Asian vegetables. They possess health benefits and exhibit increasing demand in the vegetable market. Additionally, they are easy to cultivate and fit the current trends in the European vegetable market because of their unique taste, higher value, decorative ability and small size. We strongly believe that these crops have the potential to be introduced in the European market in the future.

4.1. Korean Ginseng Sprouts (Panax Ginseng Meyer)

Korean ginseng is a perennial plant in the *Araliaceae* family. Korean ginseng has been used as a medicinal plant in East Asia [61] for more than 2,000 years [62]. Ginsenoside (triterpene glycoside saponin) is the main bioactive ingredient in Korean ginseng and is known to have various pharmacological and physiological benefits [61] such as anti-cancer [63], anti-diabetic [64,65], immunomodulatory [63,66], neuroprotective [63], radioprotective [67], anti-amnestic [63], and anti-stress properties [68,69].

Korean ginseng sprout [59] is a new medicinal vegetable recently introduced in the South Korean vegetable market with demand increasing [59]. Korean ginseng is customarily used only for its roots and should be grown for about 5–6 years before harvest [70]. Because of this length of time, it is a high-priced product for consumers (Figure 1). In contrast, Korean ginseng sprout uses all parts of the plant, not only roots, but especially the leaves due to their young tender texture. Leaves contain higher levels of certain types of ginsenosides (Rg1, Rg2+Rh1, Rd, and Rg3) than roots [71]. It can be grown in a hydroponic cultivation system either without soil [72,73] or with nursery soil [59,74] Therefore, ginseng sprouts can be produced all year round without using pesticides in appropriate conditions. It only needs to be grown for 2–8 weeks after transplanting of one-year-old-ginseng-seedlings [59,72].

Figure 1. Mountain-cultivated Korean ginseng (*Panax ginseng* Meyer).

These plants are found in mountainous regions (Gangwon-do, South Korea). Compared to intensive field-cultivated plants, mountain-cultivated ginseng is more expensive. Since wild mountain ginseng is rare and Eastern Asians strongly believe in its health benefits, its value is highly appreciated. Intensively cultivated ginseng needs 5–6 years until harvest, while wild mountain ginseng has a slow growth rate and requires a longer time. Mountain ginseng has a smaller size than intensively cultivated ginseng. Korean ginseng is not usually sold in supermarkets, but in particular markets only for Korean ginseng products or online markets (Source: Hong, personal communication)

Korean ginseng sprouts are a small sized vegetable with a mild, bitter flavor. Soil-less ginseng sprouts can provide health benefits and a pesticide-free alternative for consumers. It can be consumed in a variety of ways, including in salads, milkshakes, deep-fried, sushi, soups, and tea. Furthermore, Korean ginseng sprout can be used in health food supplements and cosmetics.

4.2. Ssamchoo (Brassica lee ssp. Namai cv. Ssamchoo) and Red Ssamchoo (Brassica koreana Lee var. Redleaf)

Ssamchoo [18] is a hybrid plant species. This handle-fan-shaped leafy vegetable was developed by aneuploidy crossing Chinese cabbage with regular cabbage as 'Ssam' leafy vegetables [75,76]. Therefore, Ssamchoo combines the advantages of Chinese cabbage and cabbage [76]. Red Ssamchoo (*Brassica koreana Lee* var. *redleaf*) is a variety of Green Ssamchoo [76].

According to a leafy vegetables-agricultural technology guide from the Rural Development Association (RDA) South Korea [76], Ssamchoo has all the benefits of Chinese cabbage and cabbage, and a very high fiber content making it easier to digest. Overall contents of calcium, iron, vitamin A and ascorbic acid in Ssamchoo are higher than Chinese cabbage, cabbage and lettuce [76].

Producers can grow Ssamchoo either in soil or hydroponic systems. The method of cultivation is similar to that of lettuce. Ssamchoo can be harvested one month after sowing [76]. However, this vegetable is intolerant to hot and humid weather [76].

Ssamchoo has a smaller leaf size (Figure 2) than those of Chinese cabbage and regular cabbage [76]. Red Ssamchoo draws attention due to its red leaf and stem color [76]. Ssamchoo brings together the slight bitterness from Chinese cabbage and nutty sweetness from regular cabbage [76]. This leafy vegetable is lightly aromatic, sweet, juicy, mini-sized, and can be used in salads and soups and may be deep-fried [76].

Figure 2. Ssamchoo (*Brassica lee* ssp. namai cv. Ssamchoo).

Leaves of Ssamchoo can be easily found in open showcases fitted with a cool-mist humidifier in supermarkets. This vegetable is displayed with other various types of fresh leafy vegetables in a 'Ssam-leafy-vegetables-section' of supermarkets. Ssamchoo can be purchased in offline or online stores. Some online stores deliver fresh Ssamchoo leaves within 24 h after placing an order, even in the early morning of the next day of purchase (Source: Hong, personal collection).

4.3. Peucedanum Japonicum

Peucedanum japonicum [18,19,58] is a perennial herb [77] that belongs to the *Umbelliferae* family and is used as a medicinal plant [78]. This herb is found in China, South Korea, Japan, Taiwan, and the Philippines [79]. Its leaves are consumed traditionally to treat coughs in the Okinawa islands in Japan [78]. *Peucedanum japonicum* is reported to possess pharmacological benefits such as anti-obesity [78,80,81], anti-oxidant [82], anti-bacterial [83], anti-diabetic [81], tyrosinase inhibition [84], and anti-platelet aggregation [77].

Peucedanum japonicum is traditionally had been foraged from the wild. However, the plant is now cultivated in South Korea for its health benefits (Figure 3). The leaves contain a fragrant aroma, a little bitter and sweet taste. This leafy vegetable can be eaten raw as a salad, blanched, pickled or deep-fried. *Peucedanum japonicum* can be cultivated in an open field, greenhouse or hydroponic cultivation system [85,86].

Figure 3. Peucedanum japonicum.

Young leaves with stems, packed in a plastic bag are sold in 'Namul-leafy vegetable section' of supermarkets. *Peucedanum japonicum* can be purchased in offline or online stores. Some online stores deliver this fresh leafy vegetable within 24 h after placing an order, even in the early morning of the day after purchase (Source: Hong, personal collection).

4.4. Aralia Elata (Miq.) Seem.

Aralia elata (Miq.) Seem. [18,19,59] is a shrub that is found across Northeastern China, Korea, Japan [87,88] and eastern Russia [89]. In Chinese traditional medicine, its root cortex has been used as a tonic, anti-arthritic and anti-diabetic agent [90]. In the leaves and root cortex, many saponins are isolated and reported to possess an anti-diabetic [90] and a cytoprotective effect [91]. Additionally, a study found specific anti-tumor activity in the extract of the shoots [92].

The young shoots of *Aralia elata* (Miq.) Seem. have been generally foraged from the wild, but it is cultivated in Japan and South Korea for its expected health benefits (Figure 4). This young shoot has a mildly bitter and sweet taste, fragrant aroma, and a crunchy texture. This shoot vegetable is enough to be an ornament itself in dishes, and thus this young shoot is used as a garnish in Japanese cuisine [93]. In South Korea, the shoots are simply blanched and consumed with a spicy-sour sauce. This aromatic vegetable can also be pickled and deep-fried.

Figure 4. *Aralia elata* (Miq.) Seem.

Young shoots of *Aralia*, packed in clear transparent disposable plastic containers can be found in the leafy vegetable section of supermarkets. The shoots are still attached to a small part of woody stems to keep the shoots fresh. Consumers should carefully remove woody stems covered with sharp thorns. Young shoots can be purchased in offline or online stores. Some online stores deliver this fresh young shoots within 24 h after placing an order, even in the early morning of the day after purchase (Source: Hong, personal collection).

Young shoots of *Aralia elata* (Miq.) Seem. can be harvested (April–May) in open field cultivation from a whole tree [94,95]. If the shoots are harvested in the greenhouse (February–April), stem cuttings must be prepared [95]. Shoots will develop naturally from the stem cuttings given appropriate humidity [95]. Producers following forcing culture practices need a simple cultivation system such as a double-layer plastic greenhouse, heater, and white greenhouse felt for insulation [95]. Stem cuttings can sprout in immersion cultivation, in which shoots can be harvested after 15 days (30–35 days long) [94]. This method results in harvest at least two weeks earlier than from forcing cultivation [94].

5. The European Market for Asian Vegetables

Fresh fruit and vegetable consumption in Europe started to recover slightly in 2014 after suffering declines since 2000 [96]. Daily consumption in the EU-28 stood at 353 g of fresh fruit and vegetables per capita in 2014 [96]. This was split, on average, between 192 g of fruits and 161 g of vegetables consumed each day [96]. The consumption of fruits and vegetables in the EU market is expected to increase in the coming decades until 2030 [97]. The main reasons for this phenomenon are increasing interests in a healthy diet and the easier access to fruits and vegetables at new marketplaces, including petrol stations and snack bars [97].

According to Florkowski [98], the market of Asian fruit and vegetables is still small in Europe. However, it is expected to expand across the EU member states, due to increasing consumer interests in diverse, balanced and healthy foods, exchange with Asian people, and contact with the Asian food culture.

When people migrate from one country to another, they have to change their consumption patterns because some vegetables are not available, i.e., the people moving from the Mediterranean region to northern Europe. However, the migrants asking for the vegetables that they were used to eating in their home country find that they cannot be found easily in the new country. In this way, they stimulated the introduction of new vegetables into Europe. An analogous mechanism can be seen during holidays abroad. In foreign countries, people are acquainted with new vegetables. After returning from holiday, they then desire to buy these products in their home country. The introduction of eggplant (*Solanum melongena*) and zucchini (*Cucurbita pepo* subsp. *pepo* convar. *giromontiina*) to the Dutch [99] and German markets are examples of this.

The small market share of Asian fruits and vegetables in Europe could be further increased through consumer education and safety assurances, as well as marketing campaigns focused on their unique taste, health benefits, ease of preparation, and reasonable price [98]. On the other hand, we believe that the total European fresh vegetable market size will increase over the long term due to strong trends, such as vegetarianism, health and well-being. There is also rising demand from consumers for organic, local, seasonal, fresh, sustainable and new vegetable species and varieties, quality labels, and specific nutrients of vegetables [100]. Thus, increasing demand for locally grown and organic vegetables can lead to the cultivation of Asian vegetables in Europe as a promising business opportunity, especially when these products possess new and additional quality attributes.

Due to stricter European import regulations concerning organic and safety standards of fruits and vegetables, locally-based producers may have comparative advantages in meeting the requirements for supplying fresh Asian vegetables to the European market. For instance, southeastern Asian countries including Malaysia, Thailand and Vietnam which export exotic fruits and vegetables to Europe will need to invest in cultivation practices and laboratory testing for certifications ensuring the hygiene standards of their products [101]. Since most southeastern Asian fruit and vegetable producers are small and live under relatively lower socio-economic conditions, the stricter European regulations are very challenging to meet [102,103].

5.1. Import from Developing and Asian Countries

The import of exotic products from developing countries into Europe is continuously increasing [104]. According to a statistical overview from Eurostat on the fruit and vegetable sector in the EU [105], the value of imported fruit and vegetables from developing countries into the EU was approximately 20.1 billion euros in 2017, while the value of exports from the EU to non-Member States was 4.6 billion euros [105]. Fruit imports accounted for 84.7%, while vegetable imports made up 15.3% of the total value [105]. The reason for this large difference is the relatively higher self-production of vegetables than fruits in Europe [106].

Among Asian countries, China is the leading exporter of vegetables and fruit to the EU. According to data sourced from Eurostat Comext, of total agri-food imports to the EU from China in 2017, fresh, chilled and dried vegetables accounted for 12% (623 million euros) and prepared vegetables, fruit or nuts accounted for 8% (416 million euros) [107]. The EU also imports vegetables from South Korea and Japan [108]. Although the European market for fresh fruit and vegetables is highly competitive, this market is large, mature, and has a relatively stable demand [106].

5.2. Exotic and Higher-Value Products

European consumers have become more interested in exotic and higher-value fruits and vegetables, with the market for these exclusive niche products growing [104]. Niche products and new varieties are expected to continuously emerge, particularly within the high-end market, due to their unique taste and consumer demands [104].

The fruit and vegetable market in Europe focused on unique produces with specific nutritional value is expected to continuously grow. For example, products with specific health benefits such as blueberries, avocados, and pomegranates known commercially as 'superfoods' have become more popular, especially in northwestern Europe [104].

According to CBI research [106], the past five years (2014–2018) have seen the highest growth rates and values from the import of certain fruits from developing countries. For instance, blueberries at 249% (449 million euro in 2018), avocados at 180% (1115 million euro) and mangoes at 51% (778 million euro) [106]. Compared to these exotic fruits, specific vegetables imported from non-EU suppliers are considered to be niche products due to the smaller market in Europe. However, these vegetables showed continuous growth in the European market from 2014 to 2018 with items as chili pepper (*Capsicum annuum*) at 33% (48,000 tons in 2018), asparagus (*Asparagus officinalis*) at 12% (42,000 tons), tamarinds (*Tamarindus indica*) at 14% (41,000 tons), sugar snaps (*Pisum sativum* var. *macrocarpon*), snow peas (*Pisum sativum* var. *saccharatum*) and other peas (*Pisum sativum*) at 12% (30,000 tons) [106].

5.3. Taste and Emotional Connection

According to Rabobank [109], the evolving tastes and preferences in developed countries drive the growing value in global trade of fresh fruits and vegetables. The top three drivers are related to the level of customer satisfaction with the quality, fresh appearance, taste, and the size and shape of the products [110]. Taste has recently become a deciding factor for purchase [110]. Additionally, taste preferences are the factors that essentially prevent the consumption of vegetables, especially for younger adults and women in Europe [111]. However, by appealing to the senses - seeing, touching, feeling and smelling - trust and loyalty can be established with customers. Building the emotional connection between consumers and vegetables plays an important role in encouraging customers to make repeated purchases [110]. Even the current Dutch Nationaal Actieplan Groenten en Fruit (National Action-Plan for Vegetables and Fruits) is designed to focus more on emotions and health issues (https://nagf.nl). Besides, sensory appeals of food products, such as taste, texture, quality, smell and appearance, influence consumers to choose an item [112], even if the taste of vegetables does not match the high market price [100,113]. Therefore, in-store demonstrations or cooking classes with simple and delicious recipes can provide excellent opportunities for new and unfamiliar Asian vegetables for a successful launch in the European market. With various consumer marketing strategies, the unique tastes and qualities of Asian vegetables can be efficiently introduced to future consumers.

5.4. Communication with Consumers

In order to increase the consumption of newly introduced Asian vegetables in the European market, potential European consumers need to aware of these unfamiliar vegetables. For instance, consumers need to know how to select, store, prepare or cook these vegetables. In other words, a successful entry of new Asian vegetables into the European market requires various ways of communication with consumers. Communication channels need to be developed based on the lifestyle and culture of targeted consumers [98]. Websites with easy recipes, such as salads or stir fry, information about health benefits or the use of vegetables can inspire consumers. In addition, social media such as Instagram where people can take, edit, and share photos and videos [114] might contribute to increased demand for new vegetables. For instance, avocado was globally one of the 20 most hash-tagged vegetables on Instagram in 2018 [115]. Avocados had been imported into the European market with the highest growth rate of 180% for 5 years (2014–2018) and in value in 2018 [106].

6. Asian Vegetables from the Perspective of Climate Change and Cultivation Method

Trending vegetables in the European market are convenient, healthy, colorful, decorative, organic, local, seasonal [100], mini-sized, and have exceptional nutritional value. Therefore, for a successful entry in the European vegetable market, the introduced Asian vegetables must be tasty, not too exotic, functional, and easy to prepare [98]. For European growers, the new Asian vegetables must be easy to cultivate and should promise a good harvest with high yield. Moreover, these vegetables must be nutritious, clean, fresh, and high-quality to fulfill consumer demands [100,116].

To ensure a holistically sustainable development trajectory, the transformation towards vegetable-rich production and food systems should support climate change adaptation and protection of human and environmental health [117]. Production and the quality of vegetables are being threatened by exposure to an extreme environment, primarily due to climate change—summer heat-waves, heavy rains, drought, possible increasing infestation of weeds, pests, diseases and changes of CO2 concentrations in the air [116–119]. Although traditional outdoor cultivation is cheaper in regards to energy and technology [120], innovative greenhouse cultivation technologies with evaporative cooling, shading, adequate water supply, and drainage systems can secure better vegetable quality [116,121,122].

7. Difficulties in the Introduction of New Vegetables and Other Ideas of Diversification

Promoting diverse vegetable supply should be accompanied by different measures—a selection of the best cultivar-genotypes to be grown in the new region, the development of cultivation techniques, avoidance of post-harvest facilities and providing consumers with information about the new ingredients and advice for cocking and preparation. However, the way of introducing new vegetables in the market is not always based on the same pattern. Introduction of kiwi and broccoli in the European market are examples. Even though kiwifruit is originally from China, it was introduced through New Zealand, where the fruit was named after the national "kiwi" bird, to Europe. Broccoli, a native European vegetable with origins in primitive cultivars grown in the Roman Empire, gained its reputation in the European market only after being re-introduced from the USA in the last century.

In many areas of the world, increased consumption of vegetables has been not only because of a growing understanding of the health benefits but also because of increasing availability of various vegetables [39]. Consumers want a continuous supply of vegetables, as well as diversity and novelty, which have become important selling points in modern markets—people are always looking for something new [123]. Introducing new Asian vegetables is just one of the ways to increase the consumption of vegetables in Europe. Introducing colorful vegetables, mini-vegetables, sprout-vegetables, and microgreens are other ways to increase consumption. The new vegetables that have been introduced to the European market include cherry and pear-shaped tomatoes (*Solanum lycopersicum*), baby-carrots (*Daucus carota* subsp. *sativus*), mini-cauliflowers (*Brassica oleracea* var. *botrytis*), mini-broccoli (*Brassica oleracea* var. *italica*), artichoke hearts (*Cynara scolymus*), fresh baby leaf curly kale (*Brassica oleracea* var. *sabellica*), pickling cucumbers (*Cucumis sativus*), beans sprouts (*Vigna radiata*), and different microgreens. The demand for lightly processed, ready-to-eat vegetables is also increasing [39].

Another way to increase vegetable consumption is the time of availability of the products. Year-round vegetable production is possible by using protected cultivation methods. Thus, colorful vegetables could be marketed in the winter period, especially in regions where the vegetable assortment is poor. Even well-known plants in the Mediterranean region, such as eggplant and sweet pepper (*Capsicum annuum* var. *grossum*), have been only recently available thanks to greenhouses. Besides, soil-less culture is popularly used because of its advantage in optimizing the root system and efficient control of pathogens without soil fumigation. Therefore, producers can achieve high-quality products as well as higher yields at a reasonable production cost using minimal pesticides [124–126]. Different cultivation methods with porous growing media and water culture systems (floating system, nutrient film technique, and aeroponics) are practiced in commercial greenhouse production [124].

In Asia, the indoor production of high-priced and high-quality vegetables is a well-known tradition. Farms use so-called plant factories as highly controlled environments. Figure 5 presents different cultivation steps and plant development stages for Korean ginseng sprouts from a farm in Korea. The vegetable growers of tomatoes, leafy greens, cucumbers, peppers, and others in Europe are progressively using hydroponics primarily because of the high-profit margins and ten times higher yields than traditional cultivation [127]. By using all-environment controlling technologies, producers in Europe can cultivate new Asian vegetables easily and efficiently, meeting food safety standards, only if they are rightly informed about the optimal cultivation conditions of these vegetables. New light processing and packaging technologies will enable fresh vegetables to be presented in a convenient form, e.g., with increased shelf life of pre-cut vegetables while maintaining nutritional value, safety, and freshness [39].

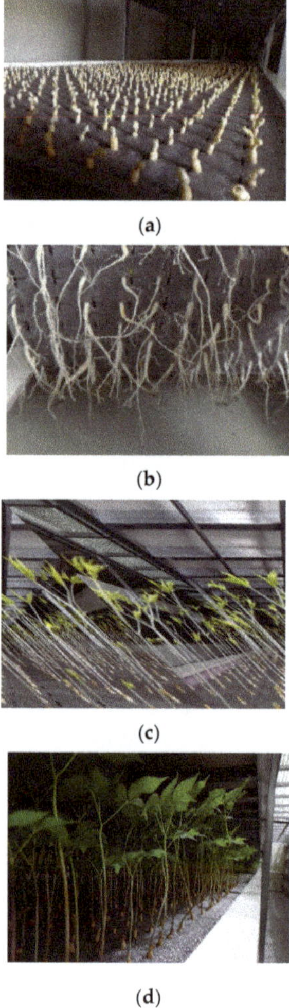

(a)

(b)

(c)

(d)

Figure 5. *Cont.*

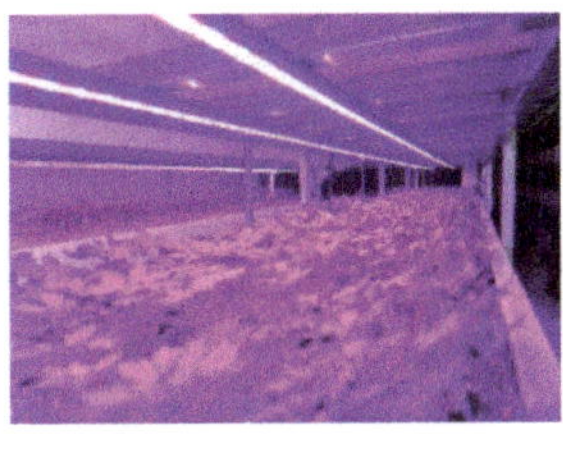

(**e**)

Figure 5. Plant development stages of Korean ginseng sprouts (*Panax ginseng* Meyer). Korean ginseng sprouts in a modern plant factory using aquaponics. (**a**) one-year-old-ginseng-seedlings, freshly transplanted in holes of sponge-hydroponic-floating-rafts; (**b**) roots of one-year-old-ginseng; (**c**) roots of about ten-day old cultivated ginseng; (**d**) at about 21–27 days cultivating ginseng; (**e**) at about 21–27 days cultivating ginseng seedlings under LED lights. Lighting time: 1–15 days (10 h exposure/day), after 15 days (13 h exposure/day). Lighting lamps: LED blending ratio (blue:red = 2:1). The distance between leaves and LED lights about 10 cm (Source: kindly provided by K. Farm Factory).

8. Conclusions

Although the volume of fresh vegetable consumption in the EU is barely increasing, the outlook for the fresh vegetable market is positive. Some of the current trends in the market are the increasing demand for diverse, healthy, and exotic vegetables. In this regard, certain Asian vegetables may have good potential to be introduced and land successfully in the European vegetable market. Here, we discussed the potential of different Asian vegetables for introduction in the European market in the future and shared practical information for the cultivation of four species.

Asian vegetables can be introduced to European consumers not only as fresh vegetables but also as processed products with new and modern ideas, for instance, as Asian vegetable salad bags, as healthy juices, or as cosmetics. In the expanding European health product market, Asian vegetables may play an essential role due to their health benefits and pharmacological properties.

The cultivation of new Asian vegetables within Europe can provide a good opportunity for European vegetable growers to open a new premium niche market with safe, local/seasonal and organic products. For a successful introduction of new Asian vegetables into the European market, appropriate marketing strategies are essential.

Although interest in Asian vegetables is increasing in the European vegetable market, there have been very few studies conducted on Asian vegetables in Europe. From a marketing point of view, the consumer needs and preferences are the key factors leading the production. In complex markets and especially in export-markets, information about consumer needs and wishes does not reach producers all by itself. Thus producers must conduct market research in order to track consumer preferences and to monitor their changes continuously [99]. Further research is required on the Asian vegetable market within Europe and also on the cultivation of potential Asian vegetables under European climatic conditions. Economic, social, and ecological aspects also have to be considered. Further research will provide the basis for all those aspects.

Author Contributions: J.H. and N.S.G. conceptualized and designed the paper, J.H. carried out the literature review and wrote the paper, N.S.G. revised and finalized the paper. All authors have read and agreed to the published version of the manuscript.

Funding: This research received no external funding.

Acknowledgments: We want to thanks to K. Farm Factory, Korea, for providing photos and information about its facility and cultivation conditions.

References

1. Suhrcke, M.; McKee, M.; Stuckler, D.; Arce, R.S.; Tsolova, S.; Mortensen, J. The contribution of health to the economy in the European Union. *Pub. Health* **2006**, *120*, 994–1001. [CrossRef]
2. Eurostat. Available online: https://ec.europa.eu/eurostat/statistics-explained/index.php/Health_statistics_at_regional_level (accessed on 9 April 2020).
3. WHO. Available online: http://www.who.int/mediacentre/news/releases/2003/pr20/en/ (accessed on 9 April 2020).
4. Danaei, G.; Ding, E.L.; Mozaffarian, D.; Taylor, B.; Rehm, J.; Murray, C.J.; Ezzati, M. The preventable causes of death in the United States: Comparative risk assessment of dietary, lifestyle, and metabolic risk factors. *PLoS Med.* **2009**, *6*, e1000058. [CrossRef]
5. Lim, S.S.; Vos, T.; Flaxman, A.D.; Danaei, G.; Shibuya, K.; Adair-Rohani, H.; Al Mazroa, M.A.; Amann, M.; Anderson, H.R.; Andrews, K.G.; et al. A comparative risk assessment of burden of disease and injury attributable to 67 risk factors and risk factor clusters in 21 regions, 1990-2010: A systematic analysis for the Global Burden of Disease Study 2010. *Lancet* **2012**, *380*, 2224–2260. [CrossRef]
6. Lozano, R.; Naghavi, M.; Foreman, K.; Lim, S.; Shibuya, K.; Aboyans, V.; Abraham, J.; Adair, T.; Aggarwal, R.; Ahn, S.Y.; et al. Global and regional mortality from 235 causes of death for 20 age groups in 1990 and 2010: A systematic analysis for the Global Burden of Disease Study 2010. *Lancet* **2012**, *380*, 2095–2128. [CrossRef]
7. Visscher, T.L.S.; Seidell, J.C. The public health impact of obesity. *Annu. Rev. Pub. Health* **2001**, *22*, 355–375. [CrossRef]
8. OECD/EU. *Health at a Glance: Europe 2016: State of Health in the EU Cycle*; OECD Publishing: Paris, France, 2016; pp. 55–112, ISBN: 9789264265646 (EPUB), 9789264265592 (PDF); Available online: https://www.oecd-ilibrary.org/docserver/9789264265592-en.pdf?expires=1586446691&id=id&accname=guest&checksum=C22AA163102617DFEF45BC2DF59DE7D4 (accessed on 9 April 2020).
9. WHO. Available online: https://www.who.int/dietphysicalactivity/childhood_consequences/en/ (accessed on 9 April 2020).
10. WHO. Available online: https://www.who.int/docs/default-source/coronaviruse/situation-reports/20200418-sitrep-89-covid-19.pdf?sfvrsn=3643dd38_2 (accessed on 19 April 2020).
11. WHO. Available online: http://www.euro.who.int/en/health-topics/health-emergencies/coronavirus-covid-19/weekly-surveillance-report (accessed on 19 April 2020).
12. Chen, N.; Zhou, M.; Dong, X.; Qu, J.; Gong, F.; Han, Y.; Qiu, Y.; Wang, J.; Liu, Y.; Wei, Y.; et al. Epidemiological and clinical characteristics of 99 cases of 2019 novel coronavirus pneumonia in Wuhan, China: A descriptive study. *Lancet* **2020**, *395*, 507–513. [CrossRef]
13. Peng, Y.D.; Meng, K.; Guan, H.Q.; Leng, L.; Zhu, R.R.; Wang, B.Y.; He, M.A.; Cheng, L.X.; Huang, K.; Zeng, Q.T. Clinical characteristics and outcomes of 112 cardiovascular disease patients infected by 2019-nCoV. *Zhonghua Xin Xue Guan Bing Za Zhi* **2020**, *48*, E004. Available online: https://europepmc.org/article/med/32120458 (accessed on 19 April 2020).
14. Saghazadeh, A.; Rezaei, N. Immune-epidemiological parameters of the novel coronavirus–A perspective. *Expert Rev. Clin. Immunol.* **2020**, *16*, 1–6. [CrossRef] [PubMed]
15. Maggini, S.; Pierre, A.; Calder, P.C. Immune Function and Micronutrient Requirements Change over the Life Course. *Nutrients* **2018**, *10*, 1531. [CrossRef] [PubMed]
16. Gasmi, A.; Noor, S.; Tippairote, T.; Dadar, M.; Menzel, A.; Bjørklund, G. Individual risk management strategy and potential therapeutic options for the COVID-19 pandemic. *Clin. Immunol.* **2019**, 108409. [CrossRef]
17. Chung, H.L.; Ripperton, J.C. *Utilization and Composition of Oriental Vegetables in Hawaii*; Hawaii Agricultural Experiment Station: Honolulu, HI, USA, 1929; pp. 1–64. Available online: http://hdl.handle.net/10125/14976 (accessed on 9 April 2020).
18. National Institute of Agricultural Sciences-RDA (Rural Development Administration). *9th Revision Korean Food Composition Table II (제9개정판 국가표준 식품성분표II)*; Functional Food and Nutrition Division-Department of Agro-food Resources-National Institute of Agricultural Sciences-RDA (Rural Development Administration): Wanju, Korea, 2016; ISBN: 978-89-480-4348-8 94520, 978-89-480-4347-2. Available online: http://www.rda.go.kr/download_file/act/bookcafe078.PDF (accessed on 9 April 2020).

19. Dongbuk Regional Statistics Office–Statistics Korea. *Illustrated Crop Book at Each Growth Stages* (생육 단계별 농작물도감); Dongbuk Regional Statistics Office: Daegu, Korea, 2015. Available online: http://kostat.go.kr/e_book/stepcrop/base/EBook/index.html#page=10 (accessed on 9 April 2020).

20. Larkcom, J. *Oriental Vegetables: The Complete Guide for Garden and Kitchen*; John Murray Ltd: London, UK, 1991; ISBN 0719547814.

21. Hutton, W. *Handy Pocket Guide to Asian Vegetables*; Periplus Editions Ltd.: Hongkong, 2004; ISBN: 0794601944 (ISBN13: 9780794601942).

22. Dixon, G.R. Origins and Diversity of Brassica and its Relatives. In *Vegetable Brassicas and Related Crucifers*; CABI: Wallingford, UK, 2007; pp. 1–33, ISBN 0-85199-395-8. Available online: https://www.cabi.org/isc/FullTextPDF/2007/20073006206.pdf (accessed on 9 April 2020).

23. Balvoll, G. Production of Chinese Cabbage in Norway: Problems and Possibilities. *J. Veg. Crop Prod.* **1995**, *1*, 3–18. [CrossRef]

24. Bastow, R.; Mylne, J.S.; Lister, C.; Lippman, Z.; Martienssen, R.A.; Dean, C. Vernalization requires epigenetic silencing of FLC by histone methylation. *Nature* **2004**, *427*, 164–167. [CrossRef] [PubMed]

25. Chun, C.; Watanabe, A.; Kim, H.H.; Kozai, T. Bolting and Growth of *Spinacia oleracea* L. can be Altered by Modifying the Photoperiod during Transplant Production. *HortScience* **2000**, *35*, 624–626. [CrossRef]

26. Larkcom, J. *Oriental Vegetables*; Kodansha International Ltd: Tokyo, Japan, 2008; pp. 17–30.

27. Du, Z.; Hou, R.; Zhu, Y.; Li, X.; Zhu, H.; Wang, Z. A random amplified polymorphic DNA (RAPD) molecular marker linked to late-bolting gene in pak-choi (*Brassica campestris* ssp. chinensis Makino, L.). *Afr. J. Biotechnol.* **2011**, *10*, 7962–7968. [CrossRef]

28. Morgan, L. Hydroponic Production of Wasabi & Japanese Vegetables. In *The Best of the Growing Edge International 2000–2005: Select Cream-of-the-Crop Articles for Soilless Growers*; Weller, T., Ed.; New Moon Publishing Inc.: Corvallis, OR, USA, 2005; pp. 53–62, ISBN 978-0-944557-05-1.

29. FreshPlaza. Available online: https://www.freshplaza.com/article/9162855/italy-baby-leaf-sector-is-growing (accessed on 28 May 2020).

30. FreshPlaza. Available online: https://www.freshplaza.com/article/2119710/italy-undisputed-leader-of-mixed-salad-sector (accessed on 28 May 2020).

31. Freshplaza. Available online: https://www.freshplaza.com/article/9151954/berlin-start-up-is-dedicated-to-cultivating-asian-vegetable-specialties (accessed on 19 June 2020).

32. Freshplaza. Available online: https://www.freshplaza.com/article/9071488/growing-asian-vegetables-in-portugal-looks-promising (accessed on 19 June 2020).

33. Agudo, A.; Cabrera, L.; Amiano, P.; Ardanaz, E.; Barricarte, A.; Berenguer, T.; Chirlaque, M.D.; Dorronsoro, M.; Jakszyn, P.; Larrañaga, N.; et al. Fruit and vegetable intakes, dietary antioxidant nutrients, and total mortality in Spanish adults: Findings from the Spanish cohort of the European Prospective Investigation into Cancer and Nutrition (EPIC-Spain). *Am. J. Clin. Nutr.* **2007**, *85*, 1634–1642. [CrossRef] [PubMed]

34. Nicklett, E.J.; Semba, R.D.; Xue, Q.L.; Tian, J.; Sun, K.; Cappola, A.R.; Simonsick, E.M.; Ferrucci, L.; Fried, L.P. Fruit and vegetable intake, physical activity, and mortality in older community-dwelling women. *J. Am. Geriatr. Soc.* **2012**, *60*, 862–868. [CrossRef]

35. Trichopoulou, A.; Costacou, T.; Bamia, C.; Trichopoulos, D. Adherence to a Mediterranean diet and survival in a Greek population. *N. Engl. J. Med.* **2003**, *348*, 2599–2608. [CrossRef]

36. Wang, X.; Ouyang, Y.; Liu, J.; Zhu, M.; Zhao, G.; Bao, W.; Hu, F.B. Fruit and vegetable consumption and mortality from all causes, cardiovascular disease, and cancer: Systematic review and dose-response meta-analysis of prospective cohort studies. *BMJ* **2014**, *349*, g4490. [CrossRef]

37. Luzi, L.; Radaelli, M.G. Influenza and obesity: Its odd relationship and the lessons for COVID-19 pandemic. *Acta Diabetol.* **2020**, 1–6. [CrossRef]

38. Shahbaz, M.; Ashraf, M.; Al-Qurainy, F.; Harris, P.J.C. Salt tolerance in selected vegetable crops. *Crit. Rev. Plant Sci.* **2012**, *31*, 303–320. [CrossRef]

39. Dias, J.S. World Importance, Marketing and Trading of Vegetables. *Acta Hort.* **2011**, *921*, 153–170. [CrossRef]

40. Caton, S.J.; Blundell, P.; Ahern, S.M.; Nekitsing, C.; Olsen, A.; Møller, P.; Hausner, H.; Remy, E.; Nicklaus, S.; Chabanet, C.; et al. Learning to Eat Vegetables in Early Life: The Role of Timing, Age and Individual Eating Traits. *PLoS ONE* **2014**, *9*, e97609. [CrossRef] [PubMed]

41. Anderson, A.S.; Porteous, L.E.G.; Foster, E.; Higgins, C.; Stead, M.; Hetherington, M.; Ha, M.-A.; Adamson, A.J. The impact of a school-based nutrition education intervention on dietary intake and cognitive and attitudinal variables relating to fruits and vegetables. *Pub. Health Nutr.* **2005**, *8*, 650–656. [CrossRef]

42. Ransley, J.K.; Greenwood, D.C.; Cade, J.E.; Blenkinsop, S.; Schagen, I.; Teeman, D.; Scott, E.; White, G.; Schagen, S. Does the school fruit and vegetable scheme improve children's diet? A nonrandomised controlled trial. *J. Epidemiol. Community Health* **2007**, *61*, 699–703. [CrossRef] [PubMed]

43. Oyebode, O.; Gordon-Dseagu, V.; Walker, A.; Mindell, J.S. Fruit and vegetable consumption and all-cause, cancer and CVD mortality: Analysis of Health Survey for England data. *J. Epidemiol. Community Health* **2014**, *68*, 856–862. [CrossRef] [PubMed]

44. Eurostat. Available online: https://ec.europa.eu/eurostat/documents/2995521/7694616/3-14102016-BP-EN.pdf/1234ac94-27fd-4640-b9be-427a42d54881 (accessed on 9 April 2020).

45. Joshipura, K.J.; Hu, F.B.; Manson, J.E.; Stampfer, M.J.; Rimm, E.B.; Speizer, F.E.; Colditz, G.; Ascherio, A.; Rosner, B.; Spiegelman, D.; et al. The effect of fruit and vegetable intake on risk for coronary heart disease. *Ann. Intern. Med.* **2001**, *134*, 1106–1114. [CrossRef] [PubMed]

46. Lustig, R.H.; Schmidt, L.A.; Brindis, C.D. The toxic truth about sugar. *Nature* **2012**, *482*, 27–29. [CrossRef] [PubMed]

47. Tappy, L.; Lê, K.A.; Tran, C.; Paquot, N. Fructose and metabolic diseases: New findings, new questions. *Nutrition* **2010**, *26*, 1044–1049. [CrossRef]

48. Lustig, R.H. Fructose: Metabolic, hedonic, and societal parallels with ethanol. *Am. Diet. Assoc.* **2010**, *110*, 1307–1321. [CrossRef]

49. Eurostat. Available online: https://ec.europa.eu/eurostat/statistics-explained/pdfscache/68501.pdf (accessed on 9 April 2020).

50. Kuehnemund, M. Evaluation of the Implementation of the Strategy for Europe on Nutrition, Overweight and Obesity Related Health Issues–Final Report. Available online: https://ec.europa.eu/health/sites/health/files/nutrition_physical_activity/docs/pheiac_nutrition_strategy_evaluation_en.pdf (accessed on 9 April 2020).

51. FAO. Available online: http://www.fao.org/3/x0171e/x0171e03.htm (accessed on 28 May 2020).

52. Meldrum, G.; Padulosi, S.; Lochetti, G.; Robitaille, R.; Diulgheroff, S. Issues and prospects for the sustainable use and conservation of cultivated vegetable diversity for more nutrition-sensitive agriculture. *Agriculture* **2018**, *8*, 112. [CrossRef]

53. Fritz, D. Starting points for crop research to promote diversification. *Acta Hortic.* **1989**, *242*, 193–202. [CrossRef]

54. OECD. *Health at a Glance 2017: OECD Indicators*; OECD Publishing: Paris, France, 2017; pp. 76–77, ISBN: 9789264282780 (HTML), 9789264280403 (PDF), 9789264280885 (EPUB). Available online: https://www.oecd-ilibrary.org/docserver/health_glance-2017-en.pdf?expires=1586736825&id=id&accname=guest&checksum=43DB9F4E2757464CEA92D340758CE043 (accessed on 9 April 2020).

55. Statista. Available online: https://www.statista.com/statistics/264662/top-producers-of-fresh-vegetables-worldwide (accessed on 9 April 2020).

56. WITS (World Integrated Trade Solution). Available online: https://wits.worldbank.org/CountryProfile/en/Country/WLD/Year/LTST/TradeFlow/Export/Partner/by-country/Product/06-15_Vegetable (accessed on 9 April 2020).

57. Our World in Data. Available online: https://ourworldindata.org/grapher/average-per-capita-vegetable-intake-vs-minimum-recommended-guidelines (accessed on 9 April 2020).

58. Korea National Arboretum. *English Names for Korean Native Plants (한반도 자생식물 영어이름 목록집)*; Korea National Arboretum: Pocheon-si, Korea, 2015; ISBN 978-89-97450-98-5-93480. Available online: http://www.forest.go.kr/kna/special/download/English_Names_for_Korean_Native_Plants.pdf (accessed on 9 April 2020).

59. Jang, I.B.; Yu, J.; Suh, S.J.; Jang, I.B.; Kwon, K.B. Growth and Ginsenoside Content in Different Parts of Ginseng Sprouts Depending on Harvest Time. *Korean J. Med. Crop Sci.* **2018**, *26*, 205–213. [CrossRef]

60. Kimchimuseum. Available online: https://www.kimchimuseum.com/kimchi-lab/kimchi-kinds/5529 (accessed on 23 April 2020).

61. Oh, J.Y.; Kim, Y.J.; Jang, M.G.; Joo, S.C.; Kwon, W.S.; Kim, S.Y.; Jung, S.K.; Yang, D.C. Investigation of ginsenosides in different tissues after elicitor treatment in *Panax ginseng*. *J. Ginseng Res.* **2014**, *38*, 270–277. [CrossRef] [PubMed]

62. Jung, M.Y.; Jeon, B.S.; Bock, J.Y. Free, esterified, and insoluble-bound phenolic acids in white and red Korean ginsengs (*Panax ginseng* C. A. Meyer). *Food Chem.* **2002**, *79*, 105–111. [CrossRef]

63. Attele, A.S.; Wu, J.A.; Yuan, C.S. Ginseng pharmacology: Multiple constituents and multiple actions. *Biochem. Pharmacol.* **1999**, *58*, 1685–1693. [CrossRef]

64. Xie, J.T.; Mehendale, S.R.; Li, X.; Quigg, R.; Wang, X.; Wang, C.Z.; Wu, J.A.; Aung, H.H.; Rue, P.A.; Bell, G.I.; et al. Anti-diabetic effect of ginsenoside Re in ob/ob mice. *Biochem. Biophys. Acta* **2005**, *1740*, 319–325. [CrossRef]

65. Shang, W.; Yang, Y.; Zhou, L.; Jiang, B.; Jin, H.; Chen, M. Ginsenoside Rb1 stimulates glucose uptake through insulin-like signaling pathway in 3T3-L1 adipocytes. *J. Endocrinol.* **2008**, *198*, 561–569. [CrossRef]

66. Wang, J.; Gao, W.Y.; Zhang, J.; Zuo, B.M.; Zhang, L.M.; Huang, L.Q. Advances in study of ginsenoside biosynthesis pathway in *Panax ginseng* C. A. Meyer. *Acta Physiol. Plant.* **2012**, *34*, 397–403. [CrossRef]

67. Lee, T.K.; Johnke, R.M.; Allison, R.R.; O'Brien, K.F.; Dobbs, L.J., Jr. Radio protective potential of ginseng. *Mutagenesis* **2005**, *20*, 237–243. [CrossRef]

68. Saito, H.; Yoshida, Y.; Takagi, K. Effect of *Panax ginseng* root on exhaustive exercise in mice. *Jpn. J. Pharmacol.* **1974**, *24*, 119–127. [CrossRef]

69. Gillis, C.N. *Panax ginseng* pharmacology: A nitric oxide link? *Biochem. Pharmacol.* **1997**, *54*, 1–8. [CrossRef]

70. Soldati, F.; Tanaka, O. *Panax ginseng*: Relation between age of plant and content of ginsenosides. *Planta Med.* **1984**, *50*, 351–352. [CrossRef] [PubMed]

71. Choi, S.Y.; Cho, C.W.; Lee, Y.M.; Kim, S.S.; Lee, S.H.; Kim, K.T. Comparison of Ginsenoside and Phenolic Ingredient Contents in Hydroponically-cultivated Ginseng Leaves, Fruits, and Roots. *J. Ginseng Res.* **2012**, *36*, 425–429. [CrossRef] [PubMed]

72. Lee, J.Y.; Yang, H.; Lee, T.K.; Lee, C.H.; Seo, J.W.; Kim, J.-E.; Kim, S.Y.; Park, J.H.Y.; Lee, K.W. A short-term, hydroponic-culture of ginseng results in a significant increase in the anti-oxidative activity and bioactive components. *Food Sci. Biotechnol.* **2020**, 1–6. [CrossRef]

73. Kim, G.S.; Lee, S.E.; Noh, H.J.; Kwon, H.; Lee, S.W.; Kim, S.Y.; Kim, Y.B. Effects of Natural Bioactive Products on the Growth and Ginsenoside Contents of *Panax ginseng* Cultured in an Aeroponic System. *J. Ginseng Res.* **2012**, *36*, 430–441. [CrossRef] [PubMed]

74. Jeong, D.H.; Lee, D.Y.; Jang, I.B.; Yu, J.; Park, K.C.; Lee, E.H.; Kim, Y.J.; Park, H.W. Growth and Ginsenoside Content of One Year Old Ginseng Seedlings in Hydroponic Culture over a Range of Days after Transplanting. *Korean J. Med. Crop Sci.* **2018**, *26*, 464–470. [CrossRef]

75. Lee, K.H. Plant Species Ssamchoo and Breeding Method Thereof. U.S. Patent 6,600,092 B2, 29 July 2003. Available online: https://patentimages.storage.googleapis.com/83/67/3a/f2fd2bf0c7a207/US6600092.pdf (accessed on 9 April 2020).

76. RDA (Rural Development Administration). *Leafy Vegetables-Agricultural Technology Guide 140* (엽채류–농업기술길잡이 140(개정판)), 3rd ed.; RDA: Jeonju-si, South Korea, 2017; pp. 142–148, ISBN: 978-89-480-5001-1 94520, 978-89-480-1772-4(set). Available online: http://lib.rda.go.kr/search/mediaView.do?mets_no=000000299932 (accessed on 9 April 2020).

77. Chen, I.S.; Chang, C.T.; Sheen, W.S.; Teng, C.M.; Tsai, I.L.; Duh, C.Y.; Ko, F.N. Coumarins and antiplatelet aggregation constituents from Formosan *Peucedanum japonicum*. *Phytochemistry* **1996**, *41*, 525–530. [CrossRef]

78. Okabe, T.; Toda, T.; Nukitrangsan, N.; Inafuku, M.; Iwasaki, H.; Oku, H. *Peucedanum japonicum* Thunb inhibits high-fat diet induced obesity in mice. *Phytother. Res.* **2011**, *25*, 870–877. [CrossRef]

79. Chen, C.C.; Agrawal, D.C.; Lee, M.R.; Lee, R.J.; Kuo, C.L.; Wu, C.R.; Tsay, H.S.; Chang, H.C. Influence of LED light spectra on in vitro somatic embryogenesis and LC-MS analysis of chlorogenic acid and rutin in *Peucedanum japonicum* Thunb.: A medicinal herb. *Bot. Stud.* **2016**, *57*, 9. [CrossRef]

80. Nugara, R.N.; Inafuku, M.; Iwasaki, H.; Oku, H. Partially purified *Peucedanum japonicum* Thunb extracts exert anti-obesity effects in vitro. *Nutrition* **2014**, *30*, 575–583. [CrossRef]

81. Nukitrangsan, N.; Okabe, T.; Toda, T.; Inafuku, M.; Iwasaki, H.; Oku, H. Antiobesity activity of *Peucednum japonicum* Thunb extract in obese diabetic animal model C57 BL/6 J Ham Slc-ob/ob mice. *Intern. J. Life Sci. Med. Res.* **2012**, *2*, 28–34. Available online: https://pdfs.semanticscholar.org/a25c/4ebb567714c0e702eb25b82cafa7c32f7224.pdf (accessed on 9 April 2020). [CrossRef]

82. Hisamoto, M.; Kikuzaki, H.; Ohigashi, H.; Nakatani, N. Antioxidant compounds from the leaves of *Peucedanum japonicum* Thunb. *J. Agric. Food Chem.* **2003**, *51*, 5255–5261. [CrossRef] [PubMed]

83. Yang, E.J.; Kim, S.S.; Oh, T.H.; Song, G.; Kim, K.N.; Kim, J.Y.; Lee, N.H.; Hyun, C.G. *Peucedenum japonicum* and *Citrus unshiu* essential oils inhibit the growth of antibiotic-resistant skin pathogens. *Ann. Microbiol.* **2009**, *59*, 623–628. [CrossRef]

84. Hisamoto, M.; Kikuzaki, H.; Nakatani, N. Constituents of the leaves of *Peucedanum japonicum* Thunb. and their biological activity. *J. Agric. Food Chem.* **2004**, *52*, 445–450. [CrossRef]

85. Lee, G.J.; Heo, J.W.; Jung, C.R.; Kim, H.H.; Yoon, J.B.; Kim, D.E.; Nam, S.Y. Effects of Plant Factory Cultural Systems on Growth, Vitamin C and Amino Acid Contents, and Yield in Hydroponically Grown *Peucedanum japonicum*. *Prot. Hortic. Plant Fact.* **2015**, *24*, 281–286. [CrossRef]

86. Lee, G.J.; Heo, J.W.; Jung, C.R.; Kim, H.H.; Kim, D.E.; Nam, S.Y. Effects of Artificial Light Sources on Growth and Yield of *Peucedanum japonicum* Hydroponically Grown in Plant Factory. *Prot. Hortic. Plant Fact.* **2016**, *25*, 16–23. [CrossRef]

87. Zhang, J.; Wang, H.; Zheng, Q. Cardioprotective effect of *Aralia elata* polysaccharide on myocardial ischemic reperfusion (IR) injury in rats. *Int. J. Biol. Macromol.* **2013**, *59*, 328–332. [CrossRef]

88. Zhang, J.; Wang, H.; Xue, Y.; Zheng, Q. Cardioprotective and antioxidant activities of a polysaccharide from the root bark of *Aralia elata* (Miq.) Seem. *Carbohydr. Polym.* **2013**, *93*, 442–448. [CrossRef]

89. Luo, Y.; Lu, S.; Ai, Q.; Zhou, P.; Qin, M.; Sun, G.; Sun, X. SIRT1/AMPK and Akt/eNOS signaling pathways are involved in endothelial protection of total aralosides of *Aralia elata* (Miq) Seem against high-fat diet-induced atherosclerosis in ApoE-/-mice. *Phytother. Res.* **2019**, *33*, 768–778. [CrossRef]

90. Yoshikawa, M.; Matsuda, H.; Harada, E.; Murakami, T.; Wariishi, N.; Yamahara, J.; Murakami, N.; Elatoside, E. A new hypoglycemic principle from the root cortex of *Aralia elata* seem.: Structure-related hypoglycemic activity of oleanolic acid glycosides. *Chem. Pharm. Bull.* **1994**, *42*, 1354–1356. [CrossRef]

91. Saito, S.; Ebashi, J.; Sumita, S.; Furumoto, T.; Nagamura, Y.; Nishida, K.; Isiguro, I. Comparison of cytoprotective effects of saponins isolated from leaves of *Aralia elata* Seem. (Araliaceae) with synthesized bisdesmosides of oleanoic acid and hederagenin on carbon tetrachloride-induced hepatic injury. *Chem. Pharm. Bull.* **1993**, *41*, 1395–1401. [CrossRef]

92. Tomatsu, M.; Ohnishi-Kameyama, M.; Shibamoto, N. Aralin, a new cytotoxic protein from *Aralia elata*, inducing apoptosis in human cancer cells. *Cancer Lett.* **2003**, *199*, 19–25. [CrossRef]

93. Yoshikawa, M.; Yoshizumi, S.; Ueno, T.; Matsuda, H.; Murakami, T.; Yamahara, J.; Murakami, N. Medicinal foodstuffs. I. Hypoglycemic constituents from a garnish foodstuff "taranome," the young shoot of *Aralia elata* SEEM.: Elatosides G, H, I, J, and K. *Chem. Pharm. Bull.* **1995**, *43*, 1878–1882. [CrossRef] [PubMed]

94. Kim, S.H.; Moon, H.K.; Song, J.H.; Son, S.K.; Lee, J.H. *Cultivation Technology for Special Purpose Trees Kalopanax Septemlobu and Aralia Elata* (특수수종 음나무 및 두릅나무 재배기술); National Institute of Forest Science: Seoul, South Korea, 2013; pp. 103–169. ISBN 978-89-8176-493-7 93520. Available online: http://know.nifos.go.kr/book/search/DetailView.ax?&cid=160540 (accessed on 9 April 2020).

95. RDA. Available online: https://www.nongsaro.go.kr/portal/ps/psb/psbk/kidofcomdtyDtl.ps;jsessionid= MFA4MfU8hjAXuTSstduwOhS6BRJbNz6nixtyVGQD21unGS9Y05622tRgnQJCXu5D.nongsaro-web_ servlet_engine1?menuId=PS00067&kidofcomdtyNo=22534 (accessed on 9 April 2020).

96. Freshfel. Available online: http://freshfel.org/freshfel-consumption-monitor (accessed on 9 April 2020).

97. European Commission. *EU Agricultural Outlook for Markets and Income 2018–2030*; DG Agriculture and Rural Development: Brussels, Belgium, 2018; pp. 4–73. Available online: https://ec.europa.eu/info/sites/info/ files/food-farming-fisheries/farming/documents/medium-term-outlook-2018-report_en.pdf (accessed on 9 April 2020).

98. Florkowski, W.J. Marketing Asian Fruits and Vegetables in Europe. *Acta Hortic.* **2008**, *804*, 39–44. [CrossRef]

99. Van der Plas, M.; Verhaegh, A.P. Marketing of Exotic and Out-of-Season Fresh Fruit and Vegetables in the European Common Market. In *Research Report*; Wageningen University and Research: Wageningen, The Netherlands, 1986. Available online: https://library.wur.nl/WebQuery/wurpubs/fulltext/264326 (accessed on 28 May 2020).

100. Gruda, N. Assessing the Impact of Environmental Factors on the Quality of Greenhouse Produce. In *Achieving Sustainable Greenhouse Cultivation*; Marcelis, L., Heuvelink, E., Eds.; Burleigh Dodds Science Publishing Limited: Sawston, UK, 2019; Chapter 16; ISBN 978-1-78676-280-1. [CrossRef]

101. United Nations. *Challenges and Opportunities Arising from Private Standards on Food Safety and Environment for Exporters of Fresh Fruit and Vegetables in Asia: Experiences of Malaysia, Thailand and Viet Nam*; United Nations

Publication: New York, NY, USA; Geneva, Switzerland, 2007. Available online: https://unctad.org/en/Docs/ditcted20076_en.pdf (accessed on 23 April 2020).

102. Faour-Klingbeil, D.; Todd, E.C.D. A Review on the Rising Prevalence of International Standards: Threats or Opportunities for the Agri-Food Produce Sector in Developing Countries, with a Focus on Examples from the MENA Region. *Foods* **2018**, *7*, 33. [CrossRef] [PubMed]

103. Grote, U.; Kirchhoff, S. Environmental and Food Safety Standards in the Context of Trade Liberalization: Issues and Options. Discussion Papers on Development Policy No. 39. Available online: https://ageconsearch.umn.edu/record/18725 (accessed on 23 April 2020).

104. CBI Ministry of Foreign Affairs. Available online: https://www.cbi.eu/node/2423/pdf (accessed on 9 April 2020).

105. Eurostat. Available online: https://ec.europa.eu/eurostat/statistics-explained/index.php/The_fruit_and_vegetable_sector_in_the_EU_-_a_statistical_overview (accessed on 9 April 2020).

106. CBI Ministry of Foreign Affairs. Available online: https://www.cbi.eu/node/2918/pdf (accessed on 9 April 2020).

107. European Commission. Available online: https://ec.europa.eu/info/sites/info/files/food-farming-fisheries/news/documents/agricultural-trade-report_map2018-1_en.pdf (accessed on 9 April 2020).

108. European Commission. Available online: https://ec.europa.eu/info/food-farming-fisheries/trade/agricultural-international-trade/bilateral-agreements/asia-and-australasia_en (accessed on 9 April 2020).

109. Rabobank. Available online: https://www.rabobank.com/en/press/search/2016/20160908_fresh_produce_trade.html (accessed on 9 April 2020).

110. Fruit Logistica. Available online: https://www.fruitlogistica.com/media/fl/fl_images/fl_images_press/6/pressemitteilungen_3/Fruit_Logistica_Trend_Report_2019.pdf (accessed on 9 April 2020).

111. Pinho, M.G.M.; Mackenbach, J.D.; Charreire, H.; Oppert, J.M.; Bárdos, H.; Glonti, K.; Rutter, H.; Compernolle, S.; De Bourdeaudhuij, I.; Beulens, J.W.J.; et al. Exploring the relationship between perceived barriers to healthy eating and dietary behaviours in European adults. *Eur. J. Nutr.* **2018**, *57*, 1761–1770. [CrossRef] [PubMed]

112. Pollard, J.; Kirk, S.F.L.; Cade, J.E. Factors affecting food choice in relation to fruit and vegetable intake: A review. *Nutr. Res. Rev.* **2002**, *15*, 373–387. [CrossRef]

113. Gruda, N. Impact of Environmental Factors on Product Quality of Greenhouse Vegetables for Fresh Consumption. *Crit. Rev. Plant Sci.* **2005**, *24*, 227–247. [CrossRef]

114. Cambridge Dictionary. Available online: https://dictionary.cambridge.org/dictionary/english/instagram (accessed on 9 April 2020).

115. Sousvidetools.com. Available online: https://www.sousvidetools.com/toolshed/the-most-hashtagged-foods-on-instagram (accessed on 23 April 2020).

116. Bisbis, M.B.; Gruda, N.; Blanke, M. Potential impacts of climate change on vegetable production and product quality–A review. *J. Clean. Prod.* **2018**, *170*, 1602–1620. [CrossRef]

117. Haddad, L.; Hawkes, C.; Webb, P.; Thomas, S.; Beddington, J.; Waage, J.; Flynn, D. A new global research agenda for food. *Nature* **2016**, *540*, 30–32. Available online: https://www.nature.com/news/polopoly_fs/1.21052!/menu/main/topColumns/topLeftColumn/pdf/540030a.pdf (accessed on 28 May 2020). [CrossRef]

118. Dong, J.; Gruda, N.; Lam, S.K.; Li, X.; Duan, Z. Effects of Elevated CO2 on Nutritional Quality of Vegetables: A Review. *Front. Plant Sci.* **2018**, *9*, 924. [CrossRef]

119. Dong, J.; Gruda, N.; Li, X.; Tang, Y.; Zhang, P.; Duan, Z. Sustainable vegetable production under changing climate: The impact of elevated CO2 on yield of vegetables and the interactions with environments-A review. *J. Clean. Prod.* **2020**, *253*, 119920. [CrossRef]

120. Eigenbrod, C.; Gruda, N. Urban vegetable for food security in cities. A review. *Agron. Sustain. Dev.* **2015**, *35*, 483–498. [CrossRef]

121. Gruda, N.; Bisbis, M.B.; Tanny, J. Impacts of protected vegetable cultivation on climate change and adaptation strategies for cleaner production—A review. *J. Clean. Prod.* **2019**, *225*, 324–339. [CrossRef]

122. Gruda, N.; Bisbis, M.; Tanny, J. Influence of climate change on protected cultivation: Impacts and sustainable adaptation strategies—A review. *J. Clean. Prod.* **2019**, *225*, 481–495. [CrossRef]

123. Van Wyk, B.E. The potential of South African plants in the development of new food and beverage products. *S. Afr. J. Bot.* **2011**, *77*, 857–868. [CrossRef]

124. Savvas, D.; Gruda, N. Application of soilless culture technologies in the modern greenhouse industry—A review. *Eur. J. Hortic. Sci.* **2018**, *83*, 280–293. [CrossRef]
125. Gruda, N.S. Increasing Sustainability of Growing Media Constituents and Stand-Alone Substrates in Soilless Culture Systems. *Agronomy* **2019**, *9*, 298. [CrossRef]
126. Gruda, N. Do soilless culture systems have an influence on product quality of vegetables? *J. Appl. Bot. Food Qual.* **2009**, *82*, 141–147. [CrossRef]
127. Businesswire. Available online: https://www.businesswire.com/news/home/20191105005561/en/Europe-Hydroponics-Aggregate-Systems-Liquid-Hydroponics-Market (accessed on 9 April 2020).

horticulturae

Article

Fruit Ripening Development of 'Valencia' Orange Trees Grafted on Different 'Trifoliata' Hybrid Rootstocks

Allan Ricardo Domingues [1], Ciro Daniel Marques Marcolini [2], Carlos Henrique da Silva Gonçalves [2], Leandro Simões Azeredo Gonçalves [1], Sergio Ruffo Roberto [1,*] and Eduardo Fermino Carlos [2]

[1] Agricultural Research Center, State University of Londrina, Celso Garcia Cid Road, km 380, P.O. Box 10.011, Londrina ZIP 86057-970, Brazil; allanrdomingues@gmail.com (A.R.D.); leandrosag@uel.br (L.S.A.G.)

[2] Development Agronomic Institute of Paraná-IDR-IAPAR, Celso Garcia Cid Road, km 375, Londrina ZIP 95700-00, Brazil; ciromarcolini@emater.pr.gov.br (C.D.M.M.); carloshg@itaipu.gov.br (C.H.d.S.G.); efcarlos@iapar.br (E.F.C.)

* Correspondence: sroberto@uel.br; Tel.: +55-43-3371-4774

Abstract: The development of fruit ripening of 'Valencia' orange trees [*Citrus sinensis* (L.) Osb.] grafted on the following 'Trifoliata' hybrid rootstocks: 'US-852', IPEACS-256 and IPEACS-264 citrandarins, and F.80-3, 'W-2' citrumelo, and 'Swingle' citrumelo (control), was assessed in three different subtropical locations: Rancho Alegre (RA); São Sebastião da Amoreira (SSA); and São Jerônimo da Serra (SJS), Parana state, Brazil. The climate of the RA and SSA locations was classified as Cfa with hot summers, whereas that of the SJS location was Cfb with temperate summers, which are located at 380, 650, and 835 m a.s.l., respectively. A completely randomized block design with four replications and four trees per plot was used as a statistical model for each location. The soluble solids (SS) content, titratable acidity (TA), and the maturation index (MI) or *ratio* (SS/TA) of the juice, as well as the citrus color index (CCI) of fruit skin, were assessed monthly, beginning 200 days after flowering until harvest, totaling seven sampling dates. The data sets of each location were analyzed independently through a two-way analysis of variance (ANOVA) involving rootstocks in a split plot array in time (days) to allow for the assessment of the significance of the main effects, complemented by regression analysis. In general, the ripening of 'Valencia' orange fruits was influenced by the different 'Trifoliata' hybrid rootstocks. At the RA location, trees on IPEACS-256 and' US-852' citrandarins had the highest SS, and on 'US-852' citrandarin had the highest MI, reaching the MI_{im} earlier than the other rootstocks. The highest CCI was achieved when trees were on IPEACS-256 citrandarin. At the SSA location, trees on 'US-852', IPEACS-256, and IPEACS-264 citrandarins had the highest SS, but on 'US–852' had the highest MI, reaching the earliest MI_{im} among the rootstocks. The highest CCI was achieved when trees were on 'US-852' and IPEACS-256 citrandarins. In the SSJ location, there was no significant effect of the rootstocks on and of the variables of 'Valencia' orange fruit. This assessment can be useful in the planning of new orchards producing high-quality fruit with desirable features for the orange juice industry.

Keywords: *Citrus sinensis* (L.) Osb.; rootstocks; maturation index; citrus color index

Citation: Domingues, A.R.; Marcolini, C.D.M.; Gonçalves, C.H.d.S.; Gonçalves, L.S.A.; Roberto, S.R.; Carlos, E.F. Fruit Ripening Development of 'Valencia' Orange Trees Grafted on Different 'Trifoliata' Hybrid Rootstocks. *Horticulturae* **2021**, 7, 3. https://doi.org/10.3390/horticulturae7010003

Received: 19 November 2020
Accepted: 22 December 2020
Published: 29 December 2020

1. Introduction

'Valencia' sweet orange [*Citrus sinensis* (L.) Osb.] is grown in the most citrus-producing countries and commonly represents the leading commercial variety [1,2]. In Brazil, this species is prominent among producers owing to high fruit yield and quality. Further, this variety has late-maturing fruit, which are intended for both fresh fruit markets as well as industrial processing into juice [3,4]. The production of 'Valencia' oranges for industrial processing in the form of frozen concentrated juice (FCOJ) and not from concentrated juice (NFC) serves as one of the main activities of the citrus industry with a broad economic impact worldwide [5]. However, the production of raw materials must be of high quality to maintain the sector's market competitiveness. The quality attributes of the fruits are

151

determined by their physicochemical characteristics, which in turn vary during the ripening period [6,7].

The ripening of citrus fruit passes through different phenological stages, and the final stage is characterized by a reduced growth rate with an increase in soluble solids (SS) content, a decrease in titratable acidity (TA), and changes in skin color [8]. The SS content and TA, along with the relationship between them, also known as the maturation index (MI) or *ratio* (SS/TA), are considered the main indicators of harvest quality [9–11]. MI represents the balance between the content of sugars and organic acids in fruit, is associated with juice taste, and is widely used by the processing industry as an indicator of maturation and internal quality of fruit [7,12]. MI is considered when adjusting harvest-to-fruit destinations, either for fresh fruit markets or processing, depending on the demand from the importing country. Thus, analyzing these indicators throughout the ripening process allows for the establishment of optimal harvesting time and, consequently, improved quality of fruit for the production sector.

In citrus and other fruit species, including grapes (*Vitis* spp.) and apples (*Malus domestica*), grafting is a well-developed technique which combines a scion and a rootstock to form a new tree with a blend of characteristics, and the interaction between the rootstock and scion in a grafted tree is based on various physiological parameters [13–16]. In addition, the ripening of orange fruit is strongly influenced by the climatic conditions of the growing region [6,8,17]. The rootstock can manipulate a grafted citrus tree in many ways. It can induce tolerance or resistance to soil-borne stresses and diseases while manipulating horticultural attributes, such as tree size, yield, nutrient uptake, water potential, and fruit quality, including the development of ripening, color of skin and juice, SS and TA contents, and other juice characteristics [18,19]. Moreover, selection of an appropriate graft combination should prioritize those that provide better fruit quality to meet the demands of the orange juice industry.

The search for new rootstocks that positively affect the characteristics of sweet oranges has led to the introduction of hybrids of 'Trifoliata' [*Poncirus trifoliata* (L.) Raf.] with mandarin (*Citrus reticulata* Blanco) or grapefruit (*Citrus paradisi* Macf.), also known as citrandarins and citrumelos, respectively. These have been evaluated as alternatives for the diversification of rootstocks in Brazil and several other countries [20–23]. Among the alternatives are 'US-852', 'IPEACS-256', and 'IPEACS-264' citrandarins and 'Swingle' and 'F.80-3' citrumelos [20,23–29].

When grafted on 'Trifoliata' hybrid rootstocks, orange trees produce high-quality fruit with desirable attributes for the juice industry, such as high SS content and juice yield [20,23,30–33]. 'Swingle' is the best-known citrumelo in Brazil and globally. It is among the main rootstocks used for diversifying orange groves, providing scions with high-quality fruit, increased juice yield, SS content and yield, and reduced scion vigor [21,30].

Analysis of orange fruit ripening development enables the identification of groups of scion/rootstock combinations with different physicochemical characteristics and harvest dates, permitting improved planning of the establishment of new orchards for producing high-quality orange juice [17,34]. Thus, it is possible to establish the optimal harvest point for each scion/rootstock combination within specific regions [35]. Therefore, we aimed to assess the ripening development of 'Valencia' oranges when trees are grafted on different 'Trifoliata' hybrid rootstocks grown in different subtropical locations.

2. Materials and Methods

2.1. Locations and Plant Material

The fruit ripening development of 'Valencia' orange tree clone IAC [*Citrus sinensis* (L.) Osb.] grafted on different 'Trifoliata' [*Poncirus trifoliata* (L.) Raf.] hybrid rootstocks was assessed, and the experiment included the following rootstocks: 'US-852', IPEACS-256, and IPEACS-264 citrandarins [*Citrus reticulata* Blanco × *P. trifoliata* (L.) Raf.]; F.80-3 and 'W-2' citrumelo; 'Swingle' citrumelo (considered as control) [*Citrus paradisi* Macf. × *P. trifoliata* (L.) Raf.].

The work was assessed concomitantly in three humid subtropical areas without a dry season in the State of Parana, Brazil, as follows: Rancho Alegre (RA) (23°03'15" S, 50°55'50" W, elevation 380 m a.s.l.); São Sebastião da Amoreira (SSA) (23°24'47" S, 50°43'45" W, elevation 650 m a.s.l.); and São Jerônimo da Serra (SJS) (23°44'04" S, 50°52'32" W, elevation 835 m a.s.l.). The three locations were chosen because they represent different citrus growing conditions in this subtropical region.

According to the Köppen classification, the climate of the RA and SSA locations is Cfa with hot summers, whereas that of the SJS location is Cfb with temperate summers [36]. The higher the elevation of the location, the greater the thermal amplitude, which is the variation of the air temperature between day and night. The average annual rainfall at the RA and SSA locations is 1300 mm, and the average temperatures are 22.1 and 21.1 °C, respectively. At the SJS location, the average rainfall is 1500 mm, and the average temperature is 20.1 °C. The relative humidity is between 75 and 80%, with higher rainfall during spring and summer [37].

The budded trees were transplanted into the trials in mid-summer 2013 at commercial field sites, with a spacing of 6.0 × 2.5 m (between-row × in-row) and were non-irrigated. A completely randomized block design with four replications of each rootstock and four trees per replicate plot was used as a statistical model for each location. In each plot, the two outer trees were considered side borders, and the two inner trees were subjected to evaluations. The rootstock trial plantings were adjacent to a commercial orchard of 'Valencia' orange trees, and the management of the trial blocks was the same as the contiguous commercial blocks.

The 'US-852' citrandarin used in these trials is a cross between the 'Changsha' mandarin and the 'English Large' trifoliate created by Joe Furr at Indio, California, USA in 1965, officially released in 1999 by Kim D. Bowman of the USDA/ARS/HRL, Fort Pierce, FL, USA [25,27]. It was introduced to Brazil in 1982 through deposition into the Active Germplasm Bank (Banco Ativo do Germoplasma-BAG) at the Sylvio Moreira Citriculture Center, Agronomic Institute of Campinas (Centro de Citricultura Sylvio Moreira, Instituto Agronômico de Campinas-CCSM/IAC), Cordeirópolis, SP under access code #1454.

The IPEACS-256 citrandarin is a cross between the 'Cleopatra' mandarin and the 'English' trifoliata, with no official record of its origin or the breeder responsible for obtaining it. It was first made available by the former Research and Agricultural Experimentation Institute of Central-South Region (Instituto de Pesquisas e Experimentação Agropecuária do Centro-Sul-IPEACS), Itaguaí, RJ, Brazil. It was introduced in 1975 through deposition into the BAG of CCSM/IAC under access code #1483. The IPEACS-264 citrandarin is a cross between the 'Sunki' mandarin and the 'English' trifoliata, and also does not have any official record of its origin. It was first made available by the former IPEACS and was introduced in 1970 through deposition to the CCSM/IAC BAG under access code #1628. However, the IPEACS-256 and IPEACS-264 citrandarins were referred to as 'Indio' and 'Riverside', respectively [38].

The F.80-3 citrumelo is a cross, originally performed in 1955 by Mortimer Cohen at the Indian River Field Laboratory, Fort Pierce, FL, USA [39], and was introduced through deposition into the BAG of CCSM/IAC in 1990 under access code #1460. The 'W-2' citrumelo is another hybrid of USA origin and was introduced into the BAG of CCSM/IAC in 1990 under the access code #1455.

The 'Swingle' citrumelo is a cross between the 'Duncan' grapefruit and 'Trifoliata', performed in 1907 by Walter T. Swingle of the USDA, Eustis, FL, USA [20]. The acronym, CPB4475, was first used and the fruit was only officially released in 1974 under the name 'Swingle' in honor of its creator [39]. It was introduced into the BAG of CCSM/IAC in 1990 under access code #401.

All citrandarins and citrumelos evaluated in the present study were introduced into the BAG of CCSM/IAC by the researcher Jorgino Pompeu Jr. The nursery trees of the different scion/rootstock combinations used in the present study were produced by Pratinha Citrus

Nurseries, a certified nursery of citrus trees located in Paranavaí, PR, and the propagation material was provided by CCSM/IAC, Cordeirópolis, SP.

2.2. Physicochemical Analysis of Fruits

The fruit samples were collected during the 2017 and 2018 seasons, and the analyses were performed at the Citrus Laboratory of Cooperativa Integrada Agroindustry processing plant, located at Uraí, PR. The fruit ripening development of 'Valencia' orange trees grafted on each rootstock was assessed monthly by collecting samples of 16 fruits from each plot. Fruit samples were juiced using an FMC citrus juice extractor, and juice quality was analyzed using standard laboratory methods. The evaluations began approximately 200 days after the beginning of flowering when the fruit were approximately 90% of their final size, which corresponds to phenological stage #79 according to the BBCH phenological growth stages and identification keys for citrus trees [40]. The zero time was defined for the first fruit sampling (late summer), extending until fruit harvest (early spring), totaling seven monthly samplings from March to September. The samples were randomly collected from the periphery of the trees between 1.0 and 2.0 m from the soil, with one fruit collected from each quadrant of each evaluated tree. At the SJS location, only the 'Swingle' and 'W-2' citrumelos were evaluated as the fruit production of trees grafted on the other rootstocks was insufficient for evaluation.

Aiming at assessing the effect of the different rootstocks on ripening development of 'Valencia' orange fruits at each location independently, the samples were submitted to physicochemical analysis, such as the SS contents, TA, and MI (or SS/TA) of the juice, as well the color of fruit skin by means of the citrus color index-CCI.

SS content was determined by direct reading of the juice on a refractometer (model PAL-ALPHA, Atago®, Tokyo, Japan) with automatic temperature compensation, and the result was expressed in °Brix. The TA was determined by titration of 25 mL of juice with 0.1 N NaOH solution, with the endpoint adjusted to pH 8.2, and the result was expressed as the percentage of citric acid [41]. The MI was calculated as SS/TA. MI = 14 was considered the ideal mean (MI_{im}) for processing 'Valencia' oranges, either for the production of FCOJ or NFC orange juice [42].

The CCI was determined using a digital color reader (model CR10 Plus, Konica-Minolta, Tokyo, Japan) to obtain the variables, L^*, a^*, and b^*. Readings were performed on the outer equatorial part of the skin of eight fruit per plot. These variables were then used to calculate the CCI using the formula: CCI = $(1000 \times a^*)/(L^* \times b^*)$. The CCI values below −7 represent green with increased intensity and more negative values. Values ranging from −7 to 0 represent shades starting from light green, passing through yellowish-green, and reaching pale yellow, and values between 0 and 7 represent shades ranging from pale yellow to orange. Values above 7 represent orange, which rose in intensity as CCI increased. The time when CCI = 0 corresponds to the color of the fruit skins beginning to change from green to pale yellow [43]. Figure 1 illustrates the development of the CCI of fruits of 'Valencia' orange according to different stages of ripening.

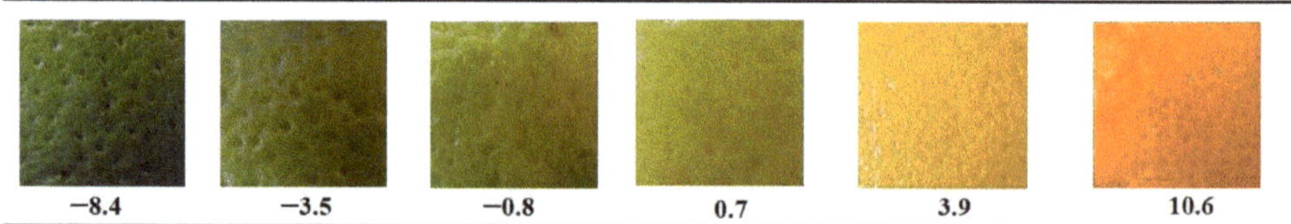

Figure 1. Representation of the development of the citrus color index (CCI) of 'Valencia' orange fruits when trees are grafted on different 'Trifoliata' hybrid rootstocks in different stages of ripening.

2.3. Statistical Analyses

The data sets of SS, TA, MI, and CCI, obtained from the monthly samplings from each of the three locations, were analyzed independently in a two-way analysis of variance (ANOVA) involving rootstocks in a split plot array in time (days) to allow the assessment of the significance of the main effects. Means were compared by Tukey's honestly significant difference (HSD) test at $p < 0.05$. To determine the effect of the 'Trifoliata' hybrid rootstocks on fruit ripening development of 'Valencia' orange at each location, a regression analysis was carried out. The model with the best fit was determined considering the coefficient of determination (R^2), which was adjusted to the number of parameters using the SISVAR software [44] and R [45]. Next, the number of days was predicted so that fruits of 'Valencia' orange trees on each rootstock reached $MI_{im} = 14$ and CCI = 0 at the three locations.

3. Results and Discussion

At the RA location, significant differences among the 'Trifoliata' hybrid rootstocks were observed for SS, MI, and CCI of 'Valencia' orange fruit (Table 1). An effect of time (in days) for all variables analyzed was also observed. For SS content, significant differences were observed after 60 days of the beginning of the evaluations, and the rootstocks IPEACS-256 and 'US-852' citrandarins had the highest means until the last evaluation (Table 2). A similar pattern was observed for MI, and during the last evaluation, the 'US-852' citrandarin exhibited the highest mean, and the 'Swingle' citrumelo the lowest (Table 2). Regarding the CCI, differences among rootstocks were also observed from the 60th day after the beginning of the evaluations, and on the last day, the IPEACS-256 citrandarin had the highest mean, and the citrumelos had the lowest (Table 2).

At the SSA location, the effect of the different 'Trifoliata' hybrid rootstocks on 'Valencia' orange fruits was observed for SS, MI, and CCI (Table 3). A significant interaction between rootstocks and time in days was only verified for SS content. For this juice characteristic, the differences were observed after 90 days from the beginning of the evaluations, and for the last one, the highest means were noted when the trees were grafted on 'US-852', IPEACS-256, and IPEACS-264 citrandarins (Table 4).

Regarding MI, the 'US-852' citrandarin had the highest mean among the evaluations (Table 4). The effect of the rootstocks on the CCI of orange skins was observed during harvest, where the 'US–852' and IPEACS-256 citrandarins resulted in the highest means, and the 'Swingle' citrumelo the lowest (Table 4).

At the SSJ location, there was no significant effects of rootstocks on 'Valencia' orange fruits for all variables evaluated (Tables 5 and 6). However, the effect of the time in days was significant, but it must be noted that, at this location, only the 'W–2' and 'Swingle' citrumelos were assessed as the fruit production of trees grafted on the other 'Trifoliata' hybrid rootstocks was inadequate for evaluation.

Table 1. Analysis of variance and mean squares of soluble solids (SS), titratable acidity (TA), maturation index (MI) or *ratio*, and citrus color index (CCI) over time of 'Valencia' orange fruits when trees are grafted on different 'Trifoliata' hybrid rootstocks at Rancho Alegre.

Sources of Variance	Df [z]	SS	TA	MI	CCI [y]
Blocks	3	0.2766	0.0343	1.64	3.47
Rootstocks (R)	5	5.11 **[x]	0.0114	4.86 **	8.82 **
Error A	15	0.364	0.032	0.95	1.28
Days (D)	6	45.43 **	7.69 **	415.07 **	925.67 **
R × D	30	0.3587 **	0.0409	0.72 **	2.81 **
Error B	108	0.1519	0.0353	0.57	1.23
CV 1 (%) [w]		6.4	12.6	11.1	4.3
CV 2 (%)		4.1	13.2	8.6	4.2

[z] df: degrees of freedom. [y] Original data transformed into x + 30. [x] **: significant ($p < 0.05$). [w] CV = coefficient of variation.

Considering the regression analysis of the fruit ripening development of 'Valencia' orange grafted on different Trifoliata hybrid rootstocks, the SS content exhibited an increasing pattern, adjusting to the quadratic model for all three locations (Figures 2A, 3A and 4A). The TA development of fruit was better adjusted by the linear regression model with decreasing values for all locations and all evaluated rootstocks. The 'Trifoliata' hybrid rootstocks induced very similar TA values throughout the ripening period (Figures 2B, 3B and 4B).

The MI of the 'Valencia' fruits of the three locations all demonstrated increasing development, and this characteristic of maturation was better adjusted to the linear model (Figures 5A, 6A and 7A). Moreover, the CCI of 'Valencia' orange fruits at all three locations assessed also increased over time and was better adjusted to the quadratic model (Figures 5B, 6B and 7B).

At the RA location, a subtropical region with a hot summer and at low elevation (380 m a.s.l.), the fruits of trees grafted on 'US-852' citrandarin reached the MI_{im} earlier than the other 'Trifoliata' hybrid rootstocks, in an estimated period of 158 days after the beginning of the evaluations, followed by IPEACS-256 citrandarin, F.80-3 citrumelo, IPEACS-264 citrandarin, 'W-2' and 'Swingle' citrumelos (162, 173, 182, 182, and 187 days, respectively) (Figure 5A). The earliest estimated period that fruits changed skin color from green to pale yellow, corresponding to a CCI = 0, was also observed when the orange trees were grafted on 'US-852' citrandarin, with an estimated period of 107 days after the beginning of the evaluations, followed by IPEACS-256 and IPEACS-264 citrandarins, and F.80-3, 'W-2', and 'Swingle' citrumelos (110, 124, 135, 136, and 141 days, respectively) (Figure 5B). At this location, considering the MI_{im}, the main characteristic for fruit processing, the evaluated rootstocks strongly influenced the ripening of 'Valencia' oranges, with a maximum variation of 29 days from the earliest ('US-852' citrandarin) to the latest ('Swingle' citrumelo) ripening.

Table 2. Soluble solids, maturation index or *ratio*, and citrus color index of 'Valencia' orange fruits when trees are grafted on different 'Trifoliata' hybrid rootstocks at Rancho Alegre at different times after the phenological stage BBCH #79.

Rootstocks	Soluble Solids–SS (°Brix)						
	Days after the Phenological Stage BBCH #79 [z]						
	0	30	60	90	120	150	180
'US–852' citrandarin	6.9b [y]	8.1a	9.1ab	10.1ab	10.8ab	11.3ab	10.7ab
IPEACS–256 citrandarin	7.9a	8.1a	9.9a	10.7a	11.5a	11.8a	11.4a
IPEACS–264 citrandarin	6.9b	8.0a	8.9b	9.8b	10.3b	10.5bc	9.6c
F.80–3 citrumelo	7.2ab	8.1a	8.9b	9.6b	10.4b	10.2c	9.3c
'W–2' citrumelo	7.1ab	7.8a	8.9b	9.3b	10.0b	10.3c	9.6c
'Swingle' citrumelo	6.7b	8.1a	9.2b	10.0ab	10.6ab	10.7bc	10.1bc
	Maturation index–MI or *ratio*						
	0	30	60	90	120	150	180
'US–852' citrandarin	2.8a	4.9a	7.1a	9.2a	12.9a	13.3ab	14.8a
IPEACS–256 citrandarin	3.1a	4.7a	6.7a	8.9ab	12.6ab	13.3a	14.4ab
IPEACS–264 citrandarin	2.9a	5.1a	5.8a	7.3b	11.9ab	11.7b	13.7ab
F.80–3 citrumelo	3.4a	4.9a	6.6a	8.8ab	12.4ab	12.4ab	13.7ab
'W–2' citrumelo	3.4a	4.7a	6.8a	7.5ab	11.0b	12.3ab	13.8ab
'Swingle' citrumelo	3.1a	5.0a	6.2a	7.9ab	11.2b	11.9ab	13.1b
	Citrus color index–CCI						
	0	30	60	90	120	150	180
'US–852' citrandarin	−11.8a	−11.3a	−5.2ab	−0.8ab	1.8a	3.2a	3.4ab
IPEACS–256 citrandarin	−12.5a	−12.5a	−3.9a	0.4a	0.8ab	3.8a	4.0a
IPEACS–264 citrandarin	−12.5a	−10.7a	−6.2b	−1.3ab	0.1ab	1.9ab	3.6ab
F.80–3 citrumelo	−10.7a	−10.9a	−4.5ab	−1.2ab	−0.4ab	1.0b	1.1c
'W–2' citrumelo	−12.2a	−11.0a	−6.5b	−2.2b	−0.7b	2.1ab	2.3bc
'Swingle' citrumelo	−12.5a	17.5a	−5.8ab	−2.2b	−0.1ab	1.6ab	1.7bc

[z] Approximately 200 days after flowering (Meier, 2001). [y] Means with the same letter are not significantly different from each other by Tukey´s honestly significant difference test at $p < 0.05$.

Table 3. Analysis of variance and mean squares of soluble solids (SS), titratable acidity (TA), maturation index (MI) or *ratio* (SS/TA), and citrus color index (CCI) over time of 'Valencia' orange fruits when trees are grafted on different 'Trifoliata' hybrid rootstocks at São Sebastião da Amoreira.

Sources of Variance	df [z]	SS	TA	MI	CCI [y]
Blocks	3	0.2887	0.0343	1.28	0.20
Rootstocks (R)	5	2.7578 **[x]	0.01114	1.77 **	14.20 **
Error A	15	0.3239	0.032	0.84	2.11
Days (D)	6	42.37 **	7.69 **	345.07 **	948.43 **
R × D	30	0.3597 **	0.0409	0.94	1.72
Error B	108	0.2270	0.0353	0.74	1.96
CV 1 (%) [w]		6.0	12.6	11.4	5.2
CV 2 (%)		5.0	13.2	10.7	5.0

[z] df: degrees of freedom. [y] Original data transformed into x + 30. [x] **: significant ($p < 0.05$). [w] CV = coefficient of variation.

At the SSA location, a subtropical region with a hot summer and at an intermediate elevation (650 m a.s.l.), the MI_{im} of 'Valencia' fruits was also reached earlier when grafted on 'US-852' citrandarin, with an estimated period of 177 days after the beginning of the evaluations (Figure 6A), followed by 'Swingle' and F.80-3 citrumelos, IPEACS-264 and IPEACS-256 citrandarins, and 'W-2' citrumelo (191, 192, 194, 194, and 202 days, respectively) (Figure 6A). The fruits of 'Valencia' orange trees grafted on the 'Trifoliata' hybrid rootstocks reached CCI = 0 in a smaller range, from 85 to 93 days (IPEACS-256 and 'US-852', respectively), except for the 'Swingle' citrumelo, which reached this index at 105 days after the beginning of the evaluations (Figure 6B).

At this location, considering the MI_{im}, the rootstocks also influenced the development of ripening of 'Valencia' orange fruit, with a variation of 25 days from the earliest ('US-852' citrandarin) to the latest ('W-2' citrumelo) ripening.

At the SJS location, a subtropical region with temperate summers and at high elevations (835 m a.s.l.), the orange fruit did not reach the MI_{im} for processing until the last evaluation. Nevertheless, according to the regression analysis, both 'Swingle' and 'W-2' citrumelos expressed a similar late ripening behavior, and the MI_{im} was estimated to be reached at 254 days after the beginning of the evaluations (Figure 7A). On the other hand, under the specific weather conditions of this location, with temperate summer and wide temperature range, with warm days and mild nights, both 'Swingle' and 'W-2' citrumelos induced very early pale-yellow skin color (CCI = 0) of 'Valencia' orange fruits at 67 days after the beginning of the evaluations (Figure 7B).

Although external and internal ripening of orange fruits in general coincide, skin and pulp behave in many respects as separate organs and thus can be considered to undergo different physiological processes. Mature citrus pulp contains a very high percentage of water (85–90%) and many different constituents, including carbohydrates; organic acids; amino acids; vitamin C; minerals; and small quantities of lipids, proteins, and secondary metabolites, such as carotenoids, flavonoids, and volatiles [46].

SS comprises 10–20% of the fresh weight of the fruit and consists mainly of carbohydrates (70–80%) and relatively minor quantities of organic acids, proteins, lipids, and minerals [46]. The accumulation of SS content during the ripening period occurs in growth stage III of sweet oranges, also known as the stage of cell expansion, and it is characterized by a rapid increase in fruit size and SS content, lasting up to 6 months according to the region and growing conditions [8]. The rootstock plays an important role in fruit ripening because it can speed up or delay citrus tree development. In this work, it was shown that the 'US-852' and IPEACS-256 citrandarins anticipated ripening of fruits. In addition, in subtropical regions with hot summers and at low elevations, the duration of stage III can be anticipated to be 3 to 4 months, whereas in subtropical regions with temperate summers, this stage tends to be prolonged, extending up to 10 months. At the end of stage III, fruits tend to accumulate considerable SS content, which is determined by the climatic conditions of the growing regions, particularly water availability.

Table 4. Soluble solids, maturation index or *ratio*, and citrus color index of 'Valencia' orange fruits when trees are grafted on different 'Trifoliata' hybrid rootstocks at São Sebastião da Amoreira at different times after the phenological stage BBCH #79.

Rootstocks	Soluble Solids–SS (°Brix)						
	Days after the Phenological Stage BBCH #79 [z]						
	0	30	60	90	120	150	180
'US–852' citrandarin	7.1a [y]	8.0a	9.7a	10.4ab	10.9a	10.9ab	10.3a
IPEACS–256 citrandarin	7.5a	8.4a	10.0a	10.7a	10.9a	11.1ab	10.6a
IPEACS–264 citrandarin	7.1a	8.2a	9.1a	10.3ab	10.5a	10.1b	10.3a
F.80–3 citrumelo	7.4a	8.0a	9.4a	9.9ab	10.3a	10.3b	9.3bc
'W–2' citrumelo	6.8a	7.8a	9.4a	9.6b	10.2a	10.6ab	8.7c
'Swingle' citrumelo	7.2a	8.2a	9.6a	10.3ab	10.8a	11.3a	9.5bc
	Maturation index–MI or *ratio*						
	0	30	60	90	120	150	180
'US–852' citrandarin	2.7a	5.2a	6.5a	7.9a	10.5a	12.6a	14.9a
IPEACS–256 citrandarin	3.5a	4.9a	5.4b	7.8a	11.1a	11.2b	13.0b
IPEACS–264 citrandarin	3.1a	4.4a	5.3b	8.0a	10.7a	11.8b	12.5b
F.80–3 citrumelo	3.1a	4.5a	5.5b	8.0a	9.4a	11.1b	13.2b
'W–2' citrumelo	3.0a	4.4a	5.2b	7.1a	10.0a	11.5b	12.4b
'Swingle' citrumelo	3.0a	4.7a	5.3b	7.6a	10.9a	10.8b	12.9b
	Citrus color index–CCI						
	0	30	60	90	120	150	180
'US–852' citrandarin	−12.0a	11.4a	−3.0ab	0.7a	2.3a	3.7a	4.6a
IPEACS–256 citrandarin	−9.8a	−9.8a	−1.2a	0.9a	2.8a	4.0a	4.4a
IPEACS–264 citrandarin	−11.2a	−10.8a	−1.4a	1.5a	2.0a	3.2a	4.0ab
F.80–3 citrumelo	−10.7a	−10.5a	−1.6a	0.8a	1.8a	3.1a	4.1ab
'W–2' citrumelo	−10.9a	−9.1a	−2.2a	0.1a	1.3a	2.7a	3.4bc
'Swingle' citrumelo	−12.0a	−10.9a	−5.5b	−0.2a	1.0a	1.8a	2.2c

[z] Approximately 200 days after flowering (Meier, 2001). [y] Means with the same letter are not significantly different from each other by Tukey's honestly significant difference test at $p < 0.05$.

Table 5. Analysis of variance and mean squares of soluble solids (SS), titratable acidity (TA), maturation index (MI) or *ratio* (SS/TA), and citrus color index (CCI) over time of 'Valencia' orange fruits when trees are grafted on different 'Trifoliata' hybrid rootstocks at São Jerônimo da Serra.

Sources of Variance	df [z]	SS	TA	MI	CCI [y]
Blocks	3	0.2196	0.1178	1.1933	2.70
Rootstocks (R)	1	0.0460	0.0240	0.3600	2.89
Error A	3	0.213	0.0393	0.4400	3.01
Days (D)	6	19.32 **[x]	2.0653 **	67.49 **	284.25 **
R × D	6	0.405	0.1057	0.35	1.63
Error B	36	0.3193	0.0651	0.76	2.38
CV 1 (%) [w]		4.6	11.9	9.8	8.9
CV 2 (%)		5.7	15.3	12.8	7.9

[z] df: degrees of freedom. [y] Original data transformed into x + 30. [x] **: significant ($p < 0.05$). [w] CV = coefficient of variation.

Table 6. Soluble solids, maturation index or *ratio*, and citrus color index of 'Valencia' orange fruits when trees are grafted on different 'Trifoliata' hybrid rootstocks at São Jerônimo da Serra-SJS location at different times after the phenological stage BBCH #79.

Rootstocks	Soluble Solids–SS (°Brix)						
	Days after the Phenological Stage BBCH #79 [z]						
	0	30	60	90	120	150	180
'W–2' citrumelo	7.5a [y]	7.8a	9.7a	10.4a	11.2a	11.8a	10.9a
'Swingle' citrumelo	7.7a	8.0a	10.0a	10.7a	11.6a	11.1a	10.4a
	Maturation index–MI or *ratio*						
	0	30	60	90	120	150	180
'W–2' citrumelo	3.7a	3.6a	5.0a	6.0a	8.4a	9.6a	10.6a
'Swingle' citrumelo	3.4a	3.7a	4.9a	7.0a	8.7a	9.8a	10.7a
	Citrus color index–CCI						
	0	30	60	90	120	150	180
'W–2' citrumelo	−10.4a	−9.4a	−0.4a	3.4a	3.5a	4.0a	4.1a
'Swingle' citrumelo	−10.1a	−8.7a	−0.8a	2.4a	3.4a	3.8a	4.0a

[z] Approximately 200 days after flowering (Meier, 2001). [y] Means with the same letter are not significantly different from each other by Tukey's honestly significant difference test at $p < 0.05$.

It is known that increasing rainfall during stage III may lead to a decrease in SS content by diluting the juice in response to excessive water absorption by trees [47,48]. Therefore, the decrease in the SS content of 'Valencia' orange fruits grafted on different rootstocks after reaching the maximum content, which is indicated by quadratic behavior, may have occurred because of the higher rainfall season during the period of SS accumulation in the three locations evaluated. This is a recurrent situation in the ripening of late orange varieties, such as 'Valencia', grown in subtropical regions, where considerable rainfall periods usually coincide with lower concentrations of these sugars.

Generally, fruits of 'Valencia' orange grown at the SJS location, with temperate summers, exhibited a slow decrease in TA during the maturation period. Some authors have reported that fruits grown in regions with hot summers tend to exhibit a rapid decrease in acidity compared with regions with temperate summers owing to the increased respiration rate [9,12,48,49].

Considering the MI_{im}, the fruits of 'Valencia' orange trees grafted on different 'Trifoliata' hybrid rootstocks grown at the RA location presented early ripeness, with an estimated time to reach this phase from 158 to 187 days after the beginning of the evaluations when grafted on 'US-852' citrandarin and 'Swingle' citrumelo, respectively (Figure 2A). The same trend was observed for both rootstocks when the 'Valencia' orange trees were grown at the SSA location, but the estimated periods that these two rootstocks took to reach this phase were 177 and 191 days, respectively (Figure 2B). According to the regression model adjusted for the SJS location, late ripeness of fruits was observed, as the estimated time to reach the MI_{im} ranged from 246 and 261 days after the beginning of the evaluations for 'Swingle' and 'W-2' citrumelos, respectively (Figure 4B).

Rootstocks have been demonstrated to change the ripening development of 'Folha Murcha' sweet orange [*C. sinensis* (L.) Osb.] by inducing different fruit harvest dates, and this difference may range from 8 to 28 days [17]. These authors also found that fruits grown in regions with higher temperatures reached full ripeness up to 92 days earlier. Under subtropical conditions with hot summers, orange fruits ripen and become marketable in a shorter period of time, whereas in milder climates, ripening takes longer [8,12]. Thus, this behavior corresponds to the 'Valencia' orange grown at the RA location, with early ripeness. Moreover, the rootstocks induced different harvest dates in 'Valencia' oranges, with emphasis on 'US-852' citrandarin grown in the RA and SSA locations.

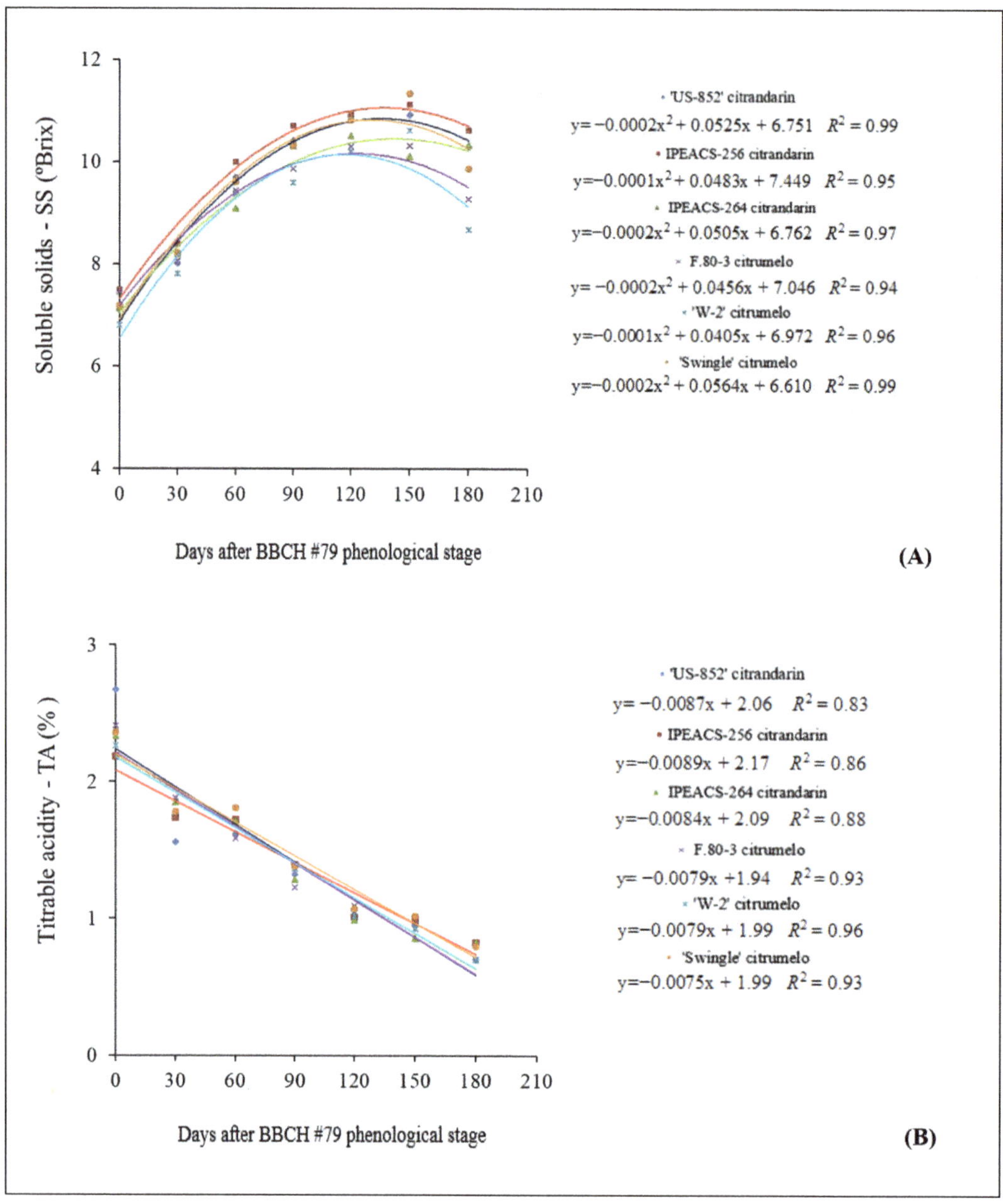

For panel (A), the fitted curves are:

$$y = -0.0002x^2 + 0.0525x + 6.751 \quad R^2 = 0.99$$
$$y = -0.0001x^2 + 0.0483x + 7.449 \quad R^2 = 0.95$$
$$y = -0.0002x^2 + 0.0505x + 6.762 \quad R^2 = 0.97$$
$$y = -0.0002x^2 + 0.0456x + 7.046 \quad R^2 = 0.94$$
$$y = -0.0001x^2 + 0.0405x + 6.972 \quad R^2 = 0.96$$
$$y = -0.0002x^2 + 0.0564x + 6.610 \quad R^2 = 0.99$$

For panel (B), the fitted lines are:

$$y = -0.0087x + 2.06 \quad R^2 = 0.83$$
$$y = -0.0089x + 2.17 \quad R^2 = 0.86$$
$$y = -0.0084x + 2.09 \quad R^2 = 0.88$$
$$y = -0.0079x + 1.94 \quad R^2 = 0.93$$
$$y = -0.0079x + 1.99 \quad R^2 = 0.96$$
$$y = -0.0075x + 1.99 \quad R^2 = 0.93$$

Figure 2. Development of 'Valencia' orange fruit ripening when trees are grafted on different 'Trifoliata' hybrid rootstocks at Rancho Alegre. (**A**): soluble solids (SS as °Brix); (**B**): titratable acidity (TA as %). Phenological stage BBCH #79 (Meier, 2001), approximately to 200 days after flowering.

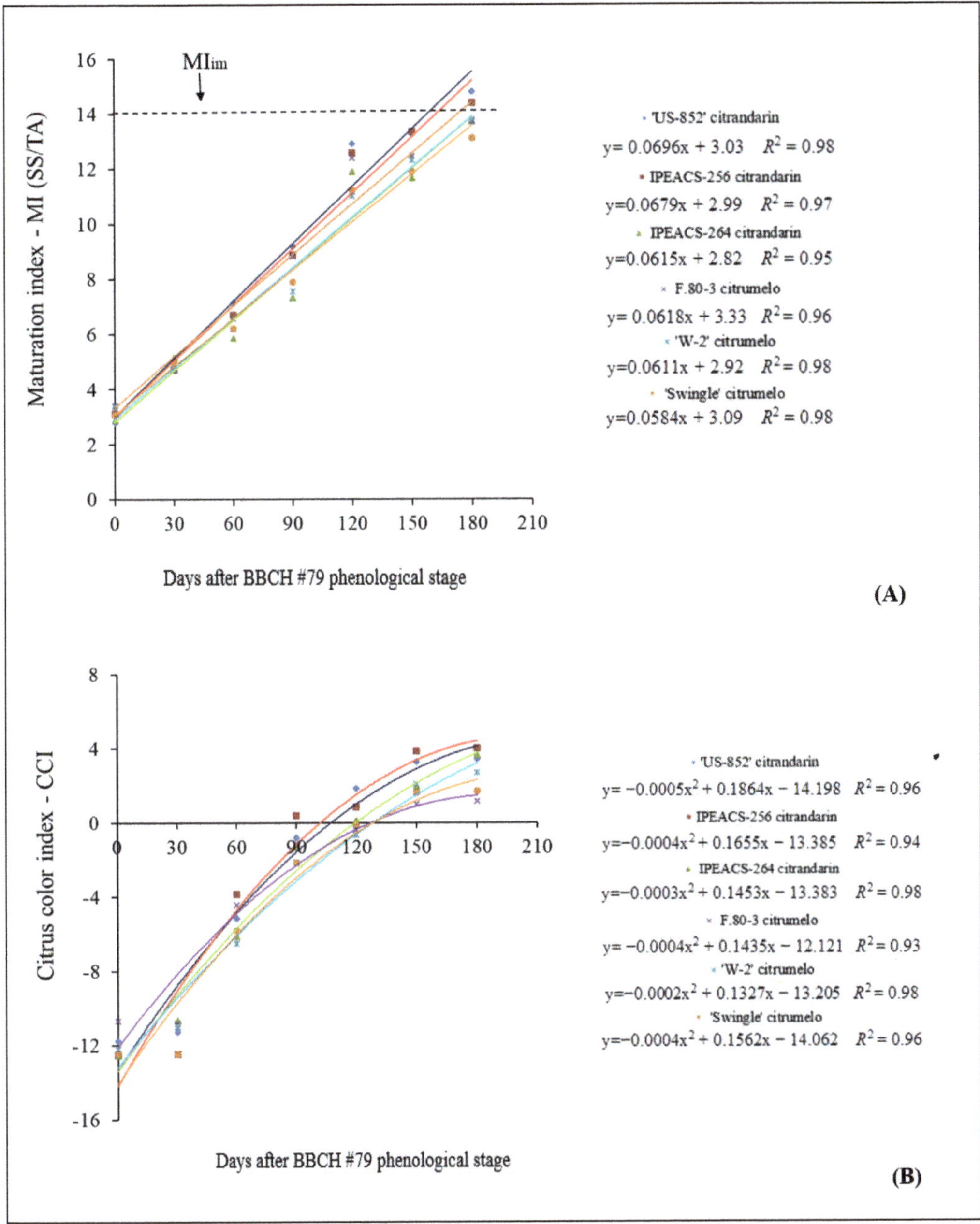

Figure 3. Development of 'Valencia' orange fruit ripening when trees are grafted on different 'Trifoliata' hybrid rootstocks at Rancho Alegre. (**A**): maturation index (Mi) or *ratio* (SS/TA); (**B**): citrus color index (CCI). MI$_{im}$ = ideal mean maturation index = 14. Phenological stage BBCH #79 (Meier, 2001), approximately to 200 days after flowering.

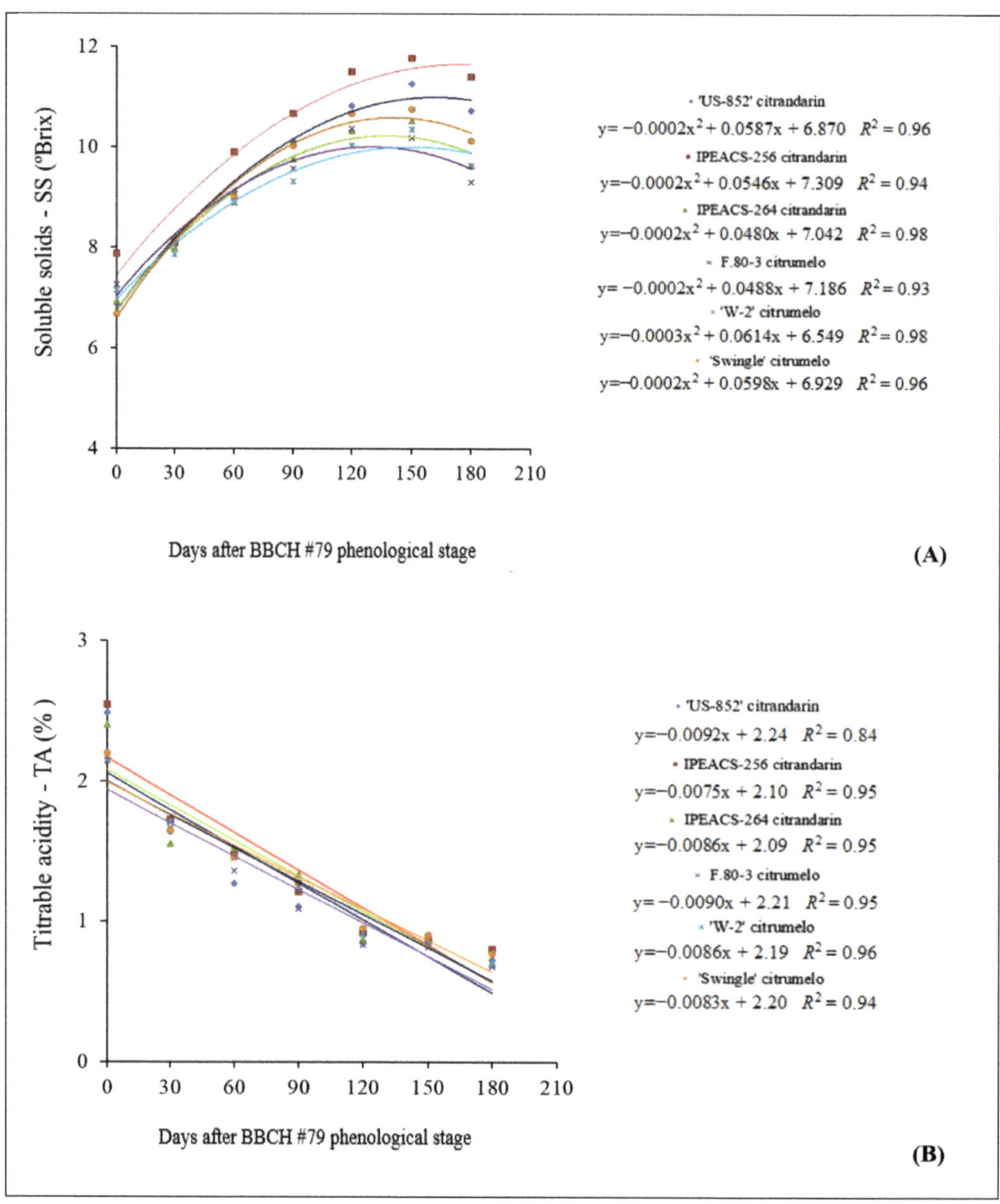

Figure 4. Development of 'Valencia' orange fruit ripening when trees are grafted on different 'Trifoliata' hybrid rootstocks at São Sebastião da Amoreira. (**A**): soluble solids (SS as °Brix); (**B**): titratable acidity (TA as %). Phenological stage BBCH #79 (Meier, 2001), approximately to 200 days after flowering.

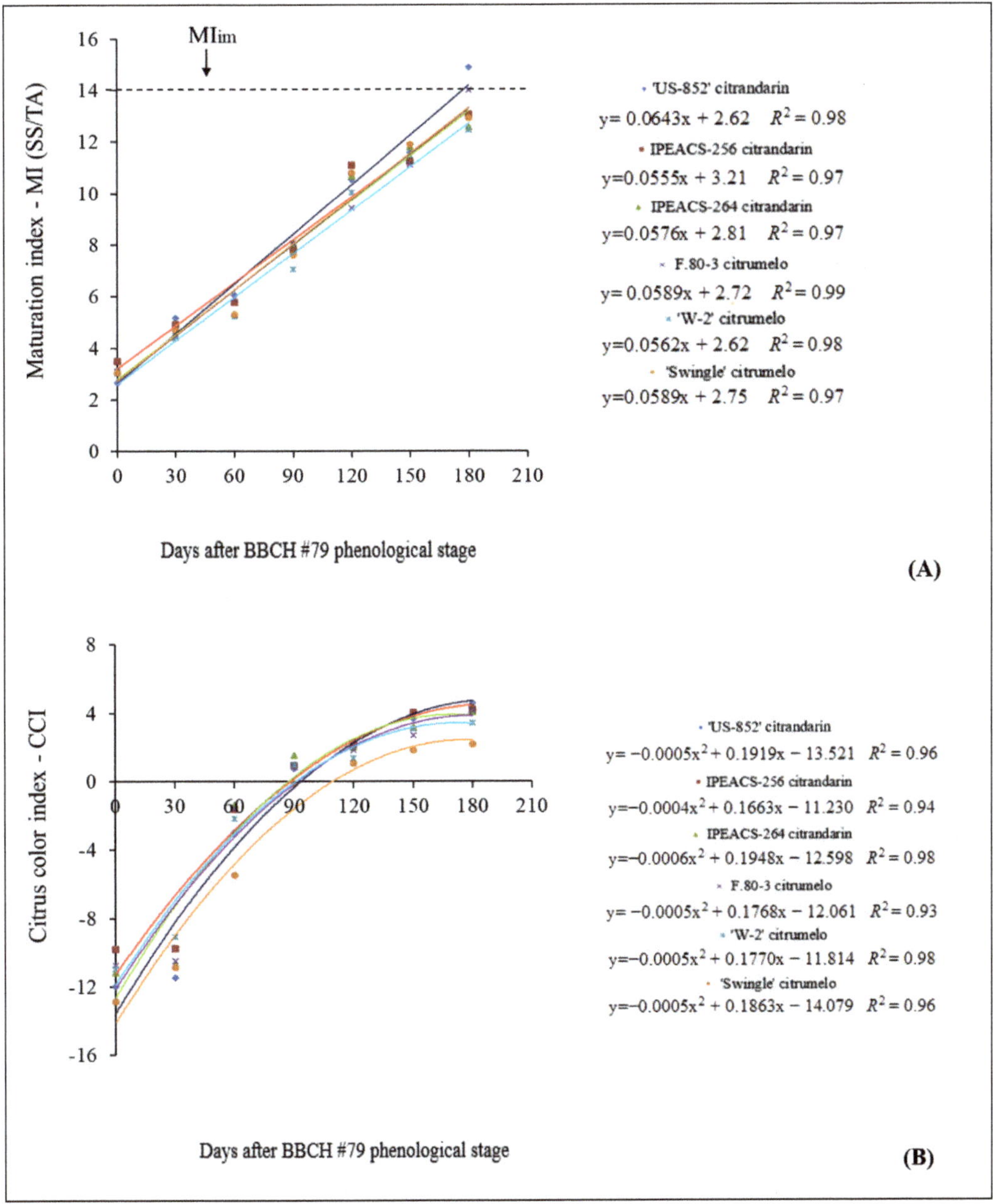

Figure 5. Development of 'Valencia' orange fruit ripening when trees are grafted on different 'Trifoliata' hybrid rootstocks at São Sebastião da Amoreira. (**A**): maturation index (MI) or *ratio* (SS/TA); (**B**): citrus color index (CCI). MI_{im} = ideal mean maturation index = 14. Phenological stage BBCH #79 (Meier, 2001), approximately to 200 days after flowering.

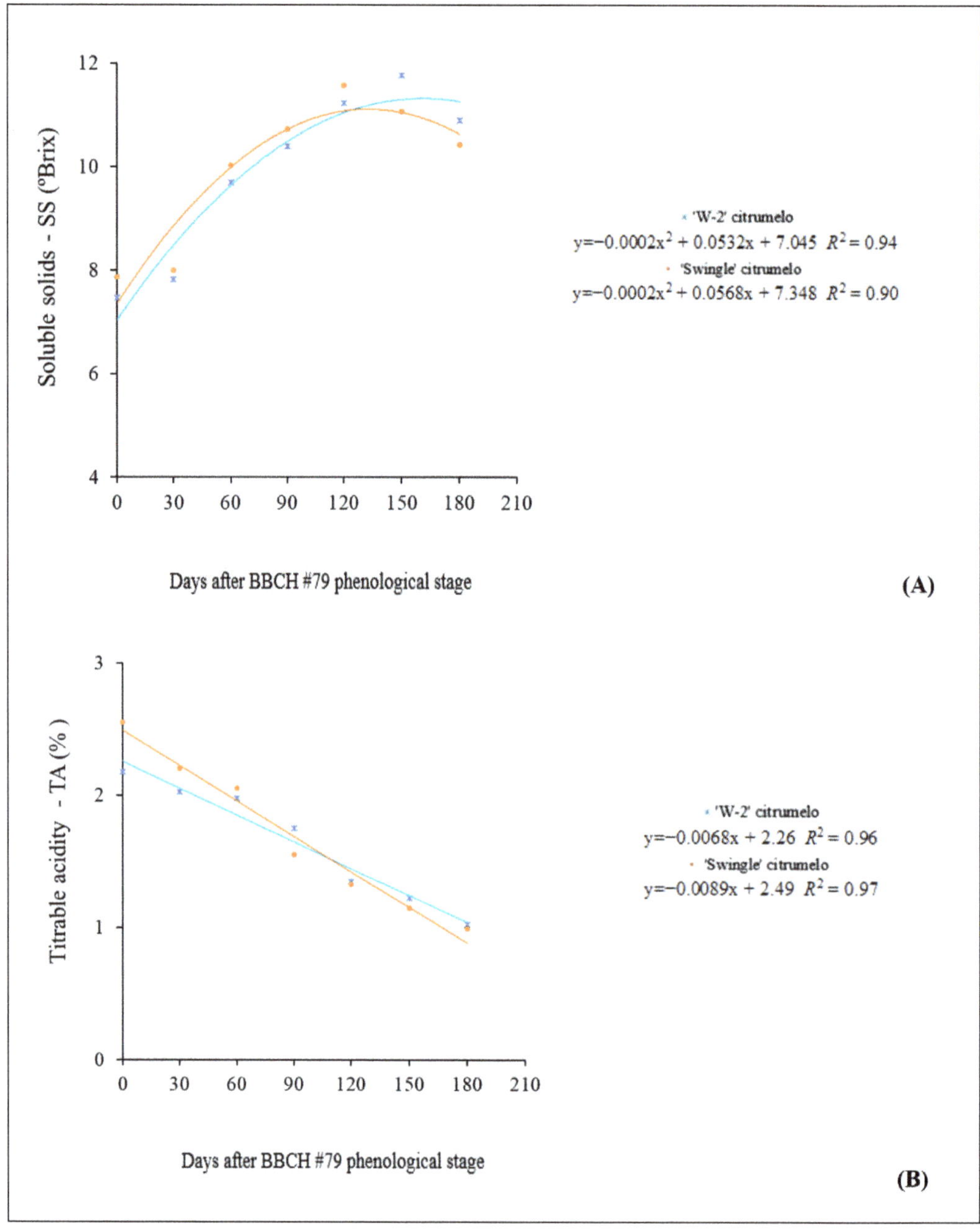

Figure 6. Development of 'Valencia' orange fruit ripening when trees are grafted on different 'Trifoliata' hybrid rootstocks at São Jerônimo da Serra. (**A**): soluble solids (SS as °Brix); (**B**): titratable acidity (TA as %). Phenological stage BBCH #79 (Meier, 2001), approximately to 200 days after flowering.

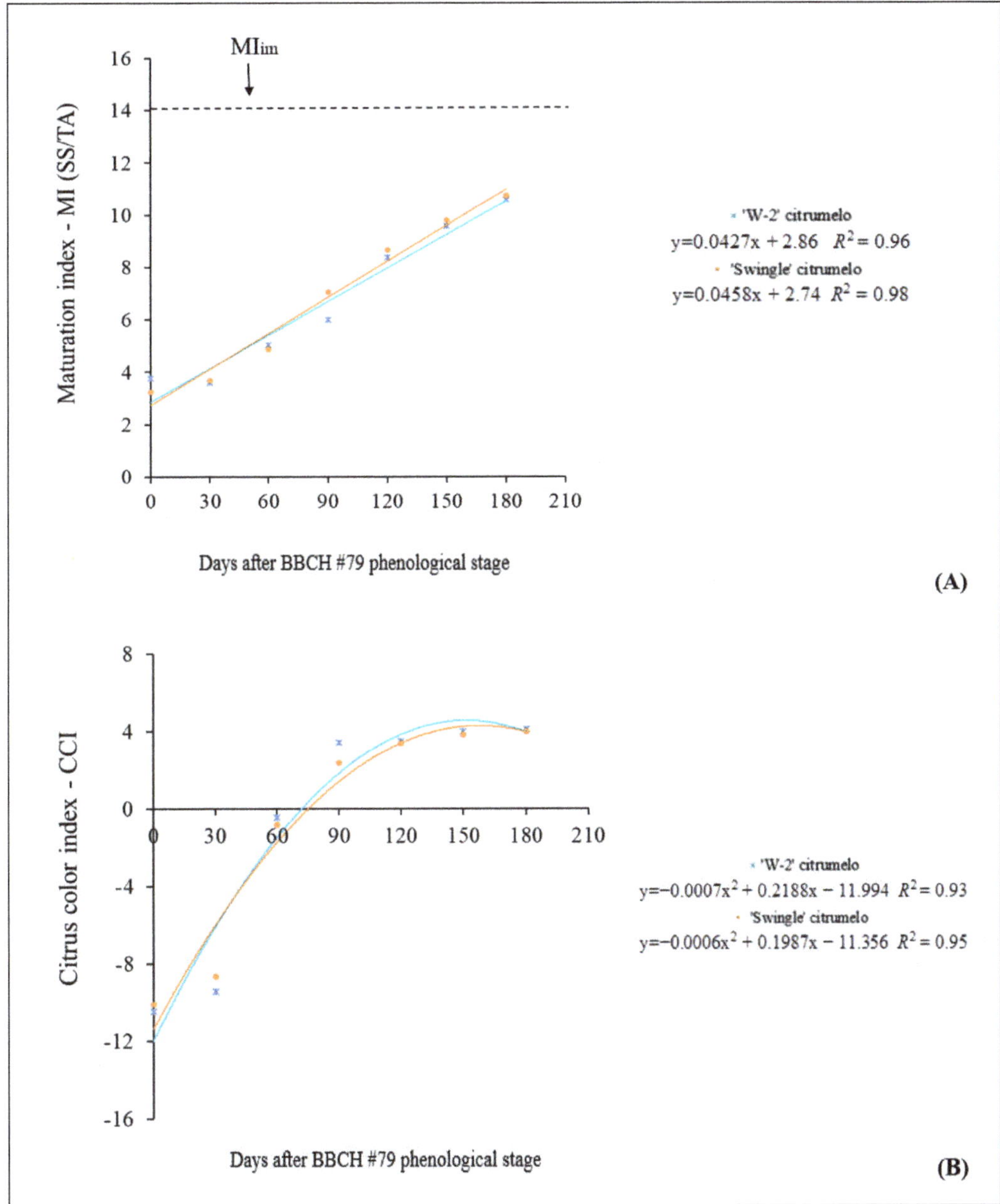

Figure 7. Development of 'Valencia' orange fruit ripening when trees are grafted on different 'Trifoliata' hybrid rootstocks at São Jerônimo da Serra. (**A**): maturation index (MI) or *ratio* (SS/TA); (**B**): citrus color index (CCI). MI_{im} = ideal mean maturation index = 14. Phenological stage BBCH #79 (Meier, 2001), approximately to 200 days after flowering.

Maximum and minimum temperatures differ even within a small interval between the latitudes of orange-producing areas, which may cause changes in fruit maturation and harvest dates as well as the behavior of scion varieties over each rootstock. Therefore, the ripening behavior found for 'Valencia' orange on 'Trifoliata' hybrid rootstocks in different subtropical growing scenarios is an important source of information for planning the marketing of fresh fruits or the industrial processing of fruits. In other words, the optimum date for harvesting can be estimated, helping to select the most appropriate rootstock genotypes according to production and processing from a commercial point of view.

The fruits of 'Valencia' orange trees grafted on 'Trifoliata' hybrid rootstocks grown at the RA location, with hot summers and low thermal amplitude during the ripening period, exhibited certain difficulty in terms of skin color change, remaining a shade of green for a long time, requiring an extended period from the beginning of the evaluations to reach CCI = 0, from 107 to 141 days (Figure 2B). When grown at the SSA and SJS locations, this period ranged from 93 to 105 days (Figure 3B) and from 64 to 69 days (Figure 3B), respectively.

The change in fruit skin color occurs due to the degradation of chlorophyll, which is responsible for the green color of fruits, and an increase in carotenoid pigments, which gives the fruit intense yellow and orange hues [50,51]. Color breaks in subtropical areas generally occur in mid-autumn when temperatures decrease, and day length diminishes. The decline in rind chlorophyll proceeds over several months, and the onset of carotenoid accumulation almost coincides with the disappearance of chlorophyll [46].

Air temperature has the greatest influence on the external color of oranges throughout maturation. Usually, the skin color of fruits grown in colder regions changes earlier. In contrast, with fruits grown in warm regions, changes occur later, and in some cases, the external color of the fruits may not develop satisfactorily [51–53]. Moreover, the genetic differences in rootstocks and their relationships with different scion varieties may influence the CCI [54–56], and it was shown in this work that citrandarins, in general, had higher CCI values than citrumelos.

Color is considered one of the most important external factors for the quality of orange fruits, as they are mainly marketed as fresh fruits, and the visual aspect of the product is considered an index of quality and maturation. However, skin maturation or morphological maturation does not always coincide with juice maturation in oranges grown in warmer tropical regions [7,12]. In addition, citrus is harvested based on internal quality (edibility), and skin changes have not been related to the best harvest times [57]. Although the 'Valencia' oranges grown at the SJS location underwent rapid development of CCI, it took a long time to reach the MI_{im}, confirming that CCI should not be used as a ripening reference alone for juice processing when growing this variety in regions with temperate summers and at high elevations. The fruits grown at the RA location, correspondingly, took less time to reach the MI_{im}, but reaching a CCI = 0 of the skin took a long time; thus, if the aim is to market fruit with a more attractive appearance, then it is necessary to extend the timing of harvest until they reach the desired skin color.

The 'Valencia' orange fruits grafted on 'US-852' or IPEACS-256 citrandarins grown in locations with hot summers presented an earliness characteristic, they reached the MI_{im} over a short period when compared with the other rootstocks evaluated. These genotypes are, therefore, a robust option for anticipating fruit harvest in order for citrus processing plants to operate earlier, over a wider period of the year.

In general, the basic quality of orange juice is determined by the fruit processor, that is, by the quality of fruit juice accepted at the reception area. Subsequent processing steps cannot improve the main quality parameters of a given production batch, except by blending a particular juice with superior quality orange juice or concentrate. In extracted juice, the concentration of sugar typically varies from 9 °Brix for early season varieties to 12 °Brix for fruit harvested late in the season. However, citrus processors usually consider 11.8 °Brix and 11.0 °Brix as the minimum grade of orange juice for FCOJ and NFC, respectively, or an MI ranging from 12.5 to 20.5 for both [42,58]. MI offers a more comprehensive analysis of orange juice quality because it also considers its organic acid content. Nevertheless, the removal of acid from the juice, also known as deacidification, would also increase MI but is not permitted for orange juice in most countries [42,59]. Thus, the SS content in fact had a major impact on MI in these trials, as the 'Trifoliata' hybrid rootstocks had no influence on this juice characteristic (Tables 1, 3 and 5). This range of MI permits processors to market different kinds of juices according to their consumer preference, but off-specification oranges are rejected to ensure robust quality juices. Thus, the monitoring of orange ripening when trees are grafted on different rootstocks at a given location is essential to improve juice quality.

Although earliness is a characteristic of the scion, the rootstock may change the ripening development of orange fruits [30,33,35]; thus, rootstock selection plays a crucial role in new citrus orchards. The rootstock is an important component of a healthy and productive citrus tree, influencing the fruit yield, fruit quality, tree size, and tolerance of diseases [60,61]. However, the selection of rootstocks in many growing countries has been based mainly on disease control.

Currently, in Sao Paulo State, Brazil, 'Rangpur' lime and 'Swingle' citrumelo are the main rootstocks, while in Florida, US, 'Carrizo' citrange and 'Swingle' citrumelo are the main rootstocks [62,63]. As demonstrated in this work, some citrandarin rootstocks, such as 'US-852' and IPEACS-256, are strong options for improving the quality of orange juice, and diversification is an important tool to avoid emerging and destructive diseases that put the citrus industry at risk.

The results of this study provide new information in terms of rootstock selection for 'Valencia' orange trees, taking into consideration the location, climatic conditions, and demands of each market, using certain important characteristics of orange fruit ripening development, such as the MI_{im} and CCI.

4. Conclusions

The aim of this work was to assess the ripening development of 'Valencia' orange fruits when trees are grafted on different 'Trifoliata' hybrid rootstocks, grown in different subtropical locations, to obtain high-quality fruits. In general, the rootstocks influenced the main features of orange ripening, such as soluble solids content, maturation index, and citrus color index. At the RA location (380 m a.s.l), trees on IPEACS-256 and 'US-852' citrandarins had the highest SS and the 'US-852' citrandarin the highest MI, reaching the MI_{im} earlier than the other rootstocks. The highest CCI was achieved when trees were on IPEACS-256 citrandarin. At the SSA location (650 m a.s.l.), trees on 'US-852', IPEACS–256, and IPEACS–264 citrandarins had the highest SS, but on 'US-852' had the highest MI, reaching the earliest MI_{im} among the rootstocks. The highest CCI was achieved when trees were on'US-852' and IPEACS-256 citrandarins. At the SSJ location (835 m a.s.l.), there was no significant effect of rootstocks on 'Valencia' orange fruits for any of the variables evaluated. This assessment can be useful in the planning of new orchards producing high-quality fruits with desirable features for the orange juice industry.

Author Contributions: E.F.C., A.R.D., C.D.M.M. and S.R.R. conceived and designed the experiments. A.R.D., E.F.C., C.D.M.M. and C.H.d.S.G. performed the experiments. A.R.D. and L.S.A.G. analyzed the data, A.R.D., S.R.R. and E.F.C. wrote the manuscript. All authors have read and agreed to the published version of the manuscript.

Funding: This work was partially supported by the Brazilian Federal Agency for Support and Evaluation of Graduate Education (CAPES) within the Ministry of Education, Finance Code 001.

Conflicts of Interest: The authors declare no conflict of interest.

References

1. Saunt, J. Citrus varieties of the world. In *An Illustrated Guide*, 1st ed.; Sinclair International Limited: Norwich, UK, 2000; p. 160.
2. Spreen, T.H. The word citrus industry. In *Soils, Plant Growth and Crop Production*, 1st ed.; Verheye, W.H., Ed.; UNESCO/Eolss Publishers: Singapore, 2010; pp. 249–269.
3. Bastos, D.C.; Ferreira, E.A.; Passos, O.S.; de Sá, J.F.; Ataíde, E.M.; Calgaro, M. Cultivares copa e porta-enxertos para a citricultura brasileira. *Inf. Agropec.* **2014**, *35*, 36–45.
4. Carvalho, S.A.; Girardi, E.A.; Mourão Filho, F.A.A.; Ferrarezi, R.S.; Filho, H.D.C. Advances in citrus propagation in Brazil. *Rev. Bras. Frutic.* **2019**, *41*, e-422. [CrossRef]

5.   USDA United States Department of Agriculture. Citrus: Markets and Trade. Foreign Agricultural Service. 2019. Available online: https://apps.fas.usda.gov/psdonline/circulars/citrus.pdf (accessed on 15 June 2019).

6.   Bermejo, A.; Cano, A. Analysis of nutritional constituents in twenty citrus cultivars from the Mediterranean area at different stages of ripening. *Food. Nutr. Sci.* **2012**, *3*, 639–650. [CrossRef]

7.   Lado, J.; Zacarías, L.; Rodrigo, M.J. Maturity indicators and citrus quality. *Stewart Postharvest Rev.* **2014**, *10*, 1–6.

8.   Davies, F.S.; Albrigo, L.G. Environmental factors affecting fruit growth, development and quality. In *Citrus*, 2nd ed.; Davies, F.S., Albrigo, L.G., Eds.; Cab International: Wallingford, SC, USA, 1994; pp. 77–82.

9.   Ladaniya, M.S. *Citrus Fruit: Biology, Technology and Evaluation*, 2nd ed.; Elsevier Inc.: Amsterdam, The Netherlands, 2008; p. 593.

10.   Sato, A.J.S.; Silva, B.S.J.; Bertolucci, R.; Carielo, M.; Guiraud, M.C.; Fonseca, I.C.B.; Roberto, S.R. Ripening evolution and physico-chemical characteristics of 'Isabel' grape on different rootstocks in North of Parana. *Semin. Agrar.* **2009**, *30*, 11–20. [CrossRef]

11.   Itakura, K.; Saito, Y.; Suzuki, T.; Kondo, N.; Hosoi, F. Estimation of citrus maturity with fluorescence spectroscopy using deep learning. *Horticulturae* **2019**, *5*, 2. [CrossRef]

12.   El-Otmani, M.; Ait-Oubahou, A.; Zacarias, L. *Citrus* spp.: Orange, mandarin, mandarin, clementine, grapefruit, pomelo, lemon and lime. In *Postharvest Physiology and Technology: Tropical and Subtropical Fruits*, 1st ed.; Yahia, E.M., Ed.; Woodhead publishing: Cambridge, UK, 2011; pp. 437–516.

13.   Tietel, Z.; Srivastava, S.; Fait, A.; Tel-Zur, N.; Carmi, N.; Raveh, E. Impact of scion/rootstock reciprocal effects on metabolomics of fruit juice and phloem sap in grafted *Citrus reticulata*. *PLoS ONE* **2020**, *15*, e0227192. [CrossRef]

14.   Hanana, M.; Hamrouni, L.; Hamed, K.; Abdelly, C. Influence of the rootstock/scion combination on the grapevines behavior under salt stress. *J. Plant Biochem. Physiol.* **2015**, *3*. [CrossRef]

15.   Naor, A.; Klein, I.; Doron, I. Stem water potential and apple size. *J. Am. Soc. Hortic. Sci.* **1995**, *120*, 577–582. [CrossRef]

16.   Shackel, K.A.; Ahmadi, H.; Biasi, W.; Buchner, R.; Goldhamer, D. Plant water status as an index of irrigation need in deciduous fruit trees. *Horttechnology* **1997**, *7*, 23–29. [CrossRef]

17.   Stenzel, N.M.C.; Neves, C.S.V.J.; Marur, C.J.; Scholz, M.B.S.; Gomes, J.C. Maturation curves and degree-days accumulation for fruits of 'Folha Murcha' orange trees. *Sci. Agric.* **2006**, *63*, 219–225. [CrossRef]

18.   Pompeu, J., Jr. Porta-enxertos. In *Citros*; Mattos, D., Jr., Ed.; Instituto Agronômico/Fundag: Campinas, Brazil, 2005; pp. 63–104.

19.   Castle, W.S. A career perspective on citrus rootstocks, their development, and commercialization. *Hortscience* **2010**, *45*, 11–15. [CrossRef]

20.   Castle, W.S.; Baldwin, J.C. Rootstock effects on 'Hamlin' and 'Valencia' orange trees growing at central ridge and flatwoods locations. *Hortscience* **2005**, *118*, 4–14.

21.   Pompeu, J., Jr.; Blumer, S. Citrumelos como porta-enxertos para laranja 'Valência'. *Pesq. Agropec. Bras.* **2011**, *46*, 105–107. [CrossRef]

22.   Pompeu, J., Jr.; Blumer, S. Híbridos de trifoliata como porta-enxertos para laranjeira Pêra. *Pesq. Agropec. Trop.* **2014**, *44*, 9–14. [CrossRef]

23.   Bowman, K.D.; McCollum, G.; Albrecht, U. Performance of 'Valencia' orange [*Citrus sinensis* (L.) Osbeck] on 17 rootstocks in a trial severely affected by huanglongbing. *Sci. Hortic.* **2016**, *201*, 355–361. [CrossRef]

24.   Hutchison, D.J. Swingle citrumelo—a promising rootstock hybrid. *Proc. Fla. State Hort. Soc.* **1974**, *87*, 89–91.

25.   Wutscher, H.K.; Hill, L.L. Performance of 'Hamlin' orange on 16 rootstocks in East-Central Florida. *Hortscience* **1995**, *30*, 41–43. [CrossRef]

26.   Bowman, K.D.; Wutscher, H.K.; Kaplan, D.T.; Chaparro, J.X. A new hybrid citrus rootstock for Florida: US-852. *Proc. Fla. State Hortic. Soc.* **1999**, *112*, 54–55.

27.   Castle, W.S.; Baldwin, J.C.; Grosser, J.W. Performance of 'Washington' navel orange trees in rootstocks trials located in Lake and St. Lucie counties. *Proc. Fla. State Hortic. Soc.* **2000**, *113*, 106–111.

28.   Schinor, E.H.; Cristofani-Yaly, M.; Bastianel, M.; Machado, M.A. Sunki Mandarin vs Poncirus trifoliata hybrids as rootstocks for Pera sweet orange. *J. Agric. Sci.* **2013**, *5*, 190–200. [CrossRef]

29.   França, N.O.; Amorim, M.S.; Girardi, E.A.; Passos, O.S.; Soares Filho, W.S. Performance of 'Tuxpan Valencia' sweet orange grafted onto 14 rootstocks in northern Bahia, Brazil. *Rev. Bras. Frutic.* **2016**, *38*, e-684. [CrossRef]

30.   Castle, W.S.; Baldwin, J.C.; Murano, R.P. Performance of 'Valencia' sweet orange trees on 12 rootstocks at two locations and an economic interpretation as a basis for rootstock selection. *Hortscience* **2010**, *45*, 523–533. [CrossRef]

31.   Castle, W.S.; Baldwin, J.C.; Murano, R.P. Rootstocks and the performance and economic returns of 'Hamlin' sweet orange trees. *Hortscience* **2010**, *45*, 875–881. [CrossRef]

32.   Cantuarias-Avilés, T.; Mourão Filho, F.A.A.; Stuchi, E.S.; Silva, S.R.; Espinoza-Nuñez, E. Horticultural performance of 'Folha Murcha' sweet orange onto twelve rootstocks. *Sci. Hortic.* **2011**, *129*, 259–265. [CrossRef]

33.   Donadio, L.C.; Lederman, I.E.; Roberto, S.R.; Stuchi, E.S. Dwarfing-canopy and rootstock cultivars for fruit trees. *Rev. Bras. Frutic.* **2019**, *41*, e997. [CrossRef]

34.   Cavalcante, I.H.L.; Martins, A.B.G.; Stuchi, E.S.; Campos, M.C.C. Fruit Maturation as a parameter for selection of sweet orange cultivars in Brazil. *J. Food Agric. Environ.* **2009**, *7*, 316–319.

35.   Carlos, E.F.; Stuchi, E.S.; Donadio, L.C. *Porta-Enxertos Para a Citricultura Paulista*; Funep: Jaboticabal, Brazil, 1997; 47p.

36.   Alvares, C.A.; Stape, J.S.; Gonçalves, J.L.M.; Sparovek, G. Köppen's climate classification map for Brazil. *Meteorol. Zeitschrift* **2013**, *22*, 711–728. [CrossRef]

37.   Nietsche, P.R.; Caramori, P.H.; da Ricce, W.S.; Pinto, L.F.D. *Atlas Climático do Estado do Paraná*; PR, IAPAR: Londrina, Brazil, 2019.

38.   Rodrigues, M.J.S.; Ledo, C.A.S.; Girardi, E.A.; Almeida, L.A.H.; Soares Filho, W.S. Caracterização de frutos e propagação de porta-enxertos híbridos de citros em ambiente protegido. *Rev. Bras. Frutic.* **2015**, *37*, 457–470. [CrossRef]

39.   Castle, W.S.; Wutscher, H.K.; Youtsey, C. Citrumelos as rootstocks for Florida citrus. *Proc. Fla. State Hortic. Soc.* **1988**, *101*, 28–33.

40. Meier, U. *Growth Stages of Mono and Dicotyledonous Plants: BBCH Monograph*, 2nd ed.; Federal Biological Research Centre for Agriculture and Forestry, Blackwell: Oxford, UK, 2001; p. 158.

41. AOAC. Association of Official Analytical Chemists. In *Official Methods of Analysis*, 15th ed.; AOAC: Arlington, VA, USA, 1990; 298p.

42. Ringblom, U. *The Orange Book*, 3rd ed.; Tetra Pak Processing Systems AB: Lund, Sweden, 2017; 208p.

43. Jimenez-Cuesta, M.; Cuquerella Cayuela, J.; Martinez-Javega, J.M. *Teoria y Practicca de la Desverdización de Los Cítricos*; INIA: Madrid, Spain, 1983; 22p.

44. Ferreira, D.F. Sisvar: A computer statistical analysis system. *Ciência Agrotecnologia (UFLA) Lavras* **2011**, *35*, 1039–1042. [CrossRef]

45. R Core Team. *R: A Language and Environment for Statistical Computing*; R Foundation for Statistical Computing: Vienna, Austria, 2017; Available online: http://www.R-project.org/ (accessed on 17 August 2020).

46. Iglesias, D.J.; Cercós, M.; Colmenero-Flores, J.M.; Naranjo, M.A.; Ríos, G.; Carrera, E.; Ruiz-Rivero, O.; Lliso, I.; Morillon, R.; Tadeo, F.R.; et al. Physiology of citrus fruiting. *Braz. J. Plant Physiol.* **2007**, *19*, 333–362. [CrossRef]

47. Perez-Perez, J.G.; Robles, J.M.; Botia, P. Influence of deficit irrigation in phase 3 of fruit growth on fruit quality in 'Lane late' sweet orange. *Agric. Water Manag.* **2009**, *96*, 969–974. [CrossRef]

48. Kallsen, C.E.; Sanden, B. Early navel orange fruit yield, quality, and maturity in response to late-season water stress. *Hortscience* **2011**, *46*, 1163–1169. [CrossRef]

49. Susanto, S.; Abdila, A.; Sulistyningrum, D. Growth and postharvest quality of mandarin (*Citrus reticulata* 'Fremont') fruit harvested from different altitudes. *Acta. Hortic.* **2013**, *975*, 421–426. [CrossRef]

50. Lee, H.S.; Castle, W.S. Seasonal changes of carotenoid pigments and colour in Hamlin, Earlygold, and Budd Blood orange juices. *J. Agric. Food Chem.* **2001**, *49*, 877–882. [CrossRef]

51. Rodrigo, M.J.; Alquezar, B.; Alos, E.; Lado, J.; Zacarias, L. Biochemical bases and molecular regulation of pigmentation in the skin of citrus fruit. *Sci. Hortic.* **2013**, *163*, 42–62. [CrossRef]

52. Mesejo, C.; Gambetta, G.; Gravina, A.; Martinez-Fuentes, A.; Reig, C.; Agusti, M. Relationship between soil temperature and fruit color development of 'Clemenpons' Clementine mandarin (*Citrus clementina* Hort. ex. Tan). *J. Sci. Food Agric.* **2012**, *92*, 520–525. [CrossRef]

53. Porras, I.; Brotons, J.M.; Conesa, A.; Manera, F.J. Influence of temperature and net radiation on the natural degreening process of grapefruit (*Citrus paradisi* Macf.) cultivars 'Rio Red' and 'Star Ruby'. *Sci. Hortic.* **2014**, *173*, 45–53. [CrossRef]

54. Legua, P.; Forner, J.B.; Hernandez, F.; Forner-Giner, M.A. Physicochemical properties of orange juice from ten rootstocks using multivariate analysis. *Sci. Hortic.* **2013**, *160*, 268–273. [CrossRef]

55. Forner-Giner, M.A.; Rodriguez-Gamir, J.; Martinez-Alcantara, B.; Quinones, A.; Iglesias, D.J.; Primo-Millo, E.; Forner, J. Performance of 'Navel' orange trees grafted onto two new dwarfing rootstocks (Forner-Alcaide 517 and Forner-Alcaide 418). *Sci. Hortic.* **2014**, *179*, 376–387. [CrossRef]

56. Continella, A.; Panniteri, C.; La Malfa, S.; Legua, P.; Distefano, G.; Nicolosi, E.; Gentile, A. Influence of different rootstocks on yield precocity and fruit quality of 'Tarocco Scirè' pigmented sweet orange. *Sci. Hortic.* **2018**, *230*, 62–67. [CrossRef]

57. Alam-Eldein, S.; Albrigo, G.; Etxeberria, E.; Burns, J.; Rouseff, R.; Tubeileh, A. Characterization of 'Valencia' orange skin maturation: Effect of water stress and growth regulators. *Acta Hortic.* **2017**, *1150*, 355–362. [CrossRef]

58. Cinquanta, L.; Di Matteo, M. The Quality of Orange Juice. In *Diet Quality Nutrition and Health*; Preedy, V., Hunter, L.A., Patel, V., Eds.; Humana Press: New York, NY, USA, 2013. [CrossRef]

59. Goboulev, V.N.; Salem, B. Traitement a'l'électrodialyse du jus d'orange. *Ind. Aliment. Agric.* **1989**, *106*, 175–177.

60. Barry, G.H.; Castle, W.S.; Davies, F.S. Soluble Solids Accumulation in 'Valencia' Sweet Orange as Related to Rootstock Selection and Fruit Size. *J. Amer. Soc. Hortic. Sci.* **2004**, *129*, 594–598. [CrossRef]

61. Bisi, R.B.; Albrecht, U.; Bowman, K.D. Seed and Seedling Nursery Characteristics for 10 USDA Citrus Rootstocks. *Hortscience* **2020**, *55*, 528–532. [CrossRef]

62. Carvalho, H.W.L.; Teodoro, A.V.; Barros, I.; Carvalho, L.M.; Soares Filho, W.S.; Girardi, E.A.; Passos, O.S.; Pinto-Zevallos, D.M. Rootstock-related improved performance of 'Pera' sweet orange under rainfed conditions of Northeast Brazil. *Sci. Hortic.* **2020**, *263*, 109–148. [CrossRef]

63. Bowman, K.D.; Joubert, J. Citrus rootstocks. In *The Genus Citrus*; Talon, M., Caruso, M., Gmitter, F.G., Eds.; Woodhead Publishing: Cambridge, UK, 2020; pp. 105–127. [CrossRef]

Article

Spatial and Temporal Enhancement of Colour Development in Apples Subjected to Reflective Material in the Southern Hemisphere

Kerstin Funke and Michael Blanke *

INRES-Horticultural Science, University of Bonn, 53121 Bonn, Germany; KerstinFunke@gmx.de
* Correspondence: mmblanke@uni-bonn.de; Tel.: +49-228-735142

Abstract: (1) *Background:* Climate change associated with a warm autumn often hampers the development of colouration of many fruits including late ripening apple varieties in New Zealand. (2) *Objective:* This study will provide detailed information on the possibility of enhancing colouration of apples under the diffuse light conditions in autumn in the southern hemisphere (SH). The aim is to obtain a larger proportion of fruit meeting the (red) colour market specifications, especially within the first picks, and to identify both the side of the fruit and its position within the tall trees canopy (3.5 m) as affected by reflective mulch on the ground spread at and over different times. (3) *Material and methods:* Reflective white textile mulch (Extenday®) was spread in the grassed alleyways 4 weeks or 2 weeks before the anticipated harvest in April on cv. Fuji and Pacific Rose apple trees without hail nets in the Northern Part of the South Island (41° S) of NZ. Fruit colour (blush) was determined by scoring and colourimeter during fruit maturation and at harvest, and fruit quality was determined at harvest by standard methods. (4) *Results:* (a) In cv. Pacific Rose apple, the reflective mulch increased the scored blush value from 1.5 (<50% blush) to 3.9 (ca. 75% blush) before the first pick, whereas the control fruit (without Extenday^R) reached a final score value of only 3.0. (b) Fruit colour improved after one week of exposure to reflective mulch in the SH. (c) The scored blush on fruit near the trunk with reflective mulch doubled (Pacific Rose) or tripled (Fuji) at harvest in comparison with trees with grass alleyways (control). (d) Two and four weeks of reflective mulch enhanced colouration of the down facing side for fruit of both cultivars, especially for fruit from the inside of the canopy near the tree trunk. However, reflective mulch significantly improved blush by 20% on fruit from the periphery of the canopies of the tall trees in both cultivars without significantly affecting fruit firmness, soluble solids, starch breakdown or ripeness. (5) *Conclusions:* The results from ca. 2000 colour measurements showed that the short exposure of at least two weeks of reflective mulch was sufficient for enhancing colouration for outside, inside and down facing sides of the fruit of both cultivars. As a result of this surprisingly short and efficient exposure time for these tall trees (3.5 m), the reflective mulch increased the portion of fruit harvested in the first pick by 8% (Fuji) and by 27% (Pacific Rose) with improved fruit storability or export quality and thereby increased financial returns to the grower in the SH.

Keywords: apple (*Malus domestica* Borkh.); anthocyanin; colouration; Envy, Extenday®; fruit quality; Fuji; Jazz; light reflection; PAL—Phenylalanine-amminia-lyase; reflective mulch; shading

check for
updates

Citation: Funke, K.; Blanke, M. Spatial and Temporal Enhancement of Colour Development in Apples Subjected to Reflective Material in the Southern Hemisphere. *Horticulturae* **2021**, *7*, 2. https://dx.doi.org/ horticulturae7010002

Received: 4 November 2020
Accepted: 23 December 2020
Published: 27 December 2020

Publisher's Note: MDPI stays neutral with regard to jurisdictional claims in published maps and institutional affiliations.

1. Introduction

For the consumer, fruit colour is one of the predominant incentives for purchasing a fruit. From apricot to persimmon and from apple to grape, the red colour is associated with ripeness, good taste and sweetness (Hamadziripi et al., 2014) [1]. Therefore, colour management is an important issue for fruit production around the world. New Zealand's apple industry provides markets all over the world with a wide range of red coloured apple varieties, such as Gala, Braeburn, Jazz, Envy and Fuji, during the Northern hemisphere (NH) spring and summer before local fruit in the NH become available. The formation of anthocyanin pigments in the fruit skin, which are responsible for the red colouration of fruits, including of the apple, depends mainly on environmental factors such as light quality (PAR, UV) and quantity and temperature (cool night), which are sensed by the MYB 10 gene (Wang et al., 2011) [2]. Fruit colouration i.e., blush can also be influenced to some extent by cultural practices such as pruning, thinning, fertilization, biostimulants and plant growth regulators (Andris and Crisoto, 1996) [3].

New Zealand exhibits record apple yields of 100–120 t/ha compared with 60–80 t/ha in Bonn, Germany. In New Zealand, late ripening apple cultivars, especially Fuji, are commonly grown on tall trees of 3.5 m height on semi-vigorous rootstocks and can fail to form red colour. This is due to the mutual shading, shorter autumn day length (photoperiod), decreasing light intensity (PAR and UV), decreasing solar angle (Meinhold et al., 2010a) [4] and the occurrence of warm autumns with warm nights. Reflective mulches, either metallised polyethylene or white woven polypropylene plastic, placed between tree rows on the ground, have been used in a number of NH countries to improve light distribution within the canopy and thus to enhance colouration of ripening fruits. In trials carried out on cv. Fuji apples under the high light conditions in California, the percentage area of skin red colour was increased by up to 65% (Andris and Crisosto, 1996) [3]; more fruit from the white reflective mulch treatment were packed in the well-coloured premium marketing categories "Fancy" and "Extra Fancy" than the control. Referring to Standard Fruit Specifications for apple cv. Fuji (T&G Specifications Manual, 2018) [5], high Grade fruits must have predominant pink or red colour ≥ 75% of the fruit surface area and Standard Grade fruit must have ≥ 50%. Therefore, using reflective mulches in New Zealand's Fuji orchards, with overcast autumn weather conditions and a large portion of diffuse light, could increase the export packout due to better colouration and would justify the cost of the material, which can be repeatedly used. First, we need to understand the underlying mechanisms and identify the target fruit within the canopy. With insufficient blush and colour (°hue) development of attached fruit in different positions on the tree, the aim of this study was to evaluate the ways in which reflective mulch influences fruit quality and colour on large and vigorous trees in New Zealand (NZ) at 41° S latitude in the Southern Hemisphere (SH), also with future hail nets to come in mind. A white woven reflective mulch was used 4 or 2 weeks before the predicted harvest date in mid-April to evaluate the necessary timeframe for sufficient colour development (anthocyanin synthesis) of these late-ripening cv. Fuji and Pacific Rose apples at Motueka (NZ). To identify how the position of the fruit on the tree was affected by light reflectance, the colour development on marked attached fruits from the inside and outer periphery of the tree canopy was monitored at regular intervals and fruit analysed for quality after harvest. This study was conducted to provide detailed information on the possibility of enhancing colouration of apples to obtain a larger proportion of fruit meeting the colour market specifications and therefore to increase the portion of the desired first pick. Additionally, the aim is to identify, which side of the fruit and its position within the tree are affected by reflective mulch. This could therefore lead to a better financial return for the fruit while satisfying both trader and consumer and providing not only more attractive, but also healthier apple fruit (Overbeck et al., 2013) [6] (Smrke et al., 2019) [7].

2. Materials and Methods

2.1. Apple Trees, Orchard Location, Management and Experimental Design

For the experiments, two orchards near Motueka were chosen. This Northern part of the South Island (latitude 40–42° S) is one of the major pipfruit growing areas of New Zealand. The treatments were carried out on Fuji/MM 106 apple trees without hail nets in a commercial orchard owned by Michael Moss and on Pacific Rose/MM 106 trees at HortResearch, Nelson Research Centre at Motueka. Due to the semi-vigorous rootstock, trees were ca. 3.5 m in height, and the spacing of the trees was relatively wide with 5 m × 3 m for Fuji and 4 × 2.5 m for Pacific Rose (Figure 1), respectively. Twelve apple trees per variety were employed in each location in one (Fuji) or two rows (Pacific Rose).

Trees planted in N–S-oriented rows were selected on the basis of uniform crop load and vigour. The girth measurements 20 cm above the graft union ranged from 39 cm to 51 cm for Fuji and 30 cm to 42 cm for Pacific Rose respectively. The design of the experiment was a randomized block with 4 replicates. Each block consisted of one tree per treatment (0, 2, 4 weeks) being separated by at least one border tree leaving at least 6 m between exposed and control trees (Figure 1). A new batch of reflective mulch (Extenday™, Extenday New Zealand Limited, Auckland) was placed in the alleyways on both sides of a tree 4 weeks (Fuji 2 March, Pacific Rose 7 March) or 2 weeks (Fuji 16 March, Pacific Rose 21 March) before the anticipated harvest in April in both locations. Trees with uncovered grass strips served as control (Figure 1).

(a)   (b)

Figure 1. Apple trees of cv. Pacific Rose on MM 106 (**a**) without (left) and (**b**) with reflective mulch (right) at HortResearch, Motueka (NZ).

2.2. Colour Scores (Blush) during Fruit Maturation

For each of the four replicates of the three durations, i.e., reflective mulch for 0, 2, 4 weeks prior to harvest, ten attached apple fruit from the inside the tree canopy close to the tree trunk and ten apple fruit from the outer periphery of the tree canopy were tagged. Blush development was visually classified every other day on the same 480 attached fruits in five groups (0 = no blush; 1 = 1–25% blush; 2 = 25–50% blush; 3 = 50–75% and 4 = 75–100% blush). Blush development on 240 tagged cv. Fuji apple fruit was scored ten times starting 6 March and ending 29 March, six days before the first pick. Similarly, 240 tagged Pacific Rose fruit were scored nine times starting 8 March and ending 30 March, 11 days before the first pick.

2.3. Colour Measurement during Fruit Maturation

Colour development was measured on half (five) of the already tagged fruit for blush scoring. Colour was measured on the same 240 fruit (120 fruit per variety) in situ non-destructively three times at weekly intervals prior to the predicted harvest date. Colour was measured on three spots on opposite sides on the fruit equator, i.e., the side of the fruit facing the trunk ("green side") and the side of the fruit facing the outside ("red side"), with a Minolta Chroma Meter CR-200 (Minolta Co., Osaka, Japan) based on CIE (1976;

L*, a*, b*) colour space. Colour was described by the parameters chroma and hue angle according to McGuire (1992) [8].

2.4. Fruit Quality and Colour Assessment at Harvest

To examine the effect of the reflective mulch on fruit quality and colour, eight fruit from the inside and outside of the tree canopy and 4 fruit from the top of the canopy were sampled for each treatment and replicated at each pick. Apples, representative of each treatment, were assessed for fruit weight, soluble solids (by refractometry), starch breakdown (after staining with potassium iodine), background colour, percentage blush on the peel and flesh firmness (with a penetrometer using standard methodology) (Meinhold et al., 2010a) [4] (Overbeck et al., 2013) [6]. Streif-Index was calculated by fruit firmness/(soluble solids x starch breakdown). This resulted in 720 fruit quality values for cv. Fuji (480 for cv. Pacific Rose).

At harvest, colour was measured as described above again on two opposite sides of the fruit ("red" and "green" side) and additionally on the down facing side of the fruit marked while on the tree before harvest.

2.5. Yield and Fruit Grading

For apple cv. Fuji (Pacific Rose), the harvest in Motueka, New Zealand, started with the first pick on 5 April (11 April) and the second pick on 17 April (no second pick for Pacific Rose). The initial start date of the first pick was selected according to NZ industry standards based on the internal quality of the fruit, i.e., the starch breakdown. The first pick was selective and based on blush and colour (area and brightness), whereas the last pick on 30 April (23 April) was a strip pick to remove all leftover fruit.

An automated Lynx Grader (LYNX Horticultural Systems, Auckland, New Zealand) was used to assess yield, fruit weight and fruit number for each pick.

2.6. Statistics

The experiment comprised 24 apple trees, i.e., 12 trees for each of the two varieties. Each treatment consisted of one tree plus border trees and was repeated four times in a randomised design. Twenty fruits from each tree were scored for blush development ten times (Fuji) or nine times (Pacific Rose) prior to harvest, so that 2400 score values could be obtained. Fruit quality and colour measurements at harvest were done on 20 fruit per tree per pick and variety.

Colour angle (° hue) and blush scores during fruit maturation were analysed by ANOVA separately for each side of the fruit and date and the two positions within the tree canopy for the effects of the treatments.

Percentage blush, colour angle (° hue) and fruit quality parameters at harvest were analysed by ANOVA separately by pick and the two positions within the tree canopy for the effects of the treatments.

Yield data were analysed by two-factorial analysis of variance (ANOVA) (treatment x pick). If there was no significant interaction, the data were analysed for the main effects of treatment and pick.

Overall, the yield of 24 apple trees, 4800 score values and 720 colour measurements during maturation as well as 1200 fruit quality assessments and 3600 colour measurements at harvest were statistically processed by analysis of variance (ANOVA) and post-hoc Tukey test at the 5% error level using SPSS Statistics package 26 (IBM, Michigan, USA).

3. Results

3.1. Temperature Course in March and April

Figure 2 shows the temperature course during the experiment and probable start of the induction of anthocyanin synthesis from 28 March with a T_{min} and ΔT of 1.7 °C and 17.5 °C on 31 March.

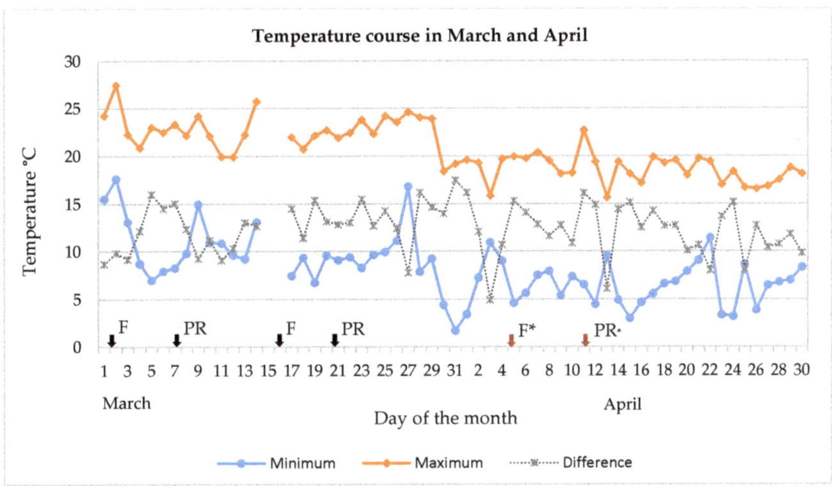

Figure 2. Temperature curve in March and April at HortResearch, Motueka, New Zealand, during the experiment- black arrows indicate the dates of laying the reflective material for Fuji (F) and Pacific Rose (PR) and the red arrows for the begin of the harvest (asterisks).

3.2. Increase in the Portion of the First Pick

In both apple cvs., Fuji and Pacific Rose, the reflective mulch increased the portion of fruit harvested in the first pick, which is relevant for storage and export quality as only the first pick goes into storage and is exported. This portion of the first pick of the late and difficult to colour cv. Fuji increased from 30% in the control to 38% with reflective mulch 4 weeks prior to harvest (Figure 3). This desired effect was more pronounced for the other later ripening cv., Pacific Rose, with an increase from 40% in the control to 62% with reflective mulch 2 weeks prior to harvest and, significantly, to 67% 4 weeks prior to harvest (Figure 4). For untreated trees, there was a significant difference of the percentage yield between the second and the third pick. There was no significant difference in mean fruit weight between the picks (result not shown), ensuring full-sized fruit in the first pick. Reflective mulches did not significantly affect overall fruit yield in either cvs., Fuji and Pacific Rose (Figures 3 and 4), thereby preventing any interference with the effects of the reflective mulch.

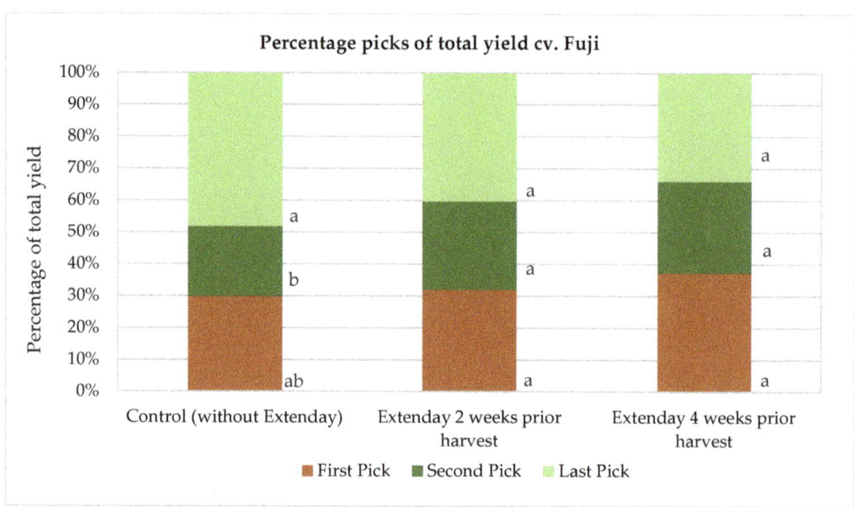

Figure 3. Summation of yield in each pick of cv. Fuji as dependent on the timing of reflective mulch in Motueka, New Zealand, (41° S). (Picks with different letter for the same treatment are statistically different ($p < 0.05$) (n = 60 fruit per treatment per position)).

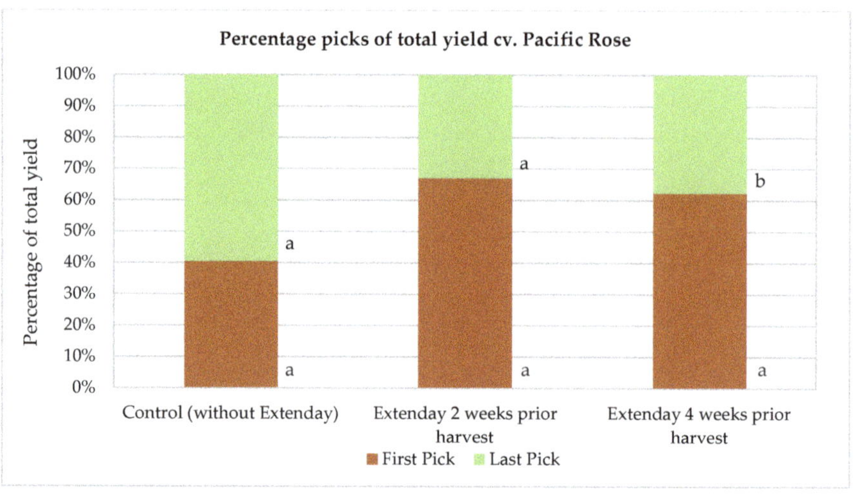

Figure 4. Summation of yield in each pick of cv. Pacific Rose as dependent on the timing of reflective mulch in Motueka, New Zealand, (41° S). (Picks with different letter for the same treatment are statistically different ($p < 0.05$) (n = 60 fruit per treatment per position)).

3.3. Colour Scores (Blush) during Fruit Maturation and at Harvest

When applied four weeks before the anticipated harvest in the late and poorly colouring cv. Fuji, the reflective mulch improved blush development of the fruit significantly compared to untreated fruit by at least one score value on a 1 to 4 colour viz., blush scale, as soon as one week after the material had been laid (Figure 5).

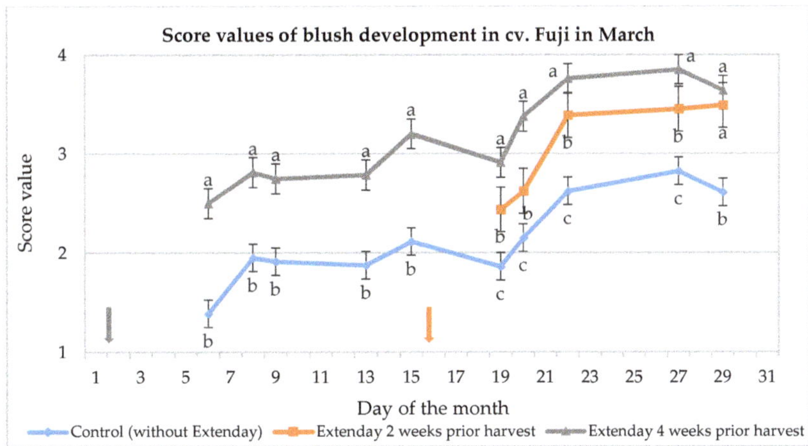

Figure 5. Score values of blush development in cv. Fuji in March as dependent on the timing of reflective mulch (average of fruit from the inside and outside of the tree canopy). (plus SEs; Different letters denote statistical differences within the same date at the $p = 0.05$ level (n = 80 fruit per treatment per pick.) Arrow indicates day of laying the mulch.

When reflective mulch was applied only two weeks before the anticipated cv. Fuji harvest, colour improvement could already be observed after one week, which has not been reported before to our knowledge. The score value of cv. Fuji fruit increased significantly from 1.5 to 3.9 before the first pick, whereas the control fruit (without mulch) reached a score value of only 3.0 (Figure 5).

In the difficult to colour cv. *Pacific Rose, the blush development also started one week after the material had been laid out* at score values 1.5 for both the control group (without Extenday^TM) and the "Extenday 4 weeks prior to harvest" group (Figure 6).

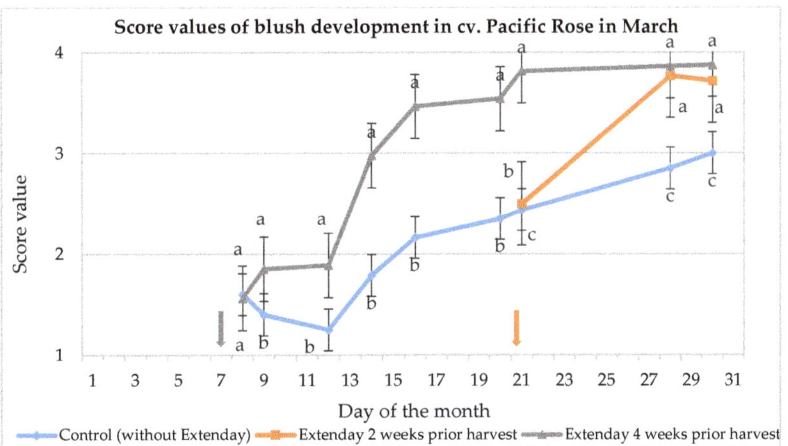

Figure 6. Score values of blush development in cv. Pacific Rose in March as dependent on the timing of reflective mulch (average of fruit from the inside and outside of the tree canopy) at Motueka (NZ). (plus SEs; Different letters denote statistical differences within the same date at the $p = 0.05$ level (n = 80 fruit per treatment per pick.) Arrow indicates day of laying the mulch.

Both exposure times for Fuji and Pacific Rose resulted in the same score values at harvest (Figures 5 and 6) being significantly different to values from fruit without reflective material.

In addition to the scored colour values recorded repeatedly on the same marked attached fruit (Figures 5 and 6), blush was also finally measured at harvest, from fruit from each of the two (Pacific Rose) or three picks (Fuji). The marked effect of the reflective mulch was on fruit at harvest of both cultivars (Fuji and Pacific Rose) from the otherwise partially shaded inside of the tree canopy (Tables 1 and 2).

Table 1. Percentage blush of cv. Fuji apple at each pick dependent on fruit position within the canopy.

	Fruit Position within the Tree Canopy					
Pick and Treatment	**Inside**		**Periphery**		**Top**	
First pick						
Control (without Extenday)	18.4	b	65.8	b	71.9	b
Extenday 2 weeks prior to harvest	62.0	a	78.1	a	74.1	b
Extenday 4 weeks prior to harvest	68.6	a	79.7	a	84.4	a
Second pick						
Control (without Extenday)	43.9	b	66.4	b	82.5	a
Extenday 2 weeks prior to harvest	77.5	a	77.0	a	72.2	a
Extenday 4 weeks prior to harvest	78.0	a	77.3	a	79.7	a
Last pick						
Control (without Extenday)	37.2	b	63.1	b	73.4	a
Extenday 2 weeks prior to harvest	80.5	a	74.4	a	76.9	a
Extenday 4 weeks prior to harvest	75.8	a	77.2	a	76.9	a

Different letters denote statistical differences within the same fruit position within the apple tree and pick at the $p = 0.05$ level (n = 80 fruit per treatment per pick). Background colours highlight / visualise significant differences in blush.

Table 2. Percentage blush of cv. Pacific Rose apple at each pick dependent on fruit position within the canopy.

	Fruit Position within the Tree Canopy					
Pick and Treatment	**Inside**		**Periphery**		**Top**	
First pick						
Control (without Extenday)	39.4	b	74.8	b	90.0	a
Extenday 2 weeks prior to harvest	86.3	a	89.4	a	95.3	a
Extenday 4 weeks prior to harvest	92.3	a	91.1	a	90.3	a
Last pick						
Control (without Extenday)	67.0	b	74.7	b	80.0	a
Extenday 2 weeks prior to harvest	84.1	a	79.1	ab	86.2	a
Extenday 4 weeks prior to harvest	89.5	a	89.7	a	88.3	a

Different letters denote statistical differences within the same fruit position within the apple tree and pick at the $p = 0.05$ level (n = 80 fruit per treatment per pick). Background colour highlights significant differences in blush.

The percentage of blush on inside fruit tripled (Fuji) or doubled (Pacific Rose) at the first pick with reflective mulch in comparison with the untreated control, i.e., grass alleyways. Similarly, reflective mulch significantly improved blush by 20% on fruit from the tree periphery for all picks in cv. Fuji and for the first pick in cv. Pacific Rose (Tables 1 and 2).

The significantly improved colouration (blush) of fruit closer to the tree trunk within the tree canopy is due to diffuse light reflection from the mulch in the alleyways into the tree canopy (Meinhold et al., 2010a) [4]. This effect was absent on fruit from the upper part and periphery of the canopy (Tables 1 and 2) due to the larger distance (3 m) from the reflective mulch and sufficient solar radiation under NZ autumn weather conditions.

3.4. Colour Development of the Apple Fruit during Maturation

Reflective mulch 2 or 4 weeks before harvest improved fruit colouration significantly during maturation of the Fuji apples, as measured by decreased hue colour angle both on the inner side of fruit (green side) inside the tree canopy (110 °hue to 75 °hue) and those in the outer tree periphery (100 °hue to 60 °hue) (Figure 7).

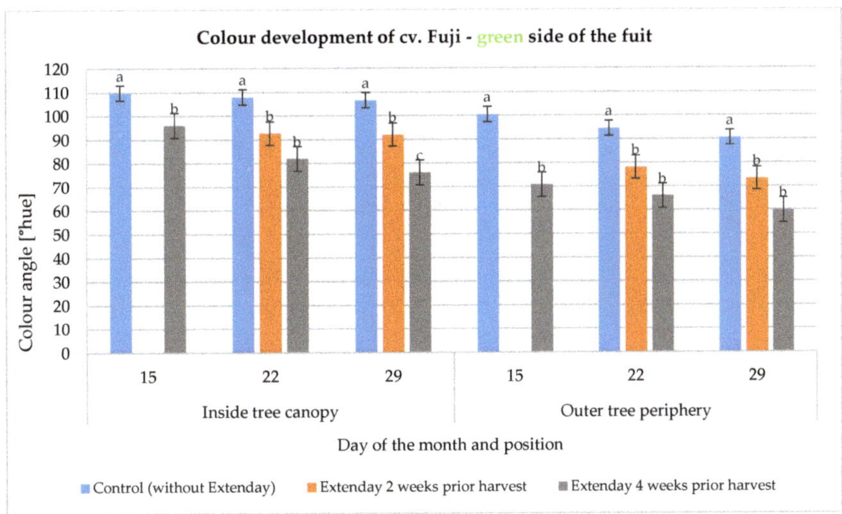

Figure 7. Colour development dependent on fruit position within the tree of cv. Fuji as measured on 15, 22 and 29 March on the inner green side of the fruit as affected by reflective mulch. (Treatments with different letter for the same position of the fruit within the same day are statistically different ($p < 0.05$) (n = 60 fruit per treatment per position)).

In contrast to the pronounced significant effect of the reflective mulch on the trunk-oriented green side of the cv. Fuji fruit (Figure 7), the colour effects of the reflective mulch were lesser on the red side of the fruit. Reflective mulches significantly improved hue colour angle of the red side of the fruit from the inside the tree canopy, irrespective of the time. By contrast, the red side of the cv. Fuji fruit in the periphery was already well coloured with colour angles of ca. 35° hue (Figure 8), so that the application of Extenday^{TM} had no significant effect.

Similarly, the effect of reflective mulch on colour development on the green inner, trunk-oriented side of the fruit was significant. Larger colour angles of 100–110 °hue in untreated fruit of cv. Pacific Rose, especially fruit from the inside of the tree canopy (Figure 9), meant greener apple fruit. Spreading Extenday^{TM} two weeks prior to harvest was sufficient under NZ weather conditions (SH at 41° S) to significantly enhance fruit colouration. On the trunk-oriented inner green side of the fruit two weeks prior to harvest, reflective mulch improved fruit colour viz., decreased the green side from 105° hue to 80° hue, whereas it remained at 100° hue in the control group. Both treatments of reflective mulch significantly enhanced colouration on the red side of the fruit in cv. Pacific Rose (Figure 10), even on the outer tree periphery (Figure 13). Overall, the effect on the green side inside the tree canopies of both varieties was more pronounced than on the red side of the fruit.

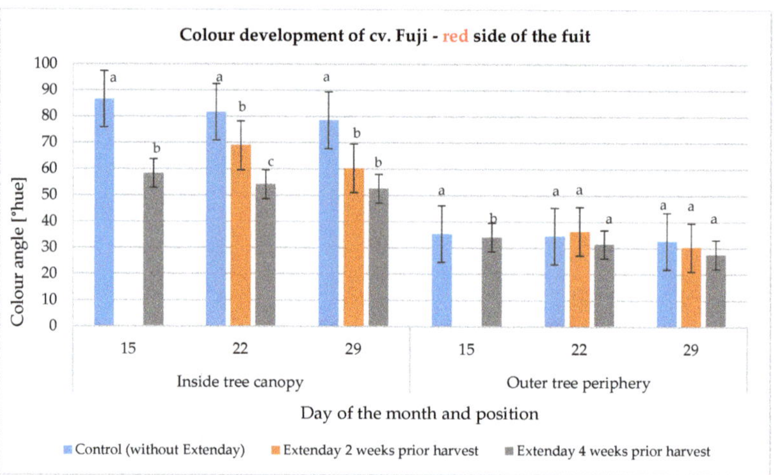

Figure 8. Colour development dependent on fruit position within the tree of cv. Fuji as measured on 15, 22 and 29 March on the red side of the fruit as affected by reflective mulch. (Treatments with different letter for the same position of the fruit within the same day are statistically different ($p < 0.05$) (n = 60 fruit per treatment per position).

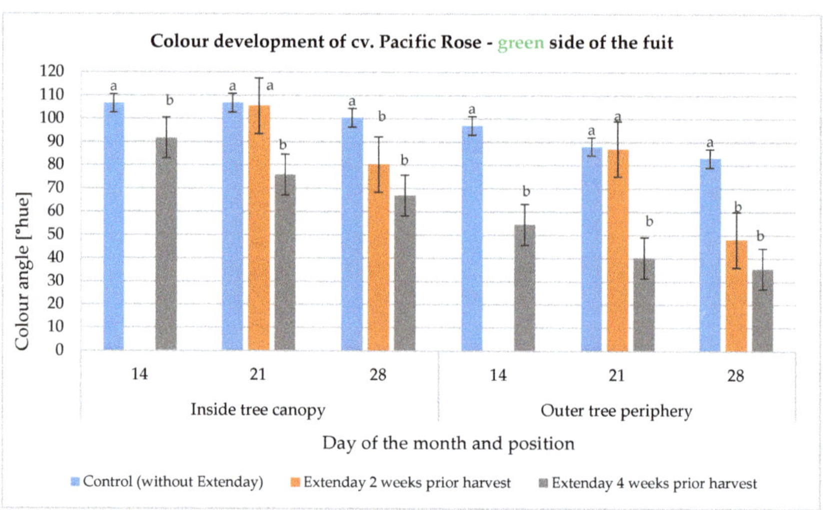

Figure 9. Colour development dependent on fruit position within the tree of cv. Pacific Rose as measured on 14, 21 and 28 March on the green side of the fruit as affected by reflective mulch spread 4 or 2 weeks prior to anticipated harvest. (Treatments with different letter for the same position of the fruit within the same day are statistically different ($p < 0.05$) (n = 60 fruit per treatment per position).

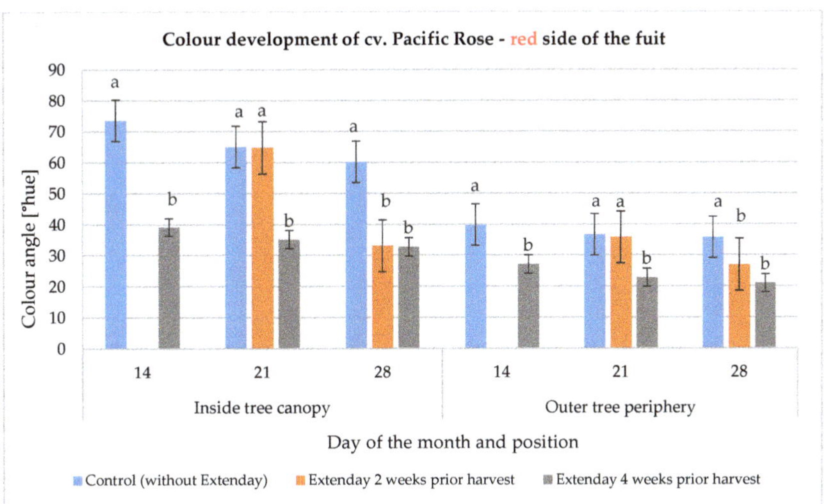

Figure 10. Colour development dependent on fruit position within the tree of cv. Pacific Rose as measured on 14, 21 and 28 March on the red side of the fruit as affected by reflective mulch. (Treatments with different letter for the same position of the fruit within the same day are statistically different ($p < 0.05$) (n = 60 fruit per treatment per position).

3.5. Colour of Apple Fruit at Harvest

Figures 11 and 12 show that two weeks of reflective mulch is sufficient for enhancing colouration of the down facing side in cv. Fuji and cv. Pacific Rose, especially for apple fruit from the inside of the canopy near the tree trunk. For Pacific Rose, these values were massively decreased from 88 °hue to 18 °hue in both cases of 2 and 4 weeks reflective mulch prior to harvest (Figure 11).

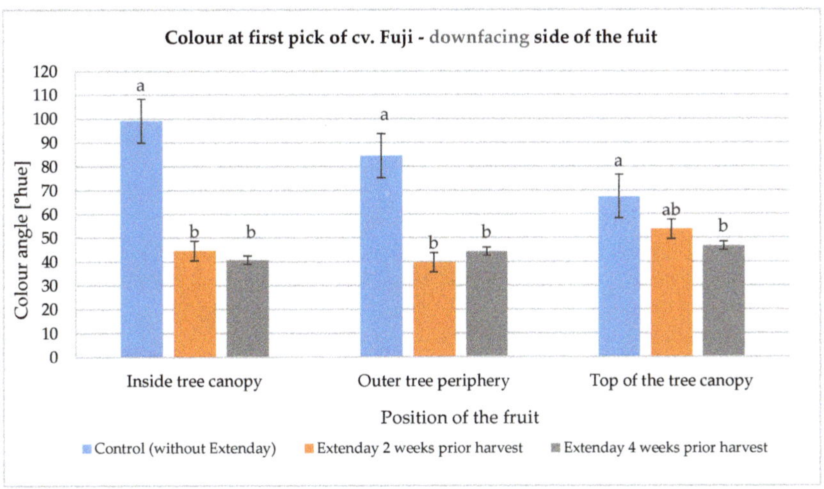

Figure 11. Colour of the down facing side of the cv. Fuji fruit at first pick as affected by the timing of reflective mulch. (Treatments with different letter for the same position of the fruit within the same pick are statistically different ($p < 0.05$) (n = 60 fruit per treatment per position).

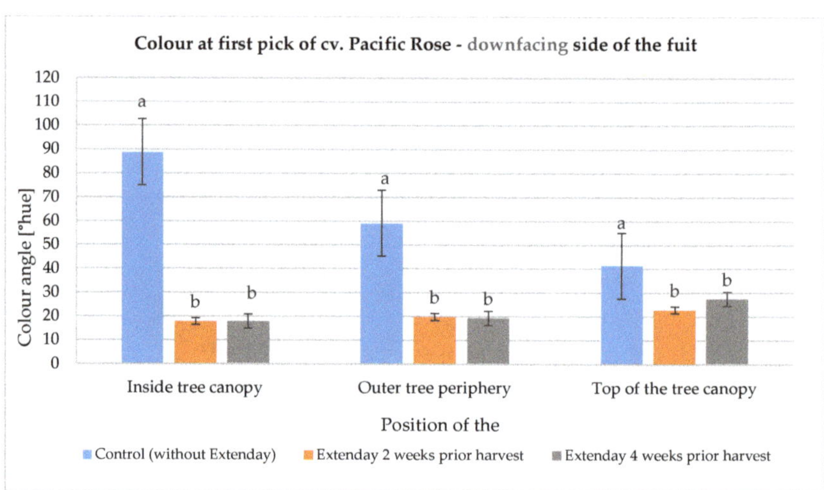

Figure 12. Colour of the down facing side of the cv. Pacific Rose fruit at first pick as affected by the timing of reflective mulch. (Treatments with different letter for the same position of the fruit within the same pick are statistically different ($p < 0.05$) (n = 60 fruit per treatment per position).

In cv. Fuji, the colour of the down facing side of the fruit was similarly reduced from nearly 100 °hue to 40–44 °hue (Figure 12). Interestingly, the down facing side of the fruit from the top of the tall trees (3.5 m) was already sufficiently coloured with 68 °hue in cv. Fuji and 40 °hue in cv. Pacific Rose (Figure 13).

Figure 13. Effect of reflective mulch on fruit colouration of cv. Pacific Rose at harvest- most conspicuous on the (down-facing) apex side of the apple fruit.

3.6. Effect of Reflective Mulches on Fruit Quality at Harvest

Fuji fruit from the periphery of the tree canopy in the first pick and fruit from the inside for the second pick showed significantly higher Streif index (Table 3). The accelerated ripeness is mainly caused by differences in starch breakdown (data not shown). These findings for apple trees with Extenday® showed a tendency to enhance starch breakdown as described by Meinhold et al. [9]. There was no significant effect on ripening (Streif index) for apple fruit in cv. Pacific Rose (Table 4).

Table 3. Quality parameter (Streif index) and ripening of cv. Fuji apple at each pick dependent on fruit position within the tree canopy.

	Fruit Position Within the Tree Canopy					
Pick and Treatment	**Inside**		**Periphery**		**Top**	
First pick						
Control (without Extenday)	0.19	a	0.15	b	0.21	a
Extenday 2 weeks prior to harvest	0.21	a	0.19	a	0.25	a
Extenday 4 weeks prior to harvest	0.20	a	0.19	a	0.24	a
Second pick						
Control (without Extenday)	0.13	b	0.12	a	0.14	a
Extenday 2 weeks prior to harvest	0.16	a	0.13	a	0.17	a
Extenday 4 weeks prior to harvest	0.16	a	0.14	a	0.13	a
Last pick						
Control (without Extenday)	0.11	a	0.10	a	0.11	a
Extenday 2 weeks prior to harvest	0.11	a	0.10	a	0.12	a
Extenday 4 weeks prior to harvest	0.11	a	0.10	a	0.10	a

Different letters denote statistical differences within the same fruit position within the apple tree and pick at the $p = 0.05$ level (n = 80 fruit per treatment per pick). Background colours highlight significant differences in fruit maturity (Streif index) between control (grassed alleyway) and reflective mulch (Extenday[TM]).

Table 4. Quality parameter (Streif index) of cv. Pacific Rose apple at each pick dependent on fruit position within the tree canopy.

	Fruit Position Within the Tree Canopy					
Pick and Treatment	**Inside**		**Periphery**		**Top**	
First pick						
Control (without Extenday)	0.44	a	0.19	a	0.40	a
Extenday 2 weeks prior to harvest	0.37	a	0.26	a	0.28	a
Extenday 4 weeks prior to harvest	0.38	a	0.29	a	0.32	a
Last pick						
Control (without Extenday)	0.39	a	0.26	a	0.20	a
Extenday 2 weeks prior to harvest	0.33	a	0.25	a	0.19	a
Extenday 4 weeks prior to harvest	0.35	a	0.21	a	0.24	a

In both cultivars, there was no significant effect of the reflective mulch on fruit weight (result not shown), as expected, given the short time of environmental microclimate modification in the orchard.

4. Discussion

The objective of the present work was to study the influence of a reflective white textile mulch (Extenday®) on colouration of fruit at various positions within the canopy of the tall trees on three sides of the fruit (inside, outside, down-facing) and quality in two late maturing apple cultivars under Southern hemisphere conditions in New Zealand at 41°S depending on the time of spreading the ground cover.

4.1. Fruit Colouration

In fruit grading, export quality (class I) depends on the overall percentage colouration of the fruit surface. The down facing side is usually the part of the apple fruit least exposed to light [10] and least coloured, and therefore, any improvement in its colouration is relevant for grading and pricing of the fruit.

Satisfactory colour of apple fruit requires a certain amount of available sunlight. Therefore, fruit or fruit parts grown in shade will not turn red during ripening due to a lack of anthocyanin synthesis, particularly under hail nets [11,12].

This may be due to providing a longer exposure to light during the time, when anthocyanin synthesis, PAL activity and MYB 1 and MYB 10 gene expression in the apple peel starts ca. 4–5 weeks prior to harvest (Lancaster, 1992, Wang et al., 2011) [2,13]. The results show the strong effects of reflective mulch on red colouring, as expressed by an increase in blush development (Figures 5 and 6) or the decline in colour angle [°hue] (Figures 9 and 10). Both effects can be shown throughout the last weeks before harvest and were more pronounced on fruit from the inside than from the outside of the canopy (Figures 7 and 8). This will result in more fruit archiving even red colouring and therefore being chosen for consumption.

Overall, the most pronounced effects of the reflective mulch on late-ripening apple cultivars were on

(a) Fruit from the inner part of the tree (in comparison with the tree periphery),
(b) The inner side of fruit in the tree (as compared with the outer side of these fruit),
(c) The down-facing side of fruit in the tree canopy (as compared to the outer side of the fruit),

which has not been studied in such detail neither in the NH nor the SH.

4.2. Fruit Picks

Fruit in the first pick are mostly from the outer and upper periphery of the tree canopy. Because the main influence of the reflective mulches can be observed and explained on fruit from the inner part of the tree canopy [12], their portion enhances in the first pick. Sometimes fewer picks have to be carried out to retrieve all fruit from a tree. Any increase of the portion of fruit in the first pick, such as 8% in cv. Fuji and 27% in cv. Pacific Rose (Figures 2 and 3), improves labour efficiency; exclusively fruit from the first pick are suitable for long term storage or export with superior financial returns. Therefore, the use of reflective mulch can be cost effective, as expensive and scarce labour is saved.

4.3. Fruit Quality

The results in Tables 3 and 4 are in line with findings in Europe, where the weather at 50° N at Bonn is less beneficial with a shortage of autumn radiation. Positive mulching effects have been shown on soluble solids and starch breakdown in cv. Elstar and Jonagold on small trees (2.8 m) on dwarfing M9 rootstock (Funke and Blanke 2011) [14].

The amount/ intensity of reflected light decreases with the square of the distance to the reflective mulch and hence hardly influences fruit in the upper part of the tree canopy.

4.4. Tree Training Size/Rootstock and Colouration

The use of semi-vigorous rootstocks such as MM106, as used in this trial, are common throughout NZ and some of the SH but can hardly be found on the European continent. The larger trees of MM106 provide a larger distance of the target fruit from the reflective mulch. Therefore, the influence of reflective mulch on fruit colouration shows a more profound effect on small trees such as Braeburn and Elstar and Jonagold all on M9 rootstock (Funke and Blanke 2005, 2011) [10,14] than on large trees (Table 5).

Table 5. Influence of latitude and tree size on minimum exposure time of reflective mulches.

Location	Latitude	Tree Height	Weeks	Yield	Time
Southern hemisphere					
Nelson, Southland, New Zealand	39–41° S	3.5 m	2+	90–120 t/ha	March–April
Northern hemisphere–North America-USA					
Washington State, Pacific Northwest, USA	46–47° N	3.5 m	3–4	60–90 t/ha	September
Geneva, New York, East Coast, USA	46° N	3.5 m	3–4	60–90 t/ha	September
Northern hemisphere -Europe					
South Tyrol, Italy	45–46.5° N	3.5m	2+	60–90t/ha	Aug.–Sept.
Lake Constance, Germany	47.4° N	3.5 m	3+	60–80 t/ha	Aug.–Sept.
Bonn/Belgium/Holland/Poland/ Somerset, East Malling, Kent, UK	50–51° N	2.8 m	4+	40–60 t/ha	September

5. Conclusions

Four weeks use of reflective mulch appeared only slightly more beneficial than 2 weeks use in the present experiment (Figures 4 and 5), which is in line with NH results with tall apple trees from South Tyrol (46°), but considerably shorter than the 4 to 6 weeks even with smaller trees (2.8 m) at 50° N in the "apple belt" along the 50–51° N, e.g., at Bonn, Belgium, Kent and Somerset etc. (Table 5). The positive results of 2 weeks less of spreading enables duplicate use in the same year starting with early ripening apple varieties such as cv. Gala and subsequent use in late ripening varieties such as cv. Fuji, Envy or Jazz and may extend the overall period of reflective mulch in the orchard. Such repeated use of these materials, textile or other, makes their use more sustainable and more economic, since the material is already out in the field and just needs to be pulled onto another plot.

Usually prices are higher in the beginning of the season with better returns for first picked fruit; long-term storage is restricted to first pick fruit. In both apple cvs Fuji and Pacific Rose, the reflective mulch increased the portion of fruit harvested in the first pick, which is the relevant portion for storage and export quality and, to our knowledge, has not been reported so far for these varieties on such large trees in the southern hemisphere (SH).

Author Contributions: Individual contributions are: Conceptualization and methodology, K.F., M.B.; investigation and data analysis, K.F.; writing—original draft preparation, K.F.; writing—review and editing M.B. All authors have read and agreed to the published version of the manuscript

Funding: There was no external funding.

Institutional Review Board Statement: Not applicable.

Informed Consent Statement: Not applicable.

Data Availability Statement: Not applicable.

Acknowledgments: We are grateful to the staff at HortResearch, Motueka, New Zealand for their assistance and support, Michael Moss, Motueka, New Zealand for the use of his orchard, Jens Wünsche and John Palmer for supervision and MDPI for the invite to this contribution and waiving the APC and Ligia Hent for editorial guidance.
The manuscript was finished under covid-19 including access restrictions to university services.

Conflicts of Interest: The authors declare no conflict of interest.

Abbreviations

NH	Northern hemisphere (NH)
PAL	Phenylalanine-ammonia-lyase
PAR	Photosynthetically active radiation
SH	Southern hemisphere

References

1. Hamadziripi, E.; Muller, M.; Theron, K.I.; Steyn, W.J. Consumer preference for apple eating quality and taste in relation to canopy position. *Acta Hortic.* **2017**, *1058*, 253–260. [CrossRef]
2. Wang, L.K.; Micheletti, D.; Palmer, J.; Volz, R.; Lozano, L.; Chagne, D.; Rowan, D.D.; Troggio, M.; Iglesias, I.; Allen, A.C. High temperature reduces apple fruit colour via regulation of the anthocyanin complex. *Plant Cell Environ.* **2011**, *34*, 1176–1190. [CrossRef] [PubMed]
3. Andris, H.; Crisoto, C.H. Reflective materials enhance 'Fuji' apple color. *Calif. Agric.* **1996**, *50*, 27–30. [CrossRef]
4. Meinhold, T.; Richters, J.P.; Damerow, L.; Blanke, M.M. Optical properties of reflection ground covers with potential for enhancing fruit colouration. *Biosyst. Eng.* **2010a**, *107*, 152–165. [CrossRef]
5. T&G Specifications Manual. FUJI (063)—Class 1. 2018. Available online: https://tandgspecmanual.helpdocsonline.com/fuji-apple-variety-specifications-class-1 (accessed on 20 October 2020).
6. Overbeck, V.; Schmitz-Eiberger, M.A.; Blanke, M.M. Reflective mulch enhances ripening and health compounds in apple fruit. *J. Sci. Food Agric.* **2013**, *93*, 2575–2579. [CrossRef] [PubMed]
7. Srmke, T.; Persic, M.; Veberic, R.; Sircelj, H.; Jakopic, J. Influence of reflective foil on persimmon (*Diospyros kaki* Thunb.) fruit peel colour and selected bioactive compounds. *Sci. Rep.* **2019**, *9*, 1989.
8. McGuire, R. Reporting of objective colour measurements. *HortScience* **1992**, *27*, 1254–1255. [CrossRef]
9. Meinhold, T.; Damerow, L.; Blanke, M.M. Reflective materials under hail net improve orchard light utilisation, fruit quality and particularly fruit colouration. *Sci. Hortic.* **2010b**, *127*, 447–451. [CrossRef]
10. Funke, K.; Blanke, M.M. Can reflective ground cover enhance fruit quality and colouration? *J. Food Agric. Environ.* **2005**, *3*, 203–206.
11. Solomakhin, A.; Blanke, M.M. Overcoming adverse effects of hail nets on fruit quality and microclimate in an apple orchard. *J. Sci. Food Agric.* **2007**, *87*, 2625–2637. [CrossRef] [PubMed]
12. Hess, P.; Kunz, A.; Blanke, M.M. Innovative strategies for the use of reflective foils for fruit colouration to reduce plastic Use in Orchards. *Sustainability* **2021**, *13*, 73. [CrossRef]
13. Lancaster, J.E. Regulation of skin colour in apples. *Crit. Rev. Plant Sci.* **1992**, *10*, 487–502. [CrossRef]
14. Funke, K.; Blanke, M.M. Mikroklima-, Farb- und Geschmacksverbesserung durch Licht reflektierende Folie zu verschiedenen Auslegeterminen bei Elstar- und Jonagold Äpfeln unter schwarzem und weißem Hagelnetz. *Erwerbs-Obstbau* **2011**, *53*, 1–10. [CrossRef]

 *horticulturae*

Article

Premature Apple Fruit Drop: Associated Fungal Species and Attempted Management Solutions

Khamis Youssef [1,*] **and Sergio Ruffo Roberto** [2,*]

[1] Agricultural Research Center, Plant Pathology Research Institute, 9 Gamaa St., Giza 12619, Egypt
[2] Agricultural Research Center, Department of Agronomy, Londrina State University, Londrina 86057-970, PR, Brazil
* Correspondence: youssefeladawy@yahoo.com (K.Y.); sroberto@uel.br (S.R.R.)

Received: 18 March 2020; Accepted: 22 May 2020; Published: 26 May 2020

Abstract: The aim of this research was to determine the incidence and possible causal pathogen(s) of premature apple fruit drop (PAFD), and also to assess some fungicides for controlling the disease organisms, in order to promote a sustainable system in orchards. The prevalence and natural incidence of apple fruit drop in cv. Anna was assessed during the 2017–2018 growing seasons in Nubaria and Cairo–Alexandria regions, Egypt. Phytopathogenic fungi were isolated from dropped fruit, and four fungicides, pyraclostrobin + boscalid, difenoconazole, carbendazim, and thiophanate methyl, were tested against the diseases in vitro and under naturally occurring infections in the field. Several phytopathogenic fungi, including *Alternaria alternata*, *Cladosporium cladosporioides*, *Fusarium semitectum*, and *Penicillium* spp., were associated with apple fruit drop. *A. alternata* was the most frequently isolated fungus occurring during the investigation. Pathogenicity tests confirmed that the maximum percentage of apple fruit drop was noted when petioles and fruits were inoculated with mixed fungal pathogens using branch sections with fruit. In vitro tests showed that the fungicides had a variable effect against the fungal isolates depending on the concentration used. All fungicides completely inhibited the growth of *A. alternata*, *C. cladosporioides*, and *F. semitectum* at 400 mg·L^{-1}. Under naturally occurring infections, thiophanate methyl applied at fruit set had the greatest effect (81.68%) against PAFD, followed by difenoconazole (73.76%), pyraclostrobin + boscalid (70.29%), and carbendazim (66.34%). The results indicated that PAFD may in part be a result of diseases caused by certain phytopathogenic fungi, which could be controlled using a number of fungicides applied at the beginning of fruit set.

Keywords: fruit drop; sustainable systems; fungicides; *Alternaria alternata*

1. Introduction

Apples (*Malus domestica* Borkh.) ranked third for worldwide fruit production, at 86.1 million tonnes and with the harvested area around 4,904,305 ha, in 2018. Egyptian apple production during 2018 was estimated to be about 704,727 tonnes and the harvested area nearly 28,085 ha, with a yield 25,092 kg·ha^{-1} [1]. Apples are susceptible to several diseases, including apple scab (*Venturia inaequalis*), bitter rot (*Colletotrichum* spp.), black rot (*Botryosphaeria obtuse*), powdery mildew (*Oidium* spp.), sooty blotch (*Gloeodes pomigena*), flyspeck (*Schizothyrium pomi*), cedar-apple rust (*Gymnosporangium juniperi-virgininae*), and fire blight (*Erwinia amylovora*) [2]. Premature fruit drop is one of the major problems for apple production, leading to substantial losses in apple yield. In Egypt, it generally appears when the fruit has reached 40–50% of their final size [3]. This problem is especially noted when fruits are left to ripen for better red color development to meet consumer expectations. In some years, losses due to premature drop can exceed 50% of the total crop yield, causing significant financial losses [4].

Previous studies have indicated an association of several fungi involved in moldy core and core rot of dropped, as well as fully mature, apple fruit on the trees, but *Alternaria alternata* is the predominant fungal pathogen responsible for moldy core and core rot of apples in different regions of the world [5,6]. Basically, in apples after fruit set, there is a period in which fruitlet drop occurs 5–6 weeks after full bloom due to physiological causes, and this is referred to as June drop. Symptoms of moldy core can also include premature ripening and fruit drop. This disease is caused by many different species of fungi that naturally occur in the orchard.

In apple trees, the integration of aminoethoxyvinylglycine (AVG) and naphthalene acetic acid (NAA) gave excellent premature drop control and did not reduce fruit firmness after controlled atmosphere storage [7]. In 'Golden Delicious' apples, AVG inhibited ethylene production, reduced premature drop, and delayed fruit maturation on the tree, and fruit ripening and softening during storage [8]. In the current research, we focused on premature fruit drop beginning around four weeks before the expected apple harvest date. To the best of our knowledge, there is a lack of pathological studies to determine if disease pathogens may be the causes of premature apple fruit drop (PAFD) in Egypt. Therefore, this research was designed to determine the incidence and possible causal pathogen(s) of PAFD, and to verify some means for controlling the diseases, in order to promote a sustainable system in orchards.

2. Materials and Methods

2.1. Disease Survey

This study was performed during the two growing seasons of 2017 and 2018 in the Nubaria region (Beheira Governorate) and on the Cairo–Alexandria desert road (Giza Governorate), Egypt. 'Anna' apple trees (*Malus domestica* Borkh.) showed natural fruit drop, and internal sections of dropped fruits showed core browning. A survey was carried out across 20 orchards at each location, four weeks before the expected harvest date. The proportion of orchards which had apple fruit drop were considered for prevalence, determined using the following formula:

$$\text{Prevalence (\%)} = \frac{\text{Number of fields showing fruit drop}}{\text{Total number of fields visited}} \times 100$$

Percentages of naturally infected apple trees, approximately 9–12 years old, showing fruit drop disease, were recorded in the above mentioned locations during April–May each year (Figure 1). Disease incidence (%) was calculated using the following Equation:

$$\text{Disease incidence (\%)} = \frac{\text{Apple trees showing fruit drop disease}}{\text{Total number of tested apple trees}} \times 100$$

2.2. Isolation and Identification of the Causal Agents

Isolation experiments were carried out on different parts of the fruit, calyx end, stem or pedicel end, and internal parts collected from commercial apple orchards showing fruit drop. Samples were collected as soon as they dropped after shaking the branches. Collected samples were smaller than healthy fruit, collapsed, and showed an abnormal appearance. To ensure that dropped fruit had not picked up the diseases from the soil and ground vegetation, the samples were washed under tap water then left to dry on folds of filter paper at room temperature. Small pieces, visually observed to contain both diseased and healthy tissues, were surface sterilized by immersing in 1% sodium hypochlorite solution for two minutes, rinsed three times in sterilized distilled water, left to dry, then transferred into 9 cm diameter Petri dishes containing potato dextrose agar (PDA) medium and incubated at 24 ± 1 °C for 3–5 days in the dark. The growing fungal colonies were purified using the hyphal tip technique [9,10]. Purified fungi were identified on the basis of their morphological characteristics, according to Barnett and Hunter (1986) [11]. Pure culture stocks of the isolated fungi were kept on

PDA slants at 5 °C for further study. The frequency (%) of each fungal species was calculated using the following Equation:

$$\text{Frequency (\%)} = \frac{\text{Number of colonies of each fungal species}}{\text{Total number of all fungi}} \times 100$$

Figure 1. An 'Anna' apple tree showing natural fruit drop (**a**), collected dropped apple fruit (**b**), internal tissues of dropped apple fruit showing core browning (**c,d**), fruit were about half their final size.

2.3. Pathogenicity Test

To fulfill Koch's postulates, pathogenicity tests were conducted on healthy 'Anna' apple fruit collected at 40–50% of full size and still attached to branches and petioles, from a private apple orchard located on the Cairo–Alexandria desert road (Giza Governorate, Egypt). Apples were thoroughly washed under tap water, surface sterilized with 0.5% sodium hypochlorite solution for 2 min, followed by washing three times in sterilized water, and then air-dried on sterilized filter paper. Finally, branches with prepared fruit were individually inserted into 1 L sterilized glass bottles, each containing 250 mL nutrient solution (1% urea + 2% K_2SO_4 + 0.5% $ZnSO_4$ + 0.3% Borax), as recommended by Khamis et al. [12].

Conidial suspensions of isolated fungi (*Alternaria alternata*, *Cladosporium cladosporioides*, *Fusarium semitectum*, and *Penicillium* spp.), individual or mixed, were prepared from seven-day-old cultures grown on PDA plates (9 cm in diameter), according to Hussien et al. [13], and adjusted to 10^6 spores·mL^{-1} following the methods of Lachhab et al. [14]. The same concentration of 10^6 spores·mL^{-1} was used for the mixed infection. Inoculation was done according to Youssef et al. [15], by injecting either fruit petioles or fruit with 50 µL of the spore suspension. A set of petioles and control fruit were injected with sterilized water only. The wounds were covered with melted (54 °C) wax [16]. Bottles were then incubated under laboratory conditions (18–25 °C) and high relative humidity for 5–7 days. Five bottles, each containing a branch with 4–5 attached fruit, were used as replicates for each treatment. The causal agents of rot were re-isolated for identity confirmation. Dropped fruit (DF) percentages were calculated using the following Equation:

$$\text{Dropped fruit (\%)} = \frac{\text{Number of dropped fruits}}{\text{Number of fruits remaining attached to branches}} \times 100$$

2.4. Effect of Different Fungicides Against Causal Agents In Vitro

Four fungicides, pyraclostrobin + boscalid (Bellis®), difenoconazole (Score®), carbendazim (Kemazed®), and thiophanate methyl (Onest®), were evaluated in vitro for their efficacy against fungal mycelial growth. Fungal mycelia plugs (5 mm in diameter), from the growing edge of one-week-old cultures, were placed in the center of Petri dishes with PDA amended with the tested fungicides at final concentrations of 100, 200, and 400 mg·L^{-1}. PDA plates without any added fungicide served as the control. Each fungicide/concentration included three plates as replicates and the whole experiment was repeated twice. The inoculated plates were incubated at 24 ± 1 °C and radial fungal growth was recorded after 5–7 days of incubation. Fungal growth (mm) was calculated as the average of the orthogonal diameter [17,18].

2.5. Field Application of Different Fungicides Against Fruit Drop Disease

Four fungicides, pyraclostrobin + boscalid (Bellis®), difenoconazole (Score®), carbendazim (Kemazed®), and thiophanate methyl (Onest®), were evaluated in the field under natural conditions of infection at recommended doses (30 g, 50 mL, 50 g, and 65 g per 100 L of water, respectively) approved by the Agricultural Pesticide Committee (APC) in the Ministry of Agriculture and Land Reclamation, Egypt. Those fungicides were already registered in Egypt to control apple scab and powdery mildew. The fungicides were applied by spraying the recommended dose at the beginning of fruit set on a private farm on the Cairo–Alexandria desert road (Giza Governorate), Egypt. Set fruits were counted before the fungicide application [15]. The orchards were divided into a randomized complete block design. Five trees, each with three branches/tree, were used as replicates for each treatment. Additionally, five trees sprayed with water only were used as the control. The percentage of dropped fruits was calculated as previously mentioned. Moreover, the efficiencies of the tested fungicides were calculated according to the following formula:

$$\text{Efficienty (\%)} = \frac{\text{Dropped fruits (\%) in the control group} - \text{Dropped fruits (\%) in the treated group}}{\text{Dropped fruits (\%) in the control group}} \times 100$$

2.6. Statistical Analysis

Percentage data were arcsine-transformed before the analyses to normalize variance. Data were subjected to one-way analysis of variance (ANOVA) using Statistica Software Ver. 6.0 (Stat Soft Inc., Tulsa, OK, USA). Fisher's protected least significant difference test was used at $p \leq 0.05$ to distinguish the differences among various treatments [19].

3. Results

3.1. Disease Survey

Disease was observed in all locations. Data presented in Table 1 show that the disease incidence (%) was higher during 2018 than 2017. The maximum disease incidence (75.3%) was recorded in the Nubaria region followed by the Cairo–Alexandria desert road (73.5%) in 2018, but reached 69.7% and 61.5%, respectively, in 2017.

Table 1. Disease incidence (%) of dropped apple fruit in the investigated locations.

Locations	Disease Incidence (%)	
	2017 Season	**2018 Season**
Nubaria region (Beheira Governorate)	69.7 ± 1.46a [z]	75.3 ± 1.13a
Cairo–Alexandria desert road (Giza Governorate)	61.5 ± 1.03a	73.5 ± 1.45b

[z] Means ± standard error followed by different letters within seasons are significantly different by Fisher's protected LSD test at $p \leq 0.05$.

3.2. Isolation and Identification of the Causal Organisms

The frequency of fungi isolated from naturally infected apple fruit collected from two different locations is shown in Table 2. Isolation and identification revealed that four different fungi, *Alternaria alternata*, *Cladosporium cladosporioides*, *Fusarium semitectum*, and *Penicillium* spp., were associated with the naturally infected apple fruit, with average frequencies of 60%, 19%, 11%, and 10%, respectively. In general, *A. alternata* was the most prevalent fungus isolated.

Table 2. Frequency (%) of fungal species from naturally infected 'Anna' apple fruit collected from different locations in Beheira and Giza Governorates, Egypt.

Fungus	Nubaria Region (Beheira Governorate)	Cairo–Alexandria Desert Road (Giza Governorate)	Mean (%)
Alternaria alternata	62.0 ± 1.15a [z]	58.0 ± 2.30a	60.0
Cladosporium cladosporioides	18.0 ± 2.03b	20.0 ± 0.88b	19.0
Fusarium semitectum	12.0 ± 1.16b	10.0 ± 0.58b	11.0
Penicillium spp.	8.0 ± 0.67b	12.0 ± 1.15b	10.0

[z] Means± standard error followed by different letters within the same column are statistically different according to Fisher's protected LSD test ($p \leq 0.05$).

3.3. Pathogenicity of the Isolated Fungi

Koch's postulates were fulfilled by artificially infecting 'Anna' apple fruit attached to branches collected from 12-year-old trees. After seven days, tests confirmed that the pathogenicity of the fungal isolates varied between them in terms of infection rate, whereas the control treatment did not show any symptoms. Generally, the maximum percentage of dropped fruit was recorded on petioles and fruits inoculated with mixed fungal pathogens, at 78% and 62%, respectively (Table 3). Comparing pathogens, the results showed that the percentages of dropped fruit were higher on petioles and fruits inoculated with *A. alternata*, followed by *C. cladosporioides*, and then *F. semitectum*. The lowest percentage of dropped fruits was observed for *Penicillium* spp., which was ignored for subsequent experiments.

Table 3. Pathogenicity of the isolated fungi on 'Anna' apple petioles and fruits expressed as the percentage of dropped fruits.

Pathogen	Dropped Fruits (%)	
	Petioles	**Fruits**
Alternaria alternata	62.0 ± 0.88b [z]	52.0 ± 2.31b
Cladosporium cladosporioides	40.0 ± 1.15c	30.0 ± 2.91c
Fusarium semitectum	22.0 ± 1.45d	18.0 ± 1.15d
Penicillium spp.	8.0 ± 1.15e	6.0 ± 1.15e
Mixed fungi	78.0 ± 0.88a	62.0 ± 2.30a
Control	0.0 ± 0.0f	0.0 ± 0.0e

[z] Means ± standard error followed by different letters within the same column are statistically different according to Fisher's protected LSD test ($p \leq 0.05$).

3.4. Effect of Different Fungicides against Causal Agents In Vitro

The effectiveness of four fungicides, each applied at three different concentrations, against the isolated pathogenic fungi is shown in Table 4. Generally, all fungicides completely inhibited the growth of *A. alternata*, *C. cladosporioides*, and *F. semitectum* at 400 mg·L^{-1}. Additionally, difenoconazole completely inhibited the growth of *A. alternata* and *C. cladosporioides* at 200 mg·L^{-1}; thiophanate methyl was able to completely inhibit the three pathogens at 200 mg·L^{-1}; carbendazim completely inhibited *C. cladosporioides* and *F. semitectum* at 200 mg·L^{-11}. None of the fungicides suppressed the pathogens at 100 mg.L^{-1}.

Table 4. Effect of four fungicides at 100, 200, and 400 mg·L^{-1} against isolated pathogenic fungi growth (mm) in vitro recovered from 'Anna' apples after premature fruit drop.

Active Ingredient	Fungal Growth (mm)								
	A. alternata			*C. cladosporioides*			*F. semitectum*		
	100	200	400	100	200	400	100	200	400
	(mg·L^{-1})								
Pyraclostrobin + boscalid (38% WG)	51.3 ± 1.33c ^{z}	26.7 ± 0.66c	0.0 ± 0.0b	58.7 ± 2.40b	18.7 ± 0.67b	0.0 ± 0.0b	49.3 ± 0.66b	28.7 ± 0.67b	0.0 ± 0.0b
Difenoconazole (25% EC)	29.3 ± 0.67d	0.0 ± 0.0d	0.0 ± 0.0b	20.7 ± 0.66d	0.0 ± 0.0c	0.0 ± 0.0b	41.3 ± 0.67c	23.3 ± 0.66c	0.0 ± 0.0b
Carbendazim (50% WP)	58.0 ± 2.0b	39.3 ± 0.66b	0.0 ± 0.0b	27.3 ± 1.3c	0.0 ± 0.0c	0.0 ± 0.0b	18.7 ± 0.66d	0.0 ± 0.0d	0.0 ± 0.0b
Thiophanate methyl (70% WP)	22.0 ± 1.15e	0.0 ± 0.0d	0.0 ± 0.0b	16.7 ± 0.67d	0.0 ± 0.0c	0.0 ± 0.0b	15.3 ± 0.67e	0.0 ± 0.0d	0.0 ± 0.0b
H$_2$O	90.0 ± 0.0a	90.0 ± 0.0a	90.0 ± 0.0a	90.0 ± 0.0a	90.0 ± 0.0a	90.0 ± 0.0a	90.0 ± 0.0a	90.0 ± 0.0a	90.0 ± 0.0a

^{z} Means ± standard error followed by different letters within the same column are statistically different according to Fisher's protected LSD test ($p \leq 0.05$).

3.5. Field Application of Different Fungicides Against Fruit Drop Disease

The effect of the four fungicides was evaluated for the control of apple fruit drop disease under natural field infection. Apple fruit drop percentages were 12.0%, 10.6%, 13.6%, and 7.4% on trees treated with pyraclostrobin + boscalid, difenoconazole, carbendazim, and thiophanate methyl, respectively (Table 5). The thiophanate methyl fungicide was the best one and recorded an efficiency of 81.6%, followed by difenoconazole, pyraclostrobin + boscalid, and carbendazim, as compared with the control group.

Table 5. Efficiency (%) of four fungicides against apple fruit drop in 'Anna' apples under naturally occurring infection in the orchard.

Active Ingredient	Applied Dose	Dropped Fruits (%)	Efficiency (%)
Pyraclostrobin + boscalid (38% WG)	30 g·100 L^{-1}	12.0 ± 0.63bc ^{z}	70.29
Difenoconazole (25% EC)	50 mL·100 L^{-1}	10.6 ± 0.98c	73.76
Carbendazim (50% WP)	50 g·100 L^{-1}	13.6 ± 1.32b	66.34
Thiophanate methyl (70% WP)	65 g·100 L^{-1}	7.4 ± 0.50d	81.68
Control	-	40.4 ± 0.75a	-

^{z} Means ± standard error followed by different letters within the same column are statistically different according to Fisher's protected LSD test ($p \leq 0.05$).

4. Discussion

The objectives of this research were to determine the incidence and possible causal pathogens of PAFD, and to verify the activity of some fungicides to control the disease. PAFD is becoming more common and severe in apple orchards in Egypt, and is inflicting substantial losses. This phenomenon, naturally common in fruit trees, usually occurs immediately prior to fruit ripening. A separation zone occurs either in the proximal end contact area of the pedicel of the fruit, or it may occur in the layer of fruit peel and skin [20]. There are several factors that can trigger early fruit drop, including overcropping, insufficient pollination, high ethylene levels, excessive summer pruning, insect damage/diseases, extremes in weather, and poor tree nutrition.

The results of this study showed that the maximum incidence of fruit drop occurred in the Nubaria region (75%) followed by the Cairo–Alexandria desert road (73.5%) area in 2018. Several factors, including the suitability of regular agricultural practices, age of the orchard, types and efficiency of sanitary methods, and chemical protectant or curative treatments against diseases and insects, were likelyinvolved, which may explain the differences between the inspected locations and growing seasons. Similarly, Raja et al. [4] demonstrated that orchard and climatic factors, including mineral nutrition, summer pruning, insects and mites, water availability, and growing season temperatures, can affect premature fruit drop.

In addition to the other factors, diseases have an important role in lowering the yield of fruit trees. In this study, various phytopathogenic fungi, i.e., *A. alternata*, *C. cladosporioides*, *F. semitectum*, and *Penicillium* spp., were recovered from dropped apple fruit. *A. alternata* was the most frequently isolated fungus. Previously, several fungi, such as *Alternaria* spp., *Penicillium* spp., and *Fusarium* spp., were isolated from apple fruit. Particularly, *Alternaria* spp. was the most common fungus associated with diseased 'Anna' apple fruit [6]. Moreover, Gao et al. [21] found that *A. alternata*, *A. tenuissima*, *A. arborescens*, *C. cladosporioides*, and *C. tenuissimum* were the main pathogens causing core browning and moldy core of 'Fuji' apple fruit in China. Furthermore, *A. alternata*, *Pleospora herbarum*, *Coniothyrium* spp., *Penicillium funiculosum*, *P. expansum*, and *P. ramulosum* were associated with core rot of 'Starking Delicious' apple fruit in South Africa [22,23]. In a previous study, four phytopathogenic fungi, *A. alternata*, *Lasiodiplodia theobromae*, *F. semitectum*, and *Pestalotia psidii*, were isolated from dropped guava fruit samples collected from different orchards [15].

Pathogenicity tests confirmed that the maximum percentage of PAFD was noted on petioles and fruits inoculated with mixed fungal pathogens. In particular, the percentages of PAFD were higher when the fruit were artificially inoculated with *A. alternata*, followed by *C. cladosporioides* and *F. semitectum*. Racskó et al. [24] confirmed that fruit drop is often caused by damage due to diseases and pests. Additionally, a fungus causing peach brown rot (*Monilinia fructicola*) was recognized as initiating fruit drop. It is polyphagous, appears in many fruit species, and is responsible for immediate premature fruit drop. A significant pathogen of larger fruit (apple, pear, apricot, peach), as well as smaller fruit (sweet and sour cherry, plum), may be due to fruit on trees overwintering as shriveled mummies [25–27]. Likewise, *A. alternata* and *F. semitectum* were associated with brown apical necrosis and caused fruit drop of English walnut [28].

In vitro tests showed that four fungicides had a variable effect against the isolated fungi, depending on the concentration used. All fungicides completely inhibited the growth of *A. alternata*, *C. cladosporioides*, and *F. semitectum* at 400 mg·L^{-1}. Under naturally occurring infection, fruit set application of thiophanate methyl had the highest efficacy (81.68%) against PAFD, followed by difenoconazole (73.76%), pyraclostrobin + boscalid (70.29%), and carbendazim (66.34%). The fungicides used in this research are registered to control apple scab and powdery mildew in several countries, including Egypt, and farmers frequently use them during the season, depending on the management program in the orchard. Thus, it is easy to integrate these fungicides in the integrated pest management program of apple. We believe that the further application of those fungicides to apple trees can play a role in reducing fruit drop. Fungicide application at fruit set may work since its early application can help to reduce or prevent primary infections/inocula by the phytopathogenic fungi identified by this research. In fact, if primary infection by phytopathogenic fungi is not well controlled, secondary infection will be a problem and be more difficult to manage. Any delay incontrolling the disease is not recommended, even if adequate fungicides are used. PAFD is variable among orchards, suggesting that cultural management can influence fruit drop. Strategies to reduce PAFD can help preserve crop yield, which is an essential factor for the economic success of an orchard [29]. In this context, a previous study showed that guava fruit drop could be controlled by applying either thiophanate methyl 50% + thiram 30% or carbendazim (50% WP) at aproportionof 70 g·100 L^{-1} water [15]. The successful control of a disease is dependent on the reliability of pathogen detection at a latent phase [30]. Raja et al. [4] summarized some effective compounds to reduce premature drop, such as naphthaleneacetic acid, naphthalene acetamide, propionic acid, propanoic acid, 2,4-dichlorophenoxy acetic acid, lactidichlor ethyl, and butyric acid. Additionally, other plant growth regulators have been investigated for their ability to reduce fruit drop, such as gibberellic acid (GA$_3$), 2,4-dichlorophenoxyacetic (2,4-D), and hydrogen peroxide [31].

Flowering and fruit set are probably the most important of allthe events occurring once an apple tree has reached reproductive maturity. Given favorable environmental conditions, the timing and intensity of flowering greatly determines when and how much fruit will be produced during the season. Prolonged flooding may lead to fruit and leaf drop, leaf chlorosis, stem dieback, and tree death. Trees are generally more tolerant of flooding during cool weather [32]. The majority of phytopathogenic fungi affecting flowers promotes and speeds up abscission [33]. In apple, the varieties exhibiting less fruit drop were 'Stayman', 'Gala', 'Melrose', 'Akane', and 'Fuji' [24]. In citrus species, fruit drop after bloom was caused by *Colletotrichum acutatum* and *C. gloeosporioides* [34]. Nature provides plants with phenolic compounds that may play an essential role in growth and resistance to pathogens, and exhibit a wide range of anti-microbial effects [35]. Racskó et al. [24] showed that fruit drop in walnuts was induced by *Gnomonia* spp., but the symptoms appeared only on the epicarp. In peaches, premature fruit drop caused by *Venturia carpophila*, the causal agent of scab, is dangerous during an extended period of drought. With a vigorous infection, fruit drop may occur in peaches when infected by *Taphrina deformans*, causing leaf-curl disease [24]. Finally, severe fruit drop could contribute to low yield in apple orchards and lead to great economic losses in many production areas. The role of different horticultural and climatic factors in apple fruit drop has not been fully investigated. This is the first research dealing with PAFD disease under Egyptian conditions.

5. Conclusions

Alternaria alternata, Cladosporium cladosporioides, Fusarium semitectum, and *Penicillium* spp. were associated with apple fruit drop, with *A. alternata* being the most prevalent. The maximum percentage of apple fruit drop in lab tests was observedwhen petioles and fruitswere inoculated with mixed fungal pathogens. The fungicides pyraclostrobin + boscalid, difenoconazole, carbendazim, and thiophanate methyl completely inhibited the growth of *A. alternata, C. cladosporioides,* and *F. semitectum* at 400 mg·L^{-1} in in vitro tests. Under natural occurring infection, thiophanate methyl had the highest efficacy in reducing premature fruit drop, followed by difenoconazole, pyraclostrobin + boscalid, and carbendazim, when applied at the beginning of fruit set. Premature apple fruit drop is a complex phenomenon and there are various factors affecting it, such as mineral nutrition, summer pruning, insects and diseases, moisture availability, and growing season temperatures. This research indicated that PAFD may also be a result of diseases caused by certain phytopathogenic fungi, which could be controlled using a number of fungicides at the fruit set stage. We recommend that growers in arid climates might need to apply fungicides more frequently than is commonly done to overcome this problem. Finally, further studies are needed to investigate the importance of disease in premature fruit drop.

Author Contributions: K.Y. conceived the research idea. S.R.R. helped to analyze the data and write the paper. All authors have read and agreed to the published version of the manuscript.

Funding: This research received no external funding.

Conflicts of Interest: The authors declare no conflict of interest.

References

1. FAOSTAT. 2019. Available online: http://www.fao.org/faostat/en/#data/QC (accessed on 10 March 2020).
2. Holb, I.J. Fungal Disease Management in Environmentally Friendly Apple Production—A Review. In *Climate Change, Intercropping, Pest Control and Beneficial Microorganisms*; Lichtfouse, E., Ed.; Springer: Dordrecht, The Netherlands, 2009; Volume 2.

3. Meier, U. *Growth Stages of Mono-and Dicotyledonous Plants*, 2nd ed.; BBCH Monograph; Federal Biological Research Centre for Agriculture and Forestry: Berlin, Germany, 2001; 158 p.

4. Raja, W.H.; UnNabi, S.; Kumawat, K.L.; Sharma, O.C. Pre harvest Fruit Drop: A Severe Problem in Apple. *Indian Farmer* **2017**, *4*, 609–614.

5. Reuveni, M.; Sheglov, D.; Sheglov, N.; Ben-Arie, R.; Prusky, D. Sensivity of Red Delicious apple fruits at various phenologic stages to infection by *Alternaria alternata* and mouldy core control. *Eur. J. Plant Pathol.* **2002**, *108*, 421–427. [CrossRef]

6. El-Mohamedy, R.S.R.; El-Sayed, H.Z. First Record of Core Rot Disease on Apple Fruit cv. Anna 106 Local Cultivar in Egypt. *Int. J. Agric. Technol.* **2015**, *11*, 1371–1380.

7. Sazo, M.M.; Robinson, T.L. The "Split" Application Strategy for Pre-Harvest Fruit Drop Control in a Super Spindle Apple Orchard in Western NY. *New York State Hort. Soc.* **2013**, *21*, 21–24.

8. Masia, A.; Ventura, M.; Gemma, H.; Sansavini, S. Effect of some plant growth regulator treatments on apple fruit ripening. *Plant Growth Reg.* **1998**, *25*, 127–134. [CrossRef]

9. Youssef, K.; Abo Rehab, M.A.; Abd El-Ghany, K. Preliminary investigation of Verticillium wilt on mango trees (*Mangifera indica* L.) in Egypt. *Am.-Eurasian J. Sustain. Agric.* **2014**, *8*, 50–58.

10. Abd-Elsalam, K.A.; Youssef, K.; Almoammar, H. First morphogenetic identification of *Fusarium solani* isolated from orange fruit in Egypt. *Phyton* **2015**, *84*, 128–131.

11. Barnett, H.L.; Hunter, B.B. *Illustrated Genera of Imperfect Fungi*, 4th ed.; Macmillan Publ. Co.: New York, NY, USA, 1986; 218 p.

12. Khamis, M.A.; Bakry, K.A.; Abd El-Moty, S.A. Improving growth and productivity of guava trees. *Minia J. Agric. Res. Dev.* **2007**, *27*, 51–70.

13. Hussien, A.; Ahmed, Y.; Al-Essawy, A.; Youssef, K. Evaluation of different salt-amended electrolysed water to control postharvest moulds of citrus. *Trop. Plant Pathol.* **2018**, *43*, 10–20. [CrossRef]

14. Lachhab, N.; Sanzani, S.M.; Fallanaj, F.; Youssef, K.; Nigro, F.; Boselli, M.; Ippolito, A. Protein hydrolysates as resistance inducers for controlling green mould of citrus fruit. *Acta Hort.* **2015**, *1065*, 1593–1598. [CrossRef]

15. Youssef, K.; Mustafa, Z.M.M.; Mounir, G.A.; Abo Rehab, M.E.A. Preliminary Studies on Fungal Species Associated with Guava Fruit Drop Disease and Possible Management. *Egypt. J. Phytopathol.* **2015**, *43*, 11–23.

16. Deverall, B.J. Biochemical changes in infection droplets containing spores of *Botrytis* spp. incubated in the seed cavities of pods of bean (*Vicia faba* L.). *Ann. Appl. Biol.* **1967**, *59*, 375–387. [CrossRef]

17. Youssef, K.; Hashim, A.F.; Margarita, R.; Alghuthaymi, M.A.; Abd-Elsalam, K.A. Fungicidal efficacy of chemically-produced copper nanoparticles against *Penicillium digitatum* and *Fusarium solani* on citrus fruit. *Philipp. Agric. Sci.* **2017**, *100*, 69–78.

18. Hashim, A.F.; Youssef, K.; Abd-Elsalam, K.A. Ecofriendly nanomaterials for controlling gray mold of table grapes and maintaining postharvest quality. *Eur. J. Plant Pathol.* **2019**, *154*, 377–388. [CrossRef]

19. Youssef, K.; de Oliveira, A.G.; Tischer, C.A.; Hussain, I.; Roberto, S.R. Synergistic effect of a novel chitosan/silica nanocomposites-based formulation against gray mold of table grapes and its possible mode of action. *Int. J. Biol. Macrom.* **2019**, *141*, 247–258. [CrossRef]

20. Arseneault, M.H.; Cline, J.A. A review of apple preharvest fruit drop and practices for horticultural management. *Sci. Hortic.* **2016**, *211*, 40–52. [CrossRef]

21. Gao, L.L.; Zhang, Q.; Sun, X.Y.; Jiang, L.; Zhang, R.; Sun, G.Y.; Zha, Y.L.; Biggs, A.R. Etiology of moldy core, core browning, and core rot of Fuji apple in China. *Plant Dis.* **2013**, *97*, 510–516. [CrossRef]

22. Combrink, J.C.; Kotzl, J.M.; Wehner, F.C.; Grobbelaar, C.J. Fungi associated with core rot of Starking apples in South Africa. *Phytophylactica* **1985**, *17*, 81–83.

23. Combrink, J.C.; Visagie, T.R.; Grobbelaar, C. Variation in the incidence and occurrence in different production areas of core rot in Starking apples. *Deciduous Fruit Grow.* **1984**, *34*, 88–89.

24. Racskó, J.; Leite, G.B.; Petri, J.L.; Zhongfu, S.; Wang, Y.; Szabó, Z.; Soltész, M.; Nyéki, J. Fruit drop: The role of inner agents and environmental factors in the drop of flowers and fruits. *Int. J. Hort. Sci.* **2007**, *13*, 13–23. [CrossRef]

25. Holb, I.J. The brown rot fungi of fruit crops (*Monilinia* spp.) I. Important features of their biology (Review). *Int. J. Hortic. Sci.* **2003**, *9*, 23–36. [CrossRef]

26. Holb, I.J. The brown rot fungi of fruit crops (*Monilinia* spp.) II. Important features of their epidemiology (Review). *Int. J. Hortic. Sci.* **2004**, *10*, 17–33. [CrossRef]

27. Holb, I.J. The brown rot fungi of fruit crops (*Monilinia* spp.) III. Important features of their disease management (Review). *Int. J. Hortic. Sci.* **2004**, *10*, 31–48. [CrossRef]
28. Belisario, A.; Santori, A.; Potente, G.; Fiorin, A.; Saphy, B.; Reigne, J.L.; Pezzini, C.; Bortolin, E.; Valier, A. Brown Apical Necrosis (BAN): A Fungal Disease Causing Fruit Drop of English Walnut. *Acta Hort.* **2010**, *861*, 449–452. [CrossRef]
29. Bravin, E.; Kilchenmann, A.; Leumann, M. Six hypotheses for profitable apple production based on the economic work-package within the ISAFRUIT Project. *J. Hortic. Sci. Biotechnol.* **2009**, *84*, 164–167. [CrossRef]
30. Alghuthaymi, M.A.; Ali, A.A.; Hashim, A.F.; Abd-Elsalam, A. A rapid method for the detection of *Ralstonia solanacerum* by isolation DNA from infested potato tubers based on magnetic nanotools. *Philipp. Agric. Sci.* **2016**, *99*, 113–118.
31. Khandaker, M.M.; Idris, N.S.; Ismail, S.Z.; Majrashi, A.; Alebedi, A.; Mat, N. Causes and Prevention of Fruit Drop of *Syzygium samarangense* (Wax Apple): A Review. *Adv. Environ. Biol.* **2016**, *10*, 112–123.
32. Crane, J.H.; Balerdi, C.F. Guava Growing in the Florida Home Landscape. Available online: https://edis.ifas.ufl.edu/pdffiles/MG/MG04500.pdf (accessed on 25 May 2020).
33. Singh, Z.; Malik, A.U.; Davenport, T.L. Fruit drop in mango. *Hort. Rev.* **2005**, *31*, 111–153.
34. Marques, J.P.R.; Amorim, L.; Spósito, M.B.; Appezzato-da-Glória, B. Histopathology of postbloom fruit drop caused by *Colletotrichum acutatum* in citrus flowers. *Eur. J. Plant Pathol.* **2013**, *135*, 783–790. [CrossRef]
35. Taha, F.S.; Wagdy, S.M.; Hassanein, M.M.M.; Hamed, S.F. Evaluation of the biological activity of sunflower hull extracts. *Grasas y Aceites* **2012**, *63*, 184–193.

 *horticulturae*

Article

Mechanical Crop Load Management (CLM) Improves Fruit Quality and Reduces Fruit Drop and Alternate Bearing in European Plum (*Prunus domestica* L.)

Sebastian Lammerich [1], Achim Kunz [1], Lutz Damerow [2] and Michael Blanke [1,*]

[1]   INRES-Horticultural Science, University of Bonn, D-53121 Bonn, Germany; slammerich@uni-bonn.de (S.L.); akunz@uni-bonn.de (A.K.)

[2]   Agricultural Engineering Department, University of Bonn, D-53121 Bonn, Germany; damerow@uni-bonn.de

*   Correspondence: mmblanke@uni-bonn.de; Tel.: +49-228735142

Received: 17 July 2020; Accepted: 26 August 2020; Published: 2 September 2020

Abstract: (1) Background: With ca. 10 million tons of annual production worldwide, the plum (*Prunus ssp.*) ranks as a major fruit crop and can suffer from small fruit size, premature fruit drop and alternate bearing, which are addressed in this paper using a range of crop load management (CLM) tools. (2) Methods: Sixty 10-year-old European plum cv. "Ortenauer" trees on dwarfing St. Julien INRA GF 655/2 rootstock (slender spindle; 4.25 × 2.80 m) in a commercial orchard near Bonn (50°N), Germany, were thinned in 2 years and flower intensity assessed in the following year. Thinning was performed either mechanically (type Bonn/Baum) or chemically, with ATS (ammonium thiosulfate) or ethephon (Flordimex), or by a combination of mechanical and chemical methods, to improve fruit quality and the proportion of Class 1 fruit. Adjacent un-thinned trees served as controls. (3) Results: Natural fruit drop in June was reduced from 290 fruits per tree in the un-thinned controls to 265 fruits after ATS blossom treatment, and to 148 fruits after mechanical thinning at 380 rpm at a 5 km/h tractor speed at full bloom. The un-thinned control trees developed a large number of small, undersized fruits. The yield of Class 1 fruits increased per tree from 47% in the un-thinned controls, up to 69% after crop load management. Sugar content and fruit firmness were unaffected. (4) Conclusions: The study has shown that fruit quality (i.e., fruit size) and financial returns could be improved by either mechanical (380 rpm at 5 km/h) or chemical thinning, or a combination of both.

Keywords: European plum (*Prunus domestica* L.); alternate bearing; crop load management (CLM); fruit drop; fruit quality; mechanical thinning; reducing chemical input; sustainability

---

1. Introduction

With 10 million tons of annual production, plums rank among the major horticultural fruit crops worldwide and in Europe, with cultivation from the Mediterranean island of Skopelos (the "plum island"; 39°N) to that of the cv. "Opal" plum in Norway (60°N). The plum is classified as a health-promoting fruit, but the tree is susceptible to alternate bearing, i.e., changes between years, with low ("OFF") or high ("ON") yields.

Crop load management (CLM) and the regulation of fruit set is considerably more difficult in plums compared with that in pome-fruit trees [1,2]. Few chemical thinning agents are effective and registered for stone fruit as a result. While fruits are scarce or absent in "OFF" years, trees in "ON" years are overloaded (Figure 1a), leading to branch breaking (Figure 1b) and an excess of small unmarketable fruits, which are often not picked, if the picking costs exceed the fruit market value. Markets, bakeries and consumers require European plums of large size, and this can be achieved by hand, chemical or mechanical thinning [3]. Plums smaller than 28 mm cannot be marketed as fresh fruit. Fruits larger than 28 mm qualify as Class 2 fruit, and those larger than 30 mm, as Class 1 (Commission Implementing

Regulation (EU) No 543/2011), with the potential for higher financial return. Alternate bearing in European plums can be overcome with blossom thinning [4,5]; however, crop load management (CLM) becomes ineffective later with progressive fruitlet development.

Figure 1. (**a**) Overloaded plum tree (top left), (**b**) which can lead to branch breakage (top right); (**c**) mechanical thinning in plum cv. "Ortenauer", using the Bonner thinning device ('BAUM') with three rotors and adapted to the spindle tree (bottom left); and (**d**) individual flowers mechanically removed from the branch (red circles, bottom right) in a commercial orchard near Bonn (Photos © M. Blanke, Bonn).

Due to the scarcity, or lack of success of, research on plum thinning, indicated by there being fewer than 10 publications worldwide, the objective of the present study was to apply a range of different CLM mechanisms to improve the fruit quality of the European plum. Field experiments were carried out in a commercial orchard over two years using the latest technology (Figure 1) for mechanical thinning, which was a result of interdisciplinary cooperation between horticulture and engineering, optimized to the needs of previous work [6].

We used this novel technology on its own or in combination with chemical thinning to meet the requirements of each production system—integrated production (IP) and organic—to reduce chemical input into agricultural systems.

2. Materials and Methods

2.1. Plum Trees

Sixty ten-year-old plum cv. "Ortenauer" trees, on dwarfing St. Julien INRA GF 655/2 rootstock, were grown in a commercial orchard near Bonn (50°N), Germany. Trees trained to slender spindles with a spacing of 4.25 × 2.80 m in two rows were employed for the trial in the years 2012 and 2013.

2.2. Crop Load Management (CLM)

The plum trees had flowered strongly, with score values of 8–9 on a scale of 0 (no flowers) to 10 (white blossom) after mechanical thinning, in the previous year [6]. The plum trees were either mechanically thinned with the Bonner thinning device (Müller Co., Eltville, Germany) with a rotor speed of 380 rpm at a tractor speed of 5 km/h (Figure 1c) at the beginning of full bloom (BBCH 63, F1) in 2012 and on 30 April 2013, or at 320 rpm combined with chemical thinner. The ethylene-releasing compound ethephon (Flordimex at 0.375 L/ha) was applied four weeks after full bloom.

At full bloom, 10 L/ha ATS was applied on 26 April 2012 and 30 April 2013 (Table 1); ATS (ammonium thiosulfate) is a foliar nitrogen fertilizer that can be used at high concentration to interrupt pollen tube growth [7]; adjacent un-thinned plum trees served as controls.

Table 1. Mechanical (green) and chemical (blue) thinning treatments on European plums in 2012 and 2013.

Treatment	Magnitude	Date	Flowering Stage
Mechanical treatments (rotor speed)			
Mechanical	380 rpm	19 April 2012	Full bloom
Mechanical (and later chemical) combination	320 rpm	30 March 2013	
Chemical treatments (concentration)			
ATS (Ammonium thiosulfate)	10 kg/ha	19 April 2012 30 March 2013	Full bloom
Flordimex (ethephon) alone	0.375 L/ha	May 2012 May 2013	4 weeks after full bloom
Flordimex (ethephon) (after mechanical thinning)			
Control (un-thinned)			

2.3. Fruit Set and Quality Assessment

Fruit set was determined on 22 May 2012, on 6 June 2013, and after (June) fruit drop on 18 July 2013. Flower clusters were counted on whole trees three days before blossom thinning on each on the trees in the experiment to calculate fruit set. Fruit set was determined after counting the fruitlets on the three largest branches of the trees in the experiment and related to the flower clusters on the respective branch. Fruit quality was assessed using 270 fruits per treatment for firmness measurement (penetrometer), sugar (refractometer type Atago 32) and acidity (titration) on 18 September 2013 as previously described [6].

2.4. Financial Returns

Financial returns from the thinning treatments were calculated as the price in Euro cents per kg of fruit, based on the relative proportions of fruit sizes obtained from the wholesale market according to fruit size: <30 mm = 4 cents; 30–32 mm = 30 cents/kg; 32–35 mm = 40 cents/kg; >35 mm = 60 cents/kg.

2.5. Statistics

The experiment comprised 60 trees in 2012 and 26 trees in 2013; three branches were employed for the flower-versus-fruit counts per tree, with nine trees per treatment. The data were subjected to statistical analysis using multifactorial analysis of variance in SPSS version 21.

3. Results

3.1. Fruit Set

Crop load regulation by mechanical blossom thinning, as an innovative technology, reduced fruit set—Viz., the number of fruit—from 75 plums in 2012 and from 90 plums per 100 flowers in 2013 to 40 fruit in 2012 and to 52 fruit in 2013 after mechanical thinning at 380 rpm (Figure 2). The chemical thinning at full bloom with ATS showed no statistically significant effect on fruit set at this early stage of fruit development.

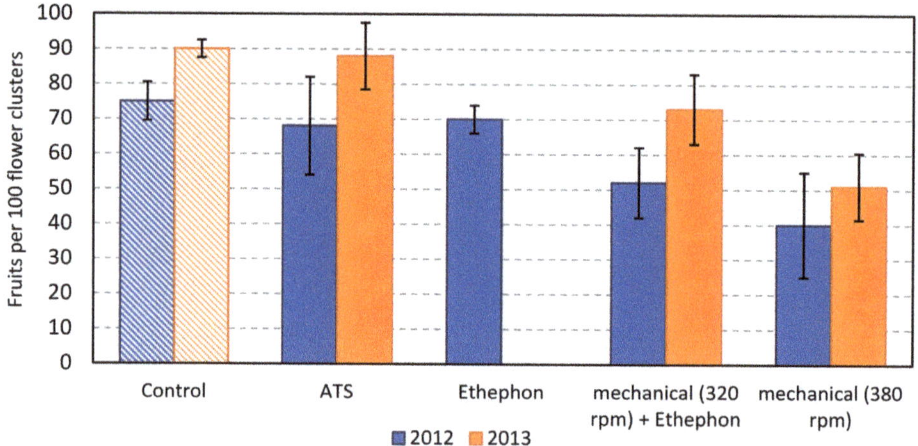

Figure 2. Effect of mechanical crop load management (CLM), chemical CLM and their combination on fruit set in European plum cv. "Ortenauer" in Grafschaft (50°N) near Bonn, Germany, in 2012 and 2013, expressed as fruits per 100 flower clusters. Error bars represent SDs at the 5% level.

3.2. Fruit Drop

Since all fruit trees undergo one or several natural fruit drop phases [3], one of the objectives of the present work was to investigate the effect and interaction of crop load management on and with the natural fruit drop of treated European plum trees. Crop load management with mechanical thinning only, at a rotor speed of 380 rpm, followed by the combination of 320 rpm at full bloom and ethephon (Flordimex) four weeks later showed the greatest effect in terms of reducing natural fruit drop in European plums (Figure 3). In 2013, the un-thinned plum trees shed 290 fruits per tree, with the largest change of 265 fruits with ATS, and lesser changes of 180 fruits and 150 fruits per tree after mechanical thinning (Figure 3), inversely reflecting the relative magnitude of the thinning motion. The greater the number of fruits removed by the thinning motion, the smaller the number of dropped fruitlets, indicating the good control and promising management of fruit set by crop load management.

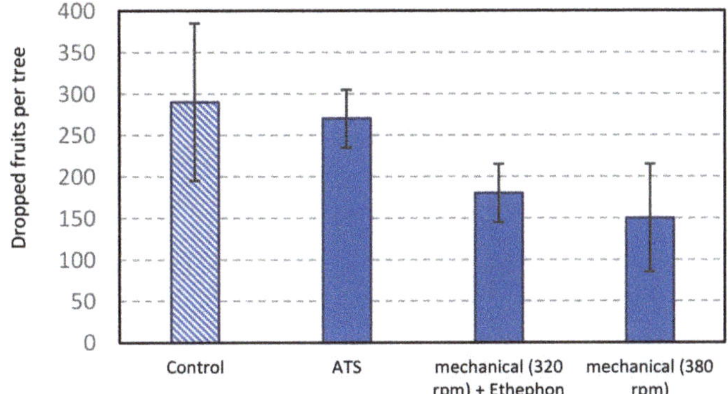

Figure 3. Effect of crop load management (CLM) on fruit drop of the European plum cv. "Ortenauer" in Bonn, Germany, in 2013; error bars represent SDs at the 5% error level.

3.3. Fruit Yield

The success of crop load management is a combination of fruit size, yield and farm gate price; the yields are shown in Figure 4 as the numbers of remaining fruits.

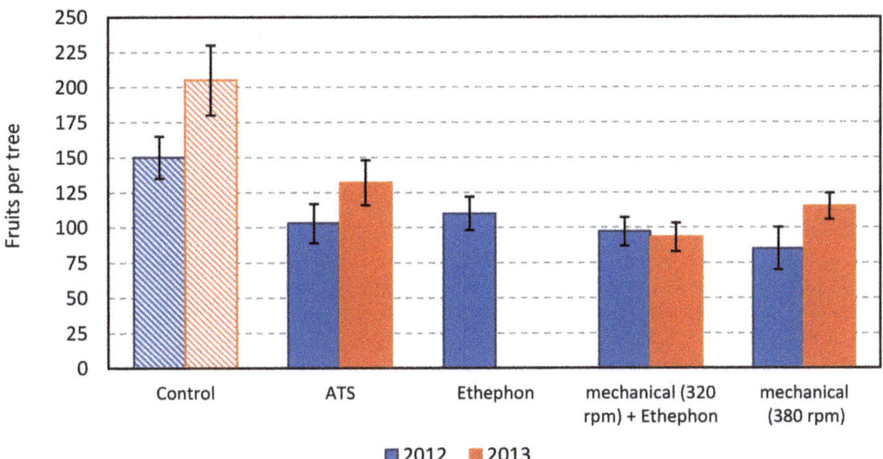

Figure 4. Effect of crop load management (CLM) on yield of the European plum cv. "Ortenauer" in Bonn, Germany, in 2012 and 2013; number of fruit per tree; error bars represent SDs at the 5% error level.

3.4. Fruit Quality

Crop load management (CLM) dramatically reduced the number of small undersized fruits at harvest: from 53% in the un-thinned control to 41% after mechanically thinning at 320 rpm, and from 34% at 380 rpm to 31% with ATS (Figure 5). The sugar content and fruit firmness were not affected by any of the treatments (results not shown).

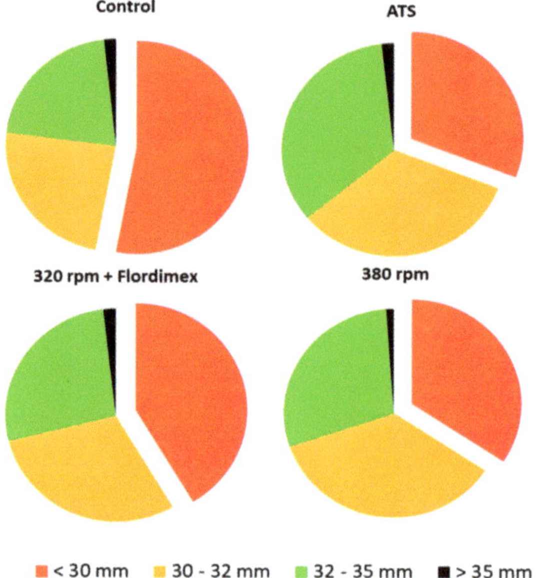

Figure 5. Relative proportions of fruit sizes and quality, as affected by mechanical and chemical thinning, in the European plum cv. "Ortenauer" in Grafschaft in 2013; unmarketable fresh fruit, <30 mm, in red; small, 30–32 mm, in orange; medium-sized fruit, 32–35 mm, in green; and large fruit, >35 mm, in black.

4. Discussion

4.1. Fruit Quality

The sugar content and fruit firmness were not affected by any of the treatments (results not shown) in contrast to previous findings [3,8–11]. One of the latter studies reported a 0.5% increase in total soluble solids TSS [12], while Weber [10] reported a tendency toward softer fruits after ethylene (Flordimex) application but—in line with Seehuber [13] and Vandal (1982) [5]—also no difference in sugar content regardless of the thinning treatment. This may be due to blossom thinning with effects on phase I of fruit growth and larger fruit rather than larger sugar/TSS content.

4.2. Economics

All the CLM treatments reduced the proportion of undersized, small fruits (Figure 5). The combination of mechanical and chemical thinning (ethephon 30 days after flowering) improved financial returns (to 23 cents per kg of fruit) relative to the un-thinned control (19 Euro cents per kg of fruit), based on the proportions of fruit sizes. Chemical thinning with ATS at flowering improved the return to 26 Euro cents per kg of fruit; mechanical thinning at 380 rpm at flowering improved the return to 24 Euro cents per kg (Figure 6). Manual thinning may improve fruit size, but the required range of 30–90 h/ha may increase labour costs, thereby decreasing the financial returns [11]. With the mechanical thinning (ca. EUR 110/ha), a portion of the labour costs in the order of EUR 300–900/ha/year can be saved, assuming a labour cost of EUR 10/h as in Western Europe.

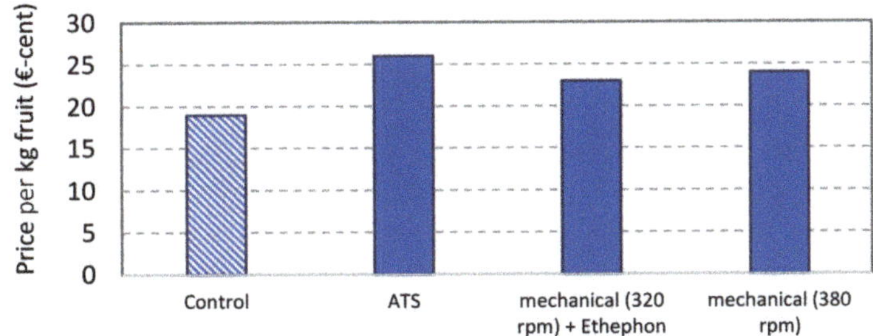

Figure 6. Financial returns from the thinning treatments, expressed as average prices in Euro cent per kg of fruit and based on the relative proportions of fruit sizes shown in Figure 5, and the following wholesale prices: <30 mm = 4 cents/kg; 30–32 mm = 30 cents/kg; 32–35 mm = 40 cents/kg; >35 mm = 60 cents/kg.

4.3. The Sustainability Aspect of Chemical Thinning

Webster and Andrews (1986) [13] used paclobutrazol (PP333, "Cultar" at 1000 and 2000 ppm) as a one-off application for fruitlet thinning 28 d after full bloom (AFB; "shuck-off phase"), and the results depended on the cultivar and weather (Table 2), but the compounds failed at full bloom.

Table 2. Features of mechanical thinning.

(1) Reduction of chemical input	(4) Improves fruit size and financial return	(7) Prevents alternate bearing
(2) Results immediately visible	(5) Prevents over-cropping and branch breakage	(8) Requires spindle trees with flexible horizontal branches
(3) Weather independent	(6) Reduction of labour cost for hand thinning	(9) Similar settings can be used for pome and stone fruit

The use of the compound PP333 has now been discontinued in horticultural food production due to residue issues in the soil, tree and fruit, similar to those in the one-off trial with hydrogen cyanamide (Fallahi et al., 1992) [9].

The gibberellin GA$_3$ was used successfully [2,8] to improve fruit size via direct hormonal effects in two consecutive years but not as a thinning agent. In table grapes, for which GA$_3$ is registered and enlarges berry size, partly through pedicel elongation, overdoses may lead to complete yield failure in the subsequent year (Gordon Hoad, 2000, personal communication).

5. Conclusions

The results of this three-year field experiment showed that the three crop load managements examined for sustainable plum production can have positive effects on the fruit load and quality of European plums without affecting sugar (TSS) and firmness, improve economic returns to the grower, mitigate alternate bearing in relevant cultivars, prevent the breakage of overloaded branches, and provide healthy fruit every year. The best results were obtained with chemical thinning with a large dose of the foliar fertiliser ATS, or solely mechanical thinning at 380 rpm (Table 2). In the absence of a registration for ATS for thinning, this latter result provides an option for where chemicals are not approved or desired, e.g., if unwanted by consumers or not permitted outside of organic cultivation. Overall, the shortage of chemicals suitable for crop load management in plums necessitates more research.

Author Contributions: Conceptualization and methodology, L.D. and M.B.; investigation, S.L. and A.K.; resources, A.K.; data analysis, S.L.; writing—Original draft preparation, S.L.; writing—Review and editing, L.D. and M.B.; visualization, S.L.; supervision, L.D. and M.B. All authors have read and agreed to the published version of the manuscript.

Funding: This research received no external funding.

Acknowledgments: We are grateful to Otto Rönn for the repeated access to his private plum orchard in Grafschaft near Bonn and continuous support throughout the experiment by helping to mark the trees and branches, counting fruit on and off the trees, weighing tree yields, taking samples for fruit quality analysis and providing his farm-gate prices for the different fruit categories. We also thank Ligia Hent for editorial guidance, MDPI for the invitation for this contribution to the special features edition of MDPI Horticulturae and waiving the APC.

Conflicts of Interest: The authors declare no conflict of interest.

References

1. Byers, R.E.; Costa, G.; Vizzotto, G. Flower and fruit thinning of peach and other Prunus species. *Hort. Rev.* **2003**, *28*, 351.

2. Meland, M.; Birken, E.M.; Kaiser, C. Managing alternate bearing of "Opal" plum (*Prunus domestica* L.) trees with GA$_3$ applications by increasing fruit size, and normalizing return bloom and yield in a Nordic climate. In *1st EUFRIN Thinning Working Group Symposia*; Blanke, M.M., Costa, G., Eds.; Acta Horticulturae: Leuven, Belgium, 2013; Volume 998, pp. 61–66.

3. Costa, G.; Blanke, M.M.; Widmer, A. Principles of thinning in fruit tree crops—Needs and novelties. In *1st EUFRIN Thinning Working Group Symposia*; Blanke, M.M., Costa, G., Eds.; Acta Horticulturae: Leuven, Belgium, 2012; Volume 998, pp. 17–26.

4. Maage, F. Fruit development of 'Victoria' plums in relation to leaf number. *Acta Hortic.* **1994**, *359*, 190–194. [CrossRef]

5. Vangdal, E. Sugars and sugar alcohols in Norwegian grown plums. *Meldinger fra Norges Lankbrukshoegskole* **1982**, *61*, 1–39.

6. Seehuber, C.; Kunz, A.; Damerow, L.; Blanke, M.M. Regulation of source: Sink relationship, fruit set, fruit growth and fruit quality in European plum (*Prunus domestica* L.)—Using thinning for crop load management. *Plant Growth Regul.* **2011**, *65*, 335–341. [CrossRef]

7. Maas, F. Control of fruit set in apple requires careful timing of ATS application. In *2nd EUFRIN Thinning Working Group Symposia*; Acta Horticulturae: Leuven, Belgium, 2016; Volume 1138, pp. 45–52.

8. Boyham, G.E.; Norton, J.B.; Abraham, B.R.; Pitts, J.A. GA$_3$ and thinning affect fruit quality and yield of 'Au-Rubrum' plum. *HortScience* **1992**, *27*, 1045. [CrossRef]

9. Fallahi, E.; Simons, B.R.; Fellman, J.K.; Colt, W.M. The use of hydrogen cyanamide in apple and plum thinning. *Plant Growth Regul.* **1992**, *11*, 435–439. [CrossRef]

10. Weber, H.J. Chemical and mechanical thinning of plums. In *1st EUFRIN Thinning Working Group Symposia*; Blanke, M.M., Costa, G., Eds.; Acta Horticulturae: Leuven, Belgium, 2013; Volume 998, pp. 51–60.

11. Walsh, K.B.; Long, R.L.; Middleton, S.G. Use of near infra-red spectroscopy in evaluation of source-sink manipulation to increase the soluble sugar content of stonefruit. *J. Hort. Sci. Biotechn.* **2007**, *82*, 316–322. [CrossRef]

12. Seehuber, C.; Damerow, L.; Blanke, M.M. Effect of mechanical thinning on source: Sink relationship, fruit quality and yield in plum [Einfluss von mechanischer Fruchtbehangsregulierung auf die Source: Sink-Verhältnisse, Fruchtqualität und Ertrag von Pflaumen.]. *Erwerbs-Obstbau* **2012**, *54*, 1–9. [CrossRef]

13. Webster, A.; Andrews, L. Flower and fruit thinning of 'Victoria' plums (*Prunus domestica* L.) with paclobutrazol. In *V International Symposium on Growth Regulators in Fruit Production*; Acta Horticulturae: Leuven, Belgium, 1986; Volume 179, pp. 703–704.

 *horticulturae*

Article

Sulphur Dioxide Pads Can Reduce Gray Mold While Maintaining the Quality of Clamshell-Packaged 'BRS Nubia' Seeded Table Grapes Grown under Protected Cultivation

Khamis Youssef [1,*], **Osmar Jose Chaves Junior** [2], **Débora Thaís Mühlbeier** [2] and **Sergio Ruffo Roberto** [2,*]

[1] Agricultural Research Center, Plant Pathology Research Institute, 9 Gamaa St., Giza 12619, Egypt

[2] Agricultural Research Center, Department of Agronomy, Londrina State University, Londrina 86057-970, PR, Brazil; osmarjcj@gmail.com (O.J.C.J.); muhlbeierdebora@gmail.com (D.T.M.)

* Correspondence: youssefeladawy@yahoo.com (K.Y.); sroberto@uel.br (S.R.R.)

Received: 3 March 2020; Accepted: 30 March 2020; Published: 1 April 2020

Abstract: The purpose of this research is to test the efficacy of different types of SO_2-generating pads on the incidence of gray mold, and on the physicochemical properties of quality of 'BRS Nubia' seeded table grapes grown under protected cultivation. Four types of SO_2-generating pads, 5 or 8 g of sodium metabisulfite dual release pads, and 4 or 7 g of sodium metabisulfite slow release pads, were used. Grapes bunches were harvested from a vineyard covered with plastic mash and stored in a cold room at $1 \pm 1\ °C$ for 45 days followed by 6 days of shelf life at $22 \pm 1\ °C$ at a high relative humidity (>95%). The results showed that SO_2-generating pads with a dual release of 5 or 8 g completely inhibited the development of gray mold at all evaluation times. Also, a high reduction of the disease incidence was achieved by using a slow release of 4 g. The study confirmed that SO_2-generating pads did not alter the physicochemical properties of 'BRS Nubia' seeded table grapes including mass loss, berry firmness, color index, total anthocyanin concentration, total soluble solids (TSS), titratable acidity (TA), and the TSS/TA ratio. Slow release pads at 4 and 7 g reduced the percentage of shattered berries by 56 and 48% as compared to control only after 6 days of shelf life. Also, all types of SO_2-generating pads reduced the stem browning score at the end of cold storage. The 5 or 8 g dual release pads and 4 g slow release pads can be considered for effective controlling of gray mold for 'BRS Nubia' table grapes grown under protected cultivation while maintaining grape quality.

Keywords: grapes; fruit quality; SO_2; *Botrytis cinerea*; rots

1. Introduction

The worldwide harvested area of grapes is estimated to be around 7,157,658 ha with a production of 79,125,982 metric tons. The Brazilian grape industry accounts for about 74,472 ha of harvested area, producing almost 1,591,986 metric tons [1]. 'BRS Nubia' (*Vitis* sp.) is a new black hybrid seeded table grape grown in the tropical and subtropical areas that requires less labor to cultivate and has a high yield, large berries, and uniform color [2]. This cultivar was obtained by crossing 'Michele Palieri' and Arkansas 2095 grapevines and has the capability of drawing the attention of customers from internal and external markets, since there is a large demand for table grapes for extended periods throughout the year globally [2,3]. 'BRS Nubia' is commonly trained on an overhead trellis system protected by a black screen with 18% shading.

During cold storage and shelf life periods, table grapes face severe postharvest problems including gray mold [4,5]. This disease is caused by *Botrytis cinerea* and causes huge losses on grapes worldwide, even when grapes are packaged in clamshells [6–8]. The control of gray mold is not easy, since postharvest applications with chemical fungicides are not allowed in many countries. As a standard preharvest treatment, fungicides such as iprodione, boscalid, chlorothalonil, captan, mancozeb, procymidone, pyrimethanil, and thiophanate-methyl are commonly used to prevent gray mold of table grapes [6,9,10]. As a postharvest treatment, the use of SO_2-generating pads during cold storage is commercially implemented to manage gray mold of table grapes [11]. Dual release SO_2 generators inside the boxes have been commonly used for grape storage and transportation for periods up to two months [7,12,13]. Grape exporters are always worried about limited long-distance transport of fresh grapes, especially when intensive production occurs in subtropical areas where two crops per year is possible. In this case, grapes can be harvested during situations highly favorable for the development of gray mold disease. The main importers of grapes, such as the EU and USA, established tolerance rates for the use of SO_2 in postharvest control.

The effectiveness of common packaging methods such as plastic bags or carton boxes on the quality of grapes has been extensively evaluated [14,15]. Nowadays, the use of clamshells has become a consequential approach for packaging of table grapes for domestic and overseas markets in a professional way to offer fresh grapes of better quality to consumers. As 'BRS Nubia' is a new hybrid table grape, there is no available data regarding the behavior of this cultivar packaged in vented clamshells during cold storage and shelf life periods in terms of gray mold incidence and grape quality, especially when vines are grown under protected cultivation. Under this situation, pathogens may find different conditions than in open-field production systems, and different postharvest techniques may be required. In this perspective, the aim of this work was to investigate the effect of different types of SO_2-generating pads on the incidence of gray mold and on physicochemical properties of 'BRS Nubia' seeded grapes grown under protected cultivation and packaged in clamshells.

2. Materials and Methods

2.1. Cultivar and Materials Used

'BRS Nubia' (*Vitis* sp.) seeded table grapes were harvested from a commercial vineyard located in Marialva, State of Parana, Brazil. The vines were 7 years old and were trained on an overhead trellis system under protected cultivation using a plastic black mesh with 18% shading. This area was chosen since it has a historical incidence of gray mold disease. At harvest, the total soluble solids (TSS) content of the grapes was 16.2 °Brix. Four types of SO_2-generating pads (Uvas Quality Grape Guard, Suragra S.A., San Bernardo, Chile) were used: (i) Slow-release pad having 4 g of sodium metabisulfite (SM) was prepared with coextruded polymer film; (ii) slow-release pad containing 7 g of SM was prepared with two polymer films including SO_2 in a solvent-free wax matrix; (iii) dual-release pad having 5 g of SM was prepared with extruded polymer film and 100% virgin paper pulp (VPP) obtained by a mechanical procedure, with amount of fast and slow phases of 1 and 4 g of SM, respectively; and (v) dual-release pad containing 8 g of SM prepared from coextruded paper with polyethylene and 100% (VPP), with fast and slow release phases of 1 and 7 g of SM, respectively. Macroperforated liners (MLs) with 0.3% of ventilation area (Suragra S.A., San Bernardo, Chile), and vent holes (70 × 90 mm), were used for all treatments including the control. Those macroperforated liners were prepared with high-density plastic and master batch (95 × 65 cm) and 12 μm thickness.

2.2. Treatments

Grape bunches free from any disorders or visual symptoms were chosen and arranged according to bunch size and shape. Grapes were accommodated in 20 × 10 cm vented clamshells (10 holes distribution) with around 0.5 kg capacity. Clamshells were wrapped in MLs (0.3% ventilation area). Then ten clamshells were located inside corrugated carton boxes (100 × 60 × 40 cm each). On the base, a unilaminar sheet of moisture-absorbing paper (33 × 46 cm dimension and 50 g m^{-2} density) was located. For each carton box, a SO_2-generating pad was placed over the clamshells. Five treatments were arranged according to the SO_2-generating pad used as follow: (i) Slow release 4 g SM; (ii) slow release 7 g SM; (iii) dual release 5 g SM; (iv) dual release 8 g SM; and (v) control treatment. All bunches were treated before harvest with iprodione at 0.2% three times during the season as commonly used by grape growers in the region. A completely randomized design including five treatments and four replicates of each was used. Each replicate consisted of 10 clamshells totaling 40 clamshells per each treatment.

2.3. Storage and Assessments

All carton boxes were stored in a cold room at 1 ± 1 °C for 45 days followed by 6 days of shelf-life at 22 ± 1 °C and high RH (>95%). MLs, absorbent paper sheets, and SO_2-generating pads were removed from the boxes at the end of cold storage. Gray mold incidence (%) was evaluated at 30 days of cold storage (DCS), 45 DCS, 3 days of shelf life (DSL), and 6 DSL according to the following formula: Disease incidence (%) = (number of affected berries/total of berries) × 100 [4]. At the end of cold storage (45 DCS) and at the end of shelf life (6 DSL), mass loss (%), shattered berries (%), stem browning, berry firmness (N), TSS, titratable acidity (TA), and TSS/TA ratio and color index (CIRG) were evaluated according to Ahmed et al. (2019) [15] and Chaves Junior et al. (2019) [8]. Also, total anthocyanin concentration was evaluated only at the end of shelf life (6 DSL) following the methods of Shahab et al., (2019) [16].

Mass loss (%) was calculated by weighing the grape bunches at the initial time of storage and at the examined time [17]: Loss of mass (%) = (initial mass-mass at examined date/initial mass) × 100. Shattered berries were evaluated as: Shattered berries (%) = (number of shattered berries inside the clamshells/total number of berries) × 100. Stem browning was measured throughout by a visual scoring system assessment according to Ngcobo et al. (2012) [18]: (1) fresh and green, (2) some light browning, (3) significant browning, and (4) severe browning. Berry firmness was determined using a texture analyzer TA.XT plus (Stable Micro Systems, Surrey, U.K.) with a cylindrical probe (35mm diameter, P35). Berries placed on the stainless steel platform were compressed in their equatorial diameter at 1 mm s^{-1} (probe speed), and firmness was measured as the force (N) needed to deform the berry by 20% at its equatorial diameter, according to Lijavetzky et al., (2012) [19].

To evaluate TSS, TA, and TSS/TA ratio, 40 berries were collected from each treatment and analyzed following the methods of Youssef et al., (2019) [5]. The berry color was investigated according to Carreño et al., (1995) [20] using a colorimeter CR-10 Plus (Konica-Minolta®, Tokyo, Japan) to get the following variables from the equatorial portion of grape berries (n = 2 per berry): L^* (lightness), C^* (chroma), and h^o (hue angle). The color index for red grapes (CIRG) was calculated following the formula CIRG = $(180 - h^o)/(L^* + C^*)$. Forty berries were collected to be analyzed from each treatment.

For total anthocyanin concentration, 3 g of berry skin were used from each plot, gently separated from the flesh using a sterile blade, and washed with distilled and deionized water. Skins were dried and added to 30 mL of acidified methanol (HCl 1% + methanol 99%) and left in the dark for 48 h. Total anthocyanin content was evaluated at 520 nm using a spectrophotometer (Genesys™ 10S UV-VIS®, Thermo Scientific, Waltham, MA, USA), and the results were expressed as mg of total anthocyanins of malvidin-3-glucoside per g of berry skin (mg/g) [16].

2.4. Statistical Analysis

Data were processed statistically for ANOVA (one-way analysis of variance) using Statistica 6.0 software (Stat Soft Inc., Tulsa, Oklahoma, USA). The experiments were repeated twice and the means were compared using Fisher's protected least significant difference (LSD) test at $p \leq 0.05$. Data (%) were arcsine transformed before analyses to normalize variance.

3. Results

3.1. Incidence of Gray Mold (%)

The natural incidence of gray mold was evaluated at 30 days of cold storage, 45 days of cold storage, followed by 3 and 6 days of shelf life (Table 1). The results showed that the dual release pads at 5 or 8 g completely inhibited the development of gray mold at all evaluation intervals. Also, the slow release pads at 4 g completely inhibited the development of gray mold at 30 days of cold storage and 45 days of cold storage and reduced the incidence by 97 and 93% at 3 and 6 days of shelf life, respectively. The results demonstrated that there was no significant difference between slow release at 7 g and the control except at 3 days of shelf life. Generally, the incidence of disease increased overtime and reached the maximum of 9.7% after 6 days of shelf life.

Table 1. Incidence of gray mold of clamshell-packaged 'BRS Nubia' seeded table grapes after 30 and 45 days of storage in cold storage (CS) at 1 ± 1 °C followed by 3 and 6 days at shelf-life (SL) at 22 ± 1 °C after the period of cold storage, packaged with different SO_2-generating pads.

Treatments	Gray Mold Incidence (%)			
	30 Days of CS	**45 Days of CS**	**3 Days of SL**	**6 Days of SL**
Slow release—4 g	0.00 ± 0.00 b [z]	0.00 ± 0.00 b	0.18 ± 0.18 c	0.60 ± 0.39 b
Slow release—7 g	0.83 ± 0.31 ab	1.84 ± 0.72 a	3.15 ± 1.26 b	10.95 ± 1.31 a
Dual release—5 g	0.00 ± 0.00 b	0.00 ± 0.00 b	0.00 ± 0.00 c	0.00 ± 0.00 b
Dual release—8 g	0.00 ± 0.00 b	0.00 ± 0.00 b	0.00 ± 0.00 c	0.00 ± 0.00 b
Control	1.81 ± 1.09 a	2.99 ± 0.68 a	6.72 ± 1.70 a	9.71 ± 2.96 a

[z] Means (± standard error) in columns marked with the same letters are not significantly different by Fisher's protected LSD test at $p \leq 0.05$.

3.2. Mass Loss (%)

The percentage of mass loss was measured at the end of cold storage (45 days) and at the end of shelf life (6 days) (Table 2). Results showed no significant difference among treatments as compared to the control after 45 days of cold storage and at 6 days of shelf life. The mass loss ranged from 3.66–5.30 and 7.44–9.56% at the end of cold storage and end of shelf life, respectively.

Table 2. Mass loss (%) and shattered berries (%) of clamshell-packaged 'BRS Nubia' seeded table grapes after 45 days of storage in cold storage (CS) at 1 ± 1 °C followed by 6 days at shelf life (SL) at 22 ± 1 °C, packaged with different SO_2-generating pads.

Treatments	Mass Loss (%)		Shattered Berries (%)	
	45 Days in CS	**6 Days of SL**	**45 Days in CS**	**6 Days of SL**
Slow release—4 g	3.66 ± 0.41 a [z]	9.56 ± 1.52 a	2.48 ± 0.91 ab	5.16 ± 0.78 b
Slow release—7 g	3.84 ± 0.14 a	9.06 ± 0.58 a	1.69 ± 0.63 b	6.12 ± 0.75 b
Dual release—5 g	4.01 ± 0.12 a	7.70 ± 1.14 a	3.51 ± 0.54 ab	10.51 ± 0.84 a
Dual release—8 g	4.23 ± 1.36 a	7.69 ± 1.19 a	5.80 ± 1.82 a	11.41 ± 1.52 a
Control	5.30 ± 1.46 a	7.44 ± 0.33 a	4.81 ± 0.71 ab	11.87 ± 1.17 a

[z] Means (± standard error) in columns marked with the same letters are not significantly different by Fisher's protected LSD test at $p \leq 0.05$.

3.3. Shattered Berries (%)

The percentage of shattered berries was evaluated at the end of cold storage (45 days) and at the end of shelf life (6 days) (Table 2). The results showed no significant difference among treatments as compared to the controlat the end of cold storage. After 6 days of shelf life, slow release pads at 4 and 7 g reduced the percentage of shattered berries by 56 and 48% as compared to the control. Overall, the percentage of shattered berries ranged from1.69–5.8% and 5.16–11.87% at the end of cold storage and end of shelf life, respectively.

3.4. Stem Browning

Stem browning, in accordance with the level of darkness, was evaluated at the end of cold storage (45 days) and at the end of shelf life (6 days) (Table 3). At the end of cold storage, all types of SO_2-generating pads significantly reduced stem browningas compared with control, and this reduction varied from15.4–23.0%. At the end of shelf life, there was no significant difference between SO_2-generating pads and the control except for dual release at 8 g which reduced stem browning by 27.8% as compared to control treatment due its antioxidant effect.

Table 3. Stem browning scores and berry firmness (N) of clamshell-packaged 'BRS Nubia' seeded table grapes after 45 days of storage in cold storage (CS) at 1 ± 1 °C followed by 6 days at shelf life (SL) at 22 ± 1 °C, packaged with different SO_2-generating pads.

Treatments	Stem Browning [z]		Berry Firmness (N)	
	45 Days in CS	**6 Days of SL**	**45 Days in CS**	**6 Days of SL**
Slow release—4 g	2.20 ± 0.14 b [y]	2.75 ± 0.70 ab	12.01 ± 0.21a	12.27 ± 0.59 a
Slow release—7 g	2.05 ± 0.05 b	2.55 ± 0.21 ab	11.73 ± 0.85a	12.06 ± 0.34 a
Dual release—5 g	2.00 ± 0.00 b	2.85 ± 0.13 a	11.61 ± 0.56a	12.77 ± 0.69 a
Dual release—8 g	2.00 ± 0.00 b	2.20 ± 0.16 b	11.33 ± 0.46a	11.77 ± 0.32 a
Control	2.60 ± 0.14 a	3.05 ± 0.27 a	11.62 ± 1.01a	12.74 ± 0.35 a

[z] Stem browning scores: (1) fresh and green, (2) some light browning, (3) significant browning, and (4) severe browning. [y] Means (± standard error) in columns marked with the same letters are not significantly different by Fisher's protected LSD test at $p \leq 0.05$.

3.5. Berry Firmness (N)

Berry firmness was evaluated at the end of cold storage (45 days) and at the end of shelf life (6 days) (Table 3). In general, no significant differences were noted between treatments with diverse SO_2-generating pads and control. Berry firmness ranged from11.3–12.0 and 11.8–12.8N at the end of cold storage and end of shelf life, respectively.

3.6. TSS, TA, and TSS/TA Ratio

TSS, TA, and TSS/TA ratio was evaluated at the end of cold storage (45 days) and at the end of shelf life (6 days) (Table 4). Generally, none of the SO_2-generating pads altered berry TSS, TA, and their ratio. TSS ranged from15.98–16.95 and 15.65–16.15 °Brix at the end of cold storage and end of shelf life, respectively. TA ranged from 0.86–0.91 and 0.80–0.86% at the end of cold storage and end of shelf life, respectively. TSS/TA ratio ranged from17.75–19.23 and 18.31–20.14 at the end of cold storage and end of shelf life, respectively.

Table 4. Total soluble solids—TSS (°Brix), titratable acidity—TA (tartaric acid %) and their ratio (TSS/TA) of clamshell-packaged 'BRS Nubia' seeded grapes berries at 45 days of cold storage (CS) at 1 ± 1 °C followed by 6 days of shelf life (SL) at 22 ± 1 °C.

Treatments	TSS (°Brix)		TA (% of Tartaric Acid)		TSS/TA	
	45 Days CS	**6 Days SL**	**45 Days CS**	**6 Days SL**	**45 Days CS**	**6 Days SL**
Slow release—4 g	16.30 ± 0.23 a [z]	16.10 ± 0.08 a	0.86 ± 0.00 a	0.83 ± 0.02 a	18.94 ± 0.27 a	19.42 ± 0.51 a
Slow release—7 g	15.98 ± 0.23 a	15.93 ± 0.15 a	0.89 ± 0.05 a	0.82 ± 0.03 a	18.06 ± 0.83 a	19.51 ± 0.76 a
Dual release—5 g	16.58 ± 0.19 a	15.78 ± 0.18 a	0.88 ± 0.01 a	0.86 ± 0.01 a	18.89 ± 0.27 a	18.31 ± 0.42 a
Dual release—8 g	16.20 ± 0.35 a	15.65 ± 0.10 a	0.91 ± 0.02 a	0.86 ± 0.01 a	17.75 ± 0.70 a	18.31 ± 0.23 a
Control	16.95 ± 0.44 a	16.15 ± 0.31 a	0.88 ± 0.02 a	0.80 ± 0.02 a	19.23 ± 0.78 a	20.14 ± 0.83 a

[z] Means (± standard error) in columns marked with the same letters are not significantly different by Fisher's protected LSD test at $p \leq 0.05$.

3.7. Color Index (CIRG)

In all cases, for grape berry color index, no significant difference was recorded among different types of SO_2-generating pads and the control at the end of cold storage and at the end of shelf life (Table 5). CIRG ranged from1.7–2.2 and 2.1–3.0 at the end of cold storage and end of shelf life, respectively.

Table 5. Color index (CIRG) of clamshell-packaged 'BRS Nubia' seeded table grapes after 45 days of storage in cold storage(CS) at $1 \pm 1\ ^{\circ}$C followed by 6 days at shelf life (SL) at $22 \pm 1\ ^{\circ}$C, packaged with different SO_2-generating pads.

Treatments	Color Index (CIRG)	
	45 Days of CS	6 Days of SL
Slow release—4 g	2.04 ± 0.25 a ^{z}	2.07 ± 0.07 b
Slow release—7 g	2.00 ± 0.26 a	2.13 ± 0.03 b
Dual release—5 g	2.16 ± 0.11 a	3.01 ± 0.55 a
Dual release—8 g	1.66 ± 0.37 a	2.43 ± 0.04 ab
Control	2.04 ± 0.04 a	2.35 ± 0.07 ab

^{z} Means ($\pm$ standard error) in columns marked with the same letters are not significantly different by Fisher's protected LSD test at $p \leq 0.05$.

3.8. Anthocyanins Concentration

At the end of shelf life, the total anthocyanin concentration of the skin was not affected by the use of different types of SO_2-generating pads. This concentration was 4.64, 4.14, 4.81, 4.95, and 4.50 mg/g for slow release pad at 4 g, slow release pad at 7 g, dual release pad at 5 g, dual release pad at 8 g, and the control, respectively.

4. Discussion

Generally, grapes are susceptible to severe losses because of some diseases, such as gray mold caused by *B. cinerea*. SO_2-generating pads areconsidered one of the most important control methods of this disease, particularly when table grapes are cold stored for extended periods [9,15]. Nowadays, the trend of 'BRS Nubia' seeded grape growers is to prolong postharvest life for both domestic and international markets.

The results obtained herein showed that SO_2-generating pads with dual release at 5 or 8 g completely inhibited the development of gray mold at all evaluation times. Also, a high reduction of the disease was achieved by using the slow release at 4 g. Those treatments were able to protect the grapes until 45 days of cold storage followed by 6 days of shelf life. The results revealed that 'BRS Nubia' seeded grapes packaged in clamshells were not affected by the SO_2 emitted in slow release pads 7 g except at 3 days of shelf life. The gas release associated with dual release at 5 or 8 g or slow release at 4 g may have worked against the initial disease progress and killed the pathogen more efficiently when those treatments were applied. Those results are in agreement with previous studies against gray mold of stored 'Calmeria', 'Red Globe', and 'BRS Vitoria' table grapes [7,21–23]. This performance of different slow release SO_2-generating pads can be variedby diverse types of polymer films, coating materials, and release forms [8,24] which provide diverse permeability and thus control altitude.

Evaluating the effectiveness of a control method is very important under naturally occurring infections, and in this research the real situation was simulated for the recent grape packing which already exists in domestic and overseas markets using clamshells. It is important to mention that the grape cultivar response to different types of SO_2-generating pads also has a significant role in terms of the control level of the disease. In a previous study, SO_2-releasing pads (7 g) were able to control gray mold of 'Benitaka' table grapes kept under the same commercial situation with high and low inoculum pressure of the pathogen [8]. The results with dual release pads at 5 and 8 g against the disease was expected because of the contact with air moisture, which leads to the release of a high quantity of gas and therefore removal of any pathogen spores around the berries. Thus, the highest concentration of gas could maintain the clamshell-packaged 'BRS Nubia' seeded table grapes completely free of mold growth during cold storage and shelf life periods. The present results are in agreement with other investigations reported previously with different grape cultivars [9,11,24–26]. Actually, low decay incidence was recorded at 30 and 45 days of cold storage even for the control treatment. In the location where the experiments were carried out, chemical fungicides such as iprodione at 0.2% is the major means to manage gray mold before harvest by applying the compound three times during the season. Moreover, when good agricultural practices are applied, low natural incidence is expected [27–29].

Because it is consumed fresh, the appearance and quality of the table grapes are crucial factors for its marketability. The current study confirmed that SO_2-generating pads did not alter the physicochemical properties of 'BRS Nubia' seeded table grapes such as mass loss, berry firmness, color index, anthocyanin content, TSS, TA, and TSS/TA ratio. Also, minor changes were noted for shattered berries and stem browning. In particular, slow release pads at 4 and 7 g reduced the percentage of shattered berries by 56 and 48% as compared to the control only after 6 days of shelf life. In addition, all types of SO_2-generating pads reduced the stem browning score at the end of cold storage. After 30 days of cold storage there was no statistical difference among the treatments in stem browning (data not shown). Additionally, the phase from grape harvest to marketing is very important in terms of the maintenance of fruit quality [15]. Our results revealed that 'BRS Nubia' grapes have a great prospect for internal and overseas markets, as high fruit quality can be obtained after 45 days of cold storage and also after 6 days of shelf life. Also, it is important to mention that a high initial quality of the grapes is essential.

The efficacy of a treatment on fruit quality is frequently disregarded, especially in laboratory experiments or small scale trials since those kinds of experiments are focused mainly on the treatment to control the disease, ignoring the quality of the final products [30–34]. In addition, table grapes grown under protection cultivation may require specific postharvest treatments, since some fungi may find a different condition to develop and cause losses [35]. However, the use of plastic mesh to cover the vineyard did not negatively affect the postharvest conditions of 'BRS Nubia' table grapes, and the use of clamshells did not interfere with SO_2 circulation inside the boxes. Recently, a strong correlation was shown between anthocyanin concentration and color index of 'Benitaka' grape berries [36].

In conclusion, 'BRS Nubia' seeded table grape is a promising cultivar for domestic and international markets. Maintaining quality during cold storage and shelf life periods is an essential requirement for the producers with respect to cultivar competition. As a postharvest treatment, dual release SO_2-generating pads at 5 or 8 g and slow release pads at 4 g can be considered as effective control of gray mold for 'BRS Nubia' table grapes while maintaining bunch quality, particularly for overseas export markets or long distance shipment.

Author Contributions: S.R.R. conceived the research idea. O.J.C.J., K.Y. and D.T.M. helped to collect the data. K.Y. and O.J.C.J. also analyzed the data and wrote the paper. All authors have read and agreed to the published version of the manuscript.

Funding: This research received no external funding.

Acknowledgments: The authors are grateful for the financial support provided by Brazilian Council for Scientific and Technological Development (CNPq) grant #315844/2018-3 to Youssef Khamis. The authors thank Suragra S.A. (Chile) for providing packing materials.

Conflicts of Interest: The authors declare no conflict of interest.

References

1. FAOSTAT. Available online: http://www.fao.org/faostat/en/#data/QC (accessed on 17 March 2020).
2. Ritschel, P.S.; Girardi, C.L.; Zanus, M.C.; Fajardo, T.V.M.; Maia, J.D.G.; Souza, R.T.; Naves, R.L.; Camargo, U.A. Novel Brazilian grape cultivars. *Acta Hortic.* **2015**, *1082*, 157–163. [CrossRef]
3. Silvestre, J.P.; Roberto, S.R.; Colombo, R.C.; Gonçalves, L.S.A.; Koyama, R.; Shahab, M.; Ahmed, S.; de Souza, R.T. Bunch sizing of 'BRS Nubia' table grape by inflorescence management, shoot tipping and berry thinning. *Sci. Hortic.* **2017**, *225*, 764–770. [CrossRef]
4. Hashim, A.F.; Youssef, K.; Abd-Elsalam, K.A. Ecofriendly nanomaterials for controlling gray mold of table grapes and maintaining postharvest quality. *Eur. J. Plant Pathol.* **2019**, *154*, 377–388. [CrossRef]

5. Youssef, K.; de Oliveira, A.G.; Tischer, C.A.; Hussain, I.; Roberto, S.R. Synergistic effect of a novel chitosan/silica nanocomposites-based formulation against gray mold of table grapes and its possible mode of action. *Int. J. Biol. Macromol.* **2019**, *141*, 247–258. [CrossRef]

6. Kishino, A.A.; Roberto, S.R.; Genta, W. Implantação do pomar. In *Viticultura Tropical: O Sistema de Produção de Uvas de Mesas do Paraná*, 2nd ed.; Kishino, A.Y., Carvalho, S.L.C., Roberto, S.R., Eds.; Instituto Agronômico do Paraná: Londrina, Brazil, 2019; pp. 161–200.

7. Domingues, A.R.; Roberto, S.R.; Ahmed, S.; Shahab, M.; Chaves Junior, O.J.; Sumida, C.H.; Souza, R.T. Postharvest techniques to prevent the incidence of botrytis mold of 'BRS Vitoria' seedless grape under cold storage. *Horticulturae* **2018**, *4*, 17. [CrossRef]

8. Chaves, O.J., Jr.; Youssef, K.; Koyama, R.; Ahmed, S.; Dominguez, A.R.; Mühlbeier, D.T.; Roberto, S.R. Control of gray mold on clamshell-packaged 'Benitaka' table grapes using sulphur dioxide pads and perforated liners. *Pathogens* **2019**, *8*, 271.

9. Stammler, G.; Brix, H.D.; Nave, B.; Gold, R.; Schoefl, U. Studies on the biological performance of boscalid and its mode of action. In *Modern Fungicides and Antifungal Compounds V, Friedrichroda*; Dehne, H.W., Deising, H.B., Gisi, U., Kuck, K.H., Russell, P.E., Lyr, H., Eds.; Deutsche PhytomedizinischeGesellschaft: Braunschweig, Germany, 2008; pp. 45–51.

10. Romanazzi, G.; Feliziani, E. Botrytis cinerea (Gray Mold). In *Post-Harvest Decay*; Academic Press: New York, NY, USA, 2014; pp. 131–146.

11. Zutahy, Y.; Lichter, A.; Kaplunov, T.; Lurie, S. Extended storage of 'Red Globe'grapes in modified SO_2 generating pads. *Postharvest Biol. Technol.* **2008**, *50*, 12–17. [CrossRef]

12. Nelson, K.E.; Ahmedullah, M. Packaging decay-control systems for storage and transit of table grapes for export. *Am. J. Enol. Vitic.* **1976**, *27*, 74–79.

13. Ahmed, S.; Roberto, S.R.; Domingues, A.R.; Shahab, M.; Chaves Junior, O.J.; Sumida, C.H.; Souza, R.T. Effects of different sulfur dioxide pads on botrytis mold in 'Italia' table grapes under cold storage. *Horticulturae* **2018**, *4*, 29. [CrossRef]

14. Lichter, A.; Zutahy, Y.; Kaplunov, T.; Lurie, S. Evaluation of table grape storage in boxes with sulfur dioxide releasing pads with either an internal plastic liner or external wrap. *HortTechnol.* **2018**, *18*, 206–214. [CrossRef]

15. Ahmed, S.; Roberto, S.R.; Youssef, K.; Colombo, R.C.; Shahab, M.; Chaves Junior, O.J.; Sumida, C.H.; Souza, R.T. Postharvest preservation of the new hybrid seedless grape, 'BRS Isis', grown under the double-cropping a year system in a subtropical area. *Agronomy* **2019**, *9*, 603. [CrossRef]

16. Shahab, M.; Roberto, S.R.; Ahmed, S.; Colombo, R.C.; Silvestre, J.P.; Koyama, R.; de Souza, R.T. Anthocyanin accumulation and color development of 'Benitaka' table grape subjected to exogenous abscisic acid application at different timings of ripening. *Agronomy* **2019**, *9*, 164. [CrossRef]

17. Youssef, K.; Roberto, S.R. Applications of salt solutions before and after harvest affect the quality and incidence of postharvest gray mold of 'Italia' table grapes. *Postharvest Biol. Technol.* **2014**, *87*, 95–102. [CrossRef]

18. Ngcobo, M.E.K.; Opara, U.L.; Thiart, G.D. Effects of packaging liners on cooling rate and quality attributes of table grape (cv. Regal seedless). *Pack. Tech. Sci.* **2012**, *25*, 73–84. [CrossRef]

19. Lijavetzky, D.; Carbonell-Bejerano, P.; Grimplet, J.; Bravo, G.; Flores, P.; Fenoll, J.; Hellín, P.; Oliveros, J.C.; Martínez-Zapater, J.M. Berry flesh and skin ripening features in *Vitis vinifera* as assessed by transcriptional profiling. *PLoS ONE* **2012**, *7*, e39547. [CrossRef]

20. Carreño, J.; Martinez, A. Proposal of an index for the objective evaluation of the color of red table grapes. *Food Res. Int.* **1995**, *28*, 373–377. [CrossRef]

21. Mustonen, H.M. The efficacy of a range of sulfur dioxide generating pads against *Botrytis cinerea* infection and on out-turn quality of Calmeria table grapes. *Aust. J. Exper. Agric.* **1992**, *32*, 389–393. [CrossRef]

22. Palou, L.; Crisosto, C.H.; Garner, D.; Basinal, L.M.; Smilanick, J.L.; Zoffoli, J.P. Minimum constant sulfur dioxide emission rates to control gray mold of cold stored table grapes. *Am. J. Enol. Vitic.* **2002**, *52*, 110–115.

23. Henríquez, J.L.; Pinochet, S. Impact of ventilation area of the liner bag, in the performance of SO_2 generator pads in boxed table grapes. *Acta Hortic.* **2016**, *1144*, 267–272. [CrossRef]

24. Fernandez-Trujillo, J.P.; Obando-Ulloa, J.M.; Baró, R.; Martinez, J.A. Quality of two table grape cultivars treated with single or dual-phase release SO_2 generators. *J. App. Bot. Food Qual.* **2008**, *82*, 1–8.

25. Zoffoli, J.P.; Latorre, B.A.; Naranjo, P. Hairline, a postharvest cracking disorder in table grapes induced by sulfur dioxide. *Postharvest Biol. Technol.* **2008**, *47*, 90–97. [CrossRef]

26. Sortino, G.; Allegra, A.; Passufiume, R.; Gianguzzi, G.; Gullo, G.; Gallota, A. Postharvest application of sulphur dioxide fumigation to improve quality and storage ability of 'Red Globe' grape cultivar during long cold storage. *Chem. Eng. Trans.* **2017**, *58*, 403–408.

27. Youssef, K.; Sanzani, S.M.; Myrta, A.; Ippolito, A. Effect of a novel potassium bicarbonate-based formulation against Penicillium decay of oranges. *J. Plant Pathol.* **2014**, *96*, 419–424.

28. Youssef, K.; Abo Rehab, M.A.; El-Ghany, K.M.A. Preliminary investigation of Verticillium wilt on mango trees (*Mangifera indica* L.) in Egypt. *Am. Eurasian J. Sust Agric.* **2014**, *8*, 50–58.

29. Garganese, F.; Sanzani, S.M.; Di Rella, D.; Schena, L.; Ippolito, A. Pre-and postharvest application of alternative means to control Alternaria Brown spot of citrus. *Crop Prot.* **2019**, *121*, 73–79. [CrossRef]

30. Lachhab, N.; Sanzani, S.M.; Fallanaj, F.; Youssef, K.; Nigro, F.; Boselli, M.; Ippolito, A. Protein hydrolysates as resistance inducers for controlling green mold of citrus fruit. *Acta Hortic.* **2015**, *1065*, 1593–1598. [CrossRef]

31. Youssef, K.; Hashim, A.F.; Margarita, R.; Alghuthaymi, M.A.; Abd-Elsalam, K.A. Fungicidal efficacy of chemically-produced copper nanoparticles against *Penicillium digitatum* and *Fusarium solani* on citrus fruit. *Philipp. Agric. Sci.* **2017**, *100*, 69–78.

32. Hussien, A.; Ahmed, Y.; Al-Essawy, A.; Youssef, K. Evaluation of different salt-amended electrolysed water to control postharvest molds of citrus. *Trop. Plant Pathol.* **2018**, *43*, 10–20. [CrossRef]

33. Roberto, S.R.; Youssef, K.; Hashim, A.F.; Ippolito, A. Nanomaterials as alternative control means against postharvest diseases in fruit crops. *Nanomaterials* **2019**, *9*, 1752. [CrossRef]

34. Youssef, K.; Hussien, A. Electrolysed water and salt solutions can reduce green and blue molds while maintain the quality properties of 'Valencia' late oranges. *Postharvest Biol. Technol.* **2020**, *159*, 111025. [CrossRef]

35. Genta, W.; Tessmann, D.J.; Roberto, S.R.; Vida, J.B.; Colombo, L.A.; Scapin, C.R.; Ricce, W.S.; Clovis, L.R. Downy mildew management in protected cultivation of table grapes 'BRS Clara'. *Pesq. Agrop. Bras.* **2010**, *45*, 1388–1395. [CrossRef]

36. Shahab, M.; Roberto, S.R.; Ahmed, S.; Colombo, R.C.; Silvestre, J.P.; Renata, K.; de Souza, R.T. Relationship between anthocyanins and skin color of table grapes treated with abscisic acid at different stages of berry ripening. *Sci. Hort.* **2020**, *259*, 108859. [CrossRef]

 *horticulturae*

Review

Influences of Postharvest Storage and Processing Techniques on Antioxidant and Nutraceutical Properties of *Rubus idaeus* L.: A Mini-Review

Ermes Lo Piccolo [1], Leani Martìnez Garcìa [2], Marco Landi [1,3], Lucia Guidi [1,3,*], Rossano Massai [1,3] and Damiano Remorini [1,3]

[1]   Department of Agriculture, Food & Environment, University of Pisa, Via del Borghetto 80, 56124 Pisa, Italy; ermes.lopiccolo@agr.unipi.it (E.L.P.); marco.landi@unipi.it (M.L.); rossano.massai@unipi.it (R.M.); damiano.remorini@unipi.it (D.R.)

[2]   Grupo Tecnico Nacional de Plantas Medicinales de la Asosiacion Cubana de Tecnicos Agricolas Agroforestales (ACTAF), Calle 98, Esquina 7ma, Playa, La Habana 11300, Cuba; leanimartinez93@gmail.com

[3]   Interdepartmental Research Center Nutrafood "Nutraceuticals and Food for Health", University of Pisa, Via del Borghetto 80, 56124 Pisa, Italy

*   Correspondence: lucia.guidi@unipi.it; Tel.: +39-50-2216613; Fax: +39-50-2216087

Received: 3 November 2020; Accepted: 14 December 2020; Published: 16 December 2020

Abstract: The growth of agricultural mechanization has promoted an increase in raspberry production, and for this reason, the best postharvest storage and processing techniques capable of maintaining the health beneficial properties of these perishable berry fruits have been widely studied. Indeed, raspberries are a rich source of bioactive chemical compounds (e.g., ellagitannins, anthocyanins, and ascorbic acid), but these can be altered by postharvest storage and processing techniques before consumption. Although there are clear differences in storage times and techniques, the content of bioactive chemical compounds is relatively stable with some minor changes in ascorbic acid or anthocyanin content during cold (5 °C) or frozen storage. In the literature, processing techniques such as juicing or drying have negatively affected the content of bioactive chemical compounds. Among drying techniques, hot air (oven) drying is the process that alters the content of bioactive chemical compounds the most. For this reason, new drying technologies such as microwave and heat pumps have been developed. These novel techniques are more successful in retaining bioactive chemical compounds with respect to conventional hot air drying. This mini-review surveys recent literature concerning the effects of postharvest storage and processing techniques on raspberry bioactive chemical compound content.

Keywords: anthocyanin; ascorbic acid; drying method; phenol; phytochemical; raspberry

---

1. Introduction

Currently, interest towards fruits and vegetables has been continually increasing due to the awareness that they are a primary source of bioactive, health-beneficial compounds. Raspberry (*Rubus idaeus* L.) belongs to the Rosaceae family and represents one of the oldest fruits that has been used for millennia for human nutrition and as a folk medicine [1]. Raspberries are a rich source of bioactive chemical compounds (e.g., ellagitannins, anthocyanins, ascorbic acid) with a high antioxidant capacity, useful for the prevention of chronic human diseases [2,3]. Indeed, there is a growing interest by consumers that the consumption of these fruit can exert positive effects on human health [2,4,5] due to their nutritional and nutraceutical content (Figure 1). Fresh raspberries have low calories, about 50 kcal 100 g^{-1}, and together with their high content in dietary fibers make them a snack that has a satiating effect [2]. The most representative bioactive chemical compounds found in red raspberry fruit belong to the polyphenol class (normally total phenolic content ranges from ~100 to 600 mg 100 g^{-1} [4,6,7]). In particular, ellagitannins, such as lambertianin C and sanguiin H-6, and flavonoids, principally anthocyanins, are the major polyphenols found in raspberry (Figure 1) [6–10]. According to recent studies, anthocyanins in raspberries are mainly represented (>90%) by cyanidin-glycosides (cyanidin 3-O-glucoside, cyanidin 3-O-rutinoside, cyanidin 3-glucosylrutinoside and cyanidin 3-sophoroside) and pelargonidin glycosides (pelargonidin 3-glucoside) [7,9]. Another important antioxidant in raspberry is ascorbic acid, which is contained in this fruit at a lower concentration as compared to orange and kiwifruit [11,12].

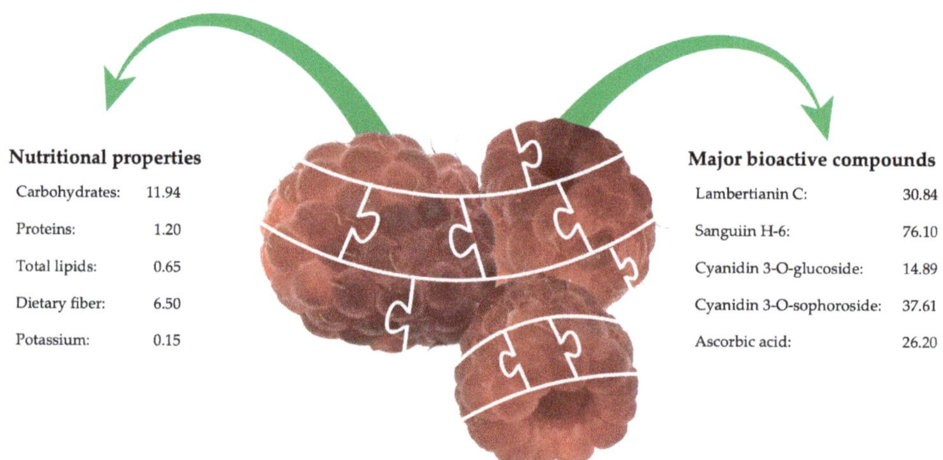

Figure 1. Nutritional (g 100 g^{-1}) and major bioactive compounds (mg 100 g^{-1}) in raw red raspberry fruits. Values were obtained from 'Phenol Explorer Database' and the literature [2,10].

The first information about raspberry cultivation comes from the Romans in the 4th century. In the last 20 years, the production of raspberry has been consistently enhanced, with a yield increase of 106% [13]. The largest production among countries is the Russian Federation with about 148 t year^{-1} (averaged data from the previous two decades), followed by Serbia (88 t year^{-1}), Poland (83 t year^{-1}), and the United States (80 t year^{-1}). However, in recent years, Mexico has become a major raspberry producer, with about 60 t year^{-1} (averaged from the last decade) [13]. The increase in the production of raspberry is likely attributable to a combination of different factors, such as globalization of berry markets, agricultural mechanization, and, last but not least, the increased interest of consumers for healthy food products.

However, raspberry is a very perishable soft fruit, and the increasing quantities produced set a double challenge for the food industry to develop processing techniques that utilize any overproduction (which results in wasted food) and preserve the fruit nutritional and nutraceutical qualities [14]. The use of postharvest storage techniques such as cold storage, a controlled atmosphere, and frozen storage can effectively prolong the postharvest quality of raspberry, making them suitable for their global commercialization [15–17]. Moreover, raspberry berries can be dry processed in order to not only preserve their nutritional properties, but also to make new foods with highly health-beneficial qualities for human consumption. In previous studies, a number of different technologies have been utilised for raspberry dehydration: freeze-drying, hot air-drying, microwave-drying, and a combination, and each technique has shown positive and negative effects [18–21]. Juice production is another industrial technique for processing fresh raspberry fruit and to extend their availability in the global market. Fruit juices are labelled as functional foods and may be introduced into the diet to increase the daily intake of functional metabolites [22]. However, the use of the abovementioned postharvest storage and processing techniques (which work with different parameters, e.g., temperatures, processing time) could cause potential variation at physical, biological and biochemical levels, which in turn may alter fruit bioactive chemical compound content, affecting the nutraceutical properties that have become appreciated by consumers [23–25]. For example, the red color of raspberries can be significantly affected by these processing techniques, since the reddish color is due to the presence of flavonoids, i.e., anthocyanins, which can be very sensitive to thermal processing techniques [26,27]. The human eye is attracted by red coloration, and, in particular, a high chromatic reddish coloration of fruit is perceived as a positive characteristic associated with fruit maturity stage, indirectly providing information about the fruit sweetness and quality [28–30]. Therefore, studies of new storage and processing/transformation technologies aimed at reducing negative changes in the content of bioactive chemical compounds of raspberry fruit represent this continuous challenge for researchers and the food industry. Moreover, as a large part of raspberry production is industrially processed, the detection of changes of the most representative bioactive compounds in raspberry fruits caused by transformation processes could provide useful information for estimating the health-related properties of a specific end product [31]. Therefore, the aim of this review is: (i) to describe the main bioactive chemical compounds contained in raspberries, highlighting their potential health benefits derived from their consumption; (ii) to report a survey of the most recent literature concerning the possible influences of postharvest storage and processing techniques (juicing and drying) on raspberry bioactive chemical compound content; and, (iii) to present novel data about the effects of a recent drying technique (heat pump) on bioactive chemical compounds of raspberry fruits.

2. Postharvest Storage

2.1. Cold/Modified Atmosphere Storage

After harvest, fresh raspberries are stored before undergoing further processing. This postharvest storage is necessary because raspberry fruits have a short postharvest life due to their high respiration rate and loss of firmness [32]. Progressive loss of firmness is associated with a loss of skin strength that leads to fruit softening, which in turn favors the onset of molds with loss of the product [33,34]. Studies conducted on the relationship between normal cold storage conditions and shelf-life of red raspberry fruits have established that raspberries can be successfully stored near 1 to 2 °C up to about 10 days, though these conditions do not reflect a realistic storage (usually 4 to 5 °C) [15,35,36]. In definition, it is difficult to preserve these small fruits, and for this reason, it is important to describe storage methods and technologies that can preserve not only the product, but also their nutritional and healthful properties.

It has been reported that the use of different storage temperatures affects the content of bioactive compounds (e.g., phenolics and ascorbic acid; Table 1) in different ways. The total phenolic content as well as the antioxidant capacity are relatively stable throughout storage by using low temperatures (1–5 °C) [35–37], even though a slight decrease in total phenolic content (with storage temperatures of 1 °C) was reported by Giuffrè et al. [35]. On the other hand, Mullen et al. [37], by simulating a realistic storage condition (4 °C), found that ellagitannin levels increased, but no changes in the total antioxidant capacity were detected.

Table 1. Changes in content of some bioactive compounds and antioxidant capacity in fresh raspberry fruit by different raspberry storage conditions. ($\uparrow$): increased contents; ($\downarrow$): decreased contents; (=): unchanged contents.

Storage Conditions	Influence on Antioxidant Compounds and Activity	References
Cold storage 1–2 °C	Phenols ($\uparrow\downarrow$)	[35,38]
	Anthocyanins ($\uparrow$)	[15,35,38,39]
	Ellagic acid (=)	[38]
	Ascorbic acid (=)	[15,35,38]
	Antioxidant capacity (=)	[35]
Cold storage ~5 °C	Phenols (=)	[40,41]
	Anthocyanins ($\uparrow$)	[40,41]
	Ellagitannins ($\uparrow$)	[37]
	Ascorbic acid ($\downarrow$)	[37,40]
	Antioxidant capacity (=$\uparrow$)	[37,40,41]
Controlled atmosphere	Anthocyanins (=)	[15]
	Ascorbic acid (=)	[15,42]
Frozen storage ~1 year	Phenols (=$\uparrow$)	[43–45]
	Anthocyanins (=$\downarrow$)	[45,46]
	Ellagic acid ($\downarrow$)	[43,47]
	Ascorbic acid ($\downarrow$)	[43,45]
	Antioxidant capacity (=)	[43,44]

Among different bioactive compounds in raspberries, anthocyanins are important for the organoleptic characteristics of these small fruits. Some have reported that total anthocyanin content increases during storage, independently of storage temperatures [15,35,39,40], but others found a decrease in total monomeric anthocyanins when raspberries were stored at 3 °C [36]. There have been contrasting results on the effect of temperature storage on phenolics in raspberry. It is important to note that, during cold storage, fruit weight loss can influence the correct estimation of bioactive compounds on a fresh weight basis and attention should be paid to this when storage effects on bioactive chemical compound content are evaluated [32]. In addition, it has been reported that during cold storage, a decrease in organic acid content can occur and this makes the carbon skeleton for the synthesis of polyphenols available [48].

Another important bioactive compound in raspberry is ascorbic acid, and it has been reported that its content in fresh raspberries stored at 1 to 2 °C was relatively stable [15,35,38]. However, its content is strictly related to the temperature used during storage. Indeed, at relatively higher temperatures (around 5 °C), ascorbic acid content slightly decreased [37,40], most likely attributed to the greater rate of oxidation of this compound that can occur at this temperature (Table 1).

Modification of storage atmosphere (at low temperatures close to 0 to 2 °C) while also changing O_2 and CO_2 concentrations (controlled atmospheres (CA)) are widely used to extend the fruit shelf-life, significantly suppressing rotting [15,49,50]. Although plenty of information is available on the effect of CA storage on fruit shelf-life for some berry fruits (e.g., blueberry and strawberry), the literature on raspberry CA storage is scarce, and the correlation between CA and red raspberry nutraceutical traits are almost non-existent. Previous studies conducted on CA effects on raspberry shelf-life confirmed an extension of berry storability (>50%), delaying fruit decay by using 10 to 35% O_2 and 15 to 45% CO_2 [15,16,42,51]. Anthocyanins and ascorbic acid content were unchanged after CA storage (Table 1), indicating that CA is effective in maintaining nutraceutical value for longer periods than normal storage [15,42]. Better results in the retention of bioactive compounds in raspberry have been related to lower O_2 concentrations, which reduce the oxidative reactions that typically can occur during fruit storage.

2.2. Frozen Storage

Certainly, lower temperatures than those described above, i.e., frozen storage, represent a way to store red raspberries for longer periods, making them available all year round. Usually, for frozen storage, temperatures around −20 °C have been used. De Ancos et al. [43] found that total phenolic content and total antioxidant capacity remained substantially unchanged after 12 months of frozen storage even though ellagic acid and ascorbic acid significantly decreased (Table 1). The decrease in free ellagic acid detected during frozen storage in red raspberry could be related to the capacity of this acid to act as a metal chelating agent with Mg^{+2} and Ca^{+2} and/or to the action of polyphenol oxidase (PPO) linked to the cell wall, which is most likely degraded by ice crystals [43,47,52] (Table 1). Others [44,45] have reported that after 12 months of frozen storage the total phenolic content increased in raspberry fruit. The authors suggested that the increase in phenolics was attributable to the degradation of cell structures by ice crystals, thereby making those compounds more easily available during extraction. Total anthocyanin content decreased in some cultivars after frozen storage [45,46] while it remained unchanged in others [46] (Table 1), indicating differing effects of frozen temperature by genotype and the type of chemical structure of the anthocyanin contained in it. For example, cyanidin 3-glucoside is much more prone to degradation during frozen storage than other cyanidin-based derivatives, especially diglycosidic anthocyanins [45,46]. More generally, it has been reported that cyanidin 3-O-glucoside is one of the most reactive anthocyanins during processing [53,54].

The loss in anthocyanin content due to oxidation and/or condensation reactions with other phenolic compounds did not interfere with antioxidant activity, as that is generally maintained or increased following frozen storage [55]. This can be attributed to the interference or association of the phenolic compounds and anthocyanin degradation products.

3. Processing Techniques

3.1. Juicing

Raspberry juice is a commercially important product, consumed pure or more usually blended with other fruit juices to enhance their color or the flavor. Red-colored juices are very appreciated by consumers; indeed, a new market-segment related to red-colored fruits and their processed products is increasing thanks to consumer association of red-colored fruit to high antioxidant properties [22,56,57]. Fruit juice is the result of a complex industrial process in which thawed fruit go through a series of operations (e.g., crushing, pressing, enzymatic treatment, thermal/non-thermal treatments, clarification/filtration) in order to obtain the final product [24]. All of these processes decrease the content of the antioxidant compounds in the juice with respect to their fresh fruit levels. Losses of bioactive chemical compounds during juice processing should be minimized to retain beneficial health effects. However, literature on this topic is scarce for raspberries, when compared to other berries such as blackberries [53,58,59] (Table 2). An interesting experiment conducted by Sojka et al. [58] with red raspberry juice showed that about 68% of total anthocyanins, 12% of ellagitannins, 31% of flavonols, and 17% of flavanols were retained in fruit juice as compared to the fresh material. This indicates that the industrial processes carried out to obtain raspberry juice resulted in a large decrease in the bioactive chemical compounds with different effects based on chemical structure. Generally, it is well known that the high temperatures used to obtain juice negatively affect nutritive quality by destroying essential nutrients and biologically active "non-essential" components such as polyphenols [24,53]. For this reason, novel technologies that can preserve nutrient quality but also food safety are desirable. For example, high hydrostatic pressure (HHP) is a relatively new technology that can be utilized instead of thermal processes such as pasteurization. However, there are no studies on the use of this technology on raspberry juice production, but it is well known that HHP allows high retention of bioactive compounds of juice [24,60].

Table 2. Changes in content of selected bioactive compounds and antioxidant capacity by different processing techniques of fresh raspberry fruit. (↑): increased content; (↓): decreased content; (=): unchanged content.

Processing Techniques	Conditions	Influence on Antioxidant Compounds and Activity	References
Juicing	Enzymatic treatment Thermal treatment	Phenols (↓) Anthocyanins (=↓) Ellagic acid (↓) Ellagitannins (↓) Antioxidant capacity (=↓)	[58,59] [53,58,59] [58] [58] [59]
Freeze-drying	~−50 °C	Phenols (=↓↑) Anthocyanins (=↓↑) Ellagic acid (↓) Ascorbic acid (=↓) Antioxidant capacity (=↓)	[18,19,21,61,62] [18,19,21,62] [20] [21,62] [18,19,62]
Hot air drying	~65 °C	Phenols (↓) Anthocyanins (↓) Ellagic acid (↓) Ascorbic acid (↓) Antioxidant capacity (↓)	[19,20,62] [19–21,59] [20] [21,62] [19,20]
Microwave	Microwave	Phenols (↓) Antioxidant capacity (↓)	[63,64] [63,64]
	Microwave/hot air	Phenols (↓) Anthocyanins (↓) Antioxidant capacity (↓)	[20,63,64] [20] [20,63,64]
	Microwave/IR	Phenols (↓) Anthocyanins (↓) Antioxidant capacity (↓)	[19] [19] [19]
	Microwave/vacuum	Phenols (↓) Anthocyanins (↓) Antioxidant capacity (↓)	[20] [20] [20]
Heat pump	30–35 °C	Phenols (↓) Anthocyanins (↓) Ascorbic acid (↓) Antioxidant capacity (↓)	[our results, see Table 3]

3.2. Freeze-Drying

Freeze-drying is considered the most effective technique for preserving food quality [65,66]. The fruit samples are dried under vacuum at low temperatures (about −50 °C) maintaining bioactive chemical compound content almost unaltered from fresh fruit in comparison to other drying techniques [65,66]. Sablani et al. [18] confirmed the effectiveness of freeze-drying in retaining bioactive chemical compounds in freeze-dried raspberry tissue, in some cases improving phenolic content compared to fresh berries (Table 2). In contrast, some authors have shown different results [19–21,61], raising doubts about its efficacy in not altering the content of bioactive compounds of raspberry fruit [21]. For example, Stamenkovic et al. [62] reported a decrease in total phenolic content, anthocyanin levels, and radical scavenging capacity after freeze drying, while ascorbic acid content remained unchanged likely due to the very low temperature and the limited oxygen availability during the drying process (Table 2). Opposite effects have also been found with other fruit and vegetable materials [67]. In general, there is no clear explanation for the reason(s) for which bioactive compound levels decrease, increase, or remain unchanged during freeze-drying since antioxidants are very sensitive to light, oxygen, temperature, and pH and can also be degraded by enzymes [67–69]. However, among drying methods, freeze drying is currently the most effective in preserving the content of bioactive compounds, but its biggest disadvantage is the high cost due to length of time and energy-costs, making it difficult to use in the food industry [67].

3.3. Hot Air-Drying

Hot air-drying technology represents the most economical and common processing technology for drying food—including raspberries—despite its high temperatures and long drying times. During the process, food samples are exposed to hot air and the presence of oxygen. This combination of oxygen and higher temperatures—even if for a short time—results in high losses of bioactive compounds [23,62]. Therefore, it is necessary to find a reasonable compromise between temperatures and duration time for the entire drying process to make it a viable option.

Many studies of the hot air-drying technique conducted with raspberry (Table 2) have used temperatures around 65 °C to assess the effects of this method on bioactive chemical compound content [19,20,23,59,62]. It has been established that high temperatures during hot air-drying negatively affect the ascorbic acid content [69], with losses of around 90% in raspberry (dried vs. fresh fruits) [21,62] (Table 2). However, the moisture of the products is involved in the negative influence on ascorbic acid content during the drying process [70].

It has also been observed that hot air-drying decreased the fruit phenolic content as compared to values for fresh raspberry fruit [19,20,62]. This phenomenon is likely due to thermal degradation of the phenolic compounds during the hot air-drying process [19,62], and not only negatively affects the health-related benefits of the final product but also affects the appearance by changing the colour (Table 2). Indeed, the products of anthocyanin thermal degradation are colourless carbinol and chalcone that cause a colour loss to the dried fruit [26,27] (Table 2).

Some authors, comparing different drying techniques, have detected a higher content in some phenols in raspberry fruits when air-dried rather than freeze-dried [19,23,62], whereas anthocyanins have always been found to decrease with air drying rather than freeze-drying samples. Nevertheless, the increase in total phenolic content in air-dried fruit compared to freeze-dried fruit was most likely linked to the release of bound phenol groups or more destruction of the tissue by higher temperatures allowing greater extraction of phenolic molecules [19,23]. Undoubtedly, the low costs of this technique are significant when compared to more sophisticated technologies (e.g., freeze-drying), but the hot air-drying approach has considerable limitations in preserving the nutraceutical properties of raspberry.

3.4. Microwave-Drying

Microwave-drying technology uses the microwaves to directly generate heat inside the fruit due to the microwave-promoted molecular oscillation. This results in decreasing the processing time [71], thereby increasing bioactive compound retention as compared to conventional hot air-drying [20]. For raspberries, microwave drying is usually combined with additional treatments such as hot air, infra-red radiation (IR), and vacuum-drying [19,20,63] (Table 2), in an attempt to increase the retention of bioactive compounds. Microwave-dried raspberries retained a high antioxidant capacity compared to microwave/hot air-dried raspberries (about 40% vs. 20%, respectively), if compared to fresh raspberry fruits [63]. Similar results were obtained by Mejia-Meza et al. [20] and Si et al. [19], in which microwave/vacuum-drying or microwave/IR-treated raspberry samples showed a higher anthocyanin level and a greater total antioxidant capacity than samples treated by microwave/hot air-drying or hot air-drying processing. The greatest reduction in antioxidant capacity occurred when hot air-drying was used, and this decrease is potentially related to the degradation of bioactive chemical compounds at high temperatures. It could be argued that even in the IR treatment, the temperatures reached inside the drying chamber were around 70 °C, but it should be taken into account that the drying time also plays a key role in the preservation of bioactive chemical compound levels. Indeed the IR treatment lasted 60 min whereas hot air-drying requires many more hours [19]. Therefore, microwave-drying (characterized by reduced time and temperatures) results are more advantageous for bioactive compound retention than conventional hot air-drying. Compared with freeze-drying, all microwave drying systems have shown high retention of total phenols, although lower levels of anthocyanins were usually detected [19,20]. The reduced drying time and costs of microwave drying processes may provide advantages when compared to freeze drying [20].

3.5. Heat Pump-Drying

A heat pump dehumidifier drying system (HP) is more environmentally friendly than conventional hot air-drying methods thanks to its high energy efficiency. This also results in lower production costs [72]. The process is relatively simple; low moisture air is blown into the drying chamber to absorb moisture from the fruit sample. The air moisture is eliminated through condensation at a low temperature (-20 to $15\,^{\circ}$C) by the dehumidifier and then heated (at 30–$35\,^{\circ}$C) to start a new cycle (Table 2). The lower drying temperatures compared to conventional hot air-drying systems potentially provide a better quality of the dried product, which retains high concentrations of phenolics and ascorbic acid [73]. However, there is a lack of knowledge about the effects of heat pump-drying on the bioactive compounds of raspberry fruit, raising the need to understand whether this new process can be a valid substitute for conventional methods. For this purpose, an experiment was conducted in our laboratory to test the effects of heat pump-drying treatment on total phenolics, total anthocyanins, and ascorbic acid content and total antioxidant capacity in comparison to hot air-drying treatment at $65\,^{\circ}$C on an ecotype of raspberry grown in Garfagnana (Tuscany), Italy, (Figure 2; Table 3). The total phenolic content decreased significantly after the dehydration with both hot air and heat pump treatments as compared with values recorded in fresh material (Tables 2 and 3). However, the decrease in phenolics caused by both drying treatments could be ascribed to two distinct causes. During the conventional hot air process, phenols can be thermally degraded as reported and explained in previous studies [19,20,74]. On the other hand, in the heat pump-drying treatment (characterized by lower temperatures), the reduction of phenolics is more likely attributable to enzymatic degradation (e.g., by PPO), since 30 to $35\,^{\circ}$C is closer to the optimum temperatures for PPO activity [75,76]. A similar process may also explain the effect on anthocyanin content; however, anthocyanins were retained more with the heat pump than the hot air treatment (87 and 66% of retention, respectively).

Table 3. Total phenolic (TP), total anthocyanin (TA), ascorbic acid (AA) content, and total antioxidant capacity (TAC) in fresh raspberry fruit grown in Garfagnana (Italy), and after conventionally hot air-drying ($65\,^{\circ}$C for 20 h) and heat pump-drying ($35\,^{\circ}$C for 2 d). Each value is the mean of three replicates $\pm$ standard deviation.

	Units	Fresh	Hot Air 65 °C	Heat Pump
TP[z]	mg GA[y] eq. g^{-1} dw	152.36 ± 5.13 a [x]	123.31 ± 6.00 b	119.27 ± 12.28 b
TA	mg Cya glu. eq. g^{-1} dw	1.18 ± 0.03 a	0.78 ± 0.07 c	1.03 ± 0.09 b
AA	mg g^{-1} dw	2.01 ± 0.13 a	1.01 ± 0.08 c	1.20 ± 0.05 b
TAC	mg Trolox eq. g^{-1} dw	83.44 ± 3.45 a	32.20 ± 1.76 b	31.38 ± 4.06 b

[z] TA was analyzed as reported in Lo Piccolo et al. [77]; TP, AA and TAC were analyzed as reported in Ceccanti et al. [78]. [y] GA: gallic acid; Cya glu: cyanidin-glucoside. [x] Within each row, means flanked by the same letter are not significantly different after a one-way ANOVA test with postharvest treatment as source of variability following a least significant difference (LSD) test ($P = 0.05$).

Our results also showed that ascorbic acid content decreased with both treatments (50 and 40% for HP and hot air-drying, respectively), although there was a difference in temperatures utilized in the two methods. Notably, though there were differences among treatments in the retention of bioactive compounds, the total antioxidant capacity decreased considerably in both treatments to about 60% of that detected in fresh fruit. The similar values in total antioxidant capacity between hot air and heat pump dried samples could be explained by the similar contents in total phenolics in both treatments. Indeed, Beekweelder et al. [79] reported that ellagitannins contributed more than 50% to the total antioxidant activity, so this may be the reason why total antioxidant activity was similar for the hot air and heat pump dried samples, even though ascorbic acid and anthocyanin contents were higher in heat pump dried samples.

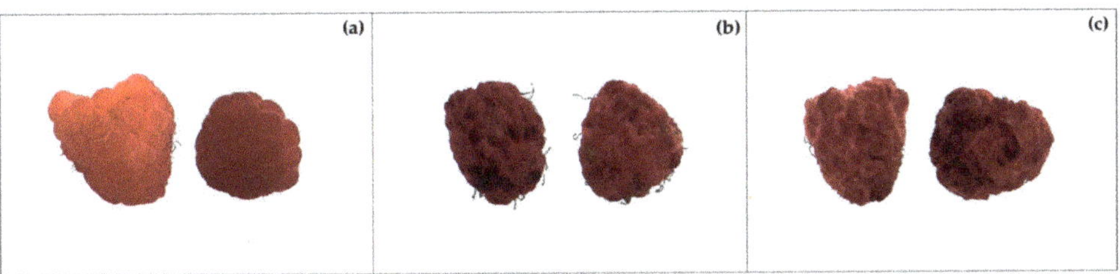

Figure 2. Fresh raspberries (**a**), ~80% moisture. Conventional hot air-dried raspberries (**b**), ~4% moisture. Heat pump-dried raspberries (**c**), ~14% moisture.

In view of the above, the use of heat pump-drying technology for the postharvest processing of raspberry fruit may allow slightly greater retention of some bioactive compounds with respect to conventional hot air drying. In addition to the benefits derived for human health, heat pump-drying technology also allows the reduction of energy consumption for processing [80].

4. Conclusions

The attention of the food industry on the preservation and improvement of the qualities of processed berries has progressively increased, with the goal of finding critical points during the postharvest processes of raspberry that affect final product quality. The storage and processing technologies of red raspberries reviewed here showed different influences specific to each processing technique on the final bioactive chemical compound content. The content of bioactive chemical compounds is relatively stable during storage, with some minor changes in ascorbic acid or anthocyanin content during cold (at 5 °C) or frozen storage.

Among drying techniques, the freeze-drying method is most likely the best processing method for retaining the content of bioactive chemical compounds, but it requires high costs for the implementation of a large-scale production of dried berries. The conventional hot air-drying method, which is the most used by the food industry, is most likely the worst drying method for bioactive chemical compound retention (especially anthocyanins and ascorbic acid).

New technologies, such as microwave- and heat pump-drying, which are cheaper than conventional methods, are emerging as promising processes to provide a higher retention of bioactive chemical compounds in dried berries. However, changes in quantity and the profile of bioactive chemical compounds in raspberry fruit subjected to these different postharvest drying processes need further investigation. In this way, the future research on this topic should be in-depth analyses at the molecular level, using different drying times, temperatures and also new possible combinations between different drying techniques (e.g., heat pump and microwave). These studies will provide useful indicators for developing new industrial processing methods.

Author Contributions: Conceived and designed the experiments, R.M., and L.G.; performed the experiments, analysed data and wrote the manuscript, E.L.P., D.R. and L.M.G.; reviewed and edited the paper, M.L., L.G., R.M. and D.R. All authors have read and agreed to the published version of the manuscript.

Funding: This research was performed in the framework of 'Erbi Boni' project, funded by FEASR Reg. UE 1305/2013-PSR 2014/2020—Regione Toscana Mis. 19—GAL MontagnAppennino Mis. 16.2.

Acknowledgments: Authors are grateful to Franca Bernardi for the technical support and Matteo Guidi, who provided us with wild raspberries.

Conflicts of Interest: The authors declare no conflict of interest.

References

1. Patel, A.; Rojas-Vera, J.; Dacke, C. Therapeutic constituents and actions of *Rubus* species. *Curr. Med. Chem.* **2004**, *11*, 1501–1512. [CrossRef] [PubMed]
2. Rao, A.V.; Snyder, D.M. Raspberries and human health: A review. *J. Agric. Food Chem.* **2010**, *58*, 3871–3883. [CrossRef] [PubMed]
3. Graham, J.; Brennan, R. *Raspberry: Breeding, Challenges and Advances*; Springer International Publishing: Cham, Switzerland, 2018; ISBN 978-3-319-99030-9.
4. De Souza, V.R.; Pereira, P.A.P.; da Silva, T.L.T.; de Oliveira Lima, L.C.; Pio, R.; Queiroz, F. Determination of the bioactive compounds, antioxidant activity and chemical composition of Brazilian blackberry, red raspberry, strawberry, blueberry and sweet cherry fruits. *Food Chem.* **2014**, *156*, 362–368. [CrossRef] [PubMed]
5. Basu, A.; Rhone, M.; Lyons, T.J. Berries: Emerging impact on cardiovascular health. *Nutr. Rev.* **2010**, *68*, 168–177. [CrossRef]
6. Pantelidis, G.; Vasilakakis, M.; Manganaris, G.; Diamantidis, G. Antioxidant capacity, phenol, anthocyanin and ascorbic acid contents in raspberries, blackberries, red currants, gooseberries and Cornelian cherries. *Food Chem.* **2007**, *102*, 777–783. [CrossRef]
7. Yang, J.; Cui, J.; Chen, J.; Yao, J.; Hao, Y.; Fan, Y.; Liu, Y. Evaluation of physicochemical properties in three raspberries (*Rubus idaeus*) at five ripening stages in northern China. *Sci. Hortic.* **2020**, *263*, 109146. [CrossRef]
8. Tosun, M.; Ercisli, S.; Karlidag, H.; Sengul, M. Characterization of red raspberry (*Rubus idaeus* L.) genotypes for their physicochemical properties. *J. Food Sci.* **2009**, *74*, C575–C579. [CrossRef]
9. Mazur, S.P.; Nes, A.; Wold, A.B.; Remberg, S.F.; Aaby, K. Quality and chemical composition of ten red raspberry (*Rubus idaeus* L.) genotypes during three harvest seasons. *Food Chem.* **2014**, *160*, 233–240. [CrossRef]
10. Neveu, V.; Perez-Jimenez, J.; Vos, F.; Crespy, V.; du Chaffaut, L.; Mennen, L.; Knox, C.; Eisner, R.; Cruz, J.; Wishart, D.; et al. Phenol-Explorer: An online comprehensive database on polyphenol contents in foods. *Database* **2010**, *2010*, bap024. [CrossRef]
11. Sivakumaran, S.; Huffman, L.; Sivakumaran, S.; Drummond, L. The nutritional composition of Zespri® SunGold kiwifruit and Zespri® Sweet Green kiwifruit. *Food Chem.* **2018**, *238*, 195–202. [CrossRef]
12. Lee, H.S.; Coates, G.A. Vitamin C in frozen, fresh squeezed, unpasteurized, polyethylene-bottled orange juice: A storage study. *Food Chem.* **1999**, *65*, 165–168. [CrossRef]
13. Statistics Division FAOSTAT—Food and Agriculture Organization of the United Nations. Available online: http://www.fao.org/faostat/en/#home (accessed on 1 July 2020).
14. FAO. *Global Food Losses and Food Waste—Extent, Causes and Prevention*; FAO: Rome, Italy, 2011.
15. Haffner, K.; Rosenfeld, H.J.; Skrede, G.; Wang, L. Quality of red raspberry *Rubus idaeus* L. cultivars after storage in controlled and normal atmospheres. *Postharvest Biol. Technol.* **2002**, *24*, 279–289. [CrossRef]
16. Forney, C.F.; Jamieson, A.R.; Munro Pennell, K.D.; Jordan, M.A.; Fillmore, S.A.E. Relationships between fruit composition and storage life in air or controlled atmosphere of red raspberry. *Postharvest Biol. Technol.* **2015**, *110*, 121–130. [CrossRef]
17. Callesen, O.; Holm, B. Storage results with red raspberry. *Acta Hortic.* **1989**, 247–254. [CrossRef]
18. Sablani, S.S.; Andrews, P.K.; Davies, N.M.; Walters, T.; Saez, H.; Bastarrachea, L. Effects of air and freeze drying on phytochemical content of conventional and organic berries. *Dry. Technol.* **2011**, *29*, 205–216. [CrossRef]
19. Si, X.; Chen, Q.; Bi, J.; Wu, X.; Yi, J.; Zhou, L.; Li, Z. Comparison of different drying methods on the physical properties, bioactive compounds and antioxidant activity of raspberry powders: Effect of drying method on properties of raspberry powders. *J. Sci. Food Agric.* **2016**, *96*, 2055–2062. [CrossRef]
20. Mejia-Meza, E.I.; Yanez, J.A.; Remsberg, C.M.; Takemoto, J.K.; Davies, N.M.; Rasco, B.; Clary, C. Effect of dehydration on raspberries: Polyphenol and anthocyanin retention, antioxidant capacity, and antiadipogenic activity. *J. Food Sci.* **2010**, *75*, H5–H12. [CrossRef]
21. Sadowska, K.; Andrzejewska, J.; Klóska, Ł. Influence of freezing, lyophilisation and air-drying on the total monomeric anthocyanins, vitamin C and antioxidant capacity of selected berries. *Int. J. Food Sci. Technol.* **2017**, *52*, 1246–1251. [CrossRef]
22. Bermúdez-Soto, M.J.; Tomás-Barberán, F.A. Evaluation of commercial red fruit juice concentrates as ingredients for antioxidant functional juices. *Eur. Food Res. Technol.* **2004**, *219*, 133–141. [CrossRef]

23. Bustos, M.C.; Rocha-Parra, D.; Sampedro, I.; de Pascual-Teresa, S.; León, A.E. The influence of different air-drying conditions on bioactive compounds and antioxidant activity of berries. *J. Agric. Food Chem.* **2018**, *66*, 2714–2723. [CrossRef]

24. Weber, F.; Larsen, L.R. Influence of fruit juice processing on anthocyanin stability. *Food Res. Int.* **2017**, *100*, 354–365. [CrossRef]

25. Michalska, A.; Łysiak, G. Bioactive compounds of blueberries: Postharvest factors influencing the nutritional value of products. *Int. J. Mol. Sci.* **2015**, *16*, 18642–18663. [CrossRef] [PubMed]

26. Furtado, P.; Figueiredo, P.; Chaves das Neves, H.; Pina, F. Photochemical and thermal degradation of anthocyanidins. *J. Photochem. Photobiol. Chem.* **1993**, *75*, 113–118. [CrossRef]

27. Patras, A.; Brunton, N.P.; O'Donnell, C.; Tiwari, B.K. Effect of thermal processing on anthocyanin stability in foods; mechanisms and kinetics of degradation. *Trends Food Sci. Technol.* **2010**, *21*, 3–11. [CrossRef]

28. Hamadziripi, E.T.; Theron, K.I.; Muller, M.; Steyn, W.J. Apple compositional and peel color differences resulting from canopy microclimate affect consumer preference for eating quality and appearance. *HortScience* **2014**, *49*, 384–392. [CrossRef]

29. Lee, S.M.; Lee, K.T.; Lee, S.H.; Song, J.K. Origin of human colour preference for food. *J. Food Eng.* **2013**, *119*, 508–515. [CrossRef]

30. Prokop, P.; Fančovičová, J. Beautiful fruits taste good: The aesthetic influences of fruit preferences in humans. *Anthropol. Anz.* **2012**, *69*, 71–83. [CrossRef]

31. Pritts, M.P. Raspberries and related fruits. In *Encyclopedia of Food Sciences and Nutrition*, 2nd ed.; Caballero, B., Ed.; Elsevier: Amsterdam, The Netherlands, 2003; pp. 4916–4921. ISBN 978-0-12-227055-0.

32. Krüger, E.; Dietrich, H.; Schöpplein, E.; Rasim, S.; Kürbel, P. Cultivar, storage conditions and ripening effects on physical and chemical qualities of red raspberry fruit. *Postharvest Biol. Technol.* **2011**, *60*, 31–37. [CrossRef]

33. Oduse, K.A.; Cullen, D. An investigation into the fruit firmness properties of some progeny and cultivars of red raspberry (*Rubus idaeus*). *J. Environ. Sci. Toxicol. Food Technol.* **2012**, *1*, 4–12. [CrossRef]

34. Zhang, D.; Quantick, P.C. Antifungal effects of chitosan coating on fresh strawberries and raspberries during storage. *J. Hortic. Sci. Biotechnol.* **1998**, *73*, 763–767. [CrossRef]

35. Giuffrè, A.M.; Louadj, L.; Rizzo, P.; De Salvo, E.; Sicari, V. The Influence of film and storage on the phenolic and antioxidant properties of red raspberries (*Rubus idaeus* L.) cv. Erika. *Antioxidants* **2019**, *8*, 254. [CrossRef] [PubMed]

36. Cortellino, G.; De Vecchi, P.; Lo Scalzo, R.; Ughini, V.; Granelli, G.; Buccheri, M. Ready-to-eat raspberries: Qualitative and nutraceutical characteristics during shelf-life. *Adv. Hortic. Sci.* **2018**, 399–406. [CrossRef]

37. Mullen, W.; Stewart, A.J.; Lean, M.E.J.; Gardner, P.; Duthie, G.G.; Crozier, A. Effect of freezing and storage on the phenolics, ellagitannins, flavonoids, and antioxidant capacity of red raspberries. *J. Agric. Food Chem.* **2002**, *50*, 5197–5201. [CrossRef] [PubMed]

38. Giovanelli, G.; Limbo, S.; Buratti, S. Effects of new packaging solutions on physico-chemical, nutritional and aromatic characteristics of red raspberries (*Rubus idaeus* L.) in postharvest storage. *Postharvest Biol. Technol.* **2014**, *98*, 72–81. [CrossRef]

39. Stavang, J.A.; Freitag, S.; Foito, A.; Verrall, S.; Heide, O.M.; Stewart, D.; Sønsteby, A. Raspberry fruit quality changes during ripening and storage as assessed by colour, sensory evaluation and chemical analyses. *Sci. Hortic.* **2015**, *195*, 216–225. [CrossRef]

40. Kalt, W.; Forney, C.F.; Martin, A.; Prior, R.L. Antioxidant capacity, vitamin C, phenolics, and anthocyanins after fresh storage of small fruits. *J. Agric. Food Chem.* **1999**, *47*, 4638–4644. [CrossRef]

41. Ali, L.; Svensson, B.; Alsanius, B.W.; Olsson, M.E. Late season harvest and storage of *Rubus* berries—Major antioxidant and sugar levels. *Sci. Hortic.* **2011**, *129*, 376–381. [CrossRef]

42. Dziedzic, E.; Błaszczyk, J.; Bieniasz, M.; Dziadek, K.; Kopeć, A. Effect of modified (MAP) and controlled atmosphere (CA) storage on the quality and bioactive compounds of blue honeysuckle fruits (*Lonicera caerulea* L.). *Sci. Hortic.* **2020**, *265*, 109226. [CrossRef]

43. Yang, M.; Ban, Z.; Luo, Z.; Li, J.; Lu, H.; Li, D.; Chen, C.; Li, L. Impact of elevated O_2 and CO_2 atmospheres on chemical attributes and quality of strawberry (*Fragaria × ananassa* Duch.) during storage. *Food Chem.* **2020**, *307*, 125550. [CrossRef]

44. Agar, I.T.; Streif, J.; Bangerth, F. Effect of high CO_2 and controlled atmosphere (CA) on the ascorbic and dehydroascorbic acid content of some berry fruits. *Postharvest Biol. Technol.* **1997**, *11*, 47–55. [CrossRef]

45. González-Orozco, B.D.; Mercado-Silva, E.M.; Castaño-Tostado, E.; Vázquez-Barrios, M.E.; Rivera-Pastrana, D.M. Effect of short-term controlled atmospheres on the postharvest quality and sensory shelf life of red raspberry (*Rubus idaeus* L.). *CyTA-J. Food* **2020**, *18*, 352–358. [CrossRef]

46. De Ancos, B.; González, E.M.; Cano, M.P. Ellagic acid, vitamin C, and total phenolic contents and radical scavenging capacity affected by freezing and frozen storage in raspberry fruit. *J. Agric. Food Chem.* **2000**, *48*, 4565–4570. [CrossRef] [PubMed]

47. Türkben, C.; Sarıburun, E.; Demir, C.; Uylaşer, V. Effect of freezing and frozen storage on phenolic compounds of raspberry and blackberry cultivars. *Food Anal. Methods* **2010**, *3*, 144–153. [CrossRef]

48. Häkkinen, S.H.; Kärenlampi, S.O.; Mykkänen, H.M.; Heinonen, I.M.; Törrönen, A.R. Ellagic acid content in berries: Influence of domestic processing and storage. *Eur. Food Res. Technol.* **2000**, *212*, 75–80. [CrossRef]

49. Šamec, D.; Piljac-Žegarac, J. Fluctuations in the levels of antioxidant compounds and antioxidant capacity of ten small fruits during one year of frozen storage. *Int. J. Food Prop.* **2015**, *18*, 21–32. [CrossRef]

50. Bobinalt, R.; Viškelis, P. Quality of Raspberry Fruits after Frozen Storage. In Proceedings of the 5th International Technical Symposium on Food Processing, Monitoring Technology in Bioprocesses and Food Quality Management, Postdam, Germany, 31 August–2 September 2009.

51. De Ancos, B.; Ibañez, E.; Reglero, G.; Cano, M.P. Frozen storage effects on anthocyanins and volatile compounds of raspberry fruit. *J. Agric. Food Chem.* **2000**, *48*, 873–879. [CrossRef]

52. Rommel, A.; Heatherbell, D.A.; Wrolstad, R.E. Red raspberry juice and wine: Effect of processing and storage on anthocyanin pigment composition, color and appearance. *J. Food Sci.* **1990**, *55*, 1011–1017. [CrossRef]

53. Boyles, M.J.; Wrolstad, R.E. Anthocyanin composition of red raspberry juice: Influences of cultivar, processing, and environmental factors. *J. Food Sci.* **1993**, *58*, 1135–1141. [CrossRef]

54. Reque, P.M.; Steffens, R.S.; Jablonski, A.; Flôres, S.H.; Rios, A.D.O.; de Jong, E.V. Cold storage of blueberry (*Vaccinium* spp.) fruits and juice: Anthocyanin stability and antioxidant activity. *J. Food Compos. Anal.* **2014**, *33*, 111–116. [CrossRef]

55. Piljac-Žegarac, J.; Šamec, D. Antioxidant stability of small fruits in postharvest storage at room and refrigerator temperatures. *Food Res. Int.* **2011**, *44*, 345–350. [CrossRef]

56. Allan, A.C.; Espley, R.V. MYBs drive novel consumer traits in fruits and vegetables. *Trends Plant Sci.* **2018**, *23*, 693–705. [CrossRef] [PubMed]

57. Espley, R.V.; Bovy, A.; Bava, C.; Jaeger, S.R.; Tomes, S.; Norling, C.; Crawford, J.; Rowan, D.; McGhie, T.K.; Brendolise, C.; et al. Analysis of genetically modified red-fleshed apples reveals effects on growth and consumer attributes. *Plant Biotechnol. J.* **2013**, *11*, 408–419. [CrossRef] [PubMed]

58. Sójka, M.; Macierzyński, J.; Zaweracz, W.; Buczek, M. Transfer and mass balance of ellagitannins, anthocyanins, flavan-3-ols, and flavonols during the processing of red raspberries (*Rubus idaeus* L.) to juice. *J. Agric. Food Chem.* **2016**, *64*, 5549–5563. [CrossRef] [PubMed]

59. Sablani, S.S.; Andrews, P.K.; Davies, N.M.; Walters, T.; Saez, H.; Syamaladevi, R.M.; Mohekar, P.R. Effect of thermal treatments on phytochemicals in conventionally and organically grown berries: Effect of thermal treatments on phytochemicals in berries. *J. Sci. Food Agric.* **2010**, *90*, 769–778. [CrossRef]

60. Jiménez-Sánchez, C.; Lozano-Sánchez, J.; Segura-Carretero, A.; Fernández-Gutiérrez, A. Alternatives to conventional thermal treatments in fruit-juice processing. Part 1: Techniques and applications. *Crit. Rev. Food Sci. Nutr.* **2017**, *57*, 501–523. [CrossRef]

61. Kamiloglu, S.; Toydemir, G.; Boyacioglu, D.; Beekwilder, J.; Hall, R.D.; Capanoglu, E. A Review on the effect of drying on antioxidant potential of fruits and vegetables. *Crit. Rev. Food Sci. Nutr.* **2016**, *56*, S110–S129. [CrossRef]

62. Ratti, C. Hot air and freeze-drying of high-value foods: A review. *J. Food Eng.* **2001**, *49*, 311–319. [CrossRef]

63. Novaković, M.M.; Stevanović, S.M.; Gorjanović, S.Ž.; Jovanovic, P.M.; Tešević, V.V.; Janković, M.A.; Sužnjević, D.Ž. Changes of hydrogen peroxide and radical-scavenging activity of raspberry during osmotic, convective, and freeze-drying. *J. Food Sci.* **2011**, *76*, C663–C668. [CrossRef]

64. Stamenković, Z.; Pavkov, I.; Radojčin, M.; Tepić Horecki, A.; Kešelj, K.; Bursać Kovačević, D.; Putnik, P. Convective drying of fresh and frozen raspberries and change of their physical and nutritive properties. *Foods* **2019**, *8*, 251. [CrossRef]

65. Harguindeguy, M.; Fissore, D. On the effects of freeze-drying processes on the nutritional properties of foodstuff: A review. *Dry. Technol.* **2020**, *38*, 846–868. [CrossRef]

66. Tomás-Barberán, F.A.; Espín, J.C. Phenolic compounds and related enzymes as determinants of quality in fruits and vegetables: Phenolics and food quality. *J. Sci. Food Agric.* **2001**, *81*, 853–876. [CrossRef]

67. Caritá, A.C.; Fonseca-Santos, B.; Shultz, J.D.; Michniak-Kohn, B.; Chorilli, M.; Leonardi, G.R. Vitamin C: One compound, several uses. Advances for delivery, efficiency and stability. *Nanomedicine Nanotechnol. Biol. Med.* **2020**, *24*, 102117. [CrossRef] [PubMed]

68. Goula, A.M.; Adamopoulos, K.G. Retention of ascorbic acid during drying of tomato halves and tomato pulp. *Dry. Technol.* **2006**, *24*, 57–64. [CrossRef]

69. Rodriguez, A.; Rodriguez, M.M.; Lemoine, M.L.; Mascheroni, R.H. Study and comparison of different drying processes for dehydration of raspberries. *Dry. Technol.* **2017**, *35*, 689–698. [CrossRef]

70. Rodriguez, A.; Bruno, E.; Paola, C.; Campañone, L.; Mascheroni, R.H. Experimental study of dehydration processes of raspberries (*Rubus Idaeus*) with microwave and solar drying. *Food Sci. Technol.* **2019**, *39*, 336–343. [CrossRef]

71. Li, F.; Chen, G.; Zhang, B.; Fu, X. Current applications and new opportunities for the thermal and non-thermal processing technologies to generate berry product or extracts with high nutraceutical contents. *Food Res. Int.* **2017**, *100*, 19–30. [CrossRef]

72. Perera, C.O.; Rahman, M.S. Heat pump dehumidifier drying of food. *Trends Food Sci. Technol.* **1997**, *8*, 75–79. [CrossRef]

73. Ong, S.P.; Law, C.L. Drying kinetics and antioxidant phytochemicals retention of salak fruit under different drying and pretreatment conditions. *Dry. Technol.* **2011**, *29*, 429–441. [CrossRef]

74. Maillard, M.-N.; Berset, C. Evolution of antioxidant activity during kilning: Role of insoluble bound phenolic acids of barley and malt. *J. Agric. Food Chem.* **1995**, *43*, 1789–1793. [CrossRef]

75. Dalmadi, I.; Rapeanu, G.; Van Loey, A.; Smout, C.; Hendrickx, M. Characterization and inactivation by thermal and pressure processing of strawberry (*Fragaria × ananassa*) polyphenol oxidase: A kinetic study. *J. Food Biochem.* **2006**, *30*, 56–76. [CrossRef]

76. Queiroz, C.; Mendes Lopes, M.L.; Fialho, E.; Valente-Mesquita, V.L. Polyphenol oxidase: Characteristics and mechanisms of browning control. *Food Rev. Int.* **2008**, *24*, 361–375. [CrossRef]

77. Beekwilder, J.; Jonker, H.; Meesters, P.; Hall, R.D.; van der Meer, I.M.; Ric de Vos, C.H. Antioxidants in raspberry: On-Line analysis links antioxidant activity to a diversity of individual metabolites. *J. Agric. Food Chem.* **2005**, *53*, 3313–3320. [CrossRef] [PubMed]

78. Taşeri, L.; Aktaş, M.; Şevik, S.; Gülcü, M.; Uysal Seçkin, G.; Aktekeli, B. Determination of drying kinetics and quality parameters of grape pomace dried with a heat pump dryer. *Food Chem.* **2018**, *260*, 152–159. [CrossRef] [PubMed]

79. Lo Piccolo, E.; Landi, M.; Massai, R.; Remorini, D.; Guidi, L. Girled-induced anthocyanin accumulation in red-leafed *Prunus cerasifera*: Effect on photosynthesis, photoprotection and sugar metabolism. *Plant Sci.* **2020**, *294*, 110456. [CrossRef] [PubMed]

80. Ceccanti, C.; Landi, M.; Antichi, D.; Guidi, L.; Manfrini, L.; Monti, M.; Tosti, G.; Frasconi, C. Bioactive Properties of fruits and leafy vegetables managed with integrated, organic, and organic no-tillage practices in the Mediterranean area: A two-year rotation experiment. *Agronomy* **2020**, *10*, 841. [CrossRef]

Publisher's Note: MDPI stays neutral with regard to jurisdictional claims in published maps and institutional affiliations.

Article

Stem Branching of Cycad Plants Informs Horticulture and Conservation Decisions

Thomas E. Marler [1,*] and **Michael Calonje [2]**

[1]   Western Pacific Tropical Research Center, University of Guam, UOG Station, Mangilao, Guam 96923, USA
[2]   Montgomery Botanical Center, 11901 Old Cutler Road, Coral Gables, Florida, FL 33156, USA; michaelc@montgomerybotanical.org
*   Correspondence: marler.uog@gmail.com; Tel.: +1-671-735-2100

Received: 22 September 2020; Accepted: 6 October 2020; Published: 8 October 2020

Abstract: The number of branches in male and female plants of *Cycas micronesica* K.D. Hill, *Cycas edentata* de Laub., *Cycas wadei* Merr., and *Zamia encephalartoides* D.W. Stev. were counted in Guam, Philippines, and Colombia, to confirm earlier reports that female plants develop fewer branches than males. *Cycas* plants produce determinate male strobili and indeterminate female strobili, but *Zamia* plants produce determinate strobili for both sexes. More than 80% of the female trees for each of the *Cycas* species were unbranched with a single stem, but more than 80% of the male trees exhibited two or more branches. The mean number of branches on male plants was more than double that of female plants. The number of branches of the *Zamia* male plants was almost triple that of female plants. Moreover, the *Zamia* plants produced 2.8-fold greater numbers of branches than the mean of the *Cycas* plants. Most of Guam's unsexed *C. micronesica* trees in 2004 were unbranched, but after 15 years of damage from non-native insect herbivores, most of the remaining live trees in 2020 contained three or more branches. The results confirm that male *Cycas* and *Zamia* plants produce more branches than female plants and suggest cycad species with determinate female strobili produce more branches on female plants than species with indeterminate female strobili. Our results indicate that the years of plant mortality on Guam due to non-native insect herbivores have selectively killed more female *C. micronesica* trees. Horticulture and conservation decisions may be improved with this sexual dimorphism knowledge.

Keywords: *Cycas*; determinate growth; dichotomous branch; isotomous branch; sexual dimorphism; *Zamia*

---

1. Introduction

Cycads are dioecious gymnosperms that are of horticultural and conservation interest [1,2], and are widely considered the most threatened plant group worldwide [3] primarily due to habitat destruction and the unsustainable trade of wild-collected plants. The unsustainable harvesting of cycads is a major concern at the local level but also internationally, where their trade is regulated by the Convention on International Trade in Endangered Species of Wild Fauna and Flora (CITES) [4,5]. Adaptive management of threatened cycad species requires the co-production of knowledge during conservation projects [6]. During horticultural projects in cultivated settings and biology or ecology projects in habitat, research results may provide crucial information for improving management decisions [7].

In wild cycad populations, only a certain proportion of plants may exhibit strobili at a given time. Consequently, most short-term field studies of these populations record sex ratio estimates that represent only what was observable at the time of fieldwork. This estimate is variously known as the coning [8], operational [9], or observed sex ratio [10], and it may differ greatly from the true genotypic sex ratio of the population and may vary widely depending on the time of year when observations were made. Therefore, the development of any protocol that would inform how to improve on the methods for determining the sex of cycads may improve decision-making.

Three publications indicated the use of stem branching of a cycad plant might be useful as secondary sex characteristics [11] for predicting sex of a plant. First, Ornduff [12] followed a cultivated group of *Zamia integrifolia* L.f. plants in Florida for 9 y to determine growth traits. The male plants produced an average of 5.7 branches per plant and the female plants produced 2.7 branches per plant. Male plants began producing cones at an earlier age and produced more cones throughout the observational period than female plants. These reproductive behavior differences were discussed as the probable cause of increased branching in male plants.

Second, Norstog and Nicholls [1] (p. 141) studied a fifty-year-old planting of *Z. integrifolia* and found marked sexual dimorphism in the cultivated cohort, with the males carrying more branches, more cones, and more leaves than female plants. The males held an average of 27 branches per plant whereas the females held an average of 8 branches per plant. The more frequent branching in male plants was attributed to the males experiencing more frequent coning events and producing more cones per reproductive episode than the female plants.

Third, Niklas and Marler [13] determined the branching traits of 483 *C. micronesica* adult plants in four research locations in Guam. The male trees exhibited a mean of 3.4 branches per tree and the female trees exhibited a mean of 1.5 branches per tree. The difference in reproductive behavior of male and female *Cycas* plants was discussed as the probable cause of increased branching in male trees.

Throughout our field work we recorded the number of stem apices on plants within numerous in situ populations as part of the plant metrics obtained to more fully understand the biology of several species. Our objectives herein were to look closely at these data to more fully understand branching behavior of four cycad species. Moreover, we observed branching behavior of *C. micronesica* before the invasions of several non-native insect herbivore species that have led to 96% mortality of Guam's population [14]. We exploited this phenomenon to look closely at branching behavior of the tree population that has survived the biological threats. Finally, we explore how this knowledge may improve horticultural and conservation decisions.

2. Materials and Methods

We collected data from numerous areas of occupancy for four cycad species in Colombia, Guam, and Philippines. These data included sympatric species, plant height and basal stem diameter, number of branches, number of leaves per plant and leaflets per leaf, length of petiole and rachis, and size of strobili if present. For the purpose of this study, we focused on the number of stem apices on each plant that we could identify as male or female. The general appearance of male and female *Cycas* trees in Guam and Philippines contrasted sharply with the appearance of male and female *Z. encephalartoides* plants in Colombia, the latter species had much shorter stems with a higher prevalence of stem branching (Figure 1).

Figure 1. The appearance of *Cycas* plants in Guam and Philippines: (**a**) male; (**b**) female. The appearance of *Zamia* plants in Colombia: (**c**) male; (**d**) female.

2.1. Cycas micronesica

The observations on *C. micronesica* trees were conducted in 2004 in Guam. Ten areas of occupancy throughout Guam were used for biochemistry studies designed to determine the biological and habitat factors that influenced neurotoxin concentrations in seeds. The plant traits were recorded in conjunction with these methods. Several non-native insect herbivore species invaded Guam during this time period, but their entry into our study sites occurred in 2005. Therefore, the 2004 data captured the population traits prior to any non-native insect herbivore damage. The number of branches on every tree that was identified as male or female was recorded. These data were combined to define traits of 1041 male and 754 female trees.

2.2. Cycas edentata

We observed 23 areas of occupancy of *C. edentata* among numerous islands within a latitudinal gradient of 9°10.655′–13°32.145′ N and longitudinal gradient of 119°52.788′–126°7.122′ E. The field work was conducted from August 2008 to March 2019. These data were combined to obtain measurements on 178 male and 234 female trees.

2.3. Cycas wadei

The only known endemic population of *C. wadei* on Culion Island was observed in March 2019 to obtain measurements on 104 male and 153 female trees.

2.4. Zamia encephalartoides

A single large grouping of *Z. encephalartoides*, restricted to a single plateau and consisting of 287 adult-sized individuals was studied in Santander, Colombia in February 2009 as part of a conservation survey for this species [15]. This population consisted of monopodial individuals or clumps containing up to 40 stems. As the stems of this cycad species may be partly subterranean and some branching may occur underground, it was sometimes difficult to ascertain with certainty whether clumps of stems belonged to the same individual plants. For the purpose of this study, only clumps exclusively containing strobili of a single sex and where all stems exhibited similar phenological stages were counted as individual plants. As the species occurs in a hot and dry environment, cone remnants can remain on plants for several months, allowing the sexual identity to be determined for over 80% of the observed plants. Measurements were obtained from 143 male and 88 female plants.

2.5. Unbiased Demography of Cycas micronesica

The methods described in the previous sub-sections employed biased methods where sexual identity of a plant was verified prior to collection of data. These methods did not allow an understanding of the branch patterns of the entire population, including plants that were mature in stature but exhibited no evidence of a strobilus. Therefore, we collected data from three high density *C. micronesica* localities with 900+ mature plants per ha along the east coast of Guam in January 2004 (13°25.674′–13°29.847′). We used four 4 × 100 m transects per locality and observed every plant greater than 100 cm in height and recorded the number of branches. A total of 432 mature plants were observed. We returned to these same localities to repeat the measurements in January 2020. By this time the density was less than 200 plants per ha. We repeated the methods with 4 × 100 m transects, but because of the plant density we increased the number of transects in each locality until we reached ≈150 plants per locality. A total of 456 mature plants were observed.

2.6. Analyses

The male versus female branch number data for each species were analyzed separately. The data did not meet parametric prerequisites, so we used the Mann–Whitney U test to determine significance (SPSS Statistics, IBM Corp., Armonk, NY, USA).

3. Results

3.1. Cycas micronesica

Most of the male *C. micronesica* trees contained two branches, accounting for about 70% of the trees (Figure 2a). The male trees that were unbranched accounted for about 10% of the trees, and the trees with three or more branches accounted for the remainder of the male trees. In contrast, the unbranched female *C. micronesica* trees accounted for about 80% of the population. The mean number of branches for male trees was almost double that for female trees (Table 1).

Table 1. The influence of sex on stem branching of four cycad species. Significance determined by the Mann–Whitney U test. Means ± standard error.

Species	Sex	N	Range	Mean	U	P
Cycas micronesica	Male	1041	1–12	2.52 ± 0.06		
	Female	754	1–8	1.36 ± 0.04	167,130	<0.001
Cycas wadei	Male	104	1–7	2.55 ± 0.12		
	Female	153	1–3	1.12 ± 0.04	1779	<0.001
Cycas edentata	Male	178	1–10	3.03 ± 0.12		
	Female	234	1–2	1.07 ± 0.02	2698	<0.001
Zamia encephalartoides	Male	143	1–40	8.14 ± 0.61		
	Female	88	1–18	2.83 ± 0.31	2649	<0.001

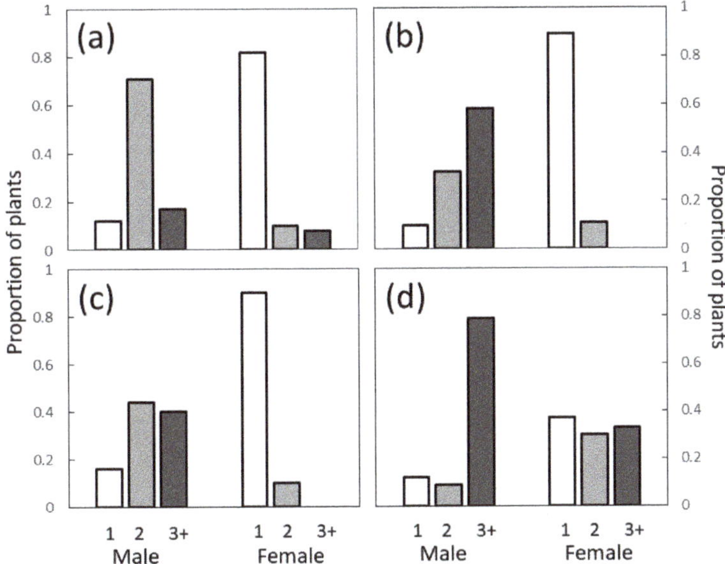

Figure 2. The influence of sex on the proportion of plants with 1, 2 or 3+ branches. (**a**) *Cycas micronesica*; (**b**) *Cycas wadei*; (**c**) *Cycas edentata*; (**d**) *Zamia encephalartoides*.

3.2. Cycas wadei

The number of male *C. wadei* trees with two branches and the number with three or more branches each accounted for about 40% of the trees (Figure 2b). Almost 20% of the male trees were monopodial. In contrast, the unbranched female *C. wadei* trees accounted for about 90% of the population, with the remainder of the trees producing two branches. The mean number of branches for male trees was 2.3-fold greater than that for female trees (Table 1).

3.3. Cycas edentata

The number of male *C. edentata* trees with three or more branches accounted for about 60% of the trees (Figure 2c). The male trees that were unbranched accounted for about 10% of the trees, and the trees with two branches accounted for the remainder of the trees. The female *C. edentata* trees exhibited branch proportions that were similar to those of *C. wadei*. The mean number of branches for male trees was 2.8-fold greater than that for female trees (Table 1).

3.4. Zamia encephalartoides

The *Z. encephalartoides* plants generally exhibited many more branches than any of the *Cycas* trees. About 80% of the male plants exhibited three or more branches (Figure 2d). The remainder of male plants were split evenly between unbranched and two-branched individuals. The female *Z. encephalartoides* plants were fairly evenly split among plants with one, two, or more than two branches. The mean number of branches for male *Z. encephalartoides* plants was more than two times greater than the mean of the three *Cycas* species and was almost triple that of female *Z. encephalartoides* plants (Table 1).

3.5. Unbiased Demography

Surveys of every *C. micronesica* tree in excess of 100 cm in height from Guam's high-density localities indicated more than 60% of the 2004 trees were unbranched (Figure 3). About 20% of these pre-invasion trees exhibited two branches, and about 20% exhibited three or more branches.

The widespread mortality that occurred prior to 2020 preferentially targeted unbranched trees, as the 2020 observations indicated less than 10% of the population was unbranched. The percentage of trees with two branches remained at about 20% of the population, but trees with three or more branches accounted for about 75% of the population. Therefore, a 7.6-fold decrease in unbranched trees accompanied a 4-fold increase in trees with more than two branches during the changes between our two years of observation.

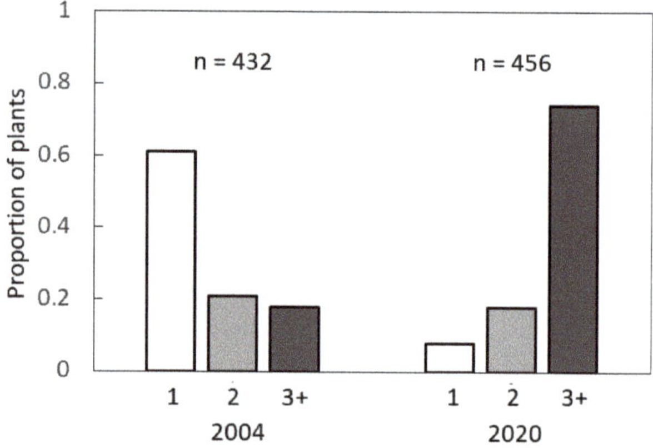

Figure 3. The influence of year on the proportion of *Cycas micronesica* plants with 1, 2 or 3+ branches. n represents the total number of trees in each census.

4. Discussion

The higher propensity for branching in male versus female cycad plants reported for cultivated *Z. integrifolia* [2,12] was also documented for in situ populations of *Cycas micronesica* [13]. The present study confirms these findings with robust in situ data sets of three *Cycas* species and one *Zamia* species. We believe the higher branch number found in male plants of the species studied may be the consequence of isotomous branching associated with the more frequent coning of male cycad plants [12,16]. Vegetative branching in cycads occurs via two known mechanisms: isotomous branching of the shoot apex or adventitious branching associated with damaged stem regions or leaf bases [17]. Adventitious branching is the most common form of branching in cycads, occurring in the ten living cycad genera and producing the offsets (known as suckers or pups) that are typically used for vegetative propagation of cycads. Isotomous vegetative branching, where the shoot apex divides dichotomously at the stem axis to form two equal branches, appears to be much rarer in cycads and has only been reported in *Cycas*, *Dioon*, and *Zamia* [1,12,17–19]. However, another dichotomous branching process, termed anisotomous branching [17] occurs in cycad genera that produce terminal strobili.

In anisotomous branching, the original apex dichotomizes and produces a fertile apex that will develop a cone (or set of cones) and another apex that goes dormant and resumes vegetative growth after the strobili mature [17]. Anisotomous branching is detectable in the anatomy of cycads stems by the presence of dome-like profiles of vascular tissue known as cone domes [1] (p. 45) but is not readily apparent externally because the vegetative apex resumes upward growth of the stem in a seemingly continuous fashion.

All cycad genera, with the exception of members of the Tribe Encephalartae (sensu [20]) produce terminal cones and therefore undergo anisotomous branching. In Encephalartae, comprised of the three related genera *Encephalartos* Lehm., *Macrozamia* Miq., and *Lepidozamia* Regel [21], cones are produced to the side of the vegetative apex with no associated apex dichotomy [22]. In *Cycas*, anisotomous branching is restricted to male plants, as female plants produce indeterminate crowns of individual sporophylls (Figure 4) that are produced similarly and alternately with crowns of leaves [23,24].

Figure 4. The reproductive behavior of a female *Cycas* plant produces a megastrobilus that is not determinate. Each sporophyll is an obvious homolog of the pinnate compound leaf. Sporophylls from a megastrobilus (blue arrows) separate the antecedent (red arrows) and subsequent (yellow arrows) leaf growth events.

Multiple studies have associated coning events with vegetative branching, suggesting a close association between anisotomous and isotomous branching. Gorelick [19] reported isotomous branching in *Zamia furfuracea* L.f. immediately following a coning event. In this case, the vegetative branches occurred on both sides of the cone axis with both branches producing simultaneous vegetative flushes. Ornduff [12] observed branching was closely associated with cone production in a cultivated cohort of *Z. integrifolia*. He found a significantly higher number of branches were produced by male plants and attributed this phenomenon to the male plants experiencing more coning events due to their reproduction starting at an earlier age than the females. Similarly, in a cultivated group of *Z. integrifolia* plants of similar age, Norstog and Nicholls [1] (p. 141) found a significantly larger number of branches on males than females. This phenomenon was attributed to increased apex subdivisions due to their more frequent cone production. Additionally, male plants were generally more robust and produced more leaves than females.

The process of isotomous vegetative branching associated with a coning event is illustrated in male *Cycas micronesica* (Figure 5). In a similar manner as described by Gorelick [19] with *Z. furfuracea*, following the production of a terminal cone, two branches of similar size are formed at either side of the cone axis and produce vegetative flushes simultaneously. In this study, isotomous vegetative branching was observed only on male *Cycas* plants, but on both sexes of *Zamia*.

The branching behaviors of the three *Cycas* species were remarkably similar, with less than 20% of male trees and more than 80% of the female trees exhibiting unbranched trunks. Most female trees were unbranched for all three *Cycas* species, but the mean and maximum number of branches observed in female *C. micronesica* trees exceeded those in the other two species. We note two factors that may explain these differences. First, the differences could be due to inter-specific genetic differences where *C. micronesica* is more prone to branching. The addition of many more *Cycas* species to this research agenda would aid in understanding the role of genetics on branching of female trees. Second, the greater branching from Guam's trees may be due to greater tropical cyclone (TC) activity. The Philippine and Guam habitats are within the most active TC basin worldwide [25]. Guam's propensity for TC occurrence must be considered when making horticultural decisions for perennial crops [26]. A major TC in 1997 caused an estimated 10% of Guam's *C. micronesica* population to become decapitated with the entire stem apex snapped off [27]. Guam's trees that suffer this form of damage during TCs develop numerous adventitious buds at positions on the stem that are proximal to the break ([28], Figure 1). After almost two years of regrowth, the trees that withstood the 1997 TC with intact apices exhibited only two vegetative growth events, but the decapitated trees exhibited as many as 25 vegetative growth events over the experimental period as a result of the numerous adventitious branches [29]. Observations since the invasions of several non-native insect herbivores indicate this new biotic threat has weakened the *C. micronesica* stems such that they are more vulnerable to TC damage [30]. This was confirmed during a 2015 TC when damage to the *C. micronesica* population was greater than expected based on historical observations of damage [31]. The third possible explanation for the greater number of apices for *C. micronesica* than for the other two *Cycas* species could be an interaction of genetic differences and TC propensity. This could be tested by observing branching patterns of mature specimens of the three *Cycas* species in a common garden setting.

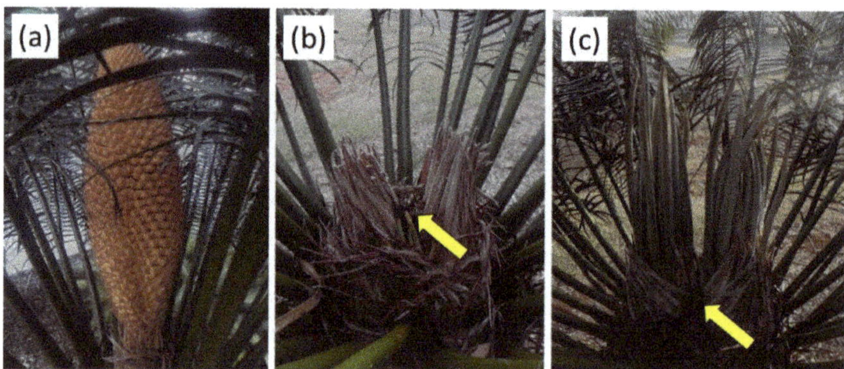

Figure 5. The reproductive behavior of a male *Cycas* plant. (**a**) July 2019 microstrobilus is determinate and terminates the stem growth; (**b**) the appearance of two adventitious apices becomes apparent five months later; (**c**) resumption of stem growth is apparent seven months after the microstrobilus. Yellow arrows point to the base of the antecedent microstrobilus.

4.1. Refinements

A look at the gestalt appearance of the entire branched *Cycas* canopy offers more refined diagnostics by using topology. The first tenet that can be employed to interpret a tree's canopy is that trees with multiple sites of bifurcation are likely males. Consider two model trees with five apices. One tree exhibits four bifurcation events leading to four stratified dichotomous branching components of the canopy (Figure 6a). The second tree exhibits five stems that converge to unite at one stratum of the basal stem (Figure 6b). The first tree is likely a male tree because of the multiple examples of dichotomous branching. The second tree could be either sex because the branching connotes adventitious branching in response to a mechanical injury at the stratum of the manifold branching. The number of *C. micronesica* trees that develop a canopy similar to Figure 6b may increase in the future due to the issues described in Section 4.1. The stems that have been weakened by non-native insect herbivory may break more often in future TCs such that more trees on average develop regrowth with multiple adventitious branches emerging from one stratum.

A second tenet that may aid in interpretation of branching is that the isotomous branching following a determinate strobilus generates two branches of a similar size, as can be seen in Figures 1c and 6a. If a cycad stem reveals branching into two stems, and one secondary stem is much larger than the other second secondary stem, this is not likely the result of a post-strobilus bifurcation event. Sometimes, two equal sized branches may develop from adventitious branching after an injury, a phenomenon referred to as pseudoisotomous branching by Stevenson [17]. Pseudoisotomous branching is distinguishable macromorphologically from isotomous branching. In isotomous branching, the leaf base pattern is continuous between the stem and the branches and no constriction is observed at the base of the branches, whereas in pseudoisotomous branching, the leaf base pattern is discontinuous and constriction is observed at the base of the branches [17]. These tenets may be most easily condensed to the simple rule that multiple examples of dichotomous branching on a *Cycas* tree likely signifies a male tree and unbranched stems likely signify a female tree. Our observations of *Cycas* trees in the field suggest that isotomous branching is most common in male plants, and that it is either very rare or inexistent in female plants. In contrast, isotomous branching was observed on both sexes in *Z. encephalartoides*, suggesting that this type of branching may be associated with the capability of producing terminal cones.

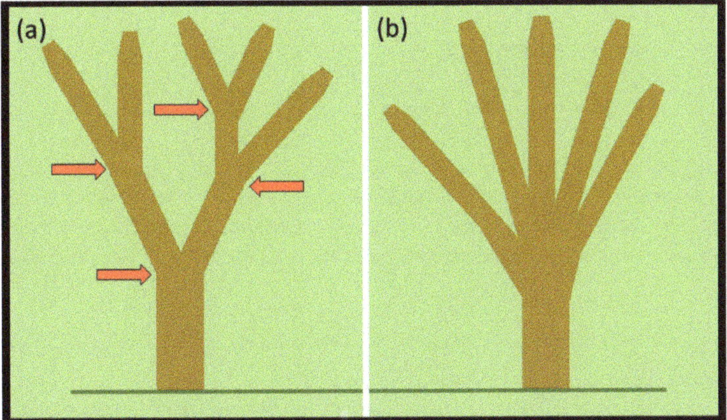

Figure 6. Two approaches for producing branches on a cycad plant. (**a**) Four bifurcations (red arrows) at different strata generate five apices; (**b**) five branches emerge from one stratum.

4.2. Horticulture and Conservation

Our unambiguous results revealing sexual dimorphism in branch number may be used by horticulturists to inform decisions. Dioecious species offer choices to horticulturists that are not available for monoecious species. For example, if a property owner has a preference for an unbranched *Cycas* tree that appears more palm-like (Figure 1b), the new knowledge that a female tree will meet that preference more often than a male tree can be exploited. Conversely, when considering highly branching, shrub-like geophilous species of *Zamia* for landscape design, male plants may make more robust specimens with a higher number of branches, leaves and cones, as demonstrated with *Z. integrifolia* by Norstog and Nicholls [1] (p. 141). Shrub-like *Zamia* species are often used in landscape designs as stand-alone specimens or in mass plantings as foreground layers or hedges. This shrub-like habit can be found in several *Zamia* species throughout the geographic range of the genus [32] but is most common in Caribbean species such as *Z. integrifolia*, and Mexican species, such as *Z. furfuracea*, one of the most abundantly cultivated cycads in the world. Small basal stem cuttings of cycad plants are highly successful in producing adventitious roots [33–35]. This knowledge may be exploited to clonally propagate a female or a male tree to obtain the desired landscape objective.

Our results also inform conservation decisions. Cycads form the most threatened group of plants worldwide [3]. For example, *C. wadei* is listed as critically endangered [36], *C. micronesica* is listed as endangered [37], *Z. encephalartoides* is listed as vulnerable [38], and *C. edentata* is listed as near threatened [39]. Our repeated visits to numerous areas of occupancy for all four of our model species have documented extensive loss of habitat due to anthropogenic land conversion actions.

This is happening in Guam due to expansive construction projects, and federally funded rescue projects have employed the collection of cuttings or excavated transplants of *C. micronesica* to rescue some of the genetic diversity of the destroyed populations. Much has been learned about the capture of in situ genetic diversity while collecting for ex situ conservation of cycad species [40,41]. The collection of clones instead of seeds improves the conservation of genetic diversity in ex situ collections [42]. The pilot study for the Guam rescue projects [43] utilized the knowledge of branch differences to ensure ≈60% of the rescued trees were unbranched, ≈20% contained two apices, and ≈20% contained three or more apices as an approach that would ensure fairly equal representation of male and female trees in the rescued population. In contrast the most recent large-scale salvage project, which employed excavation and transplantation of mature trees, did not utilize this branching knowledge, as the rescued population was heavily represented by specimens with multiple apices (personal observation, T.E.M.). This oversight ensured that most of the rescued population would be male trees, thereby reducing future recruitment and regeneration potential of the restoration site in which the plants were transplanted. These case studies illuminate the need to ensure that appropriate species expertise is guiding the decisions in conservation projects, especially when public funding is involved. Indeed, that male cycad plants produce more stem apices than female plants has been known since the 1980s [12], so if a cycad biologist had been consulted in the conservation decisions of the transplant project this mistake would not have been made.

The use of stem branching traits to inform cycad conservation decisions adds to a growing body of evidence that illuminates the importance of including conservation practitioners with a verifiable understanding of plant behavior. Other plant traits that have been discussed for use in cycad conservation decisions included stem carbon dioxide efflux [44] and stem height increment [45].

4.3. Future Directions

The collective results point toward several areas of needed research. Continued work in these directions would contribute greatly to cycad biology and conservation.

First, long-term observational studies are needed to determine the branching behavior for many cycad species. Although we believe the higher number of branches found on male plants in this study is a consequence of isotomous branching associated with more frequent terminal cone production, this is based on indirect observations, and the relative contribution of adventitious vs. isotomous branching to the stem architecture of these and other cycad species remains obscure. Direct observations over time will provide the data needed to quantify the nature of branching on different cycad species and to better understand the process of stem bifurcations following coning events. The collective data to date include only two *Zamia* and three *Cycas* species, and little is known about the branching behavior of the remaining 358 species of cycads [46]. Clearly, observations from the other eight genera and more species are required to improve our understanding of branching processes in cycads, and botanic gardens are ideal settings for this demanding long-term endeavor.

Second, future research should explore the relationship between reproductive anisotomous branching and the isotomous vegetative branching that often follows it and attempt to understand the mechanisms underlying these phenomena. While we observed this in both sexes of *Zamia*, we only observed it on male *Cycas* plants. As female *Cycas* do not form terminal cones, this observation suggests that there may be a link between the ability of cycads to produce terminal cones via anisotomous branching and their ability to produce isotomous vegetative branches. A more detailed survey of female *Cycas* plants and the plants from the three genera that do not produce terminal cones (*Encephalartos*, *Lepidozamia*, and *Macrozamia*) may confirm whether this is the case. Additionally, anatomical studies are needed in order to understand the mechanism underlying isotomous vegetative branching following cone production. For example, it is unclear if this phenomenon could be the result of trichotomous branching where the apex divides into one fertile and two vegetative apices, or of two dichotomous branching events occurring in quick succession, or of some other unknown mechanism. Careful sectioning of cycad stems undergoing this process could help resolve this conundrum.

Third, our results indicating that the recent invasions of non-native insect herbivores have killed more unbranched trees than multi-branched trees in Guam deserve further study. The results indicate more female trees have been killed than male trees, which will exert negative consequences for species recovery if future conservation interventions become successful in mitigating the ubiquitous biological threats. This disparity in vulnerability to the biological threats is likely mediated by non-structural carbohydrate relations. The greatest threat to the *C. micronesica* plants has been the armored scale *Aulacaspis yasumatsui* Takagi, and plant mortality from this herbivore is preceded by a gradual depletion of plant non-structural carbohydrates [47]. Cumulative traditional knowledge that guides the exploitation of cycad stems for the production of starch for human consumption indicates male stems yield more starch than female stems [48]. These issues collectively indicate the limited pool of non-structural carbohydrates in female trees at the time of the insect invasions may have caused the female trees to be more vulnerable than the male trees. Further studies of stem non-structural carbohydrate relations in male and female trees may improve our understanding of these historical dynamics and inform future conservation decisions for species recovery.

Fourth, illuminating clarity of how subsets of plants inform these issues may improve the predictive usage of the information in horticulture. For example, our three *Cycas* arborescent species exhibited less difference in total number of branches between male and female plants than the *Zamia* species with partly subterranean stems. Do the male-female differences in branching behave differently based on life form, rather than based on taxonomy, such that the arborescent species as a group behave differently than the subterranean species as a group? Moreover, most of our female *Cycas* individuals were unbranched. Do the indeterminate traits of *Cycas* megastrobili versus the determinate traits of the megastrobili of the other arborescent genera fully explain the increased proportion of monopodial *Cycas* individuals? For arborescent species, tree height increases with the number of primary growth events which accumulate with age [1]. Does tree height correlate positively with the number of branches for all species?

5. Conclusions

We have added more empirical evidence to earlier reports of sexual dimorphism that indicate male cycad plants produce more branches than female plants. Collectively, three *Cycas* species and two *Zamia* species have confirmed this aspect of secondary sexual behavior of the Cycadales. The changes in proportion of branched versus unbranched *C. micronesica* trees in Guam between 2004 and 2020 indicates that more female trees have died due to the damage from non-native insect herbivores that began in 2005. Horticulturists may exploit this knowledge of branching to increase the likelihood of achieving near equal numbers male and female plants during conservation rescue projects.

Author Contributions: Conceptualization, T.E.M. and M.C.; methodology, T.E.M. and M.C.; formal analysis, T.E.M.; resources, T.E.M. and M.C.; writing—original draft preparation, T.E.M.; writing—review and editing, M.C.; project administration, T.E.M. and M.C.; funding acquisition, T.E.M. and M.C. All authors have read and agreed to the published version of the manuscript.

Funding: This research was funded by the United States Department of Agriculture grant numbers 2003-05495 and 2013-31100-06057, the United States Forest Service grant numbers 09-DG-11052021-173, 13-DG-11052021-210, and 17-DG-11052021-217, and the Conservation Leadership Programme's Future Conservationist Award (project ID 130609).

Acknowledgments: We thank Nirmala Dongol, Gil Cruz, Frankie Matanane, and Michell Mooy for technical support.

Conflicts of Interest: The authors declare no conflict of interest.

References

1. Norstog, K.J.; Nicholls, T.J. *The Biology of the Cycads*; Cornell University Press: Ithaca, NY, USA, 1997; ISBN 978-0-8014-3033-6.
2. Whitelock, L.M. *The Cycads*; Timber Press: Portland, OR, USA, 2002.
3. Fragniere, Y.; Bétrisey, S.; Cardinaux, L.; Stoffel, M.; Kozlowski, G. Fighting their last stand? A global analysis of the distribution and conservation status of gymnosperms. *J. Biogeogr.* **2015**, *42*, 809–820. [CrossRef]
4. Convention on International Trade in Endangered Species of Wild Fauna and Flora. *The CITES Species.* Available online: https://www.cites.org/eng/disc/species.php (accessed on 8 October 2020).
5. Rutherford, C.; Donaldson, J.; Hudson, A.; McGough, H.N.; Sajeva, M.; Schippmann, U.; Tse-Laurence, M. *CITES and Cycads: A User's Guide*; Kew Publishing, Royal Botanic Gardens: Kew, UK, 2013.
6. Marler, T.E. Horticultural research crucial for plant conservation and ecosystem restoration. *HortScience* **2017**, *52*, 1648–1649. [CrossRef]

7.   Marler, T.E.; Lindström, A.J. The value of research to selling the conservation of threatened species: The case of *Cycas micronesica*. *J. Threat. Taxa* **2014**, *6*, 6523–6528. [CrossRef]

8.   Clark, D.A.; Clark, D.B. Temporal and environmental patterns of reproduction in *Zamia skinneri*, at tropical rain forest cycads. *J. Ecol.* **1987**, *75*, 135–149. [CrossRef]

9.   Lazcano-Lara, J.C.; Ackerman, J.D. Best in the company of nearby males: Female success in the threatened cycad, *Zamia portoricensis*. *PeerJ* **2018**, *6*, e5252. [CrossRef]

10. Calonje, C.; Calonje, M.A.; Husby, C.E. Cycad sex ratios at Montgomery Botanical Center. *Mem. N. Y. Bot. Garden* **2018**, *117*, 360–370.

11. Lloyd, D.G.; Webb, C.J. Secondary sex characteristics in seed plants. *Bot. Rev.* **1977**, *43*, 177–216. [CrossRef]

12. Ornduff, R. Gender performance in a cultivated cohort of the cycad *Zamia integrifolia* (Zamiaceae). *Am. J. Bot.* **1996**, *83*, 1006–1015. [CrossRef]

13. Niklas, K.J.; Marler, T.E. Sex and population differences in allometry of an endangered cycad species, *Cycas micronesica* (Cycadales). *Int. J. Plant Sci.* **2008**, *169*, 659–665. [CrossRef]

14. Marler, T.E.; Krishnapillai, M.V. Longitude, forest fragmentation, and plant size influence *Cycas micronesica* mortality following island insect invasions. *Diversity* **2020**, *12*, 194. [CrossRef]

15. Calonje, M.; López-Gallego, C. Cycad conservation: *Zamia encephalartoides* in Colombia. *Montgomery Bot. News* **2011**, *19*, 4.

16. Dongol, N.; Marler, T.E. Season and frequency of *Cycas micronesica* leaf and reproductive events. *Mem. N. Y. Bot. Gard.* **2018**, *117*, 497–503.

17. Stevenson, D.W. Observations on vegetative branching in cycads. *Int. J. Plant Sci.* **2020**, *181*, 564–580. [CrossRef]

18. Singh, R.; Radha, P. A new species of Cycas from Karnataka, India. *Bot. J. Linnaean Soc.* **2008**, *158*, 430–435. [CrossRef]

19. Gorelick, R. Isotomous branching at the *Zamia furfuracea* cone axis. *Cycad Newsl.* **2011**, *34*, 29.

20. Stevenson, D.W. A formal classification of the extant cycads. *Brittonia* **1992**, *44*, 220–223. [CrossRef]

21. Salas-Leiva, D.E.; Meerow, A.W.; Calonje, M.; Griffith, M.P.; Francisco-Ortega, J.; Nakamura, K.; Stevenson, D.W.; Lewis, C.E.; Namoff, S. Phylogeny of the cycads based on multiple single-copy nuclear genes: Congruence of concatenated parsimony, likelihood and species tree inference methods. *Ann. Bot.* **2013**, *112*, 1263–1278. [CrossRef]

22. Norstog, K. Cycadophyte. Encyclopedia Britanica. 2017. Available online: https://www.britannica.com/plant/cycadophyte (accessed on 8 October 2020).

23. Stevenson, D.W. Strobilar ontogeny in the Cycadales. In *Aspects of Floral Development*; Leins, S., Tucker, S.C., Endress, P.K., Eds.; Cramer: Berlin, Germany, 1988; pp. 205–224.

24. Stevenson, D.W. Morphology and systematics of the Cycadales. *Mem. N. Y. Bot. Gard.* **1990**, *57*, 8–55.

25. Marler, T.E. Pacific island tropical cyclones are more frequent and globally relevant, yet less studied. *Front. Environ. Sci.* **2014**, *2*, 42. [CrossRef]

26. Marler, T.E. Tropical cyclones and perennial species in the Mariana Islands. *Hortscience* **2001**, *36*, 264–268. [CrossRef]

27. Marler, T.E.; Hirsh, H. Guam's *Cycas micronesica* population ravaged by Supertyphoon Paka. *HortScience* **1998**, *33*, 1116–1118. [CrossRef]

28. Marler, T.E.; Lawrence, J.H. Phytophagous insects reduce cycad resistance to tropical cyclone winds and impair storm recovery. *HortScience* **2013**, *48*, 1224–1226. [CrossRef]

29. Hirsh, H.T.; Marler, T. Damage and recovery of *Cycas micronesica* after Typhoon Paka. *Biotropica* **2002**, *34*, 598–602. [CrossRef]

30. Marler, T.E. Increased threat of island endemic tree's extirpation via invasion-induced decline of intrinsic resistance to recurring tropical cyclones. *Communic. Integr. Biol.* **2013**, *6*, e22361. [CrossRef] [PubMed]

31. Marler, T.E.; Lawrence, J.H.; Cruz, G.N. Topographic relief, wind direction, and conservation management decisions influence *Cycas micronesica* K.D. Hill population damage during tropical cyclone. *J. Geogr. Nat. Disasters* **2016**, *6*, 3. [CrossRef]

32. Calonje, M.; Meerow, A.W.; Griffith, M.P.; Salas-Leiva, D.; Vovides, A.P.; Coiro, M.; Francisco-Ortega, J. A time-calibrated species tree phylogeny of the New World cycad genus *Zamia*, L. (Zamiaceae, Cycadales). *Int. J. Plant Sci.* **2019**, *180*, 286–314. [CrossRef]

33. Deloso, B.E.; Lindström, A.J.; Camacho, F.A.; Marler, T.E. Highly successful adventitious root formation of *Zamia* stem cuttings exhibits minimal response to indole-3-butyric acid. *HortScience* **2020**, *55*, 1463–1467. [CrossRef]

34. Deloso, B.E.; Paulino, C.J.; Marler, T.E. Leaf retention on stem cuttings of two *Zamia*, L. species with or without anti-transpirants does not improve adventitious root formation. *Trop. Conserv. Sci.* **2020**, *13*. [CrossRef]

35. Marler, T.E.; Deloso, B.E.; Cruz, G.N. Prophylactic treatments of *Cycas* stem wounds influence vegetative propagation. *Trop. Conserv. Sci.* **2020**, *13*, 1–6. [CrossRef]

36. Hill, K.D. *Cycas wadei. The IUCN Red List of Threatened Species 2010: E.T42097A10631294*. 2010. Available online: https://www.iucnredlist.org/species/42097/10631294 (accessed on 8 October 2020). [CrossRef]

37. Marler, T.; Haynes, J.; Lindstrom, A. *Cycas micronesica. The IUCN Red List of Threatened Species 2010: E.T61316A12462113*. 2010. Available online: https://www.iucnredlist.org/species/61316/12462113 (accessed on 8 October 2020). [CrossRef]

38. Stevenson, D.W. *Zamia encephalartoides. The IUCN Red List of Threatened Species 2010: E.T42151A10668503*. 2010. Available online: https://www.iucnredlist.org/species/42151/10668503 (accessed on 8 October 2020). [CrossRef]

39. Osborne, R.; Hill, K.D.; Nguyen, H.T.; Phan, K.L. *Cycas edentata. The IUCN Red List of Threatened Species 2010: E.T42091A10628022*. 2010. Available online: https://www.iucnredlist.org/species/42091/10628022 (accessed on 8 October 2020). [CrossRef]

40. Griffith, M.P.; Calonje, M.; Meerow, A.W.; Tut, F.; Kramer, A.T.; Hird, A.; Magellan, T.M.; Husby, C.E. Can a botanic garden cycad collection capture the genetic diversity in a wild population? *Int. J. Plant Sci.* **2015**, *176*, 1–10. [CrossRef]

41. Griffith, M.P.; Calonje, M.; Meerow, A.W.; Francisco-Ortega, J.; Knowles, L.; Aguilar, R.; Tut, F.; Sánchez, V.; Meyer, A.; Noblick, L.R.; et al. Will the same ex situ protocols give similar results for closely related species? *Biodivers. Conserv.* **2017**, *26*, 2951–2966. [CrossRef]

42. Griffith, M.P.; Clase, T.; Toribio, P.; Piñeyro, Y.E.; Jimenez, F.; Gratacos, X.; Sanchez, V.; Meerow, A.; Meyer, A.; Kramer, A.; et al. Can a botanic garden metacollection better conserve wild plant diversity? A case study comparing pooled collections with an ideal sampling model. *Int. J. Plant Sci.* **2020**, *181*, 485–496. [CrossRef]

43. Marler, T.E.; Cruz, G.N. Adventitious rooting of mature *Cycas micronesica* K.D. Hill tree stems reveals moderate success for salvage of an endangered cycad. *J. Threat. Taxa* **2017**, *9*, 10565–10570. [CrossRef]

44. Marler, T.E. Stem CO$_2$ efflux of *Cycas micronesica* is reduced by chronic non-native insect herbivory. *Plant Signal. Behav.* **2020**, *15*, e1716160. [CrossRef]

45. Marler, T.E.; Griffith, M.P.; Krishnapillai, M.V. Height increment of *Cycas micronesica* informs conservation decisions. *Plant Signal. Behav.* **2020**, *15*, e1830237. [CrossRef]

46. Calonje, M.; Stevenson, D.W.; Osborne, R. The World List of Cycads. Available online: http://cycadlist.org (accessed on 8 October 2020).

47. Marler, T.E.; Cascasan, A.N.J. Carbohydrate depletion during lethal infestation of *Aulacaspis yasumatsui* on *Cycas revoluta. Int. J. Plant Sci.* **2018**, *179*, 497–504. [CrossRef]

48. Whiting, M.G. Toxicity of cycads, a literature review. *Econ. Bot.* **1963**, *17*, 270–302. [CrossRef]

Review

Chemical Element Concentrations of Cycad Leaves: Do We Know Enough?

Benjamin E. Deloso [1], Murukesan V. Krishnapillai [2], Ulysses F. Ferreras [3], Anders J. Lindström [4], Michael Calonje [5] and Thomas E. Marler [6,*]

1   College of Natural and Applied Sciences, University of Guam, Mangilao, GU 96923, USA; delosob@triton.uog.edu
2   Cooperative Research and Extension, Yap Campus, College of Micronesia-FSM, Colonia, Yap 96943, Micronesia; muru@comfsm.fm
3   Philippine Native Plants Conservation Society Inc., Ninoy Aquino Parks and Wildlife Center, Quezon City 1101, Philippines; ulyssesferreras@gmail.com
4   Plant Collections Department, Nong Nooch Tropical Botanical Garden, 34/1 Sukhumvit Highway, Najomtien, Sattahip, Chonburi 20250, Thailand; ajlindstrom71@gmail.com
5   Montgomery Botanical Center, 11901 Old Cutler Road, Coral Gables, FL 33156, USA; michaelc@montgomerybotanical.org
6   Western Pacific Tropical Research Center, University of Guam, Mangilao, GU 96923, USA
*   Correspondence: marler.uog@gmail.com

Received: 13 October 2020; Accepted: 16 November 2020; Published: 19 November 2020

Abstract: The literature containing which chemical elements are found in cycad leaves was reviewed to determine the range in values of concentrations reported for essential and beneficial elements. We found 46 of the 358 described cycad species had at least one element reported to date. The only genus that was missing from the data was *Microcycas*. Many of the species reports contained concentrations of one to several macronutrients and no other elements. The cycad leaves contained greater nitrogen and phosphorus concentrations than the reported means for plants throughout the world. Magnesium was identified as the macronutrient that has been least studied. Only 14 of the species were represented by data from in situ locations, with most of the data obtained from managed plants in botanic gardens. Leaf element concentrations were influenced by biotic factors such as plant size, leaf age, and leaflet position on the rachis. Leaf element concentrations were influenced by environmental factors such as incident light and soil nutrient concentrations within the root zone. These influential factors were missing from many of the reports, rendering the results ambiguous and comparisons among studies difficult. Future research should include the addition of more taxa, more in situ locations, the influence of season, and the influence of herbivory to more fully understand leaf nutrition for cycads.

Keywords: *Bowenia*; *Ceratozamia*; Cycadaceae; *Cycas*; *Dioon*; *Encephalartos*; leaf element composition; leaf tissue analysis; *Lepidozamia*; *Macrozamia*; *Stangeria*; *Zamia*; Zamiaceae

---

1. Background

Effective horticultural management of economic crops or threatened plant taxa requires an adequate understanding of essential nutrient accumulation, partitioning among organs, and remobilization prior to organ senescence. These biological phenomena influence many issues such as attractiveness to herbivores, the speed of litter decomposition, and soil changes within the zone of root proliferation and leaf litterfall. Knowledge of the concentrations of essential elements in plant organs is useful for determining plant health, diagnosing the cause of an observable problem, and measuring the efficacy of a fertilizer program [1–3]. Therefore, plant tissue analysis has been part of the traditional toolbox to meet management goals in agronomy, horticulture, and silviculture or to improve knowledge about the ecology of tree species.

In any testing procedure designed to determine the presence or absence of a measurable component from a sample, adherence to protocols that were developed through verifiable research is mandatory for achieving unambiguous results [4]. Moreover, recording and reporting the biological and environmental factors which are known to influence plant nutrient concentrations are necessary for methods to become standardized and repeatable, engender trust in the results, and to justify comparisons among studies.

The group of gymnosperm plants known as cycads is comprised of the mono-generic Cycadaceae family with 117 species and the Zamiaceae family with nine genera and 241 described species [5]. Research in applied sciences such as horticulture has been insufficient for members of this plant group [6,7]. For example, the global agenda of understanding how leaf element concentrations correlate with leaf functional traits has not sufficiently included cycad species [8]. This research agenda has expanded substantially in the past six years, and the subject has never been reviewed to date.

The aim of our review is to report which taxonomic groups have been most studied, to compile a listing of the published chemical element concentration data for cycad leaves, and to establish protocols for continued research to ensure the results are comparable among the various laboratories that contribute to the agenda in the future. Moreover, we conclude with a discussion of possible future research directions with the hope of inspiring more demanding protocols to better meet horticulture and conservation goals.

2. Species Studied

The literature search identified 18 publications in the primary literature in which the concentration of at least one chemical element was reported as a constituent of leaf tissue for at least one cycad species. Our primary focus was the essential nutrients, those chemical elements that are directly involved in plant function and are required by plants to complete the life cycle. Macronutrients are required in greater quantities, and micronutrients are required in small amounts. We also report beneficial nutrients, those chemical elements that may stimulate growth in some plants but do not meet the requirements of being essential. Other chemical elements which were reported in some studies were not included in this review. The numerical concentrations of elements which were presented in figures were estimated. Some reports used logarithmic data to meet parametric statistical requirements or to smooth regression modeling. In order to standardize our reported data into one format, we transformed these data to numerical concentrations. Misspelled species names were corrected and included if the mistake was easily diagnosed. Data were not included if misspelled species names could not be determined to be a currently accepted species. Synonyms or other obsolete names with an accepted binomial [5] were reported for the currently accepted binomial. These methods identified a total of 46 cycad species from the literature search (Table 1). In addition to the taxonomic authority, we also included the countries in which each species is considered endemic or indigenous. Leaf sampling from plants growing within natural habitat were considered "in situ" and leaf sampling from plants that were growing in managed gardens were considered "ex situ."

Table 1. Forty-six cycad species with reported leaf element concentrations.

Species	Family	Taxonomic Authority	Native Range
Bowenia serrulata	Zamiaceae	(W. Bull) Chamb.	Australia
Bowenia spectabilis	Zamiaceae	Hook. ex Hook.f.	Australia
Ceratozamia mexicana	Zamiaceae	Brongn.	Mexico
Cycas armstrongii	Cycadaceae	Miq.	Australia
Cycas debaoensis	Cycadaceae	Y.C.Zhong & C.J.Chen	China
Cycas diannanensis	Cycadaceae	Z.T.Guan & G.D. Tao	China
Cycas elongata	Cycadaceae	(Leandri) D.Yue Wang	Vietnam
Cycas fairylakea	Cycadaceae	(Leandri) D.Yue Wang	China
Cycas media	Cycadaceae	R.Br.	Australia
Cycas micholitzii	Cycadaceae	Dyer	Laos, Vietnam

Table 1. *Cont.*

Species	Family	Taxonomic Authority	Native Range
Cycas micronesica	Cycadaceae	K.D. Hill	Guam, Rota, Palau, Yap
Cycas nitida	Cycadaceae	K.D.Hill & A.Lindstr.	Philippines
Cycas nongnoochiae	Cycadaceae	K.D.Hill	Thailand
Cycas panzhihuaensis	Cycadaceae	L.Zhou & S.Y.Yang	China
Cycas revoluta	Cycadaceae	Thunb.	China, Japan
Cycas rumphii	Cycadaceae	Miq.	Australia, Indonesia, Papua New Guinea
Cycas sexseminifera	Cycadaceae	F.N.Wei	China, Vietnam
Cycas siamensis	Cycadaceae	Miq.	Cambodia, Laos, Myanmar, Thailand, Vietnam
Cycas szechuanensis	Cycadaceae	W.C.Cheng & L.K.Fu	China
Cycas thouarsii	Cycadaceae	R.Br. ex Gaudich	Comoros, Kenya, Madagascar, Mozambique, Seychelles, Tanzania
Cycas wadei	Cycadaceae	Merr.	Philippines
Dioon edule	Zamiaceae	Lindl.	Mexico
Dioon mejiae	Zamiaceae	Standl. & L.O.Williams	Honduras
Dioon sonorense	Zamiaceae	(De Luca, Sabato & Vázq.Torres) Chemnick, T.J.Greg. & Salas-Mor.	Mexico
Dioon spinulosum	Zamiaceae	Dyer ex Eichler	Mexico
Encephalartos cupidus	Zamiaceae	R.A.Dyer	South Africa
Encephalartos ferox	Zamiaceae	G.Bertol	Mozambique, South Africa
Encephalartos gratus	Zamiaceae	Prain	Malawi, Mozambique
Lepidozamia hopei	Zamiaceae	Regel	Australia
Lepidozamia peroffskyana	Zamiaceae	Regel	Australia
Macrozamia communis	Zamiaceae	L.A.S.Johnson	Australia
Macrozamia lucida	Zamiaceae	L.A.S.Johnson	Australia
Macrozamia macleayi	Zamiaceae	Miq.	Australia
Macrozamia moorei	Zamiaceae	F.Muell.	Australia
Macrozamia mountperriensis	Zamiaceae	F.M.Bailey	Australia
Macrozamia parcifolia	Zamiaceae	P.I.Forst. & D.L.Jones	Australia
Macrozamia reidlei	Zamiaceae	(Gaudich.) C.A.Gardner	Australia
Macrozamia serpentina	Zamiaceae	D.L.Jones & P.I.Forst	Australia
Stangeria eriopus	Zamiaceae	(Kunze) Baill.	South Africa
Zamia erosa	Zamiaceae	O.F.Cook & G.N.Collins	Cuba, Jamaica, Puerto Rico
Zamia fischeri	Zamiaceae	Miq.	Mexico
Zamia furfuracea	Zamiaceae	L.f.	Mexico
Zamia integrifolia	Zamiaceae	L.f.	Bahamas, Cayman Islands, Cuba, United States
Zamia portoricensis	Zamiaceae	Urb.	Puerto Rico
Zamia splendens	Zamiaceae	Schutzman	Mexico
Zamia standleyi	Zamiaceae	Schutzman	Guatemala, Honduras
Zamia vazquezii	Zamiaceae	D.W.Stev., Sabato & De Luca	Mexico

3. Green Leaf Elements

3.1. The Elements

Laboratory methods have varied among the years and among laboratories. The oldest articles in our review quantified nitrogen with Kjeldahl digestion, and most contemporary articles employ dry combustion approaches for nitrogen. The other minerals and metals are digested from the tissue, with nitric acid being used most often. Quantification is done with spectrometry most common in the earliest publications and spectroscopy being used more often in recent years. Macronutrient concentrations in cycad leaf tissue were highly variable among the elements. The total carbon found in cycad leaves was less variable than the other elements and ranged from 438–566 mg·g^{-1} among taxa of nine genera (Table 2) [9–17]. The range in nitrogen concentration in the cycad leaf tissue was considerable, with a 6.9-fold difference among the species and studies and considerable overlap among the nine genera [9–23]. The phosphorus concentration of the cycad leaf tissue was less variable than nitrogen, with a 4.9-fold difference among the species and studies represented by nine genera [9–14,17–20]. Potassium concentration was highly variable with a 7.6-fold difference among the nine genera studied [9–14,17–20,23,24]. Magnesium was determined for only two genera, yet the range in concentration was substantial with a 7.5-fold difference among the studies [11–14,17,18,20,22,24]. The calcium concentration of cycad leaf tissue was more variable than the other macronutrients, with a 19.8-fold difference among the nine genera and studies [9,11–14,17,18,20,23,24]. Sulfur concentration in cycad leaf tissue was also highly variable with a 22.8-fold difference among the nine genera and studies [9,17,19,20,23].

Table 2. Published ranges in green leaf concentrations of macronutrients, micronutrients, and beneficial elements for cycad plants.

Element	Genera	Species Studied	Species in Genus	Range	Reference
Aluminum	*Cycas*	1	117	22–60 mg·kg^{-1}	[23]
Boron	*Cycas*	2	117	11.6–43.4 mg·kg^{-1}	[11–14,20]
Calcium	*Bowenia*	2	2	5.0–6.1 mg·g^{-1}	[9]
Calcium	*Ceratozamia*	1	32	7.1 mg·g^{-1}	[9]
Calcium	*Cycas*	14	117	1.2–23.7 mg·g^{-1}	[9,11–14,17,18,20,24]
Calcium	*Dioon*	3	16	7.6–8.4 mg·g^{-1}	[9]
Calcium	*Encephalartos*	3	65	4.5–14.3 mg·g^{-1}	[9]
Calcium	*Lepidozamia*	2	2	3.6–5.0 mg·g^{-1}	[9]
Calcium	*Macrozamia*	4	41	1.4–7.1 mg·g^{-1}	[9,23]
Calcium	*Stangeria*	1	1	7.1 mg·g^{-1}	[9]
Calcium	*Zamia*	5	81	3.0–7.7 mg·g^{-1}	[9]
Carbon	*Bowenia*	2	2	508–519 mg·g^{-1}	[9]
Carbon	*Ceratozamia*	1	32	514 mg·g^{-1}	[9]
Carbon	*Cycas*	13	117	463–509 mg·g^{-1}	[9–14,17]
Carbon	*Dioon*	3	16	485–496 mg·g^{-1}	[9]
Carbon	*Encephalartos*	3	65	490–505 mg·g^{-1}	[9]
Carbon	*Lepidozamia*	2	2	438–566 mg·g^{-1}	[9,16]
Carbon	*Macrozamia*	5	41	512–524 mg·g^{-1}	[9,16]
Carbon	*Stangeria*	1	1	479 mg·g^{-1}	[9]
Carbon	*Zamia*	7	81	477–491 mg·g^{-1}	[9,15]
Chloride	*Cycas*	1	117	0.5–2.3 mg·g^{-1}	[24]
Copper	*Cycas*	2	117	2.0–17.9 mg·kg^{-1}	[11–14,18,20]
Copper	*Macrozamia*	1	41	2.1–2.8 mg·kg^{-1}	[22]
Iron	*Bowenia*	2	2	189–207 mg·kg^{-1}	[9]
Iron	*Ceratozamia*	1	32	106 mg·kg^{-1}	[9]
Iron	*Cycas*	14	117	27–410 mg·kg^{-1}	[9,11–14,18–20,24]
Iron	*Dioon*	3	16	117–163 mg·kg^{-1}	[9]
Iron	*Encephalartos*	3	65	93–363 mg·kg^{-1}	[9,19]
Iron	*Lepidozamia*	2	2	166–176 mg·kg^{-1}	[9]
Iron	*Macrozamia*	3	41	83–253 mg·kg^{-1}	[9]
Iron	*Stangeria*	1	1	228 mg·kg^{-1}	[9]
Iron	*Zamia*	6	81	142–1700 mg·kg^{-1}	[9,19]
Magnesium	*Cycas*	4	117	1.4–8.2 mg·g^{-1}	[11–14,17,18,20,24]
Magnesium	*Macrozamia*	1	41	1.1–1.9 mg·g^{-1}	[22]
Manganese	*Cycas*	3	117	20–152 mg·kg^{-1}	[11–14,18,20,24]
Manganese	*Macrozamia*	1	41	6–57 mg·kg^{-1}	[22]
Nitrogen	*Bowenia*	2	2	24–41 mg·g^{-1}	[9,16]
Nitrogen	*Ceratozamia*	1	32	13 mg·g^{-1}	[9]
Nitrogen	*Cycas*	17	117	16–44 mg·g^{-1}	[9–21]
Nitrogen	*Dioon*	4	16	14–17 mg·g^{-1}	[9,22]
Nitrogen	*Encephalartos*	3	65	15–19 mg·g^{-1}	[9,19]
Nitrogen	*Lepidozamia*	2	2	17–31 mg·g^{-1}	[9,16]
Nitrogen	*Macrozamia*	8	41	8–55 mg·g^{-1}	[9,16,21,23]
Nitrogen	*Stangeria*	1	1	22 mg·g^{-1}	[9]
Nitrogen	*Zamia*	8	81	12–30 mg·g^{-1}	[9,15,19]
Phosphorus	*Bowenia*	2	2	1.0–1.1 mg·g^{-1}	[9]
Phosphorus	*Ceratozamia*	1	32	0.8 mg·g^{-1}	[9]
Phosphorus	*Cycas*	14	117	0.7–3.4 mg·g^{-1}	[9–14,17–20]
Phosphorus	*Dioon*	3	16	0.8–1.5 mg·g^{-1}	[9]
Phosphorus	*Encephalartos*	3	65	1.0–1.3 mg·g^{-1}	[9,19]

Table 2. *Cont.*

Element	Genera	Species Studied	Species in Genus	Range	Reference
Phosphorus	*Lepidozamia*	2	2	0.8–1.2 mg·g^{-1}	[9]
Phosphorus	*Macrozamia*	4	41	0.5–1.2 mg·g^{-1}	[9,21]
Phosphorus	*Stangeria*	1	1	1.1 mg·g^{-1}	[9]
Phosphorus	*Zamia*	6	81	0.7–1.3 mg·g^{-1}	[9,19]
Potassium	*Bowenia*	2	2	5.5–6.2 mg·g^{-1}	[9]
Potassium	*Ceratozamia*	1	32	4.9 mg·g^{-1}	[9]
Potassium	*Cycas*	15	117	3.1–23.7 mg·g^{-1}	[9–14,17,18,20,24]
Potassium	*Dioon*	3	16	5.7–11.5 mg·g^{-1}	[9,19]
Potassium	*Encephalartos*	3	65	6.2–8.9 mg·g^{-1}	[9]
Potassium	*Lepidozamia*	2	2	9.5–10.6 mg·g^{-1}	[9]
Potassium	*Macrozamia*	4	41	5.1–11.3 mg·g^{-1}	[9,23]
Potassium	*Stangeria*	1	1	8.0 mg·g^{-1}	[9]
Potassium	*Zamia*	6	81	4.6–18.0 mg·g^{-1}	[9]
Selenium	*Cycas*	2	117	0.41–0.58 mg·kg^{-1}	[11,12]
Sodium	*Cycas*	2	117	0.2–1.2 mg·g^{-1}	[12,24]
Sodium	*Macrozamia*	1	41	0.3–1.0 mg·g^{-1}	[23]
Sulfur	*Bowenia*	2	2	1.9 mg·g^{-1}	[9]
Sulfur	*Ceratozamia*	1	32	1.4 mg·g^{-1}	[9]
Sulfur	*Cycas*	12	117	0.8–2.6 mg·g^{-1}	[9,17,19,20]
Sulfur	*Dioon*	3	16	1.1–1.4 mg·g^{-1}	[9]
Sulfur	*Encephalartos*	3	65	0.8–2.2 mg·g^{-1}	[9,19]
Sulfur	*Lepidozamia*	2	2	1.4–1.6 mg·g^{-1}	[9]
Sulfur	*Macrozamia*	4	41	0.8–1.9 mg·g^{-1}	[9,23]
Sulfur	*Stangeria*	1	1	2.3 mg·g^{-1}	[9]
Sulfur	*Zamia*	5	81	0.6–13.7 mg·g^{-1}	[9,19]
Zinc	*Bowenia*	2	2	19–21 mg·kg^{-1}	[9]
Zinc	*Ceratoamia*	1	32	24 mg·kg^{-1}	[9]
Zinc	*Cycas*	14	117	6–70 mg·kg^{-1}	[9,11–14,18,20,24]
Zinc	*Dioon*	3	16	12–23 mg·kg^{-1}	[9]
Zinc	*Encephalartos*	3	65	11–22 mg·kg^{-1}	[9]
Zinc	*Lepidozamia*	2	2	23–25 mg·kg^{-1}	[9]
Zinc	*Macrozamia*	4	41	4–22 mg·kg^{-1}	[9,22]
Zinc	*Stangeria*	1	1	53 mg·kg^{-1}	[9]
Zinc	*Zamia*	6	81	11–38 mg·kg^{-1}	[9]

Micronutrient concentrations in cycad leaf tissue were also highly variable among the elements and studies. Iron and zinc were the only micronutrients included in numerous articles, with nine genera represented among the studies for each element (Table 2). Iron was also the only element exhibiting one extreme outlier species, with *Zamia fischeri* [9,18] exhibiting iron concentrations more than 4-fold greater than the range of the remaining 33 species that have been studied [9,11–14,18–20,24]. The remaining micronutrients have not been observed adequately. Leaf chloride concentrations were reported for a single *Cycas* species [24], boron concentrations were reported for only two species [11–14,20], copper concentrations were reported for three species [11–14,18,20,22], and manganese concentrations were reported for four species [11–14,18,20,22,24]. The cycad leaf content of the micronutrients molybdenum and nickel have not been reported for any cycad species.

Several beneficial elements have been reported from cycad leaf tissue (Table 2). Aluminum concentration has been reported for one species [23], selenium has been reported for two species [11,12], and sodium has been reported for three species [12,23,24]. The remaining beneficial nutrients have not been studied in the context of cycad leaf physiology.

3.2. The Taxa

Bowenia, *Lepidozamia*, and *Stangeria* contain only one or two species each, and every one of these species was included in the literature review (Table 2, Table A1). *Cycas* contains more species than any other cycad genus and also is the genus with most species represented in this research agenda. However, on a percentage basis only 16% of *Cycas* species have been studied, compared with 20% of

Macrozamia species. Other speciose genera are *Encephalartos* with 5% of the species studied and *Zamia* with 10% of the species studied. The monotypic *Microcycas* was the only cycad genus that has not been included in this research agenda to date. The reported ranges in nutrient concentration did not appear to be constrained within each genus. For example, the least and greatest concentrations for some nutrients were reported within a single genus (Table 2).

The number of genera and species that have been studied for each element was greatest for most of the macronutrients, as would be expected. These are the chemical elements that are needed in greatest quantity by plants, and they comprise the core constituents of most commercial fertilizers that are manufactured to increase plant growth and productivity. Nitrogen was the most studied element with nine genera and 46 of the 358 described cycad species [5] being represented among 14 reports (Table 2). For unknown reasons, the inclusion of the macronutrient magnesium in cycad leaf tissue studies has been minimal, with only five species and two genera included. The micronutrients were much less represented in the literature. Iron and zinc were the only micronutrients that received considerable attention in this agenda. The remainder of the micronutrients have been mostly ignored during past research, with one to four *Cycas* and *Macrozamia* species included for each micronutrient. The leaf concentrations for only three of the six beneficial nutrients have been reported to date (Table 2), and each of these were represented by one or two *Cycas* or *Macrozamia* species.

Only two species have had more than 10 essential or beneficial elements reported, and both were *Cycas* species (Figure 1a). Ten of the 46 species had only one or two leaf elements reported. The most heavily studied species was *Cycas micronesica*, and five of the eight studies for this species included in situ data (Figure 1b). Only 14 of the 45 species were represented with in situ data. About two-thirds of the species were represented by only one study.

The original heavily cited description of the global leaf economic spectrum known as GLOPNET [21] compiled data from 2548 species and included nitrogen and potassium among the leaf traits that were built into the model. Their global average for leaf nitrogen was 19.4 mg·g^{-1}. Our mean of leaf nitrogen concentration for cycad leaves was 22.8 mg·g^{-1}, the greater value possibly occurring because of the nitrogen-fixing cyanobacteria endosymbionts for cycads [6]. The GLOPNET data included 155 species identified as having nitrogen-fixing endosymbionts, including one *Cycas* and one *Macrozamia* species [21]. The nitrogen mean for this subset was 25.7 mg·g^{-1}, indicating cycad leaves contain less nitrogen on average than angiosperm plants that associate with nitrogen-fixing endosymbionts. The global average for leaf phosphorus was 1.1 mg·g^{-1}, less than our mean of 1.3 mg·g^{-1} for cycad species with reported phosphorus values. Overall, our findings indicated the reported values for nitrogen and phosphorus in cycad leaves were greater than the global average. However, this direct comparison suffers from procedural ambiguities. The compilers of the GLOPNET data were careful to restrict their methods to natural settings where the plants received no management of any type (Peter Reich, personal communication). Most of the published cycad reports included leaf data from managed plants in botanic gardens, and many of the studies failed to describe irrigation and fertilization protocols that preceded the sampling dates. Moreover, the explicit comparisons of cycads to leaf economic spectrum fundamentals [9,19,25] were based exclusively on managed botanic garden plants. Managed garden plants of two *Cycas* species were compared with in situ plants to indicate the managed plants produced leaves with macronutrient concentrations that were not similar to the unmanaged plants [20]. For example, *C. nongnoochiae* leaves from garden plants contained 2.6-fold greater phosphorus and 4.1-fold greater potassium than in situ plants. This species grows in one locality in central Thailand and exhibits an extreme small endemic range. Clearly, most published leaf element data from cycad species are not currently useful for comparison to GLOPNET.

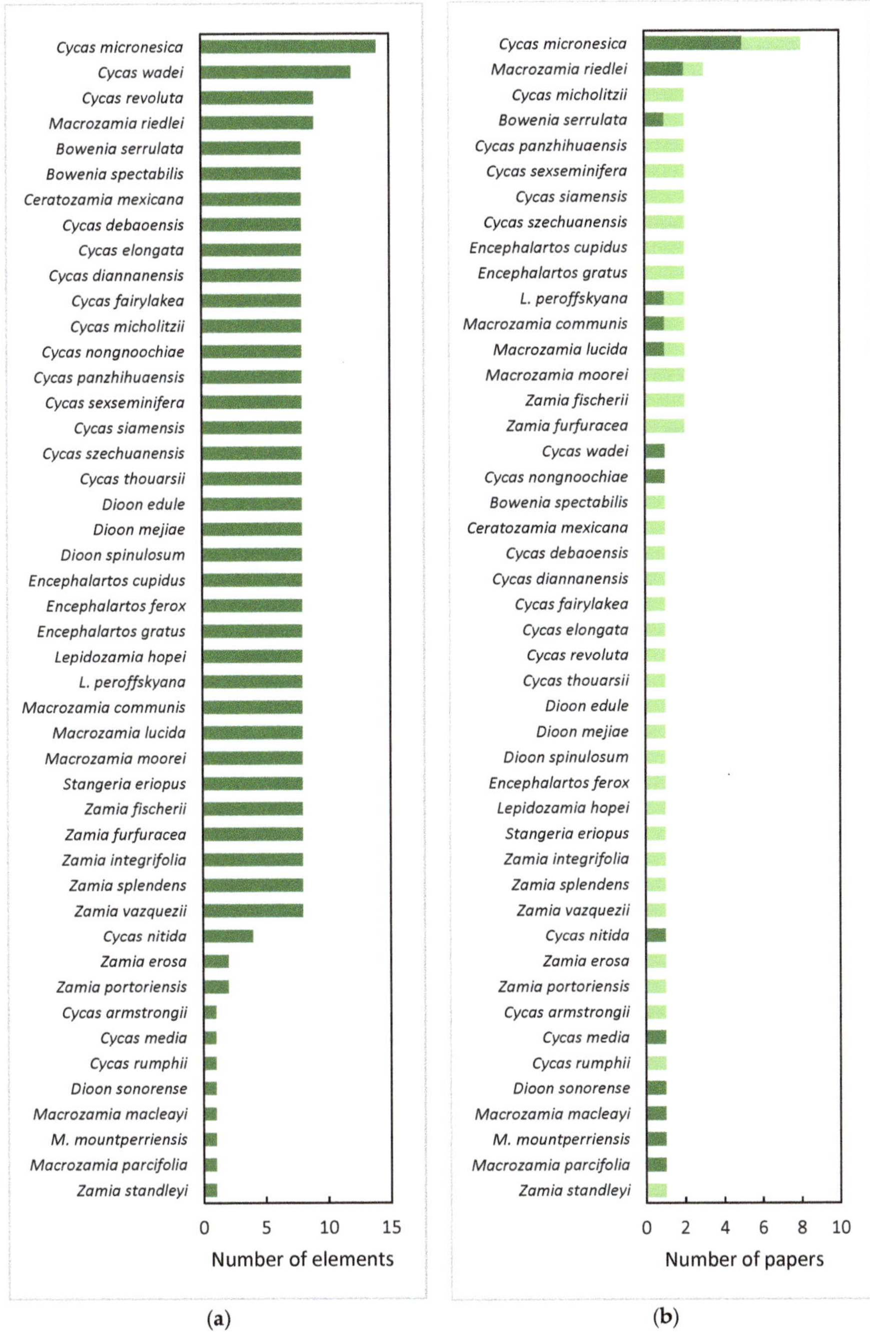

Figure 1. Statistics of forty-six cycad species. (**a**) Number of essential and beneficial elements reported from leaves. (**b**) Number of papers containing leaf element data. Dark green portions of bars depict the number of papers with in situ data.

4. Leaf Litter Elements

The elemental constituents of leaf litter interplay with many cascading ecosystem phenomena, such as plant soil feedback [26–29], the home field advantage in decomposition [30,31], and the soil food web [32–34]. Moreover, an understanding of leaf litter nitrogen is critically important for plant species in Fabaceae because these plants enter into symbiotic relationships with nitrogen-fixing bacteria (*Rhizobium*) and Cycadales because these plants enter into symbiotic relationships with nitrogen-fixing cyanobacteria (*Nostoc*) [35]. Therefore, some of the nitrogen released during litter decomposition for these plant groups represents new contributions to the bulk soil. Other plant groups that do not have nitrogen-fixing endosymbionts must absorb the required nitrogen from the edaphic substrates, then their litterfall contains that same nitrogen that is returned to the same edaphic substrates. Direct measurement of leaf litter chemistry is required for each species because translocation of green leaf elements back into the stem tissue occurs during the dismantling of a leaf's machinery as senescence commences. The percentage of resorption of each element is species-specific [36,37].

A literature review of cycad leaf litter chemistry reveals the definition of generalities is impossible because so few species have been studied. Leaf litter content of carbon and nitrogen has been determined for four *Cycas* [10,11,18,20,35,38] and two *Macrozamia* [16] species (Table 3). One to four *Cycas* species have been studied for other essential and beneficial elements (Table A2).

Table 3. Published ranges in leaf litter concentrations of macronutrients, micronutrients, and beneficial elements for cycad plants.

Element	Genera	Species Studied	Species in Genus	Range	Reference
Carbon	*Cycas*	3	117	475–534 mg·g^{-1}	[10,11,18,35,38]
Carbon	*Macrozamia*	2	41	502–546 mg·g^{-1}	[16]
Nitrogen	*Cycas*	4	117	15–22 mg·g^{-1}	[10,11,18,35,38]
Nitrogen	*Macrozamia*	2	41	11–24 mg·g^{-1}	[16]
Phosphorus	*Cycas*	4	117	0.3–2.0 mg·g^{-1}	[10,11,18,38]
Potassium	*Cycas*	4	117	1.0–14.2 mg·g^{-1}	[10,11,18,38]
Magnesium	*Cycas*	3	117	1.32–7.54 mg·g^{-1}	[11,17,38]
Calcium	*Cycas*	3	117	2.5–32.3 mg·g^{-1}	[11,18,38]
Sulfur	*Cycas*	1	117	1.20–1.38 mg·g^{-1}	[38]
Iron	*Cycas*	2	117	28–547 mg·kg^{-1}	[11,18,38]
Manganese	*Cycas*	2	117	25–141 mg·kg^{-1}	[11,18,38]
Boron	*Cycas*	2	117	29.5–51.6 mg·kg^{-1}	[11,38]
Copper	*Cycas*	2	117	1.3–5.9 mg·kg^{-1}	[11,18,38]
Zinc	*Cycas*	2	117	4.48–31.21 mg·kg^{-1}	[11,18,38]
Selenium	*Cycas*	1	117	0.48 mg·kg^{-1}	[11]

5. Biotic Factors

The direct influence of leaf age on nutrient concentration has been reported for three cycad species [13,16]. A 33% decline in leaf nitrogen occurred from youngest to oldest *C. micronesica* leaves [13], a 12% decline in leaf nitrogen occurred from youngest to oldest *M. communis* leaves [16], and a 13% increase in leaf nitrogen occurred from youngest to oldest *M. riedlei* leaves [16]. The leaf crown on a cycad plant is comprised of several cohorts of leaves with disparate age, each of which is separated by persisting cataphylls. The determination of the youngest cohort and the oldest cohort of leaves is unambiguous due to the persisting cataphylls. These contrasting results for three species were unexpected and point out the need to determine how leaf age influences leaf elements for more cycad species. The increase in nitrogen with leaf age for *M. riedlei* is in sharp contrast to the robust literature on the subject of nutrient resorption. Moreover, the description of which leaves were sampled from the plants in most cycad reports reviewed herein was not included. This oversight must be corrected in future studies. The persistence of cataphylls in cycad leaf crowns enables an unmistakable demarcation that separates the youngest leaf cohort from older leaves.

The influence of plant size on leaf nutrients has been reported for two cycad species [18,22]. Leaf nitrogen concentration declined with plant size for *C. micronesica* [18] and *D. sonorense* [22]. The results point out the need to determine the influence of plant size on leaf nutrients for more cycad species. Both of these species produce arborescent stem growth. We suggest the results were under control of allometric relations rather than height per se. Therefore, cycad species which produce stem growth that is mostly subterranean may require a different variable to quantify stem growth, such as diameter of the stem clump or number of apices per plant.

The influence of leaflet sampling position along the leaf rachis has been reported for two cycad species [14,16]. *Cycas micronesica* leaf nitrogen concentration increased linearly for young leaves and non-linearly for old leaves with distance from the petiole [14]. A non-linear increase in leaf nitrogen concentration occurred for *M. riedlei* with distance from the petiole [16]. The leaf age was not reported. The majority of papers that we reviewed did not include a description of sampling location along the pinnately compound cycad leaf rachis. As with leaf age and plant size, this oversight must be corrected in future studies.

6. Environmental Factors

The direct influence of incident light on *C. micronesica* leaf element concentrations has been reported [13]. Nitrogen, phosphorus, and potassium concentrations were greater in shaded plants than in full sun plants. Differences in *C. micronesica* leaf element concentrations were reported between homogeneous shade conditions supplied by commercial shadecloth and heterogeneous shade conditions supplied by wood slats [17]. These results reveal the dangers in relying on data from managed gardens without augmenting the results with data from natural settings. A quantification of incident light or the general level of shade has not been reported for most of the cycad studies from the literature. A comparison of two *Cycas* species between garden and in situ settings revealed the nutrient concentrations of leaves from the garden plants were dissimilar from those of leaves from habitat [20]. The benign level of competition in the gardens versus robust competition with sympatric plants in habitat was considered a causal mechanism. The use of multiple sites with contrasting soil nutrient relations has revealed that cycad leaf concentrations of some leaf nutrients track with the differences soil concentrations [10,20]. Many of the cycad studies in this review did not include a description of soil nutrient concentrations accompanying the sampled plants. Other studies reported general soil characteristics but did not include measurements of the nutrients within soils subtending the sampled cycad plants. The differences of soil chemistry directly beneath cycad plants versus away from the plants [39,40] indicate soil nutrition within the root zone of the sampled cycad plants is a metric that should be determined in order to interpret leaf nutrient results accurately.

7. Future Directions

We consider three issues as the greatest needs within this agenda as more research accumulates. First, adherence to accepted binomials for every taxon included in this research is of paramount importance. Some reports included taxa names that did not conform to any known published species names, and these data were not included herein and should not be used in future meta-analyses and reviews. Careful adherence to accepted binomials [5] in future research would mitigate this ambiguity. Moreover, as changes in cycad classification and nomenclature will continue to occur, including specific provenance or pedigree data for samples included in studies, or preparing herbarium specimens representing these samples will help researchers compiling data for future meta-analyses and reviews.

Second, more species must be added to the data before large-scale generalities will become accurate for the Cycadales. Priority should be given to taxonomic groups that have not been studied adequately. The genus *Microcycas* is missing from the published data. However, the speciose genera are also not adequately represented in the literature. For example, only 3% of *Ceratozamia*, 5% of *Encephalartos*, 10% of *Zamia*, 16% of *Cycas*, and 20% of *Macrozamia* species have been studied to date.

Third, an increase in focus on natural habitats and reduction in focus on botanic garden settings is needed. The leaf nutrient relations of only 14 of the 358 described species [5] have been determined in situ, and most of those reports included a single locality. In situ leaf sampling of *Cycas micronesica* has occurred among numerous insular habitats across four geopolitical island groups. No other species has been studied with this level of focus on in situ sampling methods. This paucity of data from natural habitats renders the current cycad literature of little value for comparing to GLOPNET. Moreover, the genetic × environmental control over leaf nutrient concentrations cannot be determined until multiple localities are included for indigenous species with an extensive native range.

Seasonal variation in leaf element concentrations may be considerable and modulated by biotic factors. For example, the influence of season on *Actinidia arguta* var. *arguta* (Siebold and Zucc.) Planch. ex Miq. leaf nutrient concentrations differed for male and female plants [41]. Moreover, the influence of season on *Olea europaea* L. leaves interacted with intraspecific genotypic variation [42]. These results indicate that research to determine the influence of season on cycad leaf nutrient relations should include multiple provenances and the distinction of male and female sampled plants. Until this is determined for numerous cycad species, the approach used by Marler and Lindström [20] is recommended for comparing more than one location, whereby one season is used to compare locations.

Zhang et al. [9,19] reported iron concentrations of *Zamia fischeri* leaves that were extreme outliers when compared with other species studied in two botanic garden locations. This observation should be confirmed in natural settings in Mexico and greater attention to iron variation among other closely related *Zamia* species may be warranted.

Marler and Lindström [20] reported that leaf magnesium concentration was constrained among *Cycas* plants from one provenance even when they were grown in different soils with substantial variation in soil magnesium concentrations. For example, *C. nongnoochiae* plants growing in Thailand habitat exhibited leaf magnesium concentration that did not differ from the plants growing in a managed cultivated garden, even though the garden soils contained magnesium that was only 14% of that in the habitat soils. Similarly, *C. micronesica* plants growing in Yap habitat exhibited leaf magnesium concentration that did not differ from the plants growing in a managed cultivated garden, even though the garden soils contained magnesium that was only 11% of that in the habitat soils. The maintenance of magnesium homeostasis in cycad leaves deserves further study. Some of the known roles of magnesium include maintenance of chlorophyll concentration, promotion of non-structural carbohydrate export from leaves, and control of ionic currents across membranes [43,44]. The observed homeostasis for two *Cycas* species is not unexpected, given this partial list of roles for this macronutrient. The observations need to be confirmed with other cycad species using multiple localities.

The nutrients which have been studied by more than one laboratory have revealed disparity in reported concentrations among the studies that may be explained by dissimilar methods. For example, green leaf nitrogen concentration reported by Kipp et al. [16] was more than double that reported by Grove et al. [23] for *Macrozamia riedlei* and almost double that reported by Zhang et al. [9] for *Bowenia serrulata*. Explanations for these differences among laboratories are difficult to consider because many of the co-varying factors discussed in Sections 5 and 6 were not reported. Effort should be made during every future study to record and report all sources of variation to improve our understanding of reported differences among studies.

Marler and Dongol [35] reported the only study that we are aware of which determined the influence of insect herbivory on cycad leaf nutrients. All three insects were invasive non-native pests. Many cycad taxa coevolved with folivorous insects, and these should be studied in a similar manner to determine how leaf nutrients are altered by the herbivory of these native sympatric insects.

The influence of *C. micronesica* leaf litter on decomposition speed, soil respiration, and mineralization dynamics has been reported [38]. This study revealed the speed of these leaf after-life phenomena was slower for the cycad leaves than for two Fabaceae species. The results indicated that the presence of cycad plants in biodiverse settings may influence community-level litter decomposition even if they are limited in incidence [45].

The long-term changes in soil nutrient concentrations beneath the canopy of cycad plants have been determined for *C. micronesica* and *Z. integrifolia* [39,40]. To our knowledge, the influences of cycad plants on the soils within the dripline of their canopy have not been studied for any other species. However, the two species that have been studied revealed that the presence of a cycad plant in unmanaged settings is valuable for introducing soil heterogeneity at the fine scale, potentially increasing biodiversity in soil organisms and increasing ecosystem health. We propose two phenomena that deserve direct study. First, rainfall rarely reaches the soil surface without first being intercepted by plant structures [46–51]. This intercepted rainfall is lost through evaporation or transferred to the soil as throughfall or stemflow. The relative proportions of these processes are affected by canopy and leaf traits, and strongly influence the spatial components of the hydrologic and chemical cycles beneath mixed stands of plants [46–50]. Throughfall is the precipitation component that drips from numerous plant surfaces, and stemflow is the precipitation component that drains along the plant stems to reach the soil. The percentage of precipitation that reaches the soil via stemflow and the concentration of solutes and suspensions of particulates in stemflow are strongly linked to leaf traits and canopy architecture traits [46–51]. Stemflow influences essential minerals and metals near the base of trees, but also influences soil carbon by the transfer of dissolved organic matter in the stem flow [51]. To our knowledge, no studies of stemflow have included a cycad representative. However, arborescent palm species exhibit stem and leaf shapes and orientations that are similar to cycads, and many palm trees are skilled at increasing soil nutrients in their root zone by maximizing stemflow [52–54]. The diameter of the *C. micronesica* leaf crown is up to 4 m for healthy trees, but the diameter of the *Z. intergrifolia* leaf crown is less than 2 m, illuminating a highly contrasting ability to intercept rainfall for the individual plant. Projected canopy area is highly influential of stemflow volume [55]. The relative diameters of leaflets and rachis surfaces are also much greater for *C. micronesica* than for *Z. integrifolia*, and these organ traits directly influence how precipitation is intercepted by an individual plant. The inclusion of a range of cycad taxa in the stemflow research agenda would add greatly to our knowledge of how cycad plants directly affect soil chemistry, but would also improve our understanding of carbon, hydrologic, and nitrogen cycles by adding this unique gymnosperm plant group to the stemflow literature.

Second, some plants may influence the biogeochemical cycle by litter trapping. The leaf and stem traits of these plants increase the volume of litterfall that is trapped in the plant's canopy, and this trapped litter becomes a privatized slow compost pile that releases nutrients over time [56]. As with stemflow, we are not aware of any cycad taxa that have been studied for litter-trapping abilities. However, palm species [52,56–58] and fern species [56,59] are highly effective at trapping litter, and the plant traits that enable this ability for palms and ferns are similar to the plant traits of cycads. Trapped litter may further magnify nutrient accumulation by attracting animals which may bring food materials and add feces directly to the litter mass [52,56]. The need to study the litter trapping traits of cycad plants is clear, as this may explain the increases in carbon and nitrogen that we have documented beneath two cycad species. Two cycad leaf traits should be considered in this line of work. First, the size, shape, and insertion angle of spines and prickles on cycad petioles vary greatly among species [6,60], and these petiole traits may directly influence how much of the incoming litterfall is trapped. Second, some cycad species produce leaves that are replaced annually, while other species produce leaves that are retained for many years. Undoubtedly, the amount of trapped litter that can accumulate over time is under the direct influence of leaf longevity, and this leaf trait should be considered in future studies on litter trapping of cycad plants.

Plants employ multiple defensive strategies against herbivores that have been studied within the context of various models [61], and plant defensive strategies are generally classified as structural or chemical. Structural defenses include leaf toughness and the construction of modified organs such as thorns, spines, and prickles. Chemical defenses include metabolites that alter the taste of the tissues to deter herbivory or that act as animal toxins. Cycads employ both defensive strategies, and cycad plants have been the subject of myriad medical and biochemical studies because of the number of known toxins that are synthesized by the plants [6,62]. Structural defenses are important after leaf expansion and maturation, but chemical defenses are important during leaf expansion [63]. The azoxyglycosides cycasin and macrozamin are among the most studied acute cycad toxins, and these nitrogenous compounds have been reported in all 10 genera and most species that have been studied [64,65]. These toxins may occur in greater concentrations in young cycad plants than in adult plants [66], which parallels the decline in leaf nitrogen concentration with plant size [18,22]. In general, elemental concentrations of plant tissues mediate defensive mechanisms [67]. These issues of secondary compounds in cycad biology suggest the individual plants with greater nutritive content are better protected with higher azoxyglycosides [66]. In consideration of the relevance of cycad toxins to human health research, continued research on element accumulation and partitioning in cycad plants may contribute substantially to toxicology research.

The elemental components of plant tissues cannot be studied in the absence of recognizing the contributions of root traits and symbionts. Cycad roots have not been adequately studied but these gymnosperms produce roots that appear typical of other seed-bearing plants, and although little is known about their general physiology, they are believed to function similarly to angiosperm roots [6 (p. 60)]. Seedlings initially produce a robust taproot which over time is augmented or replaced by similarly thick and fleshy branching secondary roots. Root hairs, which function in other plants to increase the volume of soil that plants area able to mine for nutrients, are rare in cycads and only irregularly formed in the thinnest of feeder roots. Cycads also produce specialized clusters of roots known as coralloid roots which typically grow upward above the soil surface and host nitrogen-fixing cyanobacteria which fix nitrogen for use by the plant [6,68–70]. Moreover, cycads roots are known to harbor arbuscular mycorrhizal fungi which enhance phosphorus uptake in low phosphorus soils and enhance water availability in seasonally dry habitats [71,72]. The incidence and diversity of these symbionts may contrast sharply between natural habitats where sympatric species of soil biota exist and botanic gardens where the soil biota that interacts with a cycad plant are novel to the plant More studies are needed to understand cycad root traits and to tease apart the influences of these symbiotic relationships on leaf element concentrations in various cycad taxa.

Finally, many areas of occupancy for various cycad species are characterized by edaphic characteristics that most plant species would not consider as suitable for plant growth. We highlight three examples that deserve a dedicated look during future research on cycad plant nutrition. First, multiple cycad species thrive in littoral habitats where roots are exposed to saline substrates and leaves must contend with aerosol salt deposits. Second, some cycad species flourish on limestone mountain surfaces or karst outcrops where mineral soils are scarce and drought stress is extreme. Third, cycad populations also occur on either highly acidic volcanic substrates or ultramafic habitats, where the plants must cope between the spectrum of extreme acidity and high alkalinity compounded by calcium deficiencies and metal toxicities. This group of plants is ideal for studying the mechanisms that plants exploit to compete in these extreme habitats. Moreover, some species are endemic to one of these extreme habitat types while other species are indigenous and can be found in various ecological niche habitats. Comparing these two types of cycad species may tease apart the stress physiology mechanisms that indicate facultative versus obligate approaches for tolerating extreme edaphic conditions.

8. Conclusions

Cycad species are highly prized in the horticulture trade. We have reviewed the available literature on elemental concentrations in cycad leaves. A total of six gardens were included with two in China, one in Florida, one in Thailand, one in Philippines, and one in Guam. These results were discussed along with in situ data from Australia, Guam, Mexico, Palau, Philippines, Rota, Thailand, and Yap. The review illuminates the scant research landscape of this agenda. By highlighting the unexpected results that most papers reported data from botanic gardens and the authors failed to describe the irrigation and fertilization protocols of the managed plants, we aimed to inspire an adoption of more demanding protocols for expanding this research agenda. In part, this should include measurement and reporting of plant size, leaf age, or position within the canopy, position of leaflets along the rachis, the shade level of the sampled leaves, and the soil element concentrations within the root zone of the sampled plants.

Author Contributions: Conceptualization, B.E.D. and T.E.M.; writing—original draft preparation, B.E.D. and T.E.M.; writing—review and editing, M.V.K., U.F.F., A.J.L., and M.C.; nomenclature and taxonomic clarifications, A.J.L. and M.C. All authors have read and agreed to the published version of the manuscript.

Funding: This research was funded in part by the United States Forest Service grant numbers 13-DG-11052021-210 and 17-DG-11052021-217.

Acknowledgments: The expeditions and collaborations that have enabled our joint research would not have been possible without support from our respective administrators. We thank them for their support and long-term vision.

Conflicts of Interest: The authors declare no conflict of interest. The funders had no role in the design of the study; in the collection, analyses, or interpretation of data; in the writing of the manuscript, or in the decision to publish the results.

Appendix A

Table A1. Published ranges for green leaf element concentrations of cycad plants. Misspellings of species were corrected if identity was obvious, species that were misspelled were not included if identity was not obvious. Taxonomic synonyms were corrected. Data were estimated for reports displaying data as figures and transformed if data were presented in log format.

Element	Species	Range	Reference
Carbon	*Bowenia serrulata*	519 mg·g^{-1}	[9]
Carbon	*Bowenia spectabilis*	508 mg·g^{-1}	[9]
Carbon	*Ceratozamia Mexicana*	514 mg·g^{-1}	[9]
Carbon	*Cycas debaoensis*	485 mg·g^{-1}	[9]
Carbon	*Cycas diannanensis*	463 mg·g^{-1}	[9] [1]
Carbon	*Cycas elongata*	483 mg·g^{-1}	[9]
Carbon	*Cycas fairylakea*	499 mg·g^{-1}	[9]
Carbon	*Cycas micholitzii*	475 mg·g^{-1}	[9]
Carbon	*Cycas micronesica*	479 mg·g^{-1}	[12]
Carbon	*Cycas micronesica*	484–493 mg·g^{-1}	[13]
Carbon	*Cycas micronesica*	480–505 mg·g^{-1}	[14]
Carbon	*Cycas micronesica*	475–485 mg·g^{-1}	[17]
Carbon	*Cycas nitida*	499–509 mg·g^{-1}	[10]
Carbon	*Cycas panzhihuaensis*	466–504 mg·g^{-1}	[9]
Carbon	*Cycas sexseminifera*	467 mg·g^{-1}	[9]
Carbon	*Cycas siamensis*	469 mg·g^{-1}	[9]
Carbon	*Cycas szechuanensis*	475–498 mg·g^{-1}	[9]
Carbon	*Cycas thouarsii*	497 mg·g^{-1}	[9]
Carbon	*Cycas wadei*	508 mg·g^{-1}	[11]
Carbon	*Dioon edule*	496 mg·g^{-1}	[9]
Carbon	*Dioon mejiae*	485 mg·g^{-1}	[9]
Carbon	*Dioon spinulosum*	486 mg·g^{-1}	[9]
Carbon	*Encephalartos cupidus*	490 mg·g^{-1}	[9]
Carbon	*Encephalartos ferox*	494 mg·g^{-1}	[9]
Carbon	*Encephalartos gratus*	497–505 mg·g^{-1}	[9]
Carbon	*Lepidozamia hopei*	515 mg·g^{-1}	[9]
Carbon	*Lepidozamia peroffskyana*	511 mg·g^{-1}	[9]
Carbon	*Lepidozamia peroffskyana*	473–566 mg·g^{-1}	[16]
Carbon	*Macrozamia communis*	512 mg·g^{-1}	[9]
Carbon	*Macrozamia communis*	507–560 mg·g^{-1}	[16]
Carbon	*Macrozamia lucida*	524 mg·g^{-1}	[9]
Carbon	*Macrozamia lucida*	473–522 mg·g^{-1}	[16]
Carbon	*Macrozamia macleaya*	438–508 mg·g^{-1}	[16]
Carbon	*Macrozamia moorei*	519 mg·g^{-1}	[9]

Table A1. *Cont.*

Element	Species	Range	Reference
Carbon	*Macrozamia riedlei*	455–525 mg·g^{-1}	[16]
Carbon	*Stangeria eriopus*	479 mg·g^{-1}	[9]
Carbon	*Zamia erosa*	495 mg·g^{-1}	[9] [2]
Carbon	*Zamia erosa*	481 mg·g^{-1}	[15]
Carbon	*Zamia fischeri*	458 mg·g^{-1}	[9] [3]
Carbon	*Zamia furfuracea*	477–489 mg·g^{-1}	[9]
Carbon	*Zamia integrifolia*	490–491 mg·g^{-1}	[9]
Carbon	*Zamia portoricensis*	484 mg·g^{-1}	[15]
Carbon	*Zamia splendens*	483 mg·g^{-1}	[9]
Carbon	*Zamia vazquezii*	488 mg·g^{-1}	[9]
Nitrogen	*Bowenia serrulata*	24 mg·g^{-1}	[9]
Nitrogen	*Bowenia serrulata*	41 mg·g^{-1}	[16]
Nitrogen	*Bowenia spectabilis*	24 mg·g^{-1}	[9]
Nitrogen	*Ceratozamia mexicana*	13 mg·g^{-1}	[9]
Nitrogen	*Cycas armstrongii*	21 mg·g^{-1}	[21]
Nitrogen	*Cycas debaoensis*	28 mg·g^{-1}	[9]
Nitrogen	*Cycas diannanensis*	26 mg·g^{-1}	[9] [1]
Nitrogen	*Cycas diannanensis*	26 mg·g^{-1}	[19] [1]
Nitrogen	*Cycas elongata*	28 mg·g^{-1}	[9]
Nitrogen	*Cycas fairylakea*	25 mg·g^{-1}	[9]
Nitrogen	*Cycas media*	44 mg·g^{-1}	[16]
Nitrogen	*Cycas micholitzii*	25 mg·g^{-1}	[9]
Nitrogen	*Cycas micholitzii*	25 mg·g^{-1}	[19]
Nitrogen	*Cycas micronesica*	29–30 mg·g^{-1}	[15]
Nitrogen	*Cycas micronesica*	25 mg·g^{-1}	[12]
Nitrogen	*Cycas micronesica*	14–30 mg·g^{-1}	[18]
Nitrogen	*Cycas micronesica*	18–27 mg·g^{-1}	[13]
Nitrogen	*Cycas micronesica*	18–29 mg·g^{-1}	[14]
Nitrogen	*Cycas micronesica*	23–37 mg·g^{-1}	[17]
Nitrogen	*Cycas micronesica*	17–30 mg·g^{-1}	[20]
Nitrogen	*Cycas nitida*	24–28 mg·g^{-1}	[10]
Nitrogen	*Cycas nongnoochiae*	26–30 mg·g^{-1}	[20]
Nitrogen	*Cycas panhihuaensis*	16–21 mg·g^{-1}	[9]
Nitrogen	*Cycas panhihuaensis*	16 mg·g^{-1}	[19]
Nitrogen	*Cycas rumphii*	30–31 mg·g^{-1}	[15]
Nitrogen	*Cycas sexseminifera*	19 mg·g^{-1}	[9]
Nitrogen	*Cycas sexseminifera*	19 mg·g^{-1}	[19][4]
Nitrogen	*Cycas siamensis*	18 mg·g^{-1}	[9]
Nitrogen	*Cycas siamensis*	19 mg·g^{-1}	[19]
Nitrogen	*Cycas szechuanensis*	21–25 mg·g^{-1}	[9]
Nitrogen	*Cycas szechuanensis*	21 mg·g^{-1}	[19]
Nitrogen	*Cycas thouarsii*	23 mg·g^{-1}	[9]
Nitrogen	*Cycas wadei*	21 mg·g^{-1}	[11]
Nitrogen	*Dioon edule*	15 mg·g^{-1}	[9]
Nitrogen	*Dioon mejiae*	14 mg·g^{-1}	[9]
Nitrogen	*Dioon sonorense*	14–17 mg·g^{-1}	[22]
Nitrogen	*Dioon spinulosum*	15 mg·g^{-1}	[9]
Nitrogen	*Encephalartos cupidus*	17 mg·g^{-1}	[9]
Nitrogen	*Encephalartos cupidus*	18 mg·g^{-1}	[19]
Nitrogen	*Encephalartos ferox*	15 mg·g^{-1}	[9]
Nitrogen	*Encephalartos gratus*	18–19 mg·g^{-1}	[9]
Nitrogen	*Encephalartos gratus*	18 mg·g^{-1}	[19]
Nitrogen	*Lepidozamia hopei*	17 mg·g^{-1}	[9]
Nitrogen	*Lepidozamia peroffskyana*	19 mg·g^{-1}	[9]
Nitrogen	*Lepidozamia peroffskyana*	18–31 mg·g^{-1}	[16]

Table A1. *Cont.*

Element	Species	Range	Reference
Nitrogen	*Macrozamia communis*	20 mg·g^{-1}	[9]
Nitrogen	*Macrozamia communis*	10–38 mg·g^{-1}	[16]
Nitrogen	*Macrozamia lucida*	21 mg·g^{-1}	[9]
Nitrogen	*Macrozamia lucida*	14–22 mg·g^{-1}	[16]
Nitrogen	*Macrozamia macleayi*	8–43 mg·g^{-1}	[16]
Nitrogen	*Macrozamia moorei*	20 mg·g^{-1}	[9]
Nitrogen	*Macrozamia mountperriensis*	54–55 mg·g^{-1}	[16]
Nitrogen	*Macrozamia parcifolia*	47–49 mg·g^{-1}	[16]
Nitrogen	*Macrozamia riedlei*	14 mg·g^{-1}	[21]
Nitrogen	*Macrozamia riedlei*	11–15 mg·g^{-1}	[23]
Nitrogen	*Macrozamia riedlei*	8–38 mg·g^{-1}	[16]
Nitrogen	*Macrozamia serpentina*	28–31 mg·g^{-1}	[16]
Nitrogen	*Stangeria eriopus*	22 mg·g^{-1}	[9]
Nitrogen	*Zamia erosa*	18 mg·g^{-1}	[9] [2]
Nitrogen	*Zamia erosa*	26 mg·g^{-1}	[15]
Nitrogen	*Zamia fischeri*	28 mg·g^{-1}	[9] [3]
Nitrogen	*Zamia fischeri*	28 mg·g^{-1}	[19] [3]
Nitrogen	*Zamia furfuracea*	12–14 mg·g^{-1}	[9]
Nitrogen	*Zamia furfuracea*	13 mg·g^{-1}	[19]
Nitrogen	*Zamia integrifolia*	18–21 mg·g^{-1}	[9]
Nitrogen	*Zamia portoricensis*	18 mg·g^{-1}	[15]
Nitrogen	*Zamia splendens*	15 mg·g^{-1}	[9]
Nitrogen	*Zamia standleyi*	19 mg·g^{-1}	[15]
Nitrogen	*Zamia vazquezii*	30 mg·g^{-1}	[9]
Phosphorus	*Bowenia serrulata*	1.0 mg·g^{-1}	[9]
Phosphorus	*Bowenia spectabilis*	1.1 mg·g^{-1}	[9]
Phosphorus	*Ceratozamia mexicana*	0.8 mg·g^{-1}	[9]
Phosphorus	*Cycas debaoensis*	1.4 mg·g^{-1}	[9]
Phosphorus	*Cycas diannanensis*	2.4 mg·g^{-1}	[9] [1]
Phosphorus	*Cycas diannanensis*	2.4 mg·g^{-1}	[19] [1]
Phosphorus	*Cycas elongata*	1.2 mg·g^{-1}	[9]
Phosphorus	*Cycas fairylakea*	1.1 mg·g^{-1}	[9]
Phosphorus	*Cycas micholitzii*	1.5 mg·g^{-1}	[9]
Phosphorus	*Cycas micholitzii*	1.5 mg·g^{-1}	[19]
Phosphorus	*Cycas micronesica*	2.9 mg·g^{-1}	[12]
Phosphorus	*Cycas micronesica*	1.2–2.7 mg·g^{-1}	[18]
Phosphorus	*Cycas micronesica*	0.9–2.5 mg·g^{-1}	[13]
Phosphorus	*Cycas micronesica*	0.8–2.8 mg·g^{-1}	[14]
Phosphorus	*Cycas micronesica*	2.6–2.9 mg·g^{-1}	[17]
Phosphorus	*Cycas micronesica*	1.5–2.9 mg·g^{-1}	[20]
Phosphorus	*Cycas nitida*	1.1–1.9 mg·g^{-1}	[10]
Phosphorus	*Cycas nongnoochiae*	1.3–3.4 mg·g^{-1}	[20]
Phosphorus	*Cycas panzhihuaensis*	1.0–1.1 mg·g^{-1}	[9]
Phosphorus	*Cycas panzhihuaensis*	1.1 mg·g^{-1}	[19]
Phosphorus	*Cycas sexseminifera*	1.5 mg·g^{-1}	[9]
Phosphorus	*Cycas sexseminifera*	1.2–1.5 mg·g^{-1}	[19] [4]
Phosphorus	*Cycas siamensis*	1.2 mg·g^{-1}	[9]
Phosphorus	*Cycas siamensis*	1.2 mg·g^{-1}	[19]
Phosphorus	*Cycas szechuanensis*	1.0–1.2 mg·g^{-1}	[9]
Phosphorus	*Cycas thouarsii*	1.2 mg·g^{-1}	[9]
Phosphorus	*Cycas wadei*	1.1 mg·g^{-1}	[11]
Phosphorus	*Dioon edule*	0.8 mg·g^{-1}	[9]
Phosphorus	*Dioon mejiae*	1.5 mg·g^{-1}	[9]
Phosphorus	*Dioon spinulosum*	0.8 mg·g^{-1}	[9]
Phosphorus	*Encephalartos cupidus*	1.2 mg·g^{-1}	[9]

Table A1. *Cont.*

Element	Species	Range	Reference
Phosphorus	*Encephalartos cupidus*	1.2 mg·g^{-1}	[19]
Phosphorus	*Encephalartos ferox*	1.0 mg·g^{-1}	[9]
Phosphorus	*Encephalartos gratus*	1.1–1.3 mg·g^{-1}	[9]
Phosphorus	*Encephalartos gratus*	1.1 mg·g^{-1}	[19]
Phosphorus	*Lepidozamia hopei*	0.8 mg·g^{-1}	[9]
Phosphorus	*Lepidozamia peroffskyana*	1.2 mg·g^{-1}	[9]
Phosphorus	*Macrozamia communis*	1.0 mg·g^{-1}	[9]
Phosphorus	*Macrozamia lucida*	1.2 mg·g^{-1}	[9]
Phosphorus	*Macrozamia moorei*	0.9 mg·g^{-1}	[9]
Phosphorus	*Macrozamia riedlei*	0.5 mg·g^{-1}	[21]
Phosphorus	*Stangeria eriopus*	1.1 mg·g^{-1}	[9]
Phosphorus	*Zamia erosa*	1.0 mg·g^{-1}	[9] [2]
Phosphorus	*Zamia fischeri*	1.7 mg·g^{-1}	[9] [3]
Phosphorus	*Zamia fischeri*	1.7 mg·g^{-1}	[19] [3]
Phosphorus	*Zamia furfuracea*	0.7–0.8 mg·g^{-1}	[9]
Phosphorus	*Zamia furfuracea*	0.7 mg·g^{-1}	[19]
Phosphorus	*Zamia integrifolia*	1.3 mg·g^{-1}	[9]
Phosphorus	*Zamia splendens*	0.8 mg·g^{-1}	[9]
Phosphorus	*Zamia vazquezii*	0.7 mg·g^{-1}	[9]
Potassium	*Bowenia serrulata*	5.5 mg·g^{-1}	[9]
Potassium	*Bowenia spectabilis*	6.2 mg·g^{-1}	[9]
Potassium	*Ceratozamia mexicana*	4.9 mg·g^{-1}	[9]
Potassium	*Cycas debaoensis*	4.4 mg·g^{-1}	[9]
Potassium	*Cycas diannanensis*	9.9 mg·g^{-1}	[9] [1]
Potassium	*Cycas elongata*	9.8 mg·g^{-1}	[9]
Potassium	*Cycas fairylakea*	5.8 mg·g^{-1}	[9]
Potassium	*Cycas micholitzii*	7.0 mg·g^{-1}	[9]
Potassium	*Cycas micronesica*	15.3 mg·g^{-1}	[12]
Potassium	*Cycas micronesica*	6.9–23.0 mg·g^{-1}	[18]
Potassium	*Cycas micronesica*	3.8–22.1 mg·g^{-1}	[13]
Potassium	*Cycas micronesica*	3.1–23.7 mg·g^{-1}	[14]
Potassium	*Cycas micronesica*	14.9–16.4 mg·g^{-1}	[17]
Potassium	*Cycas micronesica*	10.5–18.9 mg·g^{-1}	[20]
Potassium	*Cycas nitida*	6.4–16.6 mg·g^{-1}	[10]
Potassium	*Cycas nongnoochiae*	4.4–18.3 mg·g^{-1}	[20]
Potassium	*Cycas panzhihuaensis*	5.8–7.7 mg·g^{-1}	[9]
Potassium	*Cycas revoluta*	4.9–11.9 mg·g^{-1}	[24]
Potassium	*Cycas sexseminifera*	4.3 mg·g^{-1}	[9]
Potassium	*Cycas siamensis*	10.2 mg·g^{-1}	[9]
Potassium	*Cycas szechuanensis*	3.7–5.7 mg·g^{-1}	[9]
Potassium	*Cycas thouarsii*	8.8 mg·g^{-1}	[9]
Potassium	*Cycas wadei*	7.4 mg·g^{-1}	[11]
Potassium	*Dioon edule*	5.7 mg·g^{-1}	[9]
Potassium	*Dioon mejiae*	11.5 mg·g^{-1}	[9]
Potassium	*Dioon spinulosum*	7.9 mg·g^{-1}	[9]
Potassium	*Encephalartos cupidus*	6.2 mg·g^{-1}	[9]
Potassium	*Encephalartos ferox*	6.7 mg·g^{-1}	[9]
Potassium	*Encephalartos gratus*	7.2–8.9 mg·g^{-1}	[9]
Potassium	*Lepidozamia hopei*	9.5 mg·g^{-1}	[9]
Potassium	*Lepidozamia peroffskyana*	10.6 mg·g^{-1}	[9]
Potassium	*Macrozamia communis*	9.8 mg·g^{-1}	[9]
Potassium	*Macrozamia lucida*	11.3 mg·g^{-1}	[9]
Potassium	*Macrozamia moorei*	5.1 mg·g^{-1}	[9]
Potassium	*Macrozamia riedlei*	6.5–9.2 mg·g^{-1}	[23]
Potassium	*Stangeria eriopus*	8.0 mg·g^{-1}	[9]

Table A1. *Cont.*

Element	Species	Range	Reference
Potassium	*Zamia erosa*	10.0 mg·g^{-1}	[9] [2]
Potassium	*Zamia fischeri*	6.6 mg·g^{-1}	[9] [3]
Potassium	*Zamia furfuracea*	4.6–10.2 mg·g^{-1}	[9]
Potassium	*Zamia integrifolia*	9.3–9.5 mg·g^{-1}	[9]
Potassium	*Zamia splendens*	8.1 mg·g^{-1}	[9]
Potassium	*Zamia vazquezii*	18.0 mg·g^{-1}	[9]
Magnesium	*Cycas micronesica*	2.3 mg·g^{-1}	[12]
Magnesium	*Cycas micronesica*	1.7–8.2 mg·g^{-1}	[18]
Magnesium	*Cycas micronesica*	2.5–4.8 mg·g^{-1}	[13]
Magnesium	*Cycas micronesica*	2.9–5.1 mg·g^{-1}	[14]
Magnesium	*Cycas micronesica*	2.2–2.4 mg·g^{-1}	[17]
Magnesium	*Cycas micronesica*	3.1–7.0 mg·g^{-1}	[20]
Magnesium	*Cycas nongnoochiae*	2.4–2.6 mg·g^{-1}	[20]
Magnesium	*Cycas revoluta*	1.9–3.1 mg·g^{-1}	[24]
Magnesium	*Cycas wadei*	1.4 mg·g^{-1}	[11]
Magnesium	*Macrozamia reidlei*	1.1–1.9 mg·g^{-1}	[23]
Calcium	*Bowenia serrulata*	6.1 mg·g^{-1}	[9]
Calcium	*Bowenia spectabilis*	5.0 mg·g^{-1}	[9]
Calcium	*Ceratozamia mexicana*	7.1 mg·g^{-1}	[9]
Calcium	*Cycas debaoensis*	11.8 mg·g^{-1}	[9]
Calcium	*Cycas diannanensis*	11.4 mg·g^{-1}	[9]
Calcium	*Cycas elongata*	11.6 mg·g^{-1}	[9]
Calcium	*Cycas fairylakea*	3.9 mg·g^{-1}	[9]
Calcium	*Cycas micholitzii*	2.7 mg·g^{-1}	[9]
Calcium	*Cycas micronesica*	2.8 mg·g^{-1}	[12]
Calcium	*Cycas micronesica*	7.1–23.7 mg·g^{-1}	[18]
Calcium	*Cycas micronesica*	1.2–8.6 mg·g^{-1}	[13]
Calcium	*Cycas micronesica*	7.8–10.6 mg·g^{-1}	[14]
Calcium	*Cycas micronesica*	2.5–3.1 mg·g^{-1}	[17]
Calcium	*Cycas micronesica*	3.1–19.9 mg·g^{-1}	[20]
Calcium	*Cycas nongnoochiae*	3.2–7.0 mg·g^{-1}	[20]
Calcium	*Cycas panzhihuaensis*	6.6–7.0 mg·g^{-1}	[9]
Calcium	*Cycas revoluta*	7.7–15.6 mg·g^{-1}	[24]
Calcium	*Cycas sexseminifera*	8.6 mg·g^{-1}	[9]
Calcium	*Cycas siamensis*	9.9 mg·g^{-1}	[9]
Calcium	*Cycas szechuanensis*	1.4–2.8 mg·g^{-1}	[9]
Calcium	*Cycas thouarsii*	6.3 mg·g^{-1}	[9]
Calcium	*Cycas wadei*	2.51 mg·g^{-1}	[11]
Calcium	*Dioon edule*	7.7 mg·g^{-1}	[9]
Calcium	*Dioon mejiae*	8.4 mg·g^{-1}	[9]
Calcium	*Dioon spinulosum*	7.6 mg·g^{-1}	[9]
Calcium	*Encephalartos cupidus*	4.5 mg·g^{-1}	[9]
Calcium	*Encephalartos ferox*	14.3 mg·g^{-1}	[9]
Calcium	*Encephalartos gratus*	4.7–6.2 mg·g^{-1}	[9]
Calcium	*Lepidozamia hopei*	5.0 mg·g^{-1}	[9]
Calcium	*Lepidozamia peroffskyana*	3.6 mg·g^{-1}	[9]
Calcium	*Macrozamia communis*	1.4 mg·g^{-1}	[9]
Calcium	*Macrozamia lucida*	2.8 mg·g^{-1}	[9]
Calcium	*Macrozamia moorei*	4.7 mg·g^{-1}	[9]
Calcium	*Macrozamia riedlei*	3.1–7.1 mg·g^{-1}	[23]
Calcium	*Stangeria eriopus*	7.1 mg·g^{-1}	[9]
Calcium	*Zamia erosa*	3.0 mg·g^{-1}	[9] [2]
Calcium	*Zamia fischeri*	7.7 mg·g^{-1}	[9] [3]
Calcium	*Zamia furfuracea*	4.9–7.0 mg·g^{-1}	[9]
Calcium	*Zamia integrifolia*	4.2–4.3 mg·g^{-1}	[9]

Table A1. *Cont.*

Element	Species	Range	Reference
Calcium	*Zamia splendens*	4.4 mg·g^{-1}	[9]
Calcium	*Zamia vazquezii*	6.7 mg·g^{-1}	[9]
Chloride	*Cycas revoluta*	0.5–2.3 mg·g^{-1}	[24]
Sodium	*Cycas micronesica*	0.5 mg·g^{-1}	[12]
Sodium	*Cycas revoluta*	0.2–1.2 mg·g^{-1}	[24]
Sodium	*Macrozamia reidlei*	0.3–1.0 mg·g^{-1}	[23]
Sulfur	*Bowenia serrulata*	1.9 mg·g^{-1}	[9]
Sulfur	*Bowenia spectabilis*	1.9 mg·g^{-1}	[9]
Sulfur	*Ceratozamia mexicana*	1.4 mg·g^{-1}	[9]
Sulfur	*Cycas debaoensis*	2.6 mg·g^{-1}	[9]
Sulfur	*Cycas diannanensis*	1.6 mg·g^{-1}	[9] [1]
Sulfur	*Cycas diannanensis*	1.6 mg·g^{-1}	[19] [1]
Sulfur	*Cycas elongata*	2.0 mg·g^{-1}	[9]
Sulfur	*Cycas fairylakea*	1.7 mg·g^{-1}	[9]
Sulfur	*Cycas micholitzii*	1.4 mg·g^{-1}	[9]
Sulfur	*Cycas micholitzii*	1.4 mg·g^{-1}	[19]
Sulfur	*Cycas micronesica*	1.2–1.6 mg·g^{-1}	[17]
Sulfur	*Cycas micronesica*	1.1 mg·g^{-1}	[20]
Sulfur	*Cycas nongnoochiae*	1.4 mg·g^{-1}	[20]
Sulfur	*Cycas panzhihuaensis*	0.9–1.4 mg·g^{-1}	[9]
Sulfur	*Cycas panzhihuaensis*	0.8 mg·g^{-1}	[19]
Sulfur	*Cycas sexseminifera*	1.0 mg·g^{-1}	[9]
Sulfur	*Cycas sexseminifera*	0.9 mg·g^{-1}	[19] [4]
Sulfur	*Cycas siamensis*	1.3 mg·g^{-1}	[9]
Sulfur	*Cycas siamensis*	1.3 mg·g^{-1}	[19]
Sulfur	*Cycas szechuanensis*	1.1–1.4 mg·g^{-1}	[9]
Sulfur	*Cycas szechuanensis*	1.1 mg·g^{-1}	[19]
Sulfur	*Cycas thouarsii*	1.4 mg·g^{-1}	[9]
Sulfur	*Dioon edule*	1.4 mg·g^{-1}	[9]
Sulfur	*Dioon mejiae*	1.4 mg·g^{-1}	[9]
Sulfur	*Dioon spinulosum*	1.1 mg·g^{-1}	[9]
Sulfur	*Encephalartos cupidus*	1.2 mg·g^{-1}	[9]
Sulfur	*Encephalartos cupidus*	1.2 mg·g^{-1}	[19]
Sulfur	*Encephalartos ferox*	1.3 mg·g^{-1}	[9]
Sulfur	*Encephalartos gratus*	0.9–2.2 mg·g^{-1}	[9]
Sulfur	*Encephalartos gratus*	0.8 mg·g^{-1}	[19]
Sulfur	*Lepidozamia hopei*	1.6 mg·g^{-1}	[9]
Sulfur	*Lepidozamia peroffskyana*	1.4 mg·g^{-1}	[9]
Sulfur	*Macrozamia communis*	1.2 mg·g^{-1}	[9]
Sulfur	*Macrozamia lucida*	1.9 mg·g^{-1}	[9]
Sulfur	*Macrozamia moorei*	1.0 mg·g^{-1}	[9]
Sulfur	*Macrozamia riedlei*	0.8–1.2 mg·kg^{-1}	[23]
Sulfur	*Stangeria eriopus*	2.3 mg·g^{-1}	[9]
Sulfur	*Zamia erosa*	1.0 mg·g^{-1}	[9] [2]
Sulfur	*Zamia fischeri*	2.7 mg·g^{-1}	[9] [3]
Sulfur	*Zamia fischeri*	2.7 mg·g^{-1}	[19] [3]
Sulfur	*Zamia furfuracea*	0.6–1.5 mg·g^{-1}	[9]
Sulfur	*Zamia furfuracea*	0.6 mg·g^{-1}	[19]
Sulfur	*Zamia integrifolia*	13.6–13.7 mg·g^{-1}	[9]
Sulfur	*Zamia splendens*	1.1 mg·g^{-1}	[9]
Sulfur	*Zamia vazquezii*	2.9 mg·g^{-1}	[9]
Iron	*Bowenia serrulata*	189 mg·kg^{-1}	[9]
Iron	*Bowenia spectabilis*	207 mg·kg^{-1}	[9]
Iron	*Ceratozamia mexicana*	106 mg·kg^{-1}	[9]
Iron	*Cycas debaoensis*	114 mg·kg^{-1}	[9]

Table A1. *Cont.*

Element	Species	Range	Reference
Iron	*Cycas diannanensis*	406 mg·kg^{-1}	[9] [1]
Iron	*Cycas diannanensis*	406 mg·kg^{-1}	[19] [1]
Iron	*Cycas elongata*	149 mg·kg^{-1}	[9]
Iron	*Cycas fairylakea*	98 mg·kg^{-1}	[9]
Iron	*Cycas micholitzii*	340 mg·kg^{-1}	[9]
Iron	*Cycas micholitzii*	345 mg·kg^{-1}	[19]
Iron	*Cycas micronesica*	43.5 mg·kg^{-1}	[12]
Iron	*Cycas micronesica*	38.5–88.6 mg·kg^{-1}	[18]
Iron	*Cycas micronesica*	39.6–46.8 mg·kg^{-1}	[13]
Iron	*Cycas micronesica*	26.8–56.9 mg·kg^{-1}	[14]
Iron	*Cycas micronesica*	71.4 mg·kg^{-1}	[20]
Iron	*Cycas nongnoochiae*	76.4 mg·kg^{-1}	[20]
Iron	*Cycas panzhihuaensis*	134–215 mg·kg^{-1}	[9]
Iron	*Cycas panzhihuaensis*	225 mg·kg^{-1}	[19]
Iron	*Cycas revoluta*	31 mg·kg^{-1}	[24]
Iron	*Cycas sexseminifera*	311 mg·kg^{-1}	[9]
Iron	*Cycas sexseminifera*	300 mg·kg^{-1}	[19] [4]
Iron	*Cycas siamensis*	218 mg·kg^{-1}	[9]
Iron	*Cycas siamensis*	225 mg·kg^{-1}	[19]
Iron	*Cycas szechuanensis*	234–304 mg·kg^{-1}	[9]
Iron	*Cycas szechuanensis*	300 mg·kg^{-1}	[19]
Iron	*Cycas thouarsii*	166 mg·kg^{-1}	[9]
Iron	*Cycas wadei*	71.3 mg·kg^{-1}	[11]
Iron	*Dioon edule*	163 mg·kg^{-1}	[9]
Iron	*Dioon mejiae*	117 mg·kg^{-1}	[9]
Iron	*Dioon spinulosum*	123 mg·kg^{-1}	[9]
Iron	*Encephalartos cupidus*	363 mg·kg^{-1}	[9]
Iron	*Encephalartos cupidus*	355 mg·kg^{-1}	[19]
Iron	*Encephalartos ferox*	93 mg·kg^{-1}	[9]
Iron	*Encephalartos gratus*	121–339 mg·kg^{-1}	[9]
Iron	*Encephalartos gratus*	340 mg·kg^{-1}	[19]
Iron	*Lepidozamia hopei*	176 mg·kg^{-1}	[9]
Iron	*Lepidozamia peroffskyana*	166 mg·kg^{-1}	[9]
Iron	*Macrozamia communis*	83 mg·kg^{-1}	[9]
Iron	*Macrozamia lucida*	197 mg·kg^{-1}	[9]
Iron	*Macrozamia moorei*	253 mg·kg^{-1}	[9]
Iron	*Stangeria eriopus*	228 mg·kg^{-1}	[9]
Iron	*Zamia erosa*	142 mg·kg^{-1}	[9] [2]
Iron	*Zamia fischeri*	1697 mg·kg^{-1}	[9] [3]
Iron	*Zamia fischeri*	1700 mg·kg^{-1}	[19] [3]
Iron	*Zamia furfuracea*	194–272 mg·kg^{-1}	[9]
Iron	*Zamia furfuracea*	260 mg·kg^{-1}	[19]
Iron	*Zamia integrifolia*	211–270 mg·kg^{-1}	[9]
Iron	*Zamia splendens*	160 mg·kg^{-1}	[9]
Iron	*Zamia vazquezii*	478 mg·kg^{-1}	[9]
Manganese	*Cycas micronesica*	23.8 mg·kg^{-1}	[12]
Manganese	*Cycas micronesica*	19.5–44.7 mg·kg^{-1}	[18]
Manganese	*Cycas micronesica*	26.1–77.5 mg·kg^{-1}	[13]
Manganese	*Cycas micronesica*	25.4–95.6 mg·kg^{-1}	[14]
Manganese	*Cycas micronesica*	36.6 mg·kg^{-1}	[20]
Manganese	*Cycas micronesica*	68.6 mg·kg^{-1}	[20]
Manganese	*Cycas revoluta*	27.1–73.7 mg·kg^{-1}	[24]
Manganese	*Cycas wadei*	152 mg·kg^{-1}	[11]
Manganese	*Macrozamia riedlei*	6-57 mg·kg^{-1}	[22]
Boron	*Cycas micronesica*	13.6 mg·kg^{-1}	[12]

Table A1. *Cont.*

Element	Species	Range	Reference
Boron	*Cycas micronesica*	11.6–14.3 mg·kg^{-1}	[13]
Boron	*Cycas micronesica*	13.6–15.9 mg·kg^{-1}	[14]
Boron	*Cycas micronesica*	43.4 mg·kg^{-1}	[20]
Boron	*Cycas micronesica*	25.6 mg·kg^{-1}	[20]
Boron	*Cycas wadei*	17.2 mg·kg^{-1}	[11]
Copper	*Cycas micronesica*	4.2 mg·kg^{-1}	[12]
Copper	*Cycas micronesica*	6.5–17.9 mg·kg^{-1}	[18]
Copper	*Cycas micronesica*	3.1 mg·kg^{-1}	[13]
Copper	*Cycas micronesica*	2.0–4.0 mg·kg^{-1}	[14]
Copper	*Cycas micronesica*	7.7 mg·kg^{-1}	[20]
Copper	*Cycas micronesica*	9.7 mg·kg^{-1}	[20]
Copper	*Cycas wadei*	3.9 mg·kg^{-1}	[11]
Copper	*Macrozamia riedlei*	2.1–2.8 mg·kg^{-1}	[23]
Zinc	*Bowenia serrulata*	19.2 mg·kg^{-1}	[9]
Zinc	*Bowenia spectabilis*	21.4 mg·kg^{-1}	[9]
Zinc	*Ceratozamia mexicana*	24.4 mg·kg^{-1}	[9]
Zinc	*Cycas debaoensis*	18.6 mg·kg^{-1}	[9]
Zinc	*Cycas diannanensis*	18.9 mg·kg^{-1}	[9] [1]
Zinc	*Cycas elongata*	19.8 mg·kg^{-1}	[9]
Zinc	*Cycas fairylakea*	26.6 mg·kg^{-1}	[9]
Zinc	*Cycas micholitzii*	14.1 mg·kg^{-1}	[9]
Zinc	*Cycas micronesica*	19.0 mg·kg^{-1}	[12]
Zinc	*Cycas micronesica*	15.2–70.2 mg·kg^{-1}	[18]
Zinc	*Cycas micronesica*	20.4–45.7 mg·kg^{-1}	[13]
Zinc	*Cycas micronesica*	18.1–59.8 mg·kg^{-1}	[14]
Zinc	*Cycas micronesica*	32.5 mg·kg^{-1}	[20]
Zinc	*Cycas nongnoochiae*	28.0 mg·kg^{-1}	[20]
Zinc	*Cycas panzhihuaensis*	13.1–15.1 mg·kg^{-1}	[9]
Zinc	*Cycas revoluta*	5.7–68.5 mg·kg^{-1}	[24]
Zinc	*Cycas sexseminifera*	13.6 mg·kg^{-1}	[9]
Zinc	*Cycas siamensis*	11.1 mg·kg^{-1}	[9]
Zinc	*Cycas szechuanensis*	13.6–18.3 mg·kg^{-1}	[9]
Zinc	*Cycas thouarsii*	14.2 mg·kg^{-1}	[9]
Zinc	*Cycas wadei*	10.3 mg·kg^{-1}	[11]
Zinc	*Dioon edule*	22.6 mg·kg^{-1}	[9]
Zinc	*Dioon mejiae*	12.3 mg·kg^{-1}	[9]
Zinc	*Dioon spinulosum*	16.4 mg·kg^{-1}	[9]
Zinc	*Encephalartos cupidus*	10.5 mg·kg^{-1}	[9]
Zinc	*Encephalartos ferox*	17.8 mg·kg^{-1}	[9]
Zinc	*Encephalartos gratus*	14.9–22.2 mg·kg^{-1}	[9]
Zinc	*Lepidozamia hopei*	23.2 mg·kg^{-1}	[9]
Zinc	*Lepidozamia peroffskyana*	25.2 mg·kg^{-1}	[9]
Zinc	*Macrozamia communis*	21.5 mg·kg^{-1}	[9]
Zinc	*Macrozamia lucida*	21.0 mg·kg^{-1}	[9]
Zinc	*Macrozamia moorei*	18.2 mg·kg^{-1}	[9]
Zinc	*Macrozamia riedlei*	3.6–6.6 mg·kg^{-1}	[23]
Zinc	*Stangeria eriopus*	53.3 mg·kg^{-1}	[9]
Zinc	*Zamia erosa*	13.9 mg·kg^{-1}	[9] [2]
Zinc	*Zamia fischeri*	20.0 mg·kg^{-1}	[9] [3]
Zinc	*Zamia furfuracea*	10.5–13.7 mg·kg^{-1}	[9]
Zinc	*Zamia integrifolia*	15.5–16.1 mg·kg^{-1}	[9]
Zinc	*Zamia splendens*	13.8 mg·kg^{-1}	[9]
Zinc	*Zamia vazquezii*	38.4 mg·kg^{-1}	[9]
Aluminum	*Cycas revoluta*	22.0–59.6 mg·kg^{-1}	[24]

Table A1. *Cont.*

Element	Species	Range	Reference
Selenium	*Cycas micronesica*	0.58 mg·kg^{-1}	[12]
Selenium	*Cycas wadei*	0.41 mg·kg^{-1}	[11]

[1] Reported as *Cycas parvula* S.L. Yang ex D.Y. Wang; [2] Reported as *Zamia amblyphyllidia* D.W. Stev.; [3] The name *Z. fischeri* is widely misapplied to the species *Z. vazquezii* in cultivation. The real *Z. fischeri* is extremely rare in cultivation, and it is probable that the taxon sampled was *Z. vazquesii*; [4] Reported as *Cycas miquelii* Warb.

Table A2. Published ranges for leaf litter element concentrations of cycad plants. Misspellings of species were corrected if identity was obvious, species that were misspelled were not included if identity was not obvious. Taxonomic synonyms were corrected. Data were estimated for reports displaying data as figures and transformed if data were presented as log.

Element	Species	Range	Reference
Carbon	*Cycas micronesica*	475–486 mg·g^{-1}	[35]
Carbon	*Cycas micronesica*	501–534 mg·g^{-1}	[18]
Carbon	*Cycas micronesica*	509 mg·g^{-1}	[36]
Carbon	*Cycas nitida*	494–519 mg·g^{-1}	[10]
Carbon	*Cycas wadei*	513 mg·g^{-1}	[11]
Carbon	*Macrozamia communis*	515–546 mg·g^{-1}	[16]
Carbon	*Macrozamia riedlei*	502–534 mg·g^{-1}	[16]
Nitrogen	*Cycas micronesica*	16–22 mg·g^{-1}	[35]
Nitrogen	*Cycas micronesica*	21–22 mg·g^{-1}	[18]
Nitrogen	*Cycas micronesica*	20 mg·g^{-1}	[36]
Nitrogen	*Cycas nitida*	17–22 mg·g^{-1}	[10]
Nitrogen	*Cycas wadei*	19 mg·g^{-1}	[11]
Nitrogen	*Macrozamia communis*	11–24 mg·g^{-1}	[16]
Nitrogen	*Macrozamia riedlei*	11–20 mg·g^{-1}	[16]
Phosphorus	*Cycas micronesica*	0.5–0.9 mg·g^{-1}	[18]
Phosphorus	*Cycas micronesica*	1.3–2.0 mg·g^{-1}	[35]
Phosphorus	*Cycas nitida*	0.3–0.9 mg·g^{-1}	[10]
Phosphorus	*Cycas wadei*	0.5 mg·g^{-1}	[11]
Potassium	*Cycas micronesica*	1.0–1.9 mg·g^{-1}	[18]
Potassium	*Cycas micronesica*	2.2–14.2 mg·g^{-1}	[35]
Potassium	*Cycas nitida*	1.2–4.5 mg·g^{-1}	[10]
Potassium	*Cycas wadei*	3.2 mg·g^{-1}	[11]
Magnesium	*Cycas micronesica*	3.39–6.52 mg·g^{-1}	[18]
Magnesium	*Cycas micronesica*	3.38–5.82 mg·g^{-1}	[35]
Magnesium	*Cycas wadei*	1.32 mg·g^{-1}	[11]
Calcium	*Cycas micronesica*	4.2–15.1 mg·g^{-1}	[18]
Calcium	*Cycas micronesica*	11.9–32.3 mg·g^{-1}	[35]
Calcium	*Cycas wadei*	2.5 mg·g^{-1}	[11]
Sulfur	*Cycas micronesica*	1.20–1.38 mg·g^{-1}	[35]
Iron	*Cycas micronesica*	64–272 mg·kg^{-1}	[35]
Iron	*Cycas micronesica*	28–547 mg·kg^{-1}	[18]
Iron	*Cycas wadei*	37 mg·kg^{-1}	[11]
Manganese	*Cycas micronesica*	24.5–86.1 mg·kg^{-1}	[18]
Manganese	*Cycas micronesica*	23.0–37.3 mg·kg^{-1}	[35]
Manganese	*Cycas wadei*	141 mg·kg^{-1}	[11]
Boron	*Cycas micronesica*	29.5–51.6 mg·kg^{-1}	[35]
Boron	*Cycas wadei*	9.9 mg·kg^{-1}	[11]
Copper	*Cycas micronesica*	2.4–4.4 mg·kg^{-1}	[35]
Copper	*Cycas micronesica*	1.3–5.9 mg·kg^{-1}	[18]
Copper	*Cycas wadei*	3.3 mg·kg^{-1}	[11]
Zinc	*Cycas micronesica*	4.5–31.2 mg·kg^{-1}	[18]
Zinc	*Cycas micronesica*	11.0–23.8 mg·kg^{-1}	[35]
Zinc	*Cycas wadei*	5.9 mg·kg^{-1}	[11]
Selenium	*Cycas wadei*	0.48 mg·kg^{-1}	[11]

References

1. Schwab, G.J.; Lee, C.D.; Pearce, R. *Sampling Plant Tissue for Nutrient Analysis*; Univ. of Kentucky Cooperative Extension Service Publication AGR-92; Univ. of Kentucky: Lexington, KY, USA, 2007; p. 6.
2. Lazicki, P.; Geisseler, D. *Plant Tissue Sampling in Orchards and Vineyards*; Univ. of California: Davis, CA, USA, 2016; p. 3. Available online: https://apps1.cdfa.ca.gov/FertilizerResearch/docs/Orchard_Tissue_Sampling.pdf (accessed on 1 November 2020).
3. Vashisth, T.; Burrow, J.D.; Kadyampakeni, D.; Ferrarezi, R.S. *Citrus Leaf Sampling Procedures for Nutrient Analysis*; Univ. Florida IFAS Extension Publication #HS1355.: Gainesville, FL, USA, 2020; p. 2. Available online: http://edis.ifas.ufl.edu (accessed on 10 October 2020).
4. Campbell, C.R. (Ed.) *Reference Sufficiency Ranges for Plant Analysis in the Southern Region of the United States*; Southern Cooperative Series Bulletin #394; North Carolina Dept. of Agric.: Raleigh, NC, USA, 2000; p. 122. Available online: http://www.ncagr.gov/agronomi/saaesd/scsb394.pdf (accessed on 1 November 2020).
5. Calonje, M.; Stevenson, D.W.; Osborne, R. The World List of Cycads. Available online: http://cycadlist.org (accessed on 1 November 2020).
6. Norstog, K.J.; Nicholls, T.J. *The Biology of the Cycads*; Cornell University Press: Ithaca, NY, USA, 1997; ISBN 978-0-8014-3033-6.
7. Cascasan, A.N.; Marler, T.E. Publishing trends for the Cycadales, the most threatened plant group. *J. Threat. Taxa* **2016**, *8*, 8575–8582. [CrossRef]
8. Marler, T.E.; Lindström, A.J. Inserting cycads into global nutrient relations data sets. *Plant Signal. Behav.* **2018**, *13*, e1547578. [CrossRef] [PubMed]
9. Zhang, Y.; Cao, K.; Sack, L.; Li, N.; Wei, X.; Goldstein, G. Extending the generality of leaf economic design principles in the cycads, an ancient lineage. *New Phytol.* **2015**, *206*, 817–829. [CrossRef] [PubMed]
10. Marler, T.E.; Ferreras, U.F. Disruption of leaf nutrient remobilization in coastal *Cycas* trees by tropical cyclone damage. *J. Geogr. Nat. Disast.* **2015**, *5*, 1421–1427.
11. Marler, T.E.; Ferreras, U.F. Current status, threats and conservation needs of the endemic *Cycas wadei* Merrill. *J. Biodivers. Endanger. Species* **2017**, *5*, 3. [CrossRef]
12. Marler, T.E. Elemental profiles in *Cycas micronesica* stems. *Plants* **2018**, *7*, 94. [CrossRef]
13. Marler, T.E.; Krishnapillai, M.V. Incident light and leaf age influence leaflet element concentrations of *Cycas micronesica* trees. *Horticulturae* **2019**, *5*, 58. [CrossRef]
14. Marler, T.E.; Krishnapillai, M.V. Distribution of elements along the rachis of *Cycas micronesica* leaves: A cautionary note for sampling design. *Horticulturae* **2019**, *5*, 33. [CrossRef]
15. Krieg, C.; Watkins, J.E.; Chambers, S.; Husby, C.E. Sex-specific differences in functional traits and resource acquisition in five cycad species. *AoB Plants* **2017**, *9*, plx013. [CrossRef]
16. Kipp, M.A.; Stüeken, E.E.; Gehringer, M.M.; Sterelny, K.; Scott, J.K.; Forster, P.I.; Strömberg, C.A.; Buick, R. Exploring cycad foliage as an archive of the isotopic composition of atmospheric nitrogen. *Geobiology* **2020**, *18*, 152–166. [CrossRef]
17. Marler, T.E. Artifleck: The study of artifactual responses to light flecks with inappropriate leaves. *Plants* **2020**, *9*, 905. [CrossRef] [PubMed]
18. Marler, T.E.; Krishnapillai, M.V. Does plant size influence leaf elements in an arborescent cycad? *Biology* **2018**, *7*, 51. [CrossRef] [PubMed]
19. Zhang, Y.-J.; Sack, L.; Goldstein, G.; Cao, K.-F. Hydraulic determination of leaf nutrient concentrations in cycads. *Mem. NY Bot. Gard.* **2018**, *117*, 179–192.
20. Marler, T.E.; Lindström, A.J. Leaf nutrients of two *Cycas*, L. species contrast among in situ and ex situ locations. *J. Threat. Taxa* **2020**, *12*, 16831–16839. [CrossRef]
21. Wright, I.; Reich, P.; Westoby, M.; Ackerly, D.D.; Baruch, Z.; Bongers, F.; Cavender-Bares, J.; Chapin, T.; Cornilessen, J.H.C.; Deimer, M.; et al. The worldwide leaf economics spectrum. *Nature* **2004**, *428*, 821–827. [CrossRef]
22. Álvarez-Yépiz, J.C.; Cueva, A.; Dovčiak, M.; Teece, M.; Yepez, E.A. Ontogenetic resource-use strategies in a rare long-lived cycad along environmental gradients. *Conserv. Physiol.* **2014**, *2*. [CrossRef]
23. Grove, T.S.; O'Connell, A.M.; Malajczuk, N. Effects of fire on the growth, nutrient content and rate of nitrogen fixation of the cycad *Macrozamia riedlei*. *Austral. J. Bot.* **1980**, *28*, 271–281. [CrossRef]

24. Watanabe, T.; Broadley, M.R.; Jansen, S.; White, P.J.; Takada, J.; Satake, K.; Takamatsu, T.; Tuah, S.J.; Osaki, M. Evolutionary control of leaf element composition in plants. *New Phytol.* **2007**, *174*, 516–523. [CrossRef]
25. Zhang, Y.-J.; Sack, L.; Cao, K.-F.; Wei, X.-M.; Li, N. Speed versus endurance tradeoff in plants: Leaves with higher photosynthetic rates show stronger seasonal declines. *Sci. Rep.* **2017**, *7*, 42085. [CrossRef]
26. Kulmatiski, A.; Beard, K.H.; Stevens, J.R.; Cobbold, S.M. Plant–soil feedbacks: A meta-analytical review. *Ecol. Lett.* **2008**, *11*, 980–992. [CrossRef]
27. Van der Putten, W.H.; Bardgett, R.D.; Bever, J.D.; Bezemer, T.M.; Casper, B.B.; Fukami, T.; Kardol, P.; Klironomos, J.N.; Kulmatiski, A.; Schweitzer, J.A.; et al. Plant–soil feedbacks: The past, the present and future challenges. *J. Ecol.* **2013**, *101*, 265–276. [CrossRef]
28. Ponge, J.-F. Plant–soil feedbacks mediated by humus forms: A review. *Soil Biol. Biochem.* **2013**, *57*, 1048–1060. [CrossRef]
29. Veen, G.F.; Fry, E.L.; ten Hooven, F.C.; Kardol, P.; Morriën, E.; De Long, J.R. The role of plant litter in driving plant-soil feedbacks. *Front. Environ. Sci.* **2019**, *7*, 168. [CrossRef]
30. Gholz, H.L.; Wedin, D.A.; Smitherman, S.M.; Harmon, M.E.; Parton, W.J. Long-term dynamics of pine and hardwood litter in contrasting environments: Toward a global model of decomposition. *Glob. Chang. Biol.* **2000**, *6*, 751–765. [CrossRef]
31. Veen, G.F.; Freschet, G.T.; Ordonez, A.; Wardle, D.A. Litter quality and environmental controls of home-field advantage effects on litter decomposition. *Oikos* **2015**, *124*, 187–195. [CrossRef]
32. Veronika, G.V.; Hufnagel, L. The effect of microarthropods on litter decomposition depends on litter quality. *Eur. J. Soil Biol.* **2016**, *75*, 24–30.
33. Palozzi, J.E.; Lindo, Z. Are leaf litter and microbes team players? Interpreting home-field advantage decomposition dynamics. *Soil Biol. Biochem.* **2018**, *124*, 189–198. [CrossRef]
34. Elias, D.M.O.; Robinson, S.; Both, S.; Goodall, T.; Majalap-Lee, N.; Ostle, N.J.; McNamara, N.P. 2020 Soil microbial community and litter quality controls on decomposition across a tropical forest disturbance gradient. *Front. For. Glob. Chang.* **2020**, *3*, 81. [CrossRef]
35. Marler, T.E.; Dongol, N. Three invasive insects alter *Cycas micronesica* leaf chemistry and predict changes in biogeochemical cycling. *Communic. Integr. Biol.* **2016**, *9*, e1208324. [CrossRef]
36. Aerts, R. Nutrient resorption from senescing leaves of perennials: Are there general patterns? *J. Ecol.* **1996**, *84*, 597–608. [CrossRef]
37. Killingbeck, K.T. Nutrients in Senesced Leaves: Keys to the Search for Potential Resorption and Resorption Proficiency. *Ecology* **1996**, *77*, 1716–1727. [CrossRef]
38. Marler, T.E. Perennial trees associating with nitrogen-fixing symbionts differ in leaf after-life nitrogen and carbon release. *Nitrogen* **2020**, *1*, 111–124. [CrossRef]
39. Marler, T.E.; Krishnapillai, M.V. *Cycas micronesica* trees alter local soil traits. *Forests* **2018**, *9*, 565. [CrossRef]
40. Marler, T.E.; Calonje, M. Two cycad species affect the carbon, nitrogen, and phosphorus content of soils. *Horticulturae* **2020**, *6*, 24. [CrossRef]
41. Vance, A.J.; Strik, B.C. Seasonal changes in leaf nutrient concentration of male and female hardy kiwifruit grown in Oregon. *Eur. J. Hortic. Sci.* **2018**, *83*, 247–258. [CrossRef]
42. Pasković, I.; Lukić, I.; Žurga, P.; Majetić Germek, V.; Brkljača, M.; Koprivnjak, O.; Major, N.; Grozić, K.; Franić, M.; Ban, D.; et al. Temporal variation of phenolic and mineral composition in olive leaves is cultivar dependent. *Plants* **2020**, *9*, 1099. [CrossRef] [PubMed]
43. Shaul, O. Magnesium transport and function in plants: The tip of the iceberg. *Biometals* **2002**, *15*, 307–321. [CrossRef]
44. Cakmak, I. Magnesium in crop production, food quality and human health. *Plant Soil* **2013**, *368*, 1–4. [CrossRef]
45. Guo, C.; Cornelissen, J.H.C.; Tuo, B.; Ci, H.; Yan, E.-R. Non-negligible contribution of subordinates in community-level litter decomposition: Deciduous trees in an evergreen world. *J. Ecol.* **2020**, *108*, 1713–1724. [CrossRef]
46. Levia, D.F., Jr.; Frost, E.E. A review and evaluation of stemflow literature in the hydrologic and biogeochemical cycles of forested and agricultural ecosystems. *J. Hydrol.* **2003**, *274*, 1–29. [CrossRef]
47. Levia, D.F.; Germer, S. A review of stemflow generation dynamics and stemflow-environment interactions in forests and shrublands. *Rev. Geophys.* **2015**, *53*, 673–714. [CrossRef]

48. Van Stan, J.T.; Gordon, D.A. Mini-review: Stemflow as a resource limitation to near-stem soils. *Front. Plant Sci.* **2018**, *9*, 248. [CrossRef] [PubMed]

49. Su, L.; Zhao, C.; Xu, W.; Xie, Z. Hydrochemical fluxes in bulk precipitation, throughfall, and stemflow in a mixed evergreen and deciduous broadleaved forest. *Forests* **2019**, *10*, 507. [CrossRef]

50. Dunkerley, D. A Review of the Effects of Throughfall and Stemflow on Soil Properties and Soil Erosion. In *Precipitation Partitioning by Vegetation*; Van Stan, J., II, Gutmann, E., Friesen, J., Eds.; Springer: Cham, Switzerland, 2020; pp. 183–214.

51. Stubbins, A.; Guillemette, F.; Van Stan, J.T., II. Throughfall and Stemflow: The Crowning Headwaters of the Aquatic Carbon Cycle. In *Precipitation Partitioning by Vegetation*; Van Stan, J., II, Gutmann, E., Friesen, J., Eds.; Springer: Cham, Switzerland, 2020; pp. 121–132.

52. Raich, J.W. Understory palms as nutrient traps: A hypothesis. *Brenesia* **1983**, *21*, 119–129.

53. Scroth, G.; da Silva, L.F.; Wolf, M.-Z.; Teixeira, W.G.; Zech, W. Distribution of throughfall and stemflow in multi-strata agroforestry, perennial monoculture, fallow and primary forest in central Amazonia, Brazil. *Hydrol. Process* **1999**, *13*, 1423–1436. [CrossRef]

54. Edwards, P.J.; Fleischer-Dogley, F.; Kaiser-Bunbury, C.N. The nutrient economy of *Lodoicea maldivica*, a monodominant palm producing the world's largest seed. *New Phytol.* **2015**, *206*, 990–999. [CrossRef]

55. Zhang, Y.-F.; Wang, X.-P.; Pan, Y.-X.; Hu, R. Relative contribution of biotic and abiotic factors to stemflow production and funneling efficiency: A long-term field study on a xerophytic shrub species in Tengger Desert of northern China. *Agric. For. Meteorol.* **2020**, *280*, 107781. [CrossRef]

56. Zona, S.; Christenhusz, M.J.M. Litter-trapping plants: Filter-feeders of the plant kingdom. *Bot. J. Linn. Soc.* **2015**, *179*, 554–586. [CrossRef]

57. Rickson, F.R.; Rickson, M.M. Nutrient acquisition facilitated by litter collection and ant colonies on two Malaysian palms. *Biotropica* **1986**, *18*, 337–343. [CrossRef]

58. Alvarez-Sánchez, J.; Guevara, S. Litter interception on *Astrocaryum mexicanum* Liebm. (Palmae) in a tropical rain forest. *Biotropica* **1999**, *31*, 89–92.

59. Dearden, F.M.; Wardle, D.A. The potential for forest canopy litterfall interception by a dense fern understorey, and the consequences for litter decomposition. *Oikos* **2008**, *117*, 83–92. [CrossRef]

60. Stevenson, D.W. Spines and prickles. *Mem. NY Bot. Gard.* **2018**, *117*, 54–65.

61. Stamp, N.E. Out of the quagmire of plant defense hypotheses. *Q. Rev. Biol.* **2003**, *78*, 23–55. [CrossRef] [PubMed]

62. Whiting, M.G. Toxicity of cycads, a literature review. *Econ. Bot.* **1963**, *17*, 270–302. [CrossRef]

63. Prado, A.; Sierra, A.; Windsor, D.; Bede, J.C. Leaf traits and herbivory levels in a tropical gymnosperm, *Zamia stevensonii* (Zamiaceae). *Am. J. Bot.* **2014**, *101*, 437–447. [CrossRef]

64. Hoffmann, G.R.; Morgan, R.W. Review: Putative mutagens and carcinogens in foods. V. Cycad azoxyglycosides. *Environ. Mutagen.* **1984**, *6*, 103–116. [CrossRef] [PubMed]

65. Yagi, F. Azoxyglycoside content and beta-glycosidase activities in leaves of various cycads. *Phytochemistry.* **2004**, *65*, 3243–3247. [CrossRef]

66. Prado, A.; Rubio-Mendez, G.; Yañez-Espinosa, L.; Bede, J.C. Ontogenetic changes in azoxyglycoside levels in the leaves of *Dioon edule* Lindl. *J. Chem. Ecol.* **2016**, *42*, 1142–1150. [CrossRef]

67. Ferlian, O.; Lintzel, E.M.; Bruelheide, H.; Guerra, C.A.; Heklau, H.; Jurburg, S.; Kühn, P.; Martinez-Medina, A.; Unsicker, S.B.; Eisenhauer, N.; et al. Nutrient status not secondary metabolites drives herbivory and pathogen infestation across differently mycorrhized tree monocultures and mixtures. *Basic Appl. Ecol.* **2020**, *49*, 51282. [CrossRef]

68. Zheng, Y.; Chiang, T.; Huang, C.; Gong, X. Highly diverse endophytes in roots of *Cycas bifida* (Cycadaceae), an ancient but endangered gymnosperm. *J. Microbiol.* **2018**, *56*, 337–345. [CrossRef]

69. Chang, A.C.G.; Chen, T.; Li, N.; Duan, J. Perspectives on endosymbiosis in coralloid roots: Association of cycads and cyanobacteria. *Front. Microbiol.* **2019**, *10*, 1888. [CrossRef] [PubMed]

70. Gutiérrez-García, K.; Bustos-Díaz, E.D.; Corona-Gómez, J.A.; Ramos-Aboites, H.E.; Sélem-Mojica, N.; Cruz-Morales, P.; Pérez-Farrera, M.A.; Barona-Gómez, F.; Cibrián-Jaramillo, A. Cycad coralloid roots contain bacterial communities including cyanobacteria and *Caulobacter* spp. that encode niche-specific biosynthetic gene clusters. *Genome Biol. Evol.* **2019**, *11*, 319–334.

71. Muthukumar, T.; Udaiyan, K. Arbuscular mycorrhizas in cycads of southern India. *Mycorrhiza* **2002**, *12*, 213–217. [PubMed]

72. Fisher, J.B.; Vovides, A.P. Mycorrhizae are present in cycad roots. *Bot. Rev.* **2004**, *70*, 16–23. [CrossRef]

Publisher's Note: MDPI stays neutral with regard to jurisdictional claims in published maps and institutional affiliations.

 *horticulturae*

Review

Seed Geometry in the Arecaceae

Diego Gutiérrez del Pozo [1], José Javier Martín-Gómez [2], Ángel Tocino [3] and Emilio Cervantes [2,*]

[1] Departamento de Conservación y Manejo de Vida Silvestre (CYMVIS), Universidad Estatal Amazónica (UEA), Carretera Tena a Puyo Km. 44, Napo EC-150950, Ecuador; dgutierrez@uea.edu.ec
[2] IRNASA-CSIC, Cordel de Merinas 40, E-37008 Salamanca, Spain; jjavier.martin@irnasa.csic.es
[3] Departamento de Matemáticas, Facultad de Ciencias, Universidad de Salamanca, Plaza de la Merced 1–4, 37008 Salamanca, Spain; bacon@usal.es
* Correspondence: emilio.cervantes@irnasa.csic.es; Tel.: +34-923219606

Received: 31 August 2020; Accepted: 2 October 2020; Published: 7 October 2020

Abstract: Fruit and seed shape are important characteristics in taxonomy providing information on ecological, nutritional, and developmental aspects, but their application requires quantification. We propose a method for seed shape quantification based on the comparison of the bi-dimensional images of the seeds with geometric figures. *J* index is the percent of similarity of a seed image with a figure taken as a model. Models in shape quantification include geometrical figures (circle, ellipse, oval ...) and their derivatives, as well as other figures obtained as geometric representations of algebraic equations. The analysis is based on three sources: Published work, images available on the Internet, and seeds collected or stored in our collections. Some of the models here described are applied for the first time in seed morphology, like the superellipses, a group of bidimensional figures that represent well seed shape in species of the Calamoideae and *Phoenix canariensis* Hort. ex Chabaud. Oval models are proposed for *Chamaedorea pauciflora* Mart. and cardioid-based models for *Trachycarpus fortunei* (Hook.) H. Wendl. Diversity of seed shape in the Arecaceae makes this family a good model system to study the application of geometric models in morphology.

Keywords: circle; ellipse; lens; morphology; oval; seed shape; superellipse

---

1. Introduction

The Arecaceae Schultz Sch. (Palmae nom. cons.) is a unique family in the order Arecales, class Monocotyledoneae. It includes about 2600 species grouped in 181 genera of climbers, shrubs, and tree-like and stemless plants, all commonly known as palms, with a worldwide distribution in tropical and subtropical regions. Economically important species are the oil palm (*Elaeis oleifera* [Kunth] Cortes ex Prain), the coconut palm (*Cocos nucifera* L.), the date palm (*Phoenix dactilifera* L.) Other species in this family are the most used by local people in many tropical places like the Amazonian basin, with a great variety of applications [1–6]. It is remarkable that whole human communities, particularly in the tropics, depend on palms for their survival [1,6]. *Trachycarpus fortunei* (Hook.) H.Wendl. and *Chamaerops humilis* L. are frequently used as ornamentals in temperate zones in Mediterranean countries for their resistance to extreme temperatures.

The Arecaceae form a monophyletic group among the monocotyledons as supported by both molecular and morphological data that, in recent years, have integrated much information from DNA sequencing. Current taxonomy presents five subfamilies (Calamoideae, Nypoideae, Coryphoideae, Ceroxyloideae, and Arecoideae) divided into tribes and subtribes [1]. Table 1 shows the distribution of tribes within subfamilies and representative examples of genera. The largest subfamily, the Arecoideae, contains 106 genera grouped into 14 tribes, of which the largest, the Areceae is formed by 11 subtribes that receive their names from the respective genera (Archontophoenicinae, Arecinae, Basseliniinae, Carpoxylinae, Clinospermatinae, Dypsidinae, Linospadicinae, Oncospermatinae, Ptychospermatinae,

Rhopalostylidinae, and Verschaffeltiinae), plus an additional group of ten unplaced genera (*Bentinckia, Clinostigma, Cyrtostachys, Dictiosperma, Dransfieldia, Heterospathe, Hydriastele, Iguanura, Loxococcus,* and *Rhopaloblaste*).

Table 1. A summary of the taxonomy of the Arecaceae. Subfamilies Calamoideae, Nypoideae, Coryphoideae, Ceroxyloideae, and Arecoideae divided into tribes. The number of genera in each subfamily and tribe is given between parentheses. Some examples of genera in each of the tribes are shown. Data adapted from [1].

Subfamilies	Tribes	Genera
I. Calamoideae (21)		
	Eugeissoneae (1)	*Eugeissona*
	Lepidocaryae (7)	*Oncocalamus, Eremospatha, Laccosperma, Raphia*
	Calameae (13)	*Calamus, Retispatha, Ceratolobus*
II. Nypoideae (1)		*Nypha*
III. Coryphoideae (46)		
	Sabaleae (1)	*Sabal*
	Cryosophileae (10)	*Coccothrinax, Hemithrinax, Leucothrinax, Thrinax*
	Phoeniceae (1)	*Phoenix*
	Trachycarpeae (18)	*Chamaerops, Guihaia, Trachycarpus, Rhapis*
	Chuniophoeniceae (4)	*Livistona, Licuala, Johannesteijsmannia*
	Caryoteae (3)	*Caryota, Arenga, Wallichia*
	Corypheae (1)	*Corypha*
	Borasseae (8)	*Bismarckia, Lodoicea, Borassodendron, Borassus*
IV. Ceroxyloideae (8)		
	Cyclospatheae (1)	*Pseudophoenix*
	Ceroxyleae (4)	*Ceroxylon, Juania, Oraniopsis, Ravenea*
	Phytelepheae (3)	*Ammandra, Aphandra, Phytelephas*
V. Arecoideae (106)		
	Iriarteeae (5)	*Iriartea*
	Chamaedoreae (5)	*Chamaedorea*
	Podococceae (1)	*Podococcus*
	Oranieae (1)	*Orania*
	Sclerospermeae (1)	*Sclerosperma*
	Roystoneeae (1)	*Roystonea*
	Reinhardtieae (1)	*Reinhardtia*
	Cocoseae (18)	*Cocos, Jubaea, Bactris, Elaeis*
	Manicarieae (1)	*Manicaria*
	Euterpeae (5)	*Hyospathe, Euterpe*
	Geonomateae (6)	*Asterogyne, Geonoma*
	Leopoldiniaeae (1)	*Leopoldinia*
	Pelagodoxeae (2)	*Pelagodoxa, Sommieria*
	Areceae (58)	*Archontophoenix, Areca, Basselinia, Carpoxylon, Clinosperma, Dypsis, Linospadix, Oncosperma, Ptychosperma, Rhopalostylis, Verschaffeltia,*

Based on the utility of palms in industry at both subsistence and world market levels, the Arecaceae has been reported to be the third family in applied importance in the Plant Kingdom following the Poaceae and the Fabaceae [5]. Detailed analyses of fruit and seed morphology in this family may provide valuable information.

2. Seed Morphology in the Arecaceae

The seeds of the Arecaceae are varied in size ranging from the small size of *Prestoea* Hook., a few millimeters in length, to the largest seeds of all plants weighing up to 25 kg: The "coco de mer" (*Lodoicea maldivica* [J. F. Gmelin] Persoon). The size and shape of fruits and seeds correspond to a diversity of types of dispersion, often in water or by zoochory by a variety of animals [6–10], with a predominance of the circular, ellipsoid, and oval morphology [1,11]. The combinations of fibrous mesocarps and hard endocarps enclosing a cavity made of the fruits of *Cocos nucifera* L. (Cocoseae, Arecoideae), *L. maldivica* (Borasseae, Coryphoideae), and other species, adaptations to floating in sea water, while other fruits have fleshy mesocarps and brightly colored epicarps attractive to mammalians such as civets (*Caryota maxima* Blume; *Arenga pinnata* [Wurmb] Merr. and *Pinanga coronata* Blume), gibbons (*Arenga obtusifolia* Mart.), coyotes (*Washingtonia* spp. H. Wendl.), elephants (*Hyphaene* Gaertn. and *Borassus* L.), and agoutis (*Socratea* H. Karst) [1,6–10,12]. In the large Amazonian palms, once ripe, some of the fruits fall and are consumed by peccaries (*Tayassu pecari*) and species of rodents such as the guantas (*Cuniculus paca*), guatusas (*Dasyprocta* spp.), and coatis (*Nasua* spp.), which disperse their seeds and bury them at times of abundance [6]. Other fruits are food for birds like the Malabar hornbills (*Korthalsia* Blume) and parrots, or fishes [9]. The fruits of *Prestoea ensiformis* (Ruiz & Pav.) H.E. Moore or *Astrocaryum chambira* Burret fall by their own weight and once on the ground are eaten and their seeds transported by rodents, tapirs, and wild boar [6]. For reviews on animal-mediated seed dispersal in palms see [10,13].

In some palms, it is impossible to distinguish between the endocarp and the seed. In many species of *Licuala* (Chuniophoeniceae, Coryphoideae), such as *L. spinosa*, *L. glabra*, and *L. grandis*, the former is very thin and crustaceous. In other species, it is very thick and hard, e.g., in *Eugeissona* (Calamoideae), *Nypa* (Nypoideae), Borasseae, a few species of *Licuala* (e.g., *L. beccariana*), *Ptychococcus*, the Cocoseae (Arecoideae), and the Phytelepheae (Ceroxyloideae) [1]. Three types of endocarp have been described [14]: the coryphoid type differentiates from the inner part of the fruit wall; the chamaedoreoid type forms solely from the locular epidermis; the cocosoid-arecoid type develops from the locular epidermis, bundle sheaths, and intervening parenchyma. Different developmental patterns, described as continuous, discontinuous, and basipetal, are correlated with these endocarp types [1].

Numerous taxonomic groups in the Arecaceae received their Latin names according to characteristics of their fruits and seeds (size, hardness …). Table 2 contains examples of the names that refer to seed or fruit shape. For example, the prefix *Ptychos*, meaning folding or doubled gives the name of two genera *Ptychococcus* and *Ptychosperma* (Figure 1), both in the sub-tribe Ptychospermatinae, in the sub-family Areceae. Seedling morphology and anatomical studies have been the basis for classification in the family [15]; nevertheless, seed morphology specifically has not received much attention.

Figure 1. *Ptychococcus* species owe their name to the morphological properties of their seeds. *Ptychos* means folding or doubled seed (surface). Images: *Ptychococcus paradoxus.* Above: Two views of the seed. Below: Transversal sections in three seeds of different origins. Bar represents 1 cm. Photos courtesy of Scott Zona.

Table 2. Some of the names in the taxonomy of the Arecaceae are related to morphological characteristics of their fruits and seeds.

Name	Meaning
Actinorhytis	Radiate wrinkled (seed)
Astrocaryum	Star-like pattern of fibers around endocarp pores
Attalea amygdalina	Almond-shaped fruits
Barcella	Little boat-shaped (seed)
Calamus pycnocarpus	Thick fruit
Carpoxylon	Thick, woody fruit
Ceratolobus	Lobate horn-shaped fruit
Chelyocarpus	Turtle carapace-shaped fruit
Clinosperma, Clinospermatinae	Bent seeds (asymmetric)
Cyphosperma	Gibbous seeds
Chrysalidocarpus, Chrysalidosperma	Chrysalid-like fruit (seed)
Daemonorops oxycarpa	Sharp, pointed (fruit)
Dictyocaryum	Net of branches in the fruit
Dictyosperma	Net of branches in the seed
Eremospatha macrocarpa	Large fruit
Kentiopsis oliviformis	Olive shaped
Kentiopsis pyriformis	Pear shaped
Laccosperma	Seed with a hole or pit
Lepidocaryum	Fruit with scales
Lithocarpos cocciformis	Fruit hard and round
Lytocarium	Loose fruit
Nephrosperma	Kidney-shaped seeds
Oncosperma, Oncospermatinae	Humped or swollen seed
Pholidocarpus	Fruit with scales
Podococcus Podococceae	Foot shaped fruit
Ptychococcus Ptychosperma, Ptychospermatinae	Folding or doubled seed surface

The shapes of many seeds in the Arecaceae are often described as circular, ellipsoidal or elliptic, globose, ovoid, ovoidal, piriform, or rounded, including double adjectives such as "irregularly globose" or "broadly ovoid" [1]. A conclusion from a review of the literature is that a large proportion of species in this family have seeds resembling spheres, ellipsoids, and ovoids, but in general, the descriptions lack definition and are far from seed shape quantification. On the other hand, in regions with a great diversity of palms, such as the Amazon basin, most of the vegetative parts are difficult to access, while the seeds, due to the hardness of the endocarp, last for some time under the producing tree, being easy to find. In consequence, seed structure is often the key to the identification of palm species in the field.

Geometric models can be defined that adjust to the shapes observed making it possible to give accurate morphological descriptions of seed shape for a number of species, as well as the quantification of shape by comparison between the seed image and the model. The possibility of a model depends on the uniformity of seed shape in samples of a given species. Only when the shape is relatively constant, it is possible to describe it by means of a geometric model. Difficulties may arise from developmental aspects.

The shape is the result of a growth process and, in the same plant, or even in the same fruit, there may be differences in seed shape depending on many factors such as the developmental status, the type of inflorescence, the position of the seed in the plant or the fruit, and depending on the type of fruit. The seeds may form aggregates in the same fruit, such as *Camellia japonica* (Theaceae), *Swietenia mahogany* Jacq. (Meliaceae), *Eucalyptus* sp. L'Hér. (Myrtaceae), and *Peganum harmala* L., in the Nitrariaceae [11]. In the Arecaceae, this occurs in diverse genera (*Latania, Phytelephas, Salacca* …), and in the case of *Eremospatha macrocarpa* H. Wendl. (Lepidocaryae, Calamoideae), results in three morphological types having respectively the shape of 1/3 of a sphere, hemispherical or ellipsoidal depending on the number of seeds developing simultaneously in a fruit [1].

Shape is the tridimensional result of complex developmental processes, but the description of it is based on particular aspects. Bi-dimensional images of seeds are often similar to geometric figures that can be used as models. To find a model, the images of different seeds must be taken from a similar perspective to avoid differences in orientation and maximize similarity. The overall shape of the seed (elongated, thinner, or thicker in a different position) and anatomical points (hilium, germination pore, ventral groove) can be references for orientation. Slight changes in orientation can give different images (Figure 2).

Figure 2. Seeds of *Phoenix dactilifera* L. in three views: dorsal (**D**), lateral (**L**), ventral (**V**). A groove is visible in the ventral side, opposed to the germination pore in a dorsal position. Bar equals 1 cm.

3. Geometric Models: Definition and Application

3.1. A Conceptual Aspect

In this work, the shape of seeds will be related to geometric figures in a plane (bi-dimensional). In the literature, it can be seen that some of the adjectives used for shape description are applied to different objects: two-dimensional images of plane objects, three-dimensional objects, and their contours. This is a drawback because the same word has a different meaning when applied to each case. In other cases, different terms apply to precise, well-defined objects. For example, ovoid, or ovoidal refers to three-dimensional objects, while an oval is a bi-dimensional, plane figure. An important aspect of morphology is to define the object we work with; and a second aspect, not less important, is to define also the point of view we adopt for this object. In the description of seed shape, we consider a bi-dimensional approach easier and more straightforward than a tri-dimensional one. For example, to compare a seed with a sphere (3D) is much more difficult than to compare the image of that seed with a circle (2D). In addition, some of the models for the comparison of tri-dimensional objects are not sufficiently defined; for example, reniform means kidney-shaped, but a kidney is not an object described with precision; globose and globular mean approximatively rounded, but in a vague sense, not indicating similarity to any well-described figure.

Computer-assisted methods allow the measurements of multiple magnitudes of objects, but often the measurements are made in two dimensions [16,17]. Thus, when applying artificial vision methods [18], figures are represented by the coordinates of a set of points in the plane, that can be submitted to algebraic transformations and compared with a set of figures in databases. Plant organs, in general, and seeds in particular, have a similarity with geometric objects, and the comparison between both 2D images, the plant structure, and the geometric model, provides a direct method for the quantitative description of shapes.

3.2. J Index, a Magnitude in Seed Morphology

To be useful for classification, the geometric description of seed shape must be open to quantification. Comparing the shape of seeds (outline of seed images) to geometric figures permits a precise quantification of seed shape, and in this condition, shape description is independent of size. We have established a method based on the comparison of the bi-dimensional images of seeds with geometric figures by the calculation of *J* index [19,20]. *J* index is the percent of similarity between two plane figures: the seed image and the geometric model [19–27]. It ranges between 1 and 100 and is a measure of shape, not size. In consequence, the adjectives used to define the shapes of seeds are based on the names of well-defined geometric figures. The morphological type of seeds, corresponding to a geometric model, is a characteristic of the species, and more rarely of the genus. In some genera, such as *Acoelorrhaphe*, *Copernicia*, *Sabal*, and others, there may be a predominance of the circular type, but several observations with statistical support need to be done for each species. With this method, we can get information about seed morphology in wild species or cultivated varieties [27], as well as on aspects of seed shape related to taxonomy and life forms [28–30] or habitat [31].

3.3. Calculation of J Index

J index measures the percent of similarity of a seed image with the geometric model. To calculate *J* index in each seed image a series of graphic compositions are elaborated with Corel PHOTO-PAINT X7, in which the outline of the geometric model (circle, ellipse, or any other) is superimposed to the seed image, searching a maximum adjustment between both shapes, the seed, and the model. For each sample three graphic documents are kept: (1) a file in PSD format with the seed image and the geometric figure adapted to it, in which it is possible to make changes and corrections; (2) a file in JPG format with the geometric models in white, to discriminate between shared and total areas and thus, obtain the values of the area shared between the geometric figure and the seed image (S; Figure 3A,C), and (3) another file in JPG format with the geometric models in black, to obtain total area (T; Figure 3B,D). Representative images of (2) and (3) are shown in Figure 3 taking as an example the seed of *Geonoma congesta* H. Wendl. ex Spruce.

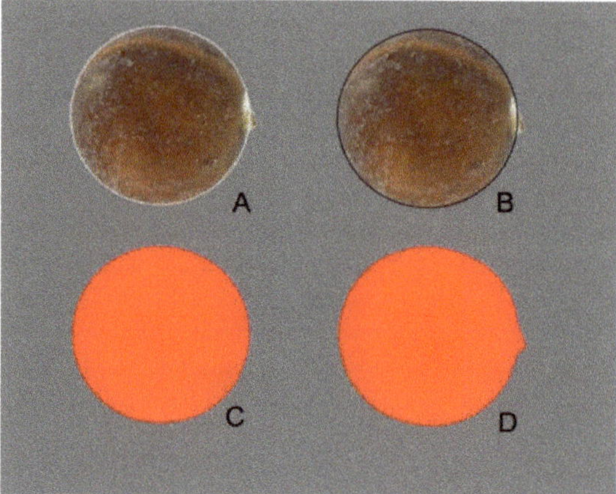

Figure 3. View of a seed of *Geonoma congesta*. (**A**) With white circle superimposed; (**B**) with a black circle superimposed; (**C**) The area shared (S) between the seed image and the geometric model for calculation with Image J. (**D**) Total area (T) for calculation in Image J. *J* index is the ratio S/T × 100. In this example, *J* index = 96.1. Image of *G. congesta* courtesy of Steven Paton. Taken from https://stricollections.org/.

It can be observed that the size of the red-figure represented in D (Total area, T) is slightly superior to the represented in C (Shared area, S). *J* index is the ratio S/T × 100.

4. Geometric Models: Definition of the Models

This section describes the geometric figures that are models in the description of seeds in the Arecaceae. Most of the descriptions are accompanied by a formula; these algebraic formulae correspond to the equation in Cartesian coordinates of the curve that delimits the figure (notice that in almost all cases the same name is used for the curve and the figure). The first groups belong to elementary geometry, while the latter groups include a set of figures constructed as graphical representations of algebraic equations. In general, we use the traditional terminology, but a particular case deserves attention. An oval is a convex figure limited by a C^2 closed curve. Convex means that any pair of its points are joined by a segment contained in the figure. The C^2 property or continuous second degree of smoothness ensures the continuity of the curvature, in particular, the absence of "peaks". In this sense, many common geometric figures are ovals (circles, ellipses ...). On the other hand, the word "oval" derives from the similarity with an egg shape (*ovus* in Latin), and this invites to its utilization in a restrictive sense, that will be used in this work, excluding, e.g., circles and ellipses.

4.1. Circle

The circle [32], the figure delimited by a circumference, is defined, according to Euclid (Book 1 Definition 15):

"A circle is a plane figure bounded by one curved line, and such that all straight lines drawn from a certain point within it to the bounding line, are equal. The bounding line is called its circumference and the point, it's center."

A circumference (centered at the origin) of radius r is represented by the expression:

$$x^2 + y^2 = r^2 \tag{1}$$

4.2. Ellipse

The ellipse [33] is "a curve surrounding two focal points, such that for all points on the curve, the sum of the two distances to the focal points is a constant". An ellipse (centered at the origin) is represented by the formula:

$$\frac{x^2}{a^2} + \frac{y^2}{b^2} = 1 \tag{2}$$

where a, $b > 0$ represent the semi-major and semi-minor axis, respectively. The circumference is a particular ellipse in which the two focal points coincide; in this case, $a = b = r$.

4.3. Oval

In our restricted sense, an oval is a curve resembling the silhouette of an egg. Unlike ellipses, ovals have only a symmetry axis [34]. Ovals can vary depending on their construction as well as on their degree of symmetry, going from figures close to ellipses to others with a remarked single symmetry. The family of ovals used in this work have the equation

$$(x^2 + ay^2)^2 = bx^3 + cxy^2 \tag{3}$$

with $a > 0$; b, c constants.

4.4. Lemniscate

A lemniscate (of Bernoulli) [35] is the locus of points whose product of distances from two fixed points (called foci) equals a constant. A lemniscate is given by the equation

$$\left(x^2 + y^2\right)^2 - 2a^2\left(x^2 - y^2\right) = 0 \tag{4}$$

where a is a real constant. By simple algebra, the equation for a half lemniscate is obtained as

$$y^2 = \sqrt{a^4 + 4a^2 x|x|} - a^2 - x|x| \tag{5}$$

4.5. Superellipse

Superellipses [36] can be seen as intermediate curves between an ellipse and its circumscribed rectangle. They are determined by the equation:

$$\left|\frac{x}{a}\right|^p + \left|\frac{y}{b}\right|^p = 1 \tag{6}$$

with $p > 2$. In particular, for $a = b$ squared circles are obtained.

4.6. Cardioid and Derivatives

The cardioid [37] has the Cartesian equation:

$$\left(x^2 + y^2 + ax\right)^2 = a^2\left(x^2 + y^2\right) \tag{7}$$

with $a > 0$. It has been used as a model for seed morphology in model plants [22,23] and it adjusts well to seeds of many species in diverse taxonomic groups [28,30,31].

4.7. Lens

A lens [38] is a convex region formed as the intersection of two circular disks (when one disk does not completely enclose the other). If the two circles that determine a lens have an equal radius, it is called a symmetric lens, otherwise, it is an asymmetric lens. The *Vesica piscis* ("fish bladder") is a special type of symmetric lens formed with circles whose centers are offset by a distance equal to the circle radii [39]. The 3D solid obtained rotating a lens around the axis through its tips is called a *lemon*.

4.8. Waterdrop

In the description of seed shape, it may be useful to modify geometric figures to obtain better adaptations to the bidimensional shape of seeds. For example, we obtained a figure resembling a water drop adapting the basis in the heart curve [40] to a circumference overlapping the maximum width of the curve. It can be seen as the joint graphical representation of the functions:

$$f(x) = -\frac{2}{3} + \sqrt{36 - x^2} + \frac{16}{3\left(2 + x^2 + |x|\right)} \; ; \; g(x) = -\frac{6}{11} - \sqrt{36 - x^2} \tag{8}$$

Derived from this combination (Model 4 in [25]), an elongated waterdrop can be obtained modifying the aspect ratio of the figure.

5. Geometric Models: Examples in Different Species in the Arecaceae

The following sections contain examples of seeds with the morphological types corresponding to the geometric models. Photographs corresponding to some of the examples are given in Figures 4–10 and a list of species for each type is provided in Table 3. Appendix A contains the sources of the images used in the Figures, while a Dataset has been published containing all the images used in this study (Supplementary Materials, doi:10.5281/zenodo.4009081). Chapter 6 discusses the relationships between morphological types and taxonomical groups. Seeds of a species can fit different models when observed from a different point of view, and for this reason, it is important to indicate the orientation in the images before proceeding to quantification with a given model.

Figure 4. Representative examples of round seeds (*J* index values are indicated between parentheses). From left to right: *Acoelorrhaphe wrightii* (93.6), *Acrocomia aculeata* (95.7), *Coccothrinax argentata* (92.6), *Geonoma congesta* (96.1), *Iriartea deltoideia* (95.3), and *Thrinax radiata* (95.4). The seed of *Iriartea deltoidea* is oriented with the attachment of the peduncle in a frontal position. In the other seeds, the place of attachment of the peduncle is in a lateral position. Bars represent 0.5 cm.

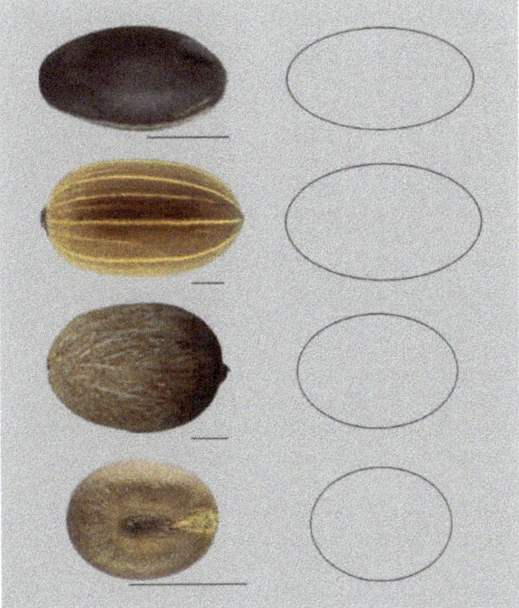

Figure 5. Representative examples of elliptic seeds with their respective models (*J* index values are indicated between parentheses): *Iriartella* sp. (94.4), *Socratea exorrhiza* (91.6), *Adonidia merrillii* (92.0), and *Washingtonia filifera* (91.9). The seed of *W. filifera* (bottom) is oriented with the attachment of the peduncle in a frontal position while in the other three seeds the attachment of the peduncle is in a lateral position. Aspect ratio values for the ellipses are (from top to bottom): 1.8, 1.7, 1.4, and 1.2. Bars represent 0.5 cm.

Figure 6. Representative examples of seeds, whose images adjust well to different ovals, (*J* index values are indicated between parentheses). From top to bottom (more to less elongated): *Serenoa repens* (90.0), *Desmoncus* sp. (93.6), *Astrocaryum standleyanum* (91.5), and *Medemia argun* (92.1). The ovals correspond to the equations given in the text. Bars represent 0.5 cm.

Figure 7. An example of the superellipse (model) and representative seeds adjusting to models in this family (between parentheses the values of *J* index with this model): *P. canariensis* (92.15 is the mean value of 25 seeds), *Welfia regia* (92.6), and *Raphia taedigera* (90.0). Bars represent 0.5 cm. The model is defined by Equation (6) with $a = 3.6$, $b = 2.2$, and $p = 2.4$.

Figure 8. The squared circle (model) and representative examples of seeds adjusting to it (between parentheses the values of *J* index with the respective models): *Clinosperma macrocarpa* (91.5), *Johannesteijsmannia altifrons* (92.4), and *Mauritia flexuosa* (92.0). In *C. macrocarpa* the seed is orientated with the stem end in a central position, while the other seeds present a side view. The model is defined by Equation (6) with $a = b = 1$, and $p = 3$.

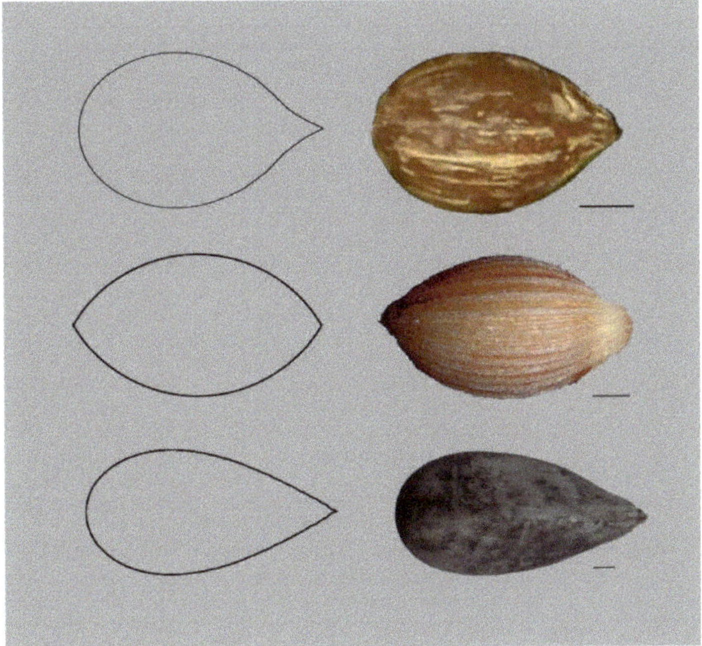

Figure 9. Water Drop, Lens, and Half Lemniscate with examples of seeds resembling each of the models (between parenthesis values of *J* index with the respective model): *Syagrus romanzoffiana* (92.3, mean of three seeds), *Asterogyne martiana* (91.8), and *Aphandra natalia (93.8)*. Bars represent 0.5 cm.

Figure 10. Seed shape quantification by an oval in *Chamaedorea pauciflora*. Left, above: Image with superimposed silhouettes of 12 seeds. Left, below: The model, an oval defined by the equation given in the text. The seed images used for seed shape quantification (comparison with the model) are in three columns and four rows. *J* index value (mean of 12 measurements) is 90.7. Bar represents 0.5 cm.

Table 3. Geometric models with some representative examples of seeds for each morphological type.

Geometric Figure	Representative Examples
Circle (42 examples)	*Acoelorrhaphe wrightii* [41], *Acrocomia aculeata, Brahea armata, B. edulis, Butia capitata, Ceroxylon parvum, Chelyocarpus chuco, Coccothrinax argentata, Copernicia alba* [42]*, C. baileyana, C. macroglossa, Corypha macropoda, C. umbraculifera, Cryosophila stauracantha, Dictyocaryum lamarckianum, Dypsis decipiens, Geonoma congesta, Guihaia argyrata, Hemithrinax ekmaniana, Hydriastele microcarpa, Iriartea deltoideia, Johannesteijsmannia altifrons, Licuala grandis, L. parviflora, L. orbicularis, Livistona chinensis* [43]*, Livistona jenkinsiana, Metroxylon warburgii, Normanbya normanbyi, Oncosperma horridum, Orania palindan, Pelagodoxa henriana, Prestoea acuminata, P. pubens, Pritchardia pacifica, Rhapis multifida, Sabal mexicana, S. minor, S. uresana, Thrinax excelsa, T.radiata, Wedlandiella* sp.
Ellipse (15 examples)	*Adonidia merrillii, Attalea butyracea, Bactris cruegeriana, B. gassipaes, Bismarckia nobilis, Calyptrogyne ghiesbreghtiana, Iriartella* sp.*, Mauritiella aculeata, Nephrosperma vanhoutteanum, Nanorhops ritchiana, Roystonea regia, Socratea exorrhiza, S. salazarii, Washingtonia filifera, W. robusta.*
Oval (34 examples)	*Actinorhytis calapparia, Archontophoenix alexandrae, Adonidia merrellii, Astrocaryum standleyanum, Attalea blepharophus, A. cohune, A. martiana, A. speciosa, Beccariophoenix fenestralis, Brahea armata, Brahea moorei, Calyptronoma occidentalis, Carpoxylon macrospermum, Chamaerops humilis, Corypha utan, Cyrtostachys renda, Desmoncus* sp.*, Dypsis bejofo, D. marojejyi, Iriartella* sp.*, Kerriodoxa elegans, Medemia argun, Nannorrhops ritchiana, Oenocarpus* sp. [44]*, Pholidostachys dactiloides, P. occidentalis, P. pulchra, P. synanthera, Retispatha dumetosa, Rhapidophyllum histrix, Serenoa repens, Voanioala gherardii, Wallichia disticha, Washingtonia robusta.*
Superellipse (5 examples)	*Cyphophoenix elegans, P. canariensis, Ph. roebelenii, Raphia taedigera, Welfia regia.*
Squared circle (4 examples)	*Johannesteijsmannia altifrons, Clinosperma macrocarpa, Mauritia flexuosa* [45]*, Ravenea rivularis*
Half lemniscate (10 examples)	*Aphandra natalia, Astrocaryum alatum, A. urostachys, Attalea dubia, Beccariophoenix fenestralis, Brahea moorei, Latania loddigesii, Pholidostachys panamensis, Phytelephas macrocarpa, Veitchia* sp.
Lens (3 examples)	*Asterogyne martiana, Butia* sp.*, Carpentaria acuminata*
Water drop (1 example)	*Syagrus romanzoffiana*

5.1. Seeds That Project Circular Images

Round seeds project circular images from any perspective. However, ellipsoidal and ovoidal seeds can also give circular images that are useful for species identification and classification. For an accurate description, the seed orientation has to be given and, in any case, values of *J* index need to be obtained in a representative number of samples. Figure 4 shows representative examples of circular seeds: *Acoelorrhaphe wrightii* (Corypheae, Coryphoideae), *Acrocomia aculeata* (Cocoseae, Arecoideae), *Coccothrinax argentata* (Cryosophileae, Coryphoideae), *Geonoma congesta* (Geonomateae, Arecoideae), *Iriartea deltoideia* (Iriarteeae, Arecoideae), and *Thrinax radiata* (Cryosophileae, Coryphoideae). Other examples are also found in diverse sub-families and tribes (Table 3).

5.2. Elliptical Seeds

Ellipsoidal seeds give bi-dimensional images that vary from circular to elliptical, depending on the selected point of view. The elliptical shape is determined by the ratio between the major and minor axes (aspect ratio). Figure 5 shows representative examples of seeds whose images adjust well to ellipses of different values. These are *Iriartella* sp. and *Socratea exorrhiza* (Iriarteae, Arecoideae), *Adonidia merrillii* (Areceae, Arecoideae), and *Washingtonia filifera* (Corypheae, Coryphoideae). Other examples are also found in diverse sub-families and tribes (Table 3).

5.3. Seeds Resembling Ovals

As was pointed out in Section 4.3, the oval has just one symmetry axis. A list of species whose seed images resemble ovals are given in Table 3. A useful mode to define the oval is by an algebraic formula, see, e.g., Equation (3). Figure 6 contains representative examples of oval seeds with their respective models: *Serenoa repens* (Corypheae, Coryphoideae), *Desmoncus* sp. (Cocoseae, Arecoideae), *Astrocaryum standleyanum* (Cocoseae, Arecoideae), and *Medemia argun* (Borasseae, Coryphoideae). Other examples are also found in diverse sub-families and tribes (Table 3).

The shape of the seeds represented in Figure 6 are ovals and all the curves representing them have been obtained with Equation (3) varying the parameters:

- *Serenoa repens*: $a = 2$; $b = 5$; $c = 2.6$;
- *Desmoncus* sp.: $a = 1.4$; $b = 4.3$; $c = 3$;
- *Astrocaryum standleyanum*: $a = 1$; $b = 4$; $c = 3.3$;
- *Medemia argun*: $a = 1$; $b = 3.4$; $c = 3.3$

Often the seeds of the different species in a genus have different morphological types, such as in *Bactris, Brahea, Butia, Chamaedorea, Geonoma,* and *Pholidostachys*. In some genera there is a considerable degree of variation in seed shape, for example, the seeds of *Pholidostachys* have ovoid shapes with varying degrees of polarity in a gradient in the direction from (acute) ovoid to ellipse (and circular):

P. panamensis > (*P. synanthera, P. dactiliodes, P. occidentalis*) > *P. pulchra* (*P. kalbreyeri*)

Once a model has been proposed for a species or a population, the proposal has to be validated statistically. *T*-test allow us to validate the hypothesis that seed lots belong to a given morphological type. We consider that a seed population belongs to a given morphological type when the hypothesis test concludes that the mean *J* index of the species is equal to or superior to 90 with a significance level of 95%. An example is later given with seeds of *Phoenix canariensis* (Figure 7).

The transitions between related forms (Circle–Ellipse–Oval) are very subtle and many species have seeds of variable shape. A mixture of shapes of the main types may be observed in images containing multiple seeds of the following species: *Actinorhytis calaparia, Aiphanes horrida, Archontophoenix alexandrae, Bismarckia nobilis, Brahea armata, Caryota maxima, Chamaedorea tuerkeimii, Chamaerops humilis, Corypha utan, Guihaya argyrata, Jubaeopsis caffra, Medemia argun, Nanorrhops ritchiana,* and *Washingtonia robusta*.

5.4. Seeds Resembling the Superellipse and Related Figures

The superellipses and squared circles [36] are figures that, to our knowledge, have not yet been mentioned in the morphological description of seeds in the Arecaceae.

The squared circle may be considered as a particular case of the superellipse (Equation (6)). Figure 7 represents the superellipse and seeds representative of this morphological type. The seeds of *P. canariensis* (Phoeniceae, Coryphoideae), *Welfia regia* (Geonomateae, Arecoideae), and *Raphia taedigera* (Lepidocaryeae, Calamoideae) adjust to a superellipse. The hypothesis test done with 25 seeds (sample mean = 92.15) concludes that the mean *J* index of the population of *P. canariensis* from which the seeds were taken is equal or superior to 90 with a significance level of 99%.

The seeds of *Clinosperma macrocarpa* (Areceae, Arecoideae), *Johannesteijsmannia altifrons* (Chuniophoeniceae, Coryphoideae), and *Mauritia flexuosa* (Lepidocaryeae, Calamoideae) and others (Table 3) adjust to a squared circle (Figure 8).

5.5. Seeds Resembling Other Figures: Half Lemniscate, Lens and Waterdrops

The figures of half lemniscate and waterdrops resemble elongated ovals with a peak in their apex. Depending on the degree of asymmetry perpendicular to the symmetry axis, the seeds resembling ovals may have shapes near to ellipses, such as *Medemia argun* or, on the other hand, be more asymmetric and divergent from the ellipse, such as *Serenoa repens* (see Figure 6). In other seeds there is still more asymmetry between the two poles, giving images similar to half lemniscate as in the case of some seeds of *Aphandra natalia* (Phytelepheae, Coryphoideae, Figure 9). Seeds of this type can be observed in other species such as *Astrocaryum urostachis*, *Syagrus romanzoffiana* and *Beccariophoenix fenestralis* (Cocoseae, Arecoideae), *Phytelephas macrocarpa* (Phytelepheae, Coryphoideae), *Pholidostachys panamensis* (Geonomeae, Arecoideae), and *Wodyetia bifurcata* (Areceae, Arecoideae).

Images of seeds of *Asterogyne martiana* (Geonomeae, Arecoideae) adjust well to a lens. This model may also be helpful in the quantification of some seeds in *Butia* sp. (Cocoseae, Arecoideae), *Carpentaria acuminata* (Areceae, Arecoideae), and others.

6. Seed Shape Quantification by Comparison with an Oval in *Chamaedorea pauciflora*

Figure 10 shows a sample of 12 seeds of *Ch. pauciflora* (Chamaedoreae, Arecoideae), with an image containing the sum of their profiles in a composed image and the model used for their quantification. The model is an oval (Equation (3) with values a = 1, b = 4, c = 3.3). *J* index value (mean of the measurements in the 12 images of Figure 10) is 90.7.

7. Morphological Aspects of the Fruits and Seeds of *Trachycarpus fortunei*

The fruits and seeds of *T. fortunei* (Trachycarpeae, Coryphoideae) adjust well to the cardioid or cardioid-related figures (Figure 11). The fruits with exocarp resemble a cardioid. Deprived of the pericarp, the seeds with endo- and mesocarps resemble a flattened cardioid. Finally, the seeds covered only by the endocarp resemble an open cardioid. The two cardioid-derived models were developed for the morphological analysis of diverse species of *Silene* L. (Caryophyllaceae). While the cardioid gave high values of *J* index with many species, a flattened cardioid was the model for *Silene latifolia* Poir., while the open cardioid was better for *Silene gallica* L. [46].

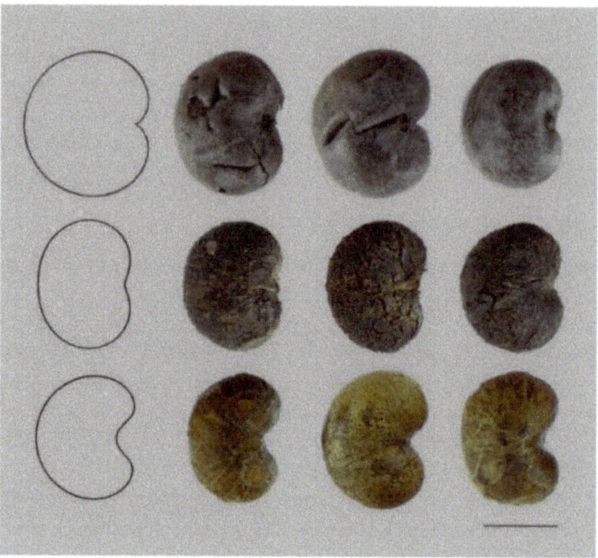

Figure 11. *T. fortunei* (between parenthesis values of *J* index with the respective model). Top: Cardioid and images of three fruits (90.4). Middle row: Modified cardioid (flattened) and images of three fruits without exocarp (the outer, visible layer is the mesocarp) (92.1). Bottom row: Modified cardioid (open) and three seeds covered only by the endocarp (exocarp and mesocarp have been removed) (91). Bar represents 0.5 cm.

8. Geometric Models in the Arecaceae: Relation with Anatomical Properties and Taxonomy

The genera with apocarpic fruits tend to have their seeds more rounded or regularly-shaped. These include *Nypa*, and many genera in the *Coryphoideae* (*Chamaerops, Chelyocarpus, Coccothrinax, Cryosophila, Guihaia, Hemithrinax, Itaya, Leucothrinax, Maxburretia, Rhapidophyllum, Rhapis, Schippia, Trachycarpus, Thrinax, Trithrinax, Zombia*) [1,47]. Nevertheless, this is not a characteristic exclusive of the sub-family because rounded or regular seeds have been observed in other taxa, and both oval and elliptical seeds are frequent throughout all the family. Oval shaped, particularly elongated ovals, half lemniscates and lens-shaped seeds are frequent in the Arecoideae, but the type is also observed in the syncarpous clade of the subfamily Coryphoideae, that includes four tribes containing high diversity in seed size and shape: *Borasseae, Corypheae, Caryoteae*, and *Chuniophoeniceae* [48].

Quantification of seed shape is interesting at the species level to describe the characteristic morphology in those species that have regular shaped seeds, but it may also be interesting in species that have variable seeds to investigate the molecular basis of morphological types and the biological meaning of geometric forms, as well as to correlate them with other morphological features such as color, surface smoothness/roughness, and striation. For example, the seeds of 17 cultivars of *Rubus* sp. L. were classified recently based on the shape of the raphe: straight, concave, or convex. Cultivars within each group could be differentiated by seed shape, size, color, and seed-coat sculpturing [49].

Species of the Arecaceae have traditionally been recognized as important components of tropical forests, and six of the 10 most common tree species in the Amazon rain forest are palms [50]. Recent work based on massive information obtained by large data-sharing networks such as RAINFOR in South America and AfriTRON in Africa [51–53] points to an even more important role of palms in the Neotropics [54], suggesting that the knowledge of palm physiology, taxonomy and ecology will contribute to predict the evolution of the ecosystem under variable climatic conditions.

Computer programs for the identification of seeds by comparison with geometric figures may develop based on the comparison of in situ photographs with information stored in databases. The development of techniques and resources in seed morphology will also have applications in archaeology as both the remains of seeds and even populations of these plants are frequently found in archaeological sites [55].

The understanding of seed geometry may be at the basis of the knowledge of structural biology to which P.B. Tomlinson referred to in 1990 when he wrote: "Palms are not then merely emblematic of the tropics, they are emblematic of how the structural biology of plants must be understood before evolutionary scenarios can be reconstructed." ([56] quoted in [1]).

9. Conclusions

The morphology of the seeds in the Arecaceae has been reviewed. This family contains a remarkable amount of generic and specific names due to seed characteristics, and in particular, morphological attributes.

New models have been proposed and a series of rules is given to accurately describe seed shape in the species of the Arecaceae. First, for those species that have seeds of regular shape, a geometric model can be identified. The seeds are round or circular when the model is the circle for any view of the seeds. Ellipsoidal and ovoid seeds or can also give images similar to circles from certain perspectives. The model for ellipsoidal seeds is an ellipse that can be defined by the Aspect Ratio. A convenient way to define an oval as a morphological 2D model is by the corresponding algebraic equation.

Additional geometric figures, other than the circle, ellipses, and ovals, can be used as models to give precise descriptions of seed shape in some species. Thus, the superellipse and the squared circle are new models useful for many seeds of the Calamoideae (*Lepidocaryum, Mauritia, Mauritiella, Korthalsia, Salacca*), as well as species in the genus *Phoenix* (Phoeniceae, Coryphoideae). The fruits of *T. fortunei* adjust well to a cardioid and the seeds to modified cardioids. Half lemniscate is a useful model for species of *Astrocaryum, Attalea, Beccariophoenix, Brahea, Lattania, Pholidostachys, Veitchia*, and others.

Lenses are good models for the description and quantification of seed shape in *Asterogyne martiana*, and most probably will be of application in other species as *Butia* sp. and *Carpentaria acuminata*.

Morphological types are varied and, for many species, there is not a clear pattern conserved in the majority of seeds that may be associated with a geometric figure. There is not an obvious relationship between the higher taxonomic divisions (subfamilies, tribes, and sub-tribes) and the models, although a trend is observed in the subfamily Coryphoideae with a high frequency of round seeds in genera with apocarpic fruits and oval, lemniscate-type seeds in the syncarpous clade of this subfamily.

Seed morphometry provides interesting tools for the identification of species in regions with a great diversity of palms, such as the Amazon basin. The seeds are critical structures in taxonomy in many cases where vegetative organs are similar among different species. In addition, thanks to their very resistant endocarp, seeds are easy to find and last a long time under the tree producer allowing differentiation up to the level of species in many instances.

The development of software of image identification based on geometric models may be an interesting contribution to future biodiversity studies.

Supplementary Materials: A Dataset has been published containing the web addresses for the images used in this study (Arecaceae. Sources of seed images in the web; https://zenodo.org/record/4009081; doi:10.5281/zenodo.4009081).

Author Contributions: Conceptualization, E.C.; methodology, J.J.M.-G.; software, J.J.M.-G, Á.T.; validation, D.G.d.P., J.J.M.-G., Á.T. and E.C.; investigation, D.G.d.P., J.J.M.-G., Á.T. and E.C.; resources D.G.d.P., J.J.M.-G., Á.T. and E.C.; data curation, D.G.d.P., J.J.M.-G and E.C.; writing—original draft preparation, E.C.; writing—review and editing, D.G.d.P., J.J.M.-G., Á.T. and E.C.; visualization, J.J.M.-G.; supervision, E.C. All authors have read and agreed to the published version of the manuscript.

Funding: This research received no external funding.

Acknowledgments: In this section, you can acknowledge any support given which is not covered by the author's contribution or funding sections. This may include administrative and technical support, or donations in kind (e.g., materials used for experiments).

Conflicts of Interest: The authors declare no conflict of interest.

Appendix A. Web Sources of the Images Used in the Figures

- Figure 1: Courtesy of Scott Zona
- Figure 2: Seed collection IRNASA-CSIC.
- Figure 3: *Geonoma congesta* seed image courtesy of Steven Paton
- Figure 4: *Acoelorrhaphe wrightii*: http://palmvrienden.net/gblapalmeraie/2017/06/27/acoelorrhaphe-wrightii/ *Acrocomia aculeata*: https://www.palmpedia.net/wiki/Acrocomia_aculeata *Coccothrinax argentata*: http://idtools.org/id/palms/palmid/factsheet.php?name=9413 *Geonoma congesta: Steven Paton* https://www.discoverlife.org/mp/20p?see=I_SP1454&res=640 *Iriartea deltoidea*: http://www.belizehank.com/IMAGES/Palm%20Seeds/Palm%20seeds%20alphabetical/Ireatea%20deltoidea.jpg *Thrinax radiata*: https://idtools.org/id/palms/palmid/factsheet.php?name=Thrinax+radiata
- Figure 5: *Iriartella* sp.: http://www.plantsoftheworldonline.org/taxon/urn:lsid:ipni.org:names:31287-1 *Socratea exorrhiza*: Steven Paton, https://www.discoverlife.org/mp/20p?see=I_SP3245&res=640 *Adonidia merrillii*: Seeds were collected at Pastaza (Ecuador) in different urban gardens of Puyo and, after being photographed, the seeds were returned to their origin. *Bismarckia nobilis*: http://idtools.org/id/palms/palmid/factsheet.php?name=9401
- Figure 6: *Serenoa repens*: http://idtools.org/id/palms/palmid/factsheet.php?name=9457 *Desmoncus* sp.: Steven Paton, https://biogeodb.stri.si.edu/bioinformatics/dfm/metas/view/8262?&lang=es *Astrocaryum standleyanum: id.* https://biogeodb.stri.si.edu/bioinformatics/dfm/metas/view/7546 *Medemia argun*: http://www.wellgrowhorti.com/Pictures/Palm%20Seeds/Thumbnail/Medemia%20argun%20seeds.jpg

- Figure 7: *Phoenix canariensis*: Seed collection IRNASA-CSIC. *Welfia regia*: http://www.belizehank.com/IMAGES/Palm%20Seeds/Palm%20seeds%20alphabetical/Welfia%20regia.jpg *Raphia taedigera*: https://www.palmpedia.net/wiki/Raphia_taedigera
- Figure 8: *Clinosperma macrocarpa*: https://www.palmpedia.net/wiki/Clinosperma_macrocarpa *Johannesteijsmannia altifrons*: https://www.rarepalmseeds.com/johannesteijsmannia-altifrons *Mauritia flexuosa*: https://www.palmpedia.net/wiki/Mauritia_flexuosa
- Figure 9: *Syagrus romanzoffiana*: http://realpalmtrees.com/palm-tree-store/queen-palm-seeds-pkg.html Available also at: https://bit.ly/3lcvFNn *Asterogyne martiana*, from www.plant.ac.cn (zhiwutong.com). *Aphandra natalia:* collected at Pastaza (Ecuador) in different urban gardens of Puyo. After being photographed, the seeds were returned to their origin.
- Figure 10: Seeds were collected at Pastaza (Ecuador) in different urban gardens of Puyo and, after being photographed, the seeds were returned to their origin.
- Figure 11: Seed collection IRNASA-CSIC.

References

1. Dransfield, J.; Uhl, N.W.; Asmussen, C.B.; Baker, W.J.; Harley, M.M.; Lewis, C.E. *Genera Palmarum: The Evolution and Classification of Palms*; Kew Publishing, Royal Botanic Gardens of Kew: London, UK, 2008.
2. Von Humboldt, A. *Personal Narrative of Travels to the Equinoctial Regions of the New Continent During the Years 1799–1804*; Longman, Hurst, Rees, Orme and Brown: London, UK, 1829; Volume 5.2, pp. 727–729.
3. Hooker, W.J. *Museum of Economic Botany: Or a Popular Guide to the Useful and Remarkable Products of the Museum of the Royal Gardens of Kew*; Longman, Brown, Green and Longmans: London, UK, 1855.
4. Ribeiro Mendes, J.C.; de Sousa Portilho, A.J.; Andrade de Aguiar-Dias, A.C.; da Silva Sampaio, K.L.; de Nazaré Farias, L. Arecaceae: Uma estratégia diferenciada para o ensino de botânica em uma escola de ensino médio na ilha de Cotijuba, Pará, Brasil. In *Enciclopédia Biosfera*; Centro Científico Conhecer: Goiânia, Brasil, 2019; Volume 16, pp. 2226–2240.
5. Meerow, A.W. Arecaceae, the palm family. In *The Encyclopedia of Fruits and Nuts*; Janick, J., Paull, R.E., Eds.; CABI: Wallingford, UK; Cambridge, MA, USA, 2006.
6. Valencia, R.; Montufar, R.; Navarrete, H.; Balslev, H. *Palmas Ecuatorianas: Biología y Uso Sostenible*; Publicaciones del Herbario QCA de la Pontificia Universidad Católica del Ecuador: Quito, Ecuador, 2013.
7. Jácome, J.; Montúfar, R. Mocora. *Astrocaryum standleyanum*. In *Palmas Ecuatorianas: Biología y Uso Sostenible*; Valencia, R., Montufar, R., Navarrete, H., Balslev, H., Eds.; Publicaciones del Herbario QCA de la Pontificia Universidad Católica del Ecuador: Quito, Ecuador, 2013; pp. 99–101.
8. Henderson, A. *Evolution and Ecology of Palms*; The New York Botanical Garden Press: Bronx, NY, USA, 2002.
9. Gottsberger, G. Seed dispersal by fish in the inundated regions of Humaita, Amazonia. *Biotropica* **1978**, *10*, 170–183. [CrossRef]
10. Zona, S.; Henderson, A. A review of animal-mediated seed dispersal of palms. *Selbyana* **1989**, *11*, 6–21.
11. Cervantes, E.; Martín Gómez, J.J.; Gutiérrez del Pozo, D.; Silva Dias, L. An Angiosperm Species Dataset Reveals Relationships between Seed Size and Two-Dimensional Shape. *Horticulturae* **2019**, *5*, 71. [CrossRef]
12. Piedade, M.T.; Parolin, P.; Junk, W.J. Phenology, fruit production and seed dispersal of *Astrocaryum jauari* (Arecaceae) in Amazonian black water floodplains. *Rev. Biol. Trop.* **2006**, *54*, 1171–1178. [CrossRef]
13. Muñoz, G.; Trøjelsgaard, K.; Kissling, W.D. A synthesis of animal-mediated seed dispersal of palms reveals distinct biogeographical differences in species interactions. *J. Biogeogr.* **2019**, *46*, 466–484. [CrossRef]
14. Murray, S.G. The formation of endocarp in palm fruits. *Principes* **1973**, *17*, 91–102.
15. Henderson, F.M.; Stevenson, D.W. A phylogenetic study of Arecaceae based on seedling morphological and anatomical data. *Aliso* **2006**, *22*, 251–264. [CrossRef]
16. Rovner, I.; Gyulai, F. Computer-assisted morphometry: A new method for assessing and distinguishing morphological variation in wild and domestic seed populations. *Econ. Bot.* **2007**, *61*, 154–172. [CrossRef]
17. Sonka, M.; Hlavac, V.; Boyle, R. *Image Processing Analysis and Machine Vision*; Thomson: Stamford, CT, USA, 2008.
18. Zheng, Y.; Guo, B.; Chen, Z.; Li, C. A Fourier Descriptor of 2D Shapes Based on Multiscale Centroid Contour Distances Used in Object Recognition in Remote Sensing Images. *Sensors* **2019**, *19*, 486. [CrossRef] [PubMed]

19. Cervantes, E.; Martín-Gómez, J.J.; Saadaoui, E. Updated Methods for Seed Shape Analysis. *Scientifica* **2016**, *2016*, 5691825. [CrossRef]

20. Cervantes, E.; Martín Gómez, J.J. Seed Shape Description and Quantification by Comparison with Geometric Models. *Horticulturae* **2019**, *5*, 60. [CrossRef]

21. Cervantes, E. Seed Morphology. Scholarly Community Encyclopedia. 2020. Available online: https://encyclopedia.pub/item/revision/7894d28ff2ebed5217b9d81a23a8c84e (accessed on 6 October 2020).

22. Cervantes, E.; Martín, J.J.; Ardanuy, R.; de Diego, J.G.; Tocino, Á. Modeling the *Arabidopsis* seed shape by a cardioid: Efficacy of the adjustment with a scale change with factor equal to the Golden Ratio and analysis of seed shape in ethylene mutants. *J. Plant Physiol.* **2010**, *67*, 408–410. [CrossRef] [PubMed]

23. Cervantes, E.; Martín, J.J.; Chan, P.K.; Gresshoff, P.M.; Tocino, Á. Seed shape in model legumes: Approximation by a cardioid reveals differences in ethylene insensitive mutants of *Lotus japonicus* and *Medicago truncatula*. *J. Plant Physiol.* **2012**, *169*, 1359–1365. [CrossRef] [PubMed]

24. Saadaoui, E.; Martín Gómez, J.J.; Tlili, N.; Khaldi, A.; Cervantes, E. Effect of Climate in Seed Diversity of Wild Tunisian *Rhus tripartita* (Ucria) Grande. *J. Adv. Biol. Biotechnol.* **2017**, *13*, 1–10. [CrossRef]

25. Martín-Gómez, J.J.; Saadaoui, E.; Cervantes, E. Seed Shape of Castor Bean (*Ricinus communis* L.) Grown in Different Regions of Tunisia. *JAERI* **2016**, *8*, 1–11.

26. Saadaoui, E.; Martín, J.J.; Bouazizi, R.; Chokri, B.R.; Grira, M.; Saad, A.; Khouja, M.L.; Cervantes, E. Phenotypic variability and seed yield of *Jatropha curcas* L Introduced to Tunisia. *Acta Bot. Mex.* **2015**, *110*, 119–134. [CrossRef]

27. Martín-Gómez, J.J.; Gutiérrez del Pozo, D.; Ucchesu, M.; Bacchetta, G.; Cabello Sáenz de Santamaría, F.; Tocino, Á.; Cervantes, E. Seed Morphology in the Vitaceae Based on Geometric Models. *Agronomy* **2020**, *10*, 739. [CrossRef]

28. Martín-Gómez, J.J.; Gutiérrez del Pozo, D.; Cervantes, E. Seed shape quantification in the Malvaceae reveals cardioid-shaped seeds predominantly in herbs. *Bot. Lith.* **2019**, *25*, 21–31. [CrossRef]

29. Cervantes, E.; Martín-Gómez, J.J. Seed shape quantification in the order Cucurbitales. *Mod. Phytomorphol.* **2018**, *12*, 1–13. [CrossRef]

30. Martín-Gómez, J.J.; Rewicz, A.; Cervantes, E. Seed Shape Diversity in families of the Order Ranunculales. *Phytotaxa* **2019**, *425*, 193–207. [CrossRef]

31. Saadaoui, E.; Martín-Gómez, J.J.; Cervantes, E. Seed morphology in Tunisian wild populations of *Capparis spinosa* L. *Acta Biol. Cracov. Bot.* **2013**, *55*, 99–106.

32. Weisstein, E.W. Circle. From MathWorld—A Wolfram Web Resource. Available online: https://mathworld.wolfram.com/Circle.html (accessed on 24 June 2020).

33. Weisstein, E.W. Ellipse. From MathWorld—A Wolfram Web Resource. Available online: https://mathworld.wolfram.com/Ellipse.html (accessed on 24 June 2020).

34. Weisstein, E.W. Oval. From MathWorld—A Wolfram Web Resource. Available online: https://mathworld.wolfram.com/Oval.html (accessed on 24 June 2020).

35. Weisstein, E.W. Lemniscate. From MathWorld—A Wolfram Web Resource. Available online: https://mathworld.wolfram.com/Lemniscate.html (accessed on 24 June 2020).

36. Weisstein, E.W. Superellipse. From MathWorld—A Wolfram Web Resource. Available online: https://mathworld.wolfram.com/Superellipse.html (accessed on 24 June 2020).

37. Weisstein, E.W. Cardioid. From MathWorld—A Wolfram Web Resource. Available online: https://mathworld.wolfram.com/Cardioid.html (accessed on 24 June 2020).

38. Weisstein, E.W. Lens. From MathWorld—A Wolfram Web Resource. Available online: https://mathworld.wolfram.com/Lens.html (accessed on 24 June 2020).

39. Weisstein, E.W. Vesica Piscis. From MathWorld—A Wolfram Web Resource. Available online: https://mathworld.wolfram.com/VesicaPiscis.html (accessed on 24 June 2020).

40. Weisstein, E.W. Heart Curve. From MathWorld—A Wolfram Web Resource. Available online: http://mathworld.wolfram.com/HeartCurve.html (accessed on 24 June 2020).

41. Clasen, A.; Kesel, A.B. Microstructural Surface Properties of Drifting Seeds—A Model for Non-Toxic Antifouling Solutions. *Biomimetics* **2019**, *4*, 37. [CrossRef] [PubMed]

42. De Oliveira Viana, E. Pós-colheita de Sementes e Produção de Mudas de *Copernicia*. Ph.D. Thesis, Universidade Federal do Ceará, Fortaleza, Brazil, 2019. Available online: http://www.repositorio.ufc.br/handle/riufc/46681 (accessed on 30 July 2020).

43. Dowe, J.L. A taxonomic account of *Livistona* R.Br. (Arecaceae). *Gard. Bull.* **2009**, *60*, 185–344.

44. De Mendonca, M.S.; de Oliveira, A.B.; de Araujo, M.G.P.; Araujo, L.M. Morpho-anatomy of the fruit and seed of *Oenocarpus* minor Mart. (Arecaceae). *Rev. Bras. Sementes* **2008**, *30*, 90–95. [CrossRef]

45. Santana Silva, R.; Monteiro Ribeiro, L.; Mercadante-Simões, M.O.; Ferreira Nunes, Y.R.; Nascimento Lopes, P.S. Seed structure and germination in buriti (*Mauritia flexuosa*), the Swamp palm. *Flora* **2014**, *209*, 674–685. [CrossRef]

46. Martín-Gómez, J.J.; Rewicz, A.; Rodríguez-Lorenzo, J.L.; Janoušek, B.; Cervantes, E. Seed morphology in *Silene* based on geometric models. *Plants* **2020**. under review.

47. Bobrov, A.V.F.C.; Lorence, D.H.; Romanov, M.S.; Romanova, E.S. Fruit Development and Pericarp Structure in *Nypa fruticans* Wurmb (Arecaceae): A Comparison with Other Palms. *Int. J. Plant Sci.* **2012**, *173*, 751–766. [CrossRef]

48. Bellot, S.; Bayton, R.P.; Couvreur, T.L.P.; Dodsworth, S.; Eiserhardt, W.L.; Guignard, M.S.; Pritchard, H.W.; Roberts, L.; Toorop, P.E.; Baker, W.J. On the origin of giant seeds: The macroevolution of the double coconut (*Lodoicea maldivica*) and its relatives (Borasseae, Arecaceae). *New Phytol.* **2020**. [CrossRef]

49. Wada, S.; Reed, B.M. Seed Coat Morphology Differentiates Blackberry Cultivars. *J. Am. Pomol. Soc.* **2010**, *64*, 152–161.

50. Ter Steege, H.; Henkel, T.W.; Helal, N.; Marimon, B.S.; Marimon-Junior, B.H.; Huth, A.; Groeneveld, J.; Sabatier, D.; de Souza Coelho, L.; de Andrade Lima Filho, D.; et al. Rarity of monodominance in hyperdiverse Amazonian forests. *Sci. Rep.* **2019**, *9*, 13822. [CrossRef]

51. Hubau, W.; De Mil, T.; Van den Bulcke, J.; Phillips, O.L.; Angoboy Ilondea, B.; Van Acker, J.; Sullivan, M.J.; Nsenga, L.; Toirambe, B.; Couralet, C.; et al. The persistence of carbon in the African forest understory. *Nat. Plants* **2019**, *5*, 133–140. [CrossRef] [PubMed]

52. Lewis, S.L.; Sonké, B.; Sunderland, T.; Begne, S.K.; Lopez-Gonzalez, G.; van der Heijden, G.M.F.; Phillips, O.L.; Affum-Baffoe, K.; Baker, T.R.; Banin, L.; et al. Above-ground biomass and structure of 260 African tropical forests. *Philos. Trans. R. Soc. B Biol. Sci.* **2013**, *368*, 20120295. [CrossRef] [PubMed]

53. Lopez-Gonzalez, G.; Lewis, S.L.; Burkitt, M.; Phillips, O.L. ForestPlots.net: A web application and research tool to manage and analyse tropical forest plot data. *J. Veg. Sci.* **2011**, *22*, 610–613. [CrossRef]

54. Muscarella, R.; Emilio, T.; Phillips, O.L.; Lewis, S.L.; Slik, F.; Baker, W.J.; Couvreur, T.L.; Eiserhardt, W.L.; Svenning, J.C.; Affum-Baffoe, K.; et al. The global abundance of tree palms. *Glob. Ecol. Biogeogr.* **2020**, *29*, 1495–1514. [CrossRef]

55. Morcote-Rios, G.; Bernal, R. Remains of palms (Palmae) at archaeological sites in the New World: A review. *Bot. Rev.* **2001**, *67*, 309–350. [CrossRef]

56. Tomlinson, P.B. *The Structural Biology of Palms*; Clarendon Press: Oxford, UK, 1990.

 *horticulturae*

Article

Rapid In Vitro Multiplication of Non-Runnering *Fragaria vesca* Genotypes from Seedling Shoot Axillary Bud Explants

Babul C. Sarker [1,2,*], Douglas D. Archbold [1], Robert L. Geneve [1] and Sharon T. Kester [1]

[1] Department of Horticulture, University of Kentucky, Lexington, KY 40546, USA; darchbol@uky.edu (D.D.A.); rgeneve@uky.edu (R.L.G.); stkest2@uky.edu (S.T.K.)
[2] Chief Scientific Officer, Pomology Division, Horticulture Research Centre, Bangladesh Agricultural Research Institute, Gazipur 1701, Bangladesh
* Correspondence: bsarker_64@yahoo.com

Received: 17 July 2020; Accepted: 26 August 2020; Published: 1 September 2020

Abstract: *Fragaria vesca* L. has become a model species for genomic studies relevant to important crop plant species in the Rosaceae family, but generating large numbers of plants from non-runner-producing genotypes is slow. To develop a protocol for the rapid generation of plants, leaf explants were compared to single axillary bud shoot explants, both from in vitro-grown *Fragaria vesca* seedlings, as sources of shoots for new plant production in response to benzyladenine (BA) or thidiazuron (TDZ) combined with indolebutyric acid (IBA) on Murashige and Skoog's Basal Salt (MS) medium. BA at 2.0 and 4.0 mg L^{-1} and TDZ at 1.5 mg L^{-1} promoted the greatest number of shoots produced per shoot explant. There were no IBA effects or IBA interactions with BA or TDZ. Significant interactions between BA and IBA, but not TDZ and IBA, occurred in leaf explant callus formation and % explants with callus at 6 and 9 weeks of culture and on shoots per leaf explant at 9 weeks. TDZ treatments produced uniformly high levels of callus but low numbers of shoots. The treatment generating the most shoot production was BA at 4.0 mg L^{-1} plus IBA at 0.50 mg L^{-1}. After 9 weeks of culture, leaf explants of the non-runner-producing genotype Baron Solemacher had generated 4.6 shoots per explant with the best treatment, while axillary bud explants had generated 30.8 shoots with the best treatment. Thus, in vitro culture of shoot axillary bud explants can generate high numbers of clonal shoots from a single seedling plant in vitro.

Keywords: in vitro multiplication; alpine strawberry; TDZ; BA; IBA; non-runnering; shoot explant

---

1. Introduction

Fragaria vesca is a self-pollinating diploid species that has become a model species for the commercial strawberry (*F. X ananassa* Duch.) and other members of the Rosaceae family because it has a small genome (240 Mb), a short generation time, and an available full genome sequence [1–3]. In order to facilitate research in ameliorating the qualitative (i.e., flavonoid biosynthesis and other polyphenols) and quantitative characters of the plant with *F. vesca* [4], a large number of replicate clonal plants are required for individual experiments. Most *Fragaria* species produce clonal plants on stolons (commonly called runners) that develop from axillary buds [5], but there are some important non-runnering genotypes within *F. vesca* [6], notably the *F. vesca semperflorens*, that flower constantly, with progeny that are primarily seed-derived and do not produce any runners (stolon, vegetative self-propagating unit).

Studies of in vitro micropropagation of *F. vesca* have successfully regenerated shoots from leaf and petiole explants using combinations of benzyladenine (BA) and indolebutryic acid (IBA) following transformation by *Agrobacterium* [7–11]. *F. vesca* leaf explants placed abaxial side up regenerated shoots more rapidly than those placed adaxial side up after an *Agrobacterium* transformation treatment [11]. No difference was observed in the regeneration response of leaf explants for *F. vesca* versus *F. vesca semperflorens*, runnering versus non-runnering phenotypes, respectively, although differences in petiole responses did differ as more shoots were produced by the latter form at comparable levels of BA plus IBA [10]. However, genotypic variation in post-transformation shoot regeneration was noticed in *F. vesca* genotypes [11]. Thidiazuron (TDZ) was shown to replace BA and promote callus and shoot initiation from *F. vesca* leaf explants [12,13], although a *F. vesca* × *F. vesca semperflorens* interspecific hybrid produced fewer shoots than *F. vesca* in response to TDZ [14], suggesting some genotype sensitivity to TDZ. TDZ may also reduce subsequent shoot elongation [2]. Within the majority of these studies, it was not clear how the transformation and subsequent selection protocols may have determined shoot regeneration potential separately from the inherent capacity for such potential within each genotype. Regeneration of plants after callus formation, as is common in transformation studies, leads to much greater somaclonal variation among progeny than after meristem micropropagation [15], and this lack of uniformity is a problem for subsequent genetic studies.

Even though *F. vesca* may be self-pollinated, seed-derived populations of genotypes that do not produce stolons exhibited some level of variability [2], necessitating lengthy periods of controlled intraspecific pollination of each genotype of interest to create reasonably uniform homozygous populations for subsequent research. Even without transformation, reliable and rapid in vitro propagation techniques for substantially and rapidly increasing the number of clonal plants from non-runnering genotypes for physiological and molecular studies are desirable. To date, there have been no protocols for shoot regeneration from shoot explants (i.e., a shoot with an axillary meristem) of non-runnering *F. vesca* genotypes. Thus, the present study was performed to establish such a protocol by (1) comparing the rates of new shoot production from seedling shoot axillary bud explants versus leaf explants, (2) determining the effective concentrations of BA, TDZ, and IBA for in vitro shoot regeneration from both leaf and shoot explants of *Fragaria vesca*, (3) determining if genotype has an effect on the responses, and (4) assessing if adaxial versus abaxial placement affects leaf explant response.

2. Materials and Methods

2.1. Plant Material

Seeds of four *F. vesca* genotypes, the non-runnering Baron Solemacher (*F. vesca semperflorens*), rarely runnering Pineapple Crush, and runnering types Ivory and Yellow Wonder, were collected from several self-pollinating, individual plants of each genotype grown in a greenhouse at the University of Kentucky, Lexington, KY, USA, washed under tap water, and air-dried.

2.2. Seed Germination and Culture

Seeds were dipped into 30% Clorox bleach (*v/v*; 2.5% sodium hypochlorite) plus 10% sodium dodecyl sulfate (SDS) (*v/v*) solution for 20 min. Seed was then rinsed with sterilized water 3 times under a laminar flow hood. After surface sterilization, 20 seeds were placed in sterile 20 mL Petri dishes containing 4.4 g L^{-1} Murashige and Skoog's Basal Salt (MS) (Sigma® M5524) [16], 30 g L^{-1} sucrose, and 7 g L^{-1} Bacto agar (BD-Difco®) for germination. The medium was prepared by adjusting the pH to 5.7 prior to autoclaving at 121 °C and 105 kPa for 70 min. After 14 days of germination, 3 seedlings with little growth were transferred to each of the 50 mL jars containing the same medium to allow for better growth. Seed germinated in two to three weeks.

2.3. Explant Culture

Leaf lamina (36 mm^2) and shoot (6–8 mm) explants were excised from 5 week-old sterile, in vitro Baron Solemacher seedlings in order to regenerate shoots. The shoot explant consisted of a piece of the main stem, a petiole base, and an axillary meristem at the petiole base. To assess the effect of different combinations and concentrations of plant growth regulators (PGRs) on shoot regeneration, two excised shoot explants were placed in each sterile 20 mL Petri plate containing MS Basal Salt with the following treatments: benzyladenine (BA) at 2 or 4 mg L^{-1} or thidiazuron (TDZ) at 1 or 1.5 mg L^{-1}, each with indole-3-butyric acid (IBA) at 0.125, 0.25, or 0.50 mg L^{-1}. Thus, there were 12 treatment combinations (BA+IBA or TDZ+IBA) in total and each was replicated 4 times. Possible genotypic variation was compared using four shoot explants from each of the 4 genotypes on 2 mg L^{-1} BA versus 1.5 mg L^{-1} TDZ, each combined with 0.25 mg L^{-1} IBA. There were 4 replicates of each genotype by treatment combination. In a third experiment, leaf explants were placed adaxial side up versus abaxial side up using the set of BA+IBA or TDZ+IBA treatments described above. With leaf explant placement as an additional treatment, there were 24 total treatments, each replicated 4 times. The Petri dishes were held in a laboratory at 22 °C temperature under fluorescent lighting with an 18 h daylength.

After 6 and 9 weeks of culture, callus production by each explant was rated as: 0 = no callus, 1 = low quantity of callus, 2 = medium quantity of callus, or 3 = high quantity of callus. After 10 weeks, regenerated shoots were transferred to Petri plates containing half-strength MS medium without growth regulator for rooting. The mean value of each set consisting of two explants in each Petri plate was considered as a replication. All the experiments were conducted in a completely randomized design (CRD).

Shoot number per shoot explant and the number of shoot explants producing new shoots were recorded after 6 and/or 9 weeks of culture. Shoot explants did not produce visible callus. With leaf explants, relative callus production and the number of new shoots per explant were recorded after 6 and 9 weeks of culture, and the % of explants producing callus was calculated. Analyses of variance (ANOVA) were performed (SigmaPlot 12.0, Systat Software, Inc., San Jose, CA, USA), and means were compared using the Student–Newman–Keuls method at $P < 0.05$. Results were expressed as least squares means ± standard error of the mean.

3. Results and Discussion

3.1. Shoot Axillary Bud Explants

BA at 2.0 and 4.0 mg L^{-1} and TDZ at 1.5 mg L^{-1} produced more shoots per axillary bud explant than with TDZ at 1 mg L^{-1} (Table 1). Preliminary work indicated that BA at 2 mg L^{-1} plus IBA at 0.125 mg L^{-1} had no effect compared to BA alone (data not shown), and IBA concentration up to 0.5 mg L^{-1} had no main effect and did not interact with BA or TDZ (data not shown). Thus, BA or TDZ alone were sufficient to generate new shoots from shoot explants. BA at 2 mg L^{-1} had a higher % of explants producing shoots than TDZ at 1 mg L^{-1}, with the remaining treatments at intermediate values (Figure 1).

Table 1. Effect of 6-benzylaminpurine (BA) and thidiazuron (TDZ) on shoot regeneration from shoot explants of *Fragaria vesca*. Data were collected 6 weeks after initiating the study.

BA or TDZ	Concentration (mg L^{-1})	Shoots per Explant	% Explants Producing Shoots
BA	BA 2.0	6.4 a ^{z}	100 a
BA	BA 4.0	7.3 a	86 ab
TDZ	TDZ 1.0	2.9 b	67 b
TDZ	TDZ 1.5	5.6 a	88 ab

^{z} Mean separation by the Student–Newman–Keuls method at $P < 0.05$.

There was a genotype-by-treatment interaction on the number of shoots produced per shoot axillary bud explant (Table 2). With IBA at 0.25 mg L^{-1}, Baron Solemacher on TDZ at 1.5 mg L^{-1} produced more shoots per explant after 9 weeks of culture than those on BA at 2.0 mg L^{-1}. Baron Solemacher shoot explants on BA and TDZ produced equal numbers of shoots by 6 weeks of culture (Table 1), although the total numbers were considerably lower than those of the shoots shown in Table 2. The longer 9-week culture period in the latter experiment led to more total shoot production and perhaps to the treatment difference, although the other cultivars in the latter experiment did not show the same BA versus TDZ difference observed with Baron Solemacher. It has been reported that the best shoot proliferation was obtained with 1 mg L^{-1} TDZ and 0.2 mg L^{-1} IBA from leaf explants of *F. vesca* cultivars (43.9% explants formed shoots) [14].

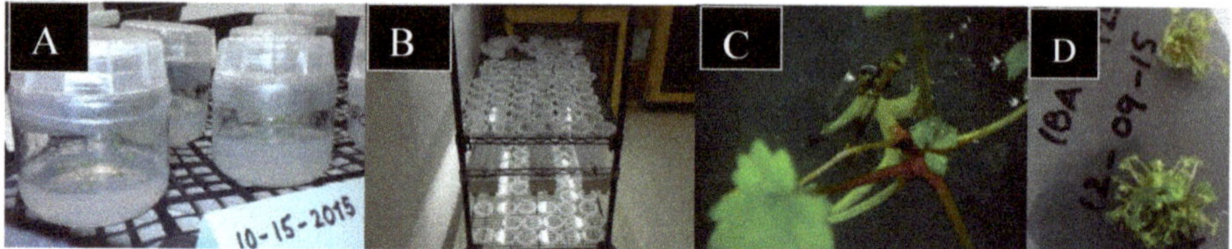

Figure 1. (**A**) Transfer of seedlings cv. Baron Solemacher to bottles with MS medium. (**B**) Setup of experiment. (**C**) Baron Solemacher single axillary bud shoot explants producing new shoots. (**D**) Baron Solemacher with single axillary bud shoot explants produced the maximum number of shoots at BA 2.00 (mg L^{-1}) + IBA 0.125 (mg L^{-1}).

Table 2. Effect of genotype and BA at 2.0 mg L^{-1} plus IBA at 0.25 mg L^{-1}, or TDZ at 1.5 mg L^{-1} plus IBA at 0.25 mg L^{-1}, on shoot regeneration from shoot explants of *Fragaria vesca*. Data were collected after 9 weeks of culture.

Genotype	BA or TDZ	Shoots per Explant
Baron Solemacher	BA	15.0 b ^{z}
Baron Solemacher	TDZ	30.8 a
Pineapple Crush	BA	22.4 ab
Pineapple Crush	TDZ	26.0 ab
Ivory	BA	27.0 a
Ivory	TDZ	26.9 ab
Yellow Wonder	BA	22.6 ab
Yellow Wonder	TDZ	29.6 a

^{z} Mean separation by the Student–Newman–Keuls method at $P < 0.05$.

3.2. Leaf Explants

Starting at 4 weeks of culture, the cut edge of leaf explants exhibited the start of callus formation. From the callus, reddish-colored shoots and light-green leaves then developed. There were significant interactions between BA and IBA on the relative amount of callus formation and % explants with callus at 6 and 9 weeks of culture, and on shoots per leaf explant at 9 weeks (Figure 2, Table 3). Except for an interaction of TDZ at 1 mg L^{-1} with IBA on relative callus formation, there was no other TDZ–IBA interaction at 6 or 9 weeks. A high level of callus production did not always result in high shoot production, as the TDZ treatments produced uniformly high levels of callus but low numbers of shoots. The treatment generating the most shoot production was BA at 4.0 mg L^{-1} combined with IBA at 0.50 mg L^{-1}. Although IBA did not have an effect on shoot regeneration by shoot explants, as noted above, it did have an increasing effect on regeneration from leaf explants with BA. Only the treatment combination of BA at 2 and 4 mg L^{-1} and IBA at 0.125 mg L^{-1} did not produce any shoots through 9 weeks of culture.

Figure 2. Shoot formation in 'Baron Solemacher' leaf discs treated with 6-benzylaminpurine (BA; 4.0 mg L^{-1}) plus indolebutyric acid (IBA; 0.5 mg L^{-1}) (**a**) Callus formation after six weeks in culture, (**b**) shoot formation after nine weeks, (**c**) abaxial side up with BA (4.0 mg L^{-1}) plus IBA (0.5 mg L^{-1}), (**d**) adaxial side up with BA (4.0 mg L^{-1}) plus IBA (0.5 mg L^{-1}), (**e**) plantlets, and (**f**) rooted plantlets.

Table 3. Effects of BA, TDZ, and IBA on relative callus formation [z], % explants with callus, and shoots per explant from seedling leaf explants of *Fragaria vesca* cv. Baron Solemacher.

PGR (mg L^{-1})			Six Weeks of Culture			Nine Weeks of Culture		
BA	TDZ	IBA	Relative Callus Production [z]	Explants with Callus (%)	Shoots per Explant	Relative Callus Production	Explants with Callus (%)	Shoots per Explant
2.0	0	0.125	0.25 c [y]	25 c	0 NS	0.33 bc	33 b	0 c
2.0	0	0.25	1.33 b	100 a	0.4	1.83 ab	100 a	0.85 b
2.0	0	0.50	1.42 b	83 ab	0	1.83 b	92 a	0.5 b
4.0	0	0.125	0 c	0 c	0	0 c	0 c	0 c
4.0	0	0.25	1.13 b	81 ab	0.9	1.38 b	81 a	1.1 b
4.0	0	0.50	2.31 a	100 a	2.9	2.5 a	100 a	4.6 a
0	1.0	0.125	1.13 b	100 a	0	1.5 b	100 a	0.4 b
0	1.0	0.25	1.67 ab	100 a	0.4	2.08 ab	100 a	0.9 b
0	1.0	0.50	1.75 a	92 ab	0.4	1.92 ab	100 a	1.0 b
0	1.5	0.125	1.25 b	92 ab	0.2	1.5 b	92 a	1.4 b
0	1.5	0.25	0.83 bc	67 b	0	1 bc	75 a	0.4 b
0	1.5	0.50	1.50 b	100 a	0	1.94 ab	100 a	0.3 b

[z] Ratings were as follows: 1 = explant with low callus production; 2 = explant with medium callus formation; 3 = explant with high callus formation. [y] Mean separation within columns by the Student–Newman–Keuls method at $P < 0.05$. PGR: plant growth regulators.

F. vesca leaf explants cultured on MS media containing 1.5 mg L^{-1} TDZ + 0.5 mg L^{-1} IBA showed the best shoot production across a set of TDZ by IBA concentrations, but overall IBA had no effect, as noted in the present study when combined with TDZ, or even reduced shoot production as the concentration increased [13]. In another study, better shoot proliferation was recorded with TDZ plus IBA than without IBA in two *F. vesca* genotypes from leaf explants [14]. Neither of these cited studies compared TDZ to BA levels.

The placement of the leaf explants exerted a significant influence on explants with callus and % explants with callus, with leaf explants placed abaxial side up producing more callus, but not shoots, than those placed adaxial side up after 6 and 9 weeks of culture (Table 4, Figure 2). Placement did not interact with BA, TDZ, or IBA (data not shown). When young *F. vesca* leaf explants were placed abaxial side up on MS medium, shoot regeneration occurred in all the treatments with BA and IBA [8].

Table 4. Effect of leaf placement—adaxial or abaxial side up—on relative callus formation, % explants with callus, and shoots per explant from leaf explants of *Fragaria vesca* cv. Baron Solemacher.

Placement of Leaf Explant	Six Weeks of Culture			Nine Weeks of Culture		
	Relative Callus Production	Explants with Callus (%)	Shoots per Explant	Relative Callus Production	Explants with Callus (%)	Shoots per Explant
Adaxial side up	1.03 b [z]	75 b	0.52	1.34 b	79	1.29
Abaxial side up	1.40 a	82 a	1.42	1.63 a	83	2.55

[z] Mean separation within columns by the Student–Newman–Keuls method at $P < 0.05$.

In vitro placing of regenerated shoots from all tissue sources, genotypes, and treatment conditions on half-strength MS medium without PGRs resulted in 100% rooting (data not shown) (Figure 2). A high level of root formation on MS basal medium in *F. vesca* has been noted [13]. All plants in these studies were successfully acclimatized to a greenhouse environment via a mist bed and culture in containers (Figure 3). There was no visible evidence of phenotypic variation within genotypes following one year of plant growth (data not shown).

Figure 3. (**a**) Rooted plantlets placed in ProMix BX in a plastic pot in a mist chamber; (**b**) rooted plantlets in the greenhouse after 2 weeks in the mist chamber; (**c**) plants after being acclimatized in the greenhouse; (**d**) acclimatized plants with very good growth in the greenhouse.

4. Conclusions

An effective in vitro shoot regeneration protocol for *Fragaria vesca* was demonstrated from both shoot axillary bud and leaf explants. For shoot explants, BA at 2 and 4 mg L^{-1} and TDZ at 1.5 mg L^{-1} were the best treatments for shoot regeneration. However, there were no effects of IBA concentration on shoot regeneration. Only Baron Solemacher among the four *F. vesca* genotypes (including Pineapple Crush, Ivory, and Yellow Wonder) showed a difference between use of TDZ versus BA. From leaf explants, 4 mg L^{-1} BA plus 0.5 mg L^{-1} IBA resulted in the maximum callus production and number of shoots per explant, and the IBA effect increased from 0.125 to 0.5 mg L^{-1}. In contrast, TDZ promoted high callus production but resulted in reduced numbers of shoots, and IBA concentration had no effect on the production of callus and shoots. Leaf explants placed abaxial side up produced more callus but no more shoots than those placed adaxial side up. After 9 weeks of culture, Baron Solemacher leaf explants had generated 4.6 shoots per explant with the best treatment (Table 3), while shoot explants from Baron Solemacher had generated 30.8 shoots with the best treatment (Table 2). Thus, the protocol for using shoot axillary bud explants is a better alternative to leaf explants for generating high numbers of clonal shoots from a single seedling plant in vitro, avoiding callus production and with the possibility of a large homozygous population of plants for the non-runnering *F. vesca* genotypes.

Author Contributions: B.C.S. performed all of the work. D.D.A., R.L.G., and S.T.K. advised during the planning and execution of the study and provided technical support. B.C.S. wrote the manuscript with input from D.D.A. and R.L.G. All authors have read and agreed to the published version of the manuscript.

Funding: Fulbright authority (CIES/IIE) provided the fellowship under Fulbright Visiting Scholar Program to perform the research.

Acknowledgments: This is publication No. 17-11-064 of the Kentucky Agricultural Experiment Station and is published with the approval of the Director. Babul C. Sarker would like to express his gratitude to the Fulbright authority (CIES/IIE) for providing a Fellowship to carry out research under the Fulbright Scholars Program at the Department of Horticulture, University of Kentucky, Lexington, Kentucky, USA. Technical as well as other support and cooperation by numerous faculty and staff of the Department of Horticulture, University of Kentucky, Lexington, Kentucky, USA are gratefully acknowledged. Thanks to the Bangladesh Agricultural Research Institute, Gazipur, Bangladesh, for approving leave for the period of study and research. This project was supported by the National Institute of Food and Agriculture, U.S. Department of Agriculture, Hatch project number KY011042, and the Fulbright Scholars Program.

References

1. Folta, K.M.; Davis, T.M. Strawberry Genes and Genomics. *Crit. Rev. Plant Sci.* **2006**, *25*, 399–415. [CrossRef]
2. Slovin, J.P.; Schmitt, K.; Folta, K.M. An inbred line of the diploid strawberry *Fragaria vesca f. semperflorens* for genomic and molecular genetic studies in the Rosaceae. *Plant Methods* **2009**, *5*, 15. [CrossRef] [PubMed]
3. Shulaev, V.; Sargent, D.J.; Crowhurst, R.N.; Mockler, T.C.; Folkerts, O.; Delcher, A.L.; Jaiswal, P.; Mockaitis, K.; Liston, A.; Mane, S.P.; et al. The genome of woodland strawberry (*Fragaria vesca*). *Nat. Genet.* **2011**, *43*, 109–116. [CrossRef] [PubMed]
4. Roy, S.; Wu, B.; Liu, W.; Archbold, D.D. Comparative analyses of polyphenolic composition of *Fragaria* spp. *Plant Physiol. Biochem.* **2018**, *125*, 255–261. [CrossRef] [PubMed]
5. Galletta, G.J.; Bringhurst, R.S. Strawberry Management. In *Small Fruit Crop Management*; Galletta, G.J., Himelrick, D.G., Eds.; Prentice Hall: New Jersey, NJ, USA, 1990; pp. 83–156.
6. Brown, T.; Wareing, P.F. The genetical control of the everbearing habit and three other characters in varieties of *Fragaria vesca*. *Euphytica* **1965**, *14*, 97–112.
7. Greene, A.E.; Davis, T.M. Regeneration of *Fragaria vesca* plants from leaf tissue. In *The Strawberry into the 21st Century*; Dale, A., Luby, J.J., Eds.; Timber Press: Portland, OR, USA, 1991; pp. 124–125.
8. El Mansouri, I.E.; Mercado, J.A.; Valpuesta, V.; López-Aranda, J.M.; Pliego-Alfaro, F.; Quesada, M.A. Shoot regeneration and *Agrobacterium*-mediated transformation of *Fragaria vesca* L. *Plant Cell Rep.* **1996**, *15*, 642–646. [CrossRef] [PubMed]
9. Haymes, K.M.; Davis, T.M. *Agrobacterium*-mediated transformation of 'Alpine' *Fragaria vesca*, and transmission of transgenes to R1 progeny. *Plant Cell Rep.* **1998**, *17*, 279–283. [CrossRef] [PubMed]

10. Alsheikh, M.K.; Suso, H.-P.; Robson, M.; Battey, N.H.; Wetten, A. Appropriate choice of antibiotic and *Agrobacterium* strain improves transformation of antibiotic-sensitive *Fragaria vesca* and *F.v. semperflorens*. *Plant Cell Rep.* **2002**, *20*, 1173–1180. [CrossRef]

11. Oosumi, T.; Gruszewski, H.A.; Blischak, L.A.; Baxter, A.J.; Wadl, P.A.; Shuman, J.L.; Veilleux, R.E.; Shulaev, V. High-efficiency transformation of the diploid strawberry (*Fragaria vesca*) for functional genomics. *Planta* **2006**, *223*, 1219–1230. [CrossRef] [PubMed]

12. Zhao, Y.; Liu, Q.; Davis, R.E. Transgene expression in strawberries driven by a heterologous phloem-specific promoter. *Plant Cell Rep.* **2004**, *23*, 224–230. [CrossRef] [PubMed]

13. Yildirim, A.B.; Turker, A.U. Effects of regeneration enhancers on micropropagation of *Fragaria vesca* L. and phenolic content comparison of field-grown and in vitro-grown plant materials by liquid chromatography-electrospray tandem mass spectrometry LC–ESI-MS/MS. *Sci. Hortic.* **2014**, *169*, 169–178. [CrossRef]

14. Landi, L.; Mezzetti, B. TDZ, auxin and genotype effects on leaf organogenesis in Fragaria. *Plant Cell Rep.* **2006**, *25*, 281–288. [CrossRef] [PubMed]

15. Morozova, T. Genetic stability of pure lines of *Fragaria vesca* L. in micropropagation and long-term storage. *Acta Hortic.* **2002**, *567*, 85–88. [CrossRef]

16. Murashige, T.; Skoog, F. A revised medium for rapid growth and bioassays with tobacco tissue cultures. *Physiol. Plant.* **1962**, *15*, 473–497. [CrossRef]

MDPI

St. Alban-Anlage 66

4052 Basel

Switzerland

Tel. +41 61 683 77 34

Fax +41 61 302 89 18

www.mdpi.com

Horticulturae Editorial Office

E-mail: horticulturae@mdpi.com

www.mdpi.com/journal/horticulturae